Systems Metabolic Engineering

Christoph Wittmann • Sang Yup Lee
Editors

Systems Metabolic Engineering

Springer

Editors
Christoph Wittmann
Institute of Biochemical Engineering
Braunschweig Integrated Center for Systems Biology (BRICS)
Center for Pharmaceutical Engineering
Technische Universität Braunschweig
Braunschweig, Germany

Sang Yup Lee
Metabolic and Biomolecular Engineering National Research Laboratory
Department of Chemical and Biomolecular Engineering (BK21 Program)
Center for Systems and Synthetic Biotechnology
Institute for the BioCentury KAIST
Daejeon, Republic of Korea

ISBN 978-94-007-4533-9 ISBN 978-94-007-4534-6 (eBook)
DOI 10.1007/978-94-007-4534-6
Springer Dordrecht Heidelberg New York London

Library of Congress Control Number: 2012941228

Printed on acid-free paper

Springer is part of Springer Science+Business Media (www.springer.com)

To Heike, Isabelle, Felix and Florian from Christoph and to Hyejean and Gina from Sang Yup for their love, support and inspiration

Preface

The integration of systems-wide omics approaches, genome-scale modeling and simulation, synthetic biological approaches, and even evolutionary engineering is opening a new era of industrial strain engineering – the design-based creation of tailor-made overproducers that are optimized at the global level. This integrated approach of metabolic engineering is now called systems metabolic engineering. At the entry into a new millennium, facing strong needs for a novel bio-economy due to global warming and shortage of fossil fuels, this seems one of the most relevant and promising areas of research and industrial application. The present book is devoted to this fascinating area of systems-wide analysis and engineering of cellular metabolism. Through a series of exciting chapters, world leading experts provide us with up-front approaches to analyze, model and re-design biological systems towards desired properties and enrich this by real-case applications for the most relevant workhorses in industrial bio-production.

The book starts with computational and experimental methods on the systems-wide analysis of biological systems, the entry and basis to create understanding and enable knowledge-based systems metabolic engineering. In Chap. 1, Professors Lee and Palsson together with their teams combine their pioneering expertise on computational modeling of genome-scale networks, the tin-opener in many of the successful projects on systems metabolic engineering reported today. They provide the full picture, touching metabolic networks, transcriptional networks and cell-signaling networks and include interesting case studies on the biological systems picked up in the later application chapters of the book. Chapter 2 by Professor Palsson and colleagues extends the stoichiometric modeling of Chap. 1 to kinetic models of metabolism, crucial to describe systems dynamics. Their contribution gives valuable hands-on advice for the creation of kinetic models, provides the fundamental mathematics and closes with practical application examples. Chapter 3 by Professor Shimizu's group complements the dry lab analysis of networks via wet lab approaches, covering state-of-art omics technologies. They describe how thoroughly designed experimental studies deliver deep understanding of industrial microorganisms and how metabolic engineers can efficiently exploit this towards optimized strains.

The above approaches on systems-wide computational and experimental analysis of biological systems, which form the initial part of the book, provide design strategies for superior cell factories that have to be implemented on the DNA level. The more we want to shape and create, the larger the genetic changes necessary. In this regard, Chap. 4 by Professor Panke and co-workers discusses how to efficiently translate design concepts into synthetic DNA sequences. The authors especially focus on novel large-scale synthetic engineering approaches and provide us with an interesting view on completely design-based synthetic systems.

Chapters 5, 6, 7, 8, 9 and 10 pickup most relevant industrial workhorses from the groups of bacteria, yeasts and fungi and illustrate how systems metabolic engineering is used today for next-level strain and bioprocess development. Chapter 5 by the group of Professor Lee focuses on *Escherichia coli*, probably the most deeply studied microorganism on the systems level. Their set of examples on a wide set of products underline how advanced we are in creating tailor-made *E. coli* cell factories and what is needed to go even further. In Chap. 6, Professor Wittmann and his team review systems metabolic engineering of *Corynebacterium glutamicum*. They explain how this gram-positive soil bacterium can be tailored to convert a broad spectrum of renewable raw materials into various chemicals, fuels, materials or therapeutics and thus, similarly, to *E.* coli, is becoming a successful bio-production platform. Chapter 7 by Professor Papoutsakis and colleagues deals with *Clostridium acetobutylicum*, a famous bacterium for production of solvents since the very beginning of industrial biotechnology almost a 100 years ago. Their contribution focuses on improved tolerance from a systems view point, a key target of superior strains, and especially valid towards high-level production of the often unnatural chemicals toxic for the cell. Chapters 8 and 9 highlight eukaryotic production systems. In Chap. 8, Professor Heijnen's group describe systems-level design of *Penicillium chrysogenum*, well-known for its high relevance for antibiotics production and a model system for the rich set of industrial processes with other filamentous fungi. Their contribution nicely recruits modeling to elucidate function and control of metabolism for strain design. Chapter 9 by Professor Nielsen and co-workers deals with yeast and illustrates how omics technologies can be integrated with synthetic biology for rational DNA modification into a knowledge-based framework for systems metabolic engineering. They complement this by two case studies from biofuel production, which are among the most relevant bioprocesses in yeast industrial biotechnology. In Chap. 10, Professor Kondo and his team provide us with interesting examples on systems metabolic engineering of cellular properties that are crucial to successfully integrate cell factories into the rising concept of biorefinery. In this regard their review discusses improved utilization of renewable raw materials – direct conversion of the mainly mixed, polymeric substrates as well as improved tolerance to toxic ingredients.

Chapter 11 closes the book by opening a new door. Professor Stephanopoulos and colleagues illustrate how we can exploit ideas and tools of systems metabolic engineering and systems biology to address key questions in medicine. Their contribution on cancer as a metabolic disease might stimulate to further extend the application of the engineering concepts described throughout the book towards a new field of research.

As compiled in this book, we are now reaching the level of global analysis, design and engineering of biological systems. This provides a cornucopia of novel possibilities – sustainable supply of chemicals, materials and fuels in a new era of bio-production as well as tailor-made therapies of threatening diseases. We hope that the book is interesting and valuable for researchers and engineers from the various disciplines that are all integrated into the field. Thanks to the worldwide experts and their excellent contributions, which are greatly appreciated, this book hopefully sets a milestone with perpetual value. We would like to deeply thank the members of our labs, led by Dr. Judith Becker, for their great efforts in editing and formatting the book. Finally, we would like to thank the people at Springer for their assistance in the production. Admittedly, we are still away from the immaculate cell factory, but the way towards it has become visible – and it is a privilege to walk on and share it with you.

TU Braunschweig, Germany — Christoph Wittmann
KAIST, Daejeon, Republic of Korea — Sang Yup Lee

Contents

Chapter 1
Genome-Scale Network Modeling

Sang Yup Lee, Seung Bum Sohn, Hyun Uk Kim, Jong Myoung Park, Tae Yong Kim, Jeffrey D. Orth, and Bernhard Ø. Palsson

Contents

S.Y. Lee (✉)
Metabolic and Biomolecular Engineering National Research Laboratory, Department of Chemical and Biomolecular Engineering (BK21 Program), Center for Systems and Synthetic Biotechnology, Institute for the BioCentury, KAIST, Daejeon, Republic of Korea
e-mail: leesy@kaist.ac.kr

S.B. Sohn • J.M. Park
Metabolic and Biomolecular Engineering National Research Laboratory, Department of Chemical and Biomolecular Engineering (BK21 Program), Center for Systems and Synthetic Biotechnology, Institute for the BioCentury, KAIST, Daejeon, Republic of Korea

Bioinformatics Research Center, KAIST, Daejeon, Republic of Korea

H.U. Kim •
T.Y. Kim
Bioinformatics Research Center, KAIST, Daejeon, Republic of Korea

J.D. Orth • B.Ø. Palsson
Department of Bioengineering, University of California, San Diego, La Jolla, CA, USA

C. Wittmann and S.Y. Lee (eds.), *Systems Metabolic Engineering*,
DOI 10.1007/978-94-007-4534-6_1, © Springer Science+Business Media Dordrecht 2012

Abstract Genome-scale models have garnered considerable interest for their ability to elucidate cellular characteristics and lead to a better understanding of biological systems. Metabolic models in particular have been widely used to study complex metabolic pathways in order to better understand microbial systems and to design strategies for engineering various biotechnological applications. Similar to metabolic networks, transcriptional and signaling network models have also been reconstructed to elucidate regulatory interactions and to further understand the response of systems to various environmental stimuli. However, a true genome-scale model that integrates all these characteristics into one comprehensive model has not yet been constructed. For the time being, the existing network models have individually contributed to the knowledge of their respective fields and to our understanding of biological systems. In selected cases they have provided design strategies for systems wide engineering of metabolism. There have been several attempts to integrate these networks to realize the full potential of a complete cellular network model, although there is still room for further development. Here, we review the different network types and highlight their contributions to biotechnological applications via illustrative examples.

Keywords Genome-scale model • Metabolic network • Transcriptional network • Signaling network • *Escherichia coli* • Network reconstruction • Genome annotation • *In silico* • Stoichiometric matrix • Steady-state • Automatic reconstruction • Single-input module • Top-down • Bottom-up • Boolean formalism • Flux analysis • Stoichiometric formalism • Kinetic formalism

1.1 Introduction

Understanding and visualizing of biological networks has become an important aspect of systems biology as more information and knowledge are being generated. The availability of a network describing a particular aspect of the biological system, whether it is metabolism or transcriptional regulation, allows the user to better understand how the system can respond to the ever-changing external environment. With the advent of the full genome sequence, the reconstruction of a full genome-scale model of a cellular system has become feasible. Currently, there are genome-scale metabolic models [1, 2] and recently genome-scale transcriptional networks have begun to appear [3, 4]. However, a true genome-scale model which integrates the metabolism, the transcriptional regulatory network, and all other networks that are found in biological systems into one comprehensive model is still being developed.

Current models of biological systems have provided researchers with a wealth of knowledge regarding their respective scopes. Metabolic network models have aided in the design of new strategies for systems metabolic engineering of host strains for the production of high-value compounds in cell factories of *Escherichia coli* [5] or *Corynebacterium glutamicum* [6], as outlined later for the respective microorganisms

Table 1.1 Overview of the different networks

Type	Metabolic networks	Transcriptional networks	Signaling networks
Definition	Network of biochemical reactions which reflect the metabolic state of the cell	Network reflecting the expression state of the genome	Network of proteins that transduce information that changes the transcriptional state of the cell
Components	Metabolites Metabolic reactions	Promoters Transcription factors Metabolites	Proteins Protein-protein interactions
Source of data	Genome annotation ^{13}C flux data Enzyme analysis Biochemical databases Other curated databases Literature	Gene expression data Location analysis Predictive algorithms for promoters Curated databases Literature	Signaling databases Protein-protein interactome Gene expression data RNAi knockdown of signaling pathways Proteome Fluorescent localization data

throughout this book. Moreover, transcriptional network models have aided in the identification of a number of new transcription factors or binding sites [7]. Despite the incompleteness of these models, they continue to provide valuable knowledge in filling in gaps in our understanding of these biological networks [8].

Here, we discuss the characteristics of three biological networks that have been extensively studied in recent years (Table 1.1). Metabolic networks have been the most studied of the three, and have been utilized in a large number of applications from drug discovery to industrial production of high-valued biochemicals opening a new era of design-based metabolic engineering as illustrated throughout this book. Transcriptional networks and signaling networks are still in their infancy with regard to large-scale network reconstructions. However, there have been major advances in each field, which will be discussed in their respective sections.

1.2 Metabolic Networks

A metabolic network model describes the metabolic state of the cell. It is composed of biochemical reactions that are constrained by the laws of thermodynamics and mass action. The metabolic network can be modeled on different levels of complexity based on what is being examined. The uppermost level is the cellular level, where only the cellular inputs and outputs are of concern and the mechanics within the cell are not. Below that is the functional level of the metabolic network where the network is divided based on the functions performed by a particular section of the network, e.g., catabolic or anabolic. The next level examines the pathways in functional groups, such as glycolysis or amino acid biosynthesis. Finally, at the foundation of the metabolic network are the individual biochemical reactions.

In the rest of this chapter, we will concern ourselves with this level of the metabolic network as most current genome-scale metabolic networks are reconstructed as lists of biochemical reactions.

To rebuild the metabolic network of a particular organism, a draft reconstruction is first assembled based on the organism's genome annotation. All known metabolic reactions in the organism of interest must be collected and incorporated into the network reconstruction. The reactions associated with each metabolic gene can come from sources such as annotated gene names, EC numbers [9], and GO terms [10]. Multispecies metabolic databases such as KEGG [11] can also be used to match genes with their reactions. The assembly of a draft metabolic reconstruction can be performed manually, or it can be automated [12, 13]. The draft metabolic reconstruction will certainly contain errors and missing information, particularly if the draft reconstruction was automated, and so it must be manually curated.

Confidence levels for the presence of each biochemical reaction in the network should be determined. These confidence levels are based on evidence of the existence of the reaction in the organism according to literature and experimental studies. For instance, biochemical data indicating a reaction's presence in the organism, such as an assay of a purified enzyme, would have the highest confidence level and would be included in the metabolic reconstruction. Unfortunately, specific biochemical data for every biochemical reaction for every species does not exist. Therefore other data sources are utilized. With the availability of full genome sequences of many other organisms, the function of genes can be determined through the use of genetic data from previously characterized organisms indicating the corresponding metabolic reactions. Data on gene knockouts and their effect on metabolism and homology with genes with known functions from other species are some examples of genetic data, and account for most of the information in metabolic networks of less well-characterized species.

Physiological data, such as secreted metabolite concentrations or glucose and oxygen uptake rates, are also useful. When *in vivo* physiology does not match the *in silico* results, it is usually due to an incomplete annotation or insufficient characterization of the organism. For example, if a species is known to produce a certain metabolite and yet the genome does not include known genes encoding the required metabolic enzymes, those reactions are included in the metabolic network with a lower confidence level to ensure consistency with known physiology. Finally there are *in silico* simulation data, which gives the lowest confidence level for those biochemical reactions included in the metabolic network. Reactions of this sort are usually included to ensure that the target objective is generated, such as biomass synthesis reactions.

A completely curated metabolic network reconstruction must be converted to a mathematical model to make computational predictions. Constraint-based modeling is most often used to represent metabolic networks [14–17]. The reactions are encoded in a stoichiometric matrix in which each row represents a metabolite and each column represents a reaction. The elements of the matrix are the stoichiometric coefficients of each metabolite in the reaction. Upper and lower bound constraints on each reaction can be imposed. This model can then be used in different mathematical analyses, the most common being flux balance analysis (FBA) [15]. Here, the steady-state metabolic flux distribution of the cell is determined using linear programming where an objective function, such as maximum cell growth rate, is selected, giving a particular state of the metabolic network. The assumption of steady-state is made possible by the time scale difference between the cellular level growth and the reaction level fluxes. Once the mathematical model of the metabolic network is constructed, the accuracy of the network can be validated through comparison with experimental data. Phenotypic data for growth on different substrates [18] or with gene knockouts [19] can be directly compared to FBA predicted phenotypes. These comparisons continue in an iterative fashion, where the network is tested and updated based on the results. Other analyses that can be performed on the metabolic model include pathway analysis [20], elementary flux mode analysis [21], gene expression analysis [22], and adaptive evolution analysis [23].

1.2.1 Metabolic Network Case Studies

The metabolic network model of *Haemophilus influenza* was the first reconstructed genome-scale metabolic network, published in 1999 [24]. Since then, more than 70 genome-scale metabolic networks have been reconstructed and published (Fig. 1.1). Although the majority of published metabolic networks are of bacterial systems, metabolic networks for archaea and eukaryotes, including *Homo sapiens* [25, 26], have been reconstructed and studied. These reconstructed metabolic networks have been utilized to analyze and characterize their respective organisms. With these metabolic networks, researchers have been able to investigate physiological characteristics of the organism of interest or suggest engineering strategies for improving target organisms for the overproduction of value-added substances.

The *E. coli* metabolic network has been at the center of metabolic network reconstruction due to the important role it plays in microbiology. It is the best-characterized microorganism with a wealth of literature support, and its veteran status in microbial studies has created well established tools for genetic manipulation. Because of these resources, the *E. coli* metabolic model is perhaps the most easily validated metabolic model available. It has undergone continuous updates since its initial publication in 2000 [1, 27, 28]. Metabolic models of *E. coli* have been utilized in various biotechnological applications, particularly in metabolic engineering, where the *E. coli* metabolic network has been engineered to produce high quantities of value-added substances. Examples include the use of *E. coli* to produce lycopene [29, 30], L-valine [5], and biopolymers such as poly-lactate [31]. In all of these cases, the metabolic network was analyzed through the use of algorithms, such as MOMA (which uses quadratic programming to identify gene knockout targets) [19] or FSEOF (which identifies gene amplification targets) [30], to identify the best target for modification in the metabolic network such that production of the target compound increases.

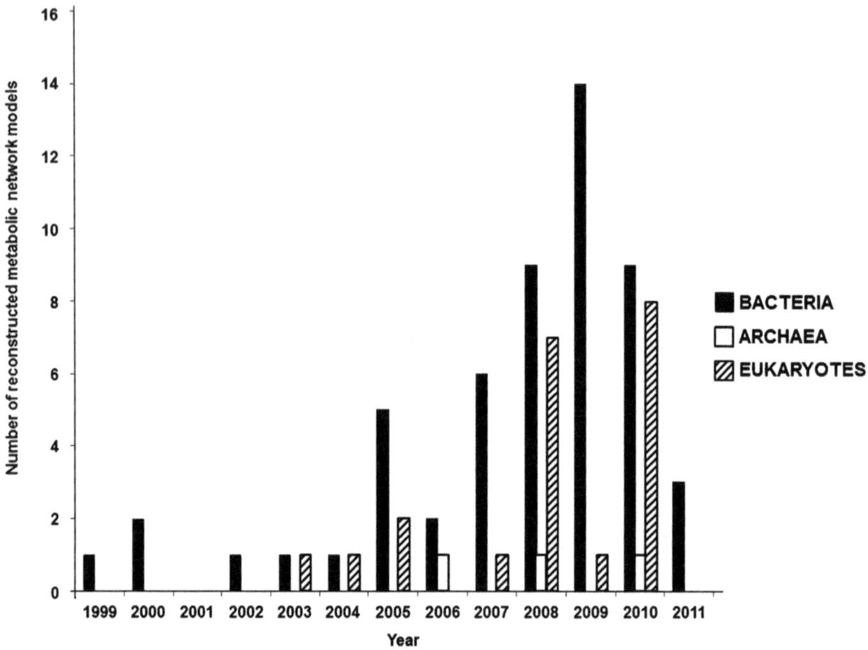

Fig. 1.1 Number of publications of metabolic networks over the last 10 years broken down for the number of publications for bacteria, archaea, and eukarya

Several algorithms now exist for the developing recombinant strains with improved production of high-valued compounds using metabolic network models. OptKnock [32] is an algorithm that uses bi-level linear programming to identify the optimal set of gene knockouts to couple production of a target compound to growth. OptGene [33] is a related method that uses a genetic algorithm to identify gene targets. The algorithm OptStrain [34] predicts new reactions to add to the metabolic network in conjunction with gene knockouts to improve production of the target compound. In addition to gene knockouts, the approach Flux Design [35] identifies targets for amplification as well as down regulation through the use of elementary flux mode analysis. Other strategies have been predicted for *E. coli* [36], and designs for the production of L-lactate have been constructed and optimized by adaptive evolution [37] and through media design for increased L-methionine production [38].

Other metabolic networks have been utilized in a similar capacity for the metabolic engineering of new strains with improved performance. *Mannheimia succiniciproducens* [39] has been studied for the production of succinic acid, *C. glutamicum* [40] for the production of various amino acids, and *Streptomyces coelicolor* A3(2) [41] for the production of antibiotics. Other industrial species with available metabolic network reconstructions include *Pseudomonas putida* [42], *Clostridium acetobutylicum* [43], and *Zymomonas mobilis* [44] to name a few. As many of these species are not as extensively characterized as *E. coli*, their genome-scale metabolic networks have had limited use in metabolic engineering. However, they have been helpful in furthering our understanding of the species' metabolism and have provided a platform for further experimental studies of their metabolism.

Pathogenic organisms such as *Helicobacter pylori* [45], *Vibrio vulnificus* [46], *Pseudomonas aeruginosa* PAO1 [47], and *Acinetobacter baumannii* [48] have had their metabolic networks reconstructed so that drug targets could be identified for the treatment of infections. These targets are usually at points of fragility in the metabolic network of the pathogen [48, 49]. Robustness analysis of metabolism identifies these fragile points, but also identifies possible alternate routes the organism can use to counteract treatments that are suggested [50, 51]. On a related note, the human metabolic network [25] is also utilized when searching for possible drug targets against pathogens. To prevent possible side-effects of a newly introduced drug in the human host, the human metabolic network is analyzed to determine if human metabolism will be affected. The drug must target components of the pathogenic metabolism that are not found in the human metabolic network.

The field of metabolic network reconstruction has mainly focused on prokaryotes due to the relative simplicity of prokaryotic cellular systems. However, there have been a number of eukaryotic organisms for which metabolic networks have been reconstructed. The most extensively studied eukaryotic metabolic network is that of the budding yeast *Saccharomyces cerevisiae* [52]. Like the prokaryote *E. coli, S. cerevisiae* is the best-characterized eukaryotic biological system with a wide array of tools available for in depth studies. The *S. cerevisiae* metabolic network has undergone multiple upgrades and revisions to incorporate new information and data into the metabolic network [52, 53]. However, because of this network's complexity, including compartmentalization of the metabolic network to represent organelles, there have been limited practical applications. Several recent updates to the metabolic network have improved the accuracy of the *S. cerevisiae* metabolic model [54, 55]. Zomorrodi and Maranas improved upon the previously reconstructed metabolic network of *S. cerevisiae*, *i*MM904 [52], and Dobson and coworkers improved upon the consensus reconstruction Yeast 1.0 [56]. The difference between the two reconstructions lies in the level of confidence the authors are willing to attribute to the metabolic reactions included in the network. These slightly different reconstructions of the same organism give different results to an identical problem, allowing researchers to perform comparative analyses, which in turn lead to the identification of previously unknown characteristics of the metabolic network and thereby improve our knowledge of the network. Other eukaryotic systems for which metabolic networks have been reconstructed include the methylotrophic yeast *Pichia pastoris* [2], several fungi from the genus *Aspergillus* [57], the mouse, *Mus musculus* [58], the plant model organism, *Arabidopsis thaliana* [59], and human [25, 26].

The human metabolic network has many complexities that prevent it from being analyzed in the same way as metabolic networks of unicellular organisms. The human biological system is composed of many different tissues with different cell types, each having their own unique metabolic profiles. As a result, use of the metabolic network typically requires additional information regarding the metabolism of the cell type of interest. Studies have been performed using the human metabolic network by mapping gene expression data from specific tissues types, such as brain cells, to the metabolic network [60, 61]. From these studies, the authors were able to elucidate specific aspects of the metabolism of a particular cell type under the given conditions from which the expression data was taken.

Recently, another emerging trend related to metabolic networks is the process of automatic reconstruction [12, 62]. Generally, the process of reconstructing genome-scale metabolic networks is performed manually in more than 90 steps [63]. This process is extremely complex and explains the slow pace of the construction of new metabolic networks. To speed up the process, Henry et al. created the Model SEED, a web-based platform that can, in a high-throughput manner, generate, optimize, and analyze reconstructed metabolic networks [12]. The Model SEED integrates existing methods; draft reconstruction of metabolic networks, gap-filling, analysis of metabolic networks, comparison of metabolic networks with phenotypic data, and manual curation. It introduces techniques to automate the steps of the reconstruction process, taking less than 2 days to reconstruct a metabolic network from an assembled genome sequence. Another automatic platform, MEMOSys (MEtabolic MOdel research and development System) supports the development of a new metabolic network by providing a version control system that can show the complete developmental history. Utilizing MEMOSys, existing models can be researched through the use of search systems, references to external databases, and feature-rich comparison mechanism to verify and refine pre-reconstructed metabolic networks [62]. While the tools for the reconstruction of metabolic networks still have limitations, particularly pertaining to poorly characterized organisms and complex metabolic networks, the automatic reconstruction pipeline is yet another advancement in the field of genome-scale metabolic networks.

1.3 Transcriptional Networks

A transcriptional network reconstruction is a representation of the network that controls the gene expression state of the cell. Not all genes in the genome are expressed at the same time in the cell, and the transcriptional network provides a blueprint for how the cell controls the timing of which genes are expressed under specific conditions. Biochemically, transcription is only partially understood, but it is known that the interactions in the transcriptional network include protein-protein interactions and protein-DNA interactions.

There are two fundamental building blocks needed to reconstruct the transcriptional network: the promoters of the genes and the transcription factors (TFs) that bind to each promoter. Identifying promoter regions of the genome is relatively straightforward, but identifying which TF binds to which promoter is more complex. In the case of higher organisms, the promoters possess sites for multiple TFs, resulting in an increasing number of combinations that can bind to each promoter. This greater complexity, and therefore a greater number of possibilities in transcriptional states, allows for more possibilities in functional states of the cell.

Information on promoters, TFs, and DNA binding proteins is usually taken from different types of experimental data which are classified by their increasing complexity: component data, interaction data, and network state data. Component data contains details about the individual components, such as TFs, in the transcription network system. Interaction data can show which TFs are active and with which promoters the TFs are interacting. Network state data shows the transcriptional state of the entire system at a specific time, which can be utilized to determine connectivity in the transcriptional network.

These components are then assembled into an interactive network representing the transcriptional state of the genome. With the construction of the transcriptional network, it was found that several basic motifs commonly occur, such as the feed-forward loop, in which components activate sections of the network downstream to speed up the response towards the input; the single input module (SIM), where a single input leads to the activation of multiple outputs; and dense overlapping regulons (DOR), which are composed of regions of complex interactions and TFs that are involved in multiple interactions in the network [64]. The classification of the different motifs simplifies the complexities of the transcriptional network and allows the user to better visualize the roles of the various components of the network.

There are two main approaches to reconstruct the transcriptional network: Top-down and Bottom-up (Fig. 1.2). The top-down approach usually utilizes high-throughput data that simultaneously measure large numbers of data points to identify the individual components of the transcriptional network. Some examples of top-down approaches include identification of the expression status of the genome, identification of all promoter sites computationally, and the experimental identification of all protein binding sites on the DNA. As with all high-throughput data, each type requires detailed curation before being used in the reconstruction. From the opposite end, the bottom-up approach involves the individual components, which are studied, characterized, and connected to the network. This method attributes high confidence to the data being used in the reconstruction of the transcriptional network. However, the process of obtaining all the necessary data for each individual component is time intensive. Therefore, a combination of the top-down approach and the bottom-up approach is usually preferred for transcriptional network reconstruction.

With the reconstruction of the transcriptional network, one can examine various properties of the system's transcriptional intricacies. First, the user can utilize the transcriptional network to obtain a better understanding of the transcription patterns through analysis of the network to reveal new information or explain observed effects. Second, causal relationships can be better understood and new relationships between previously unrelated components can be identified. Third, a reaction mechanism can be suggested based on the analysis of the transcriptional model. Finally, kinetic constants can be better estimated with the help of the transcriptional model through tuning and refinement [65].

Various methods have been developed to reconstruct transcriptional regulatory networks based on gene expression profiling data, literature data, and databases. These methods yield directed graph networks [64], Boolean networks [66–68], Boolean networks in a matrix format [69], dynamic modeling of the network [70], and probabilistic modeling of the network using Bayesian network analysis [71].

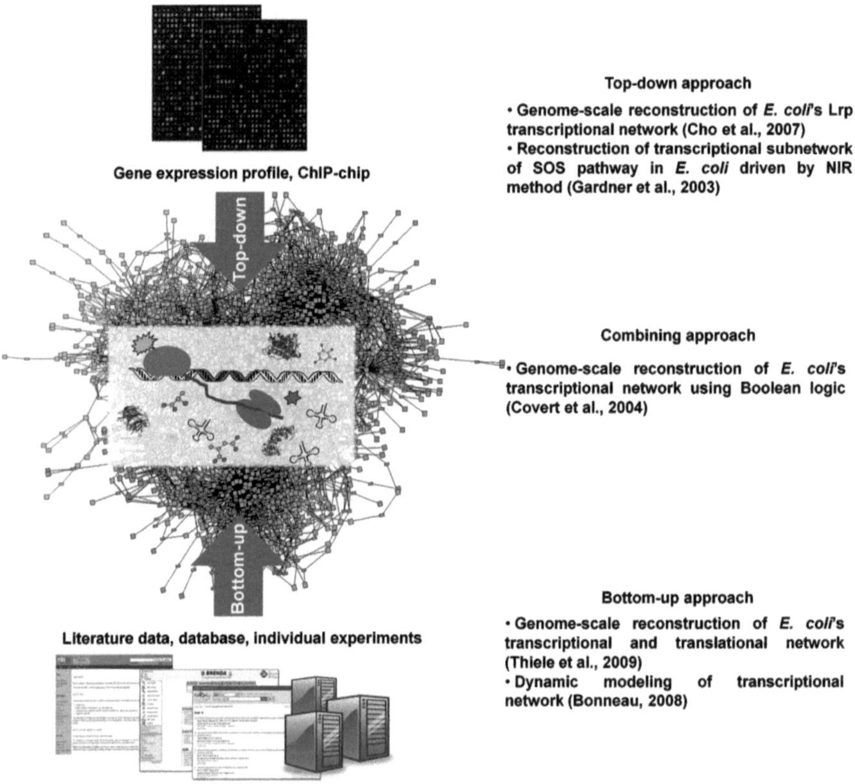

Fig. 1.2 Breakdown of the different approaches towards transcriptional network reconstruction. These two approaches are also utilized together in a combinatorial approach seen in Covert et al. [66]

Furthermore, the reconstructed transcriptional networks can be integrated with other network types, including metabolic [66, 68, 72, 73] and signaling networks [74, 75]. These integrated models allow the accurate prediction on the effects the transcriptional regulatory perturbations has for a given condition on the metabolic network. The various networks can be used to integrate omics data to analyze cellular phenotypes and more accurately predict the phenotypes of a cell for multiple conditions, and therefore can be useful for systems metabolic engineering.

1.3.1 Transcriptional Network Case Studies

Transcriptional network reconstruction can utilize high-throughput experimental data for large-scale measurement of transcriptional interactions and components, such as genome-wide expression profiling and chromatin immunoprecipitation followed by microarray hybridization (ChIP-chip) [76]. Based on combinations of high-throughput experimental data, several top-down approaches for the reconstruction of transcriptional networks have been developed [77, 78]. The reconstruction of the genome-scale transcriptional network of the leucine-responsive protein (Lrp) TF in *E. coli* K-12 MG1655 is one example of the top-down approach [77]. Lrp is a global transcriptional regulator and its regulon includes genes involved in pili synthesis, amino acid biosynthesis and degradation, among other cellular functions [79, 80]. To reconstruct the network, a systems approach integrating genome-wide data from ChIP-chip for Lrp and RNA polymerase and from gene expression profiling was employed. A four-step process to reconstruct the Lrp transcriptional network was performed. First, high-resolution ChIP-chip data and expression profiles were obtained to determine the Lrp-binding regions of the genome and to measure the changes in RNA polymerase occupancies of promoters. mRNA transcript levels were used to classify the binding states under multiple environmental conditions. Second, six distinct regulatory modes were determined, including independent, concerted, and reciprocal mode, all controlled by Lrp. Third, regulatory network motifs for metabolites that are affected by the corresponding gene products were identified. Fourth, the amino acids and metabolites with the same regulatory motifs were classified, and it was determined how leucine was able to affect the regulatory motifs for the metabolites. The physiological role of the Lrp regulon was thus understood comprehensively through the reconstruction of this transcriptional network.

Another example of the top-down approach is the transcriptional network reconstruction strategy called Network Identification by multiple Regression (NIR) [78]. In this strategy, genes in a nine transcript subnetwork of the SOS pathway in *E. coli* were perturbed for down- or up-regulation, and the resulting expression profiles for all genes were measured. Then, the NIR method, using a first-order model, was applied to infer a model of the perturbed network using the expression profiles. As a result of this analysis, a first-order model of regulatory interactions in this nine transcript subnetwork of the SOS pathway was reconstructed. The inferred network provides values between genes, called connection strengths, that indicate transcriptional relationships and interactions between genes.

Bottom-up approaches have been performed to reconstruct transcriptional networks from individual experiments, databases, and literature data [65, 70]. The genome-scale network of *E. coli*'s transcriptional and translational machinery was reconstructed using information from databases, literature, and the revised *E. coli* K-12 MG1655 genome annotation [65]. The mathematical representation of the reconstruction was designated the Expression-matrix (E-matrix), representing the expression of mRNA and proteins. By implementing the stoichiometric E-matrix from the transcriptional and the translational machinery, the quantitative integration of omics data into the transcriptional and translational network is possible. This reconstructed network can also be used to compute functional states of the network. For example, the network model accurately predicted the ribosome production in *E. coli* without any parameterization, as well as the effects of the deletion of single or multiple rRNA operons [65]. To understand transcriptional regulatory networks, transcriptional interactions and dynamics can also be modeled by differential equations and stochastic models based on individual experiments and analysis of subnetworks, which provide detailed descriptions of regulatory systems and require accurate measurement of a large number of parameters for each condition [70].

Accordingly, achieving full genome-scale analysis with dynamic modeling of transcriptional networks has significant limitations [70].

The combination of top-down and bottom-up approaches has also been utilized in the reconstruction of transcriptional networks [66, 68]. An integrated genome-scale model of transcriptional regulatory and metabolic networks in *E. coli* was reconstructed based on information from literature and databases. Gene expression profiling data was also used to reconstruct the transcriptional network [66, 68]. The model was then validated and upgraded by comparing computational predictions with experimental data from growth phenotypes for multiple gene knockouts and growth on different substrates, and with gene expression data from microarray experiments [66, 68]. To incorporate a metabolic network with a transcriptional regulatory network, Boolean logic was used to represent the availability (ON) or unavailability (OFF) of genes, proteins, and reactions as binary values [69, 81]. The transcriptional regulatory network was then combined with the genome-scale metabolic network of *E. coli* in order to determine which open reading frames (ORFs) are transcribed under given conditions and aid the accurate predictions of cellular physiology and model-driven discovery [1, 22]. Methods for the prediction of gene expression, metabolic fluxes, and steady-state regulatory flux balance analysis (SR-FBA) were developed [72]. In addition, other methods, including iFBA [74] and idFBA [75], integrating metabolic, transcriptional regulatory, and signal transduction were developed.

1.4 Cell Signaling Networks

A cell signaling network is a communication network that transduces information regarding the external conditions of the cell, allowing the cell to adjust its transcriptional state accordingly. When a cell receives a signal at the external membrane, it activates a cascade of events and information flow through the cell that ultimately ends at the nucleus or the genome. It is here that the information is integrated to affect transcription. Signals from the environment are received by the cells through various means, such as chemical (e.g. chemotaxis) and physical (e.g. pressure) stimuli. Radiation can also serve as a signal to cells, as in the case of phototaxis. In multicellular organisms, cells in different parts of the body require means to send signals to each other to ensure proper function of the body. This can be accomplished through chemicals, such as hormones, dissolved in blood or other circulatory media, physical changes to the extracellular matrix, or through direct cell to cell communication. Input from these extracellular stimuli is one of the three main components of the signaling network. The other two components are the reactions that make up the signaling network from the membrane to the nucleus and the events in the nucleus affecting transcription. Through these steps, information is transduced through the cell to the nucleus so that it can be processed and allow the cell to appropriately respond to external stimuli from the environment.

Detailed mechanisms of the complete signaling network are not fully known. However, advancements in the reconstruction of signaling networks are being

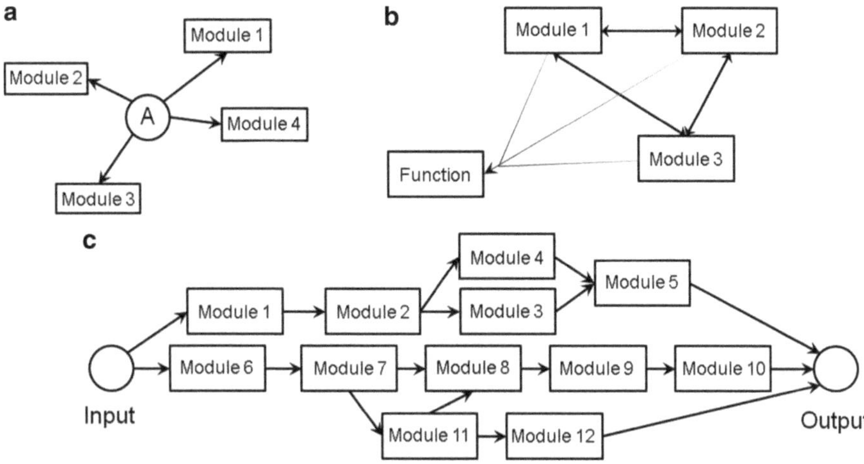

Fig. 1.3 Different strategies in the approach towards signaling network modeling. (**a**) Single signaling molecule represented by the sole node linked to various modules in the signaling network. (**b**) Different modules working together towards a common function. (**c**) Single input initiating a cascade of events that lead to an output

made. Studies have found that signaling network structure is similar to that of the metabolic network with respect to the interconnectivity and interactions between the various nodes. For instance, the degree of interconnectivity of metabolic networks is similar to that seen in the *S. cerevisiae* signaling network where there is an average of more than five protein interactions for a given protein [82–84]. One significant property of the signaling network is the combinatorial control of the components, where a few proteins can form combinations with other proteins to create receptors that can respond to a wide range of environmental stimuli [20, 85]. Therefore, while detailed studies characterizing the components of the network are being performed and have allowed for a limited level of construction of the signaling network, the full potential of the signaling network has yet to be realized.

Due to our limited knowledge of the complete signaling network, there are three different strategies for modeling signaling networks (Fig. 1.3). The first strategy involves the reconstruction of a network centered on a specific node, for example, all pathways in which the neurotransmitter acetylcholine is involved. This would encompass all paths regardless of functionality and include all roles that the node could play in the signaling network. The second strategy is the grouping of different cellular signaling components that function together under certain conditions [86, 87]. This would incorporate the interactions between different components under the specified conditions and usually includes kinetic parameters. The third strategy is the reconstruction of a signaling network consisting of a single given input and output [88].

The levels of detail in the network can also be incorporated based on the available information. The connectivity of the nodes can be either simple or complex, depending on the level of information on the mechanisms of the reactions in the signaling network (e.g. $A \rightarrow B$ as opposed to $A \rightarrow C \rightarrow B$). The reactions in

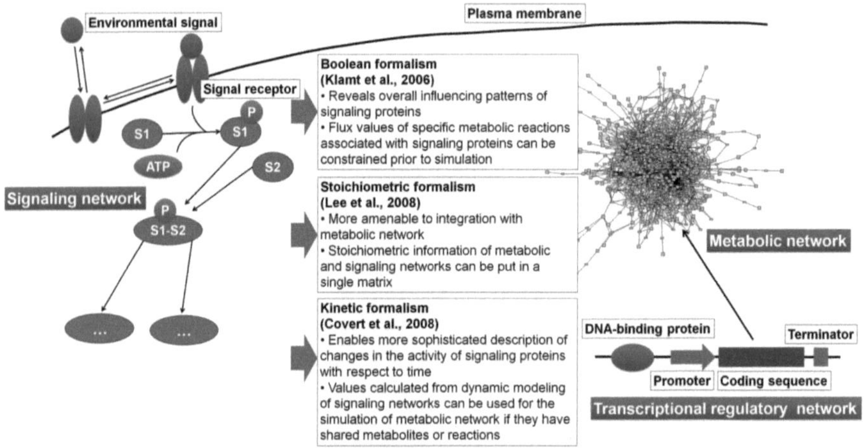

Fig. 1.4 Overview of the signaling network and the different approaches utilized in the signaling network reconstruction. There is the Boolean formalism, the stoichiometric formalism and the kinetic formalism

the network can also be further detailed by the inclusion of kinetic information. With the incorporation of kinetic parameters, an additional time dimension is added to the network allowing for a better dynamic representation of the transcriptional network. Without the kinetic information, the network would consist of reactions represented by simple causal relationships. Finally, mechanistic information on the signaling reactions can be incorporated in the form of stoichiometric coefficients (e.g. 2 A + 2 B $\rightarrow$ 1 AB_BA).

High-throughput techniques for the characterization of signaling components allowing their incorporation into a signaling network reconstruction fall under two categories: (1) biochemical techniques used to characterize protein-protein interactions, and (2) assays that elucidate functional characteristics. Some examples of protein-protein interaction studies include two-hybrid systems and mass spectrometry [89, 90]. Assays include perturbation analysis [91], RNAi knockdown [92, 93], proteome analysis [94], and fluorescence labeling [95]. These methods all have their advantages and disadvantages. Therefore, combining methods to compensate for disadvantages is suggested. Success has been achieved by integrating various data types to generate a systems-level hypotheses on the nature of the interactions between several essential proteins [96].

Modeling methodologies for signaling networks deserve further discussion (Fig. 1.4). Thus far, signaling networks have been constructed based on stoichiometric and Boolean formalisms, as they do not necessitate intricate kinetic parameters, and can be easily scaled to a large size [67, 96]. In addition to these methods, dynamic or kinetic modeling and network inference using machine learning algorithms can be applied. The question then becomes, what is the best option for modeling the signaling network? Our belief is that there is no 'one-size-fits-all' solution. Each approach has its unique strengths such that they should be considered simultaneously in order to complement one another. This would help assemble separate pieces and reveal the whole picture of the signaling network as well as its interaction with other layers of biological networks, namely metabolic and transcriptional networks.

1.4.1 Signaling Network Case Studies

Despite the relative lack of detailed information on signaling networks, there have been several attempts to model them. Most of the existing signaling network models are focused on mammalian and human cells because of their sophisticated sensing systems and cell-to-cell communication. Palsson and colleagues reconstructed the largest signaling network so far for toll-like receptors, comprised of 909 reactions and 752 components [97]. Similar to procedures used in metabolic network modeling, this signaling network was reconstructed based on a stoichiometric formalism, such that flux balance analysis (FBA) could be used for simulations. Distinct input–output pathways were calculated and control points that are specific for the target pathways while not affecting other parts of the signaling network were identified.

The two-component regulatory system for signal transduction has been modeled for bacterial systems wherein sensor proteins embedded in the cell membrane sense an external signal from the environment and are phosphorylated, transmitting the information to the response regulator proteins [98]. The response regulator protein ultimately binds DNA to accordingly regulate transcription. Because of the dynamic behavior of signal transduction, stoichiometric network modeling of this system under pseudo-steady state has not been reported to our knowledge. Instead, most mathematical modeling of the two-component system resorts to kinetic modeling. Examples include phototaxis and chemotaxis of the archaeon *Halobacterium salinarum* [99], chemotaxis of *E. coli* with emphasis on the phosphatase CheZ [100], bacterial chemotaxis focused on the histidine kinase CheA [101], and the KdpD/KdpE system of *E. coli* that regulates expression of the high affinity K^+ uptake system [102]. Successful descriptions of such signaling pathways in production hosts will be of great importance in systems metabolic engineering because they may contribute to identification and optimization of unnoticed bioprocess parameters.

Aside from these studies, relatively few studies have been conducted on signaling network modeling for organisms appropriate for systems metabolic engineering when only the stoichiometric formalism and optimization-based simulations are considered. One reason would be that bacterial signaling networks are still not fully understood, despite the relative simplicity of their intracellular networks compared to eukaryotic signaling networks. Many bacterial signaling networks include signaling between other organisms in their environment, and thus can be just as complex as eukaryotes. Furthermore, the links between metabolic networks and signaling networks are not fully established. Current research has limited the integration between the two networks to specific regions, and not full networks.

Therefore, in the case of microorganisms, the focus has been on the development of integrative network models rather than independent signaling networks.

Examples of integrative modeling include the models of *E. coli* [74] and *S. cerevisiae* [75]. Both models simultaneously account for metabolic, transcriptional regulatory, and signal transduction information. In the study of *E. coli*, integrated FBA (iFBA) was developed [74], in which the stoichiometric metabolic network model of *E. coli* was integrated with a Boolean regulatory model [81] and an ordinary differential equation (ODE)-based kinetic model of *E. coli* describing the phosphotransferase (PTS) catabolite repression mechanism [103]. In this algorithm, a Boolean model of transcriptional regulation is used to constrain reactions to be active or inactive, under given condition. Then an ODE model of PTS catabolite repression is used to calculate numerical values which are passed to the model through common metabolites. iFBA was demonstrated for wild-type *E. coli* and single gene mutants for diauxic growth on glucose/lactose and glucose/glucose-6-phosphate. A significant improvement in predictive capability was found compared to individual FBA and ODE models.

Likewise, integrated dynamic FBA (idFBA), was developed and applied to the high-osmolarity glycerol response (HOG) pathway in *S. cerevisiae*, a crucial signaling pathway for adaptation to high external osmolarity [75]. In contrast to prokaryotic hosts such as *E. coli*, the consideration of signaling networks becomes more important in yeast as it has a more developed and complex signaling system. Here, unlike iFBA, signaling information was incorporated into the metabolic network through a stoichiometric formalism, thereby enabling simultaneous simulation via optimization. Another important distinction is the use of the incidence matrix with binary parameters that indicate activation or inactivation of reactions, 1 or 0 respectively, at each discretized time point. This matrix is updated progressively, producing a time-dependent dynamic simulation.

1.5 Concluding Remarks

Biological networks are complex systems that are not fully understood at our current level of knowledge. However, as compartmentalized networks such as metabolic, transcriptional, or signaling networks, become more sophisticated, we move one step closer to achieving a fully reconstructed genome-scale cellular network. While transcriptional and signaling networks do not encompass the same level of information as metabolic networks, recent studies have elucidated many characteristics of these networks that were then reincorporated into the metabolic network to aid in further understanding. Networks incorporating limited information from other networks have also been valuable in the study of biological systems. They also provide hypotheses for designing strategies to fully incorporate the different types of networks into a single network representing a complete biological system.

There should be consistent feedback between experimental and network modeling to gain better insight into biological systems. Experimental data no doubt lays the foundation for reconstructing initial versions of biological networks, which in turn generate new hypotheses that must be experimentally validated. Once validated, these hypotheses would then contribute to updating the biological network. Although we did not discuss experimental techniques in detail in this chapter, various high-throughput techniques at different biological levels, including genome, transcriptome, proteome, metabolome, and fluxome levels, deserve close attention. Pieces of information from this modeling effort, in combination with experimental data, should help to elucidate the big picture of biological systems.

Acknowledgments This work was supported by the Intelligent Synthetic Biology Center (2011-0031963) through the Global Frontier Project of Ministry of Education, Science and Technology.

References

1. Feist AM, Henry CS, Reed JL, Krummenacker M, Joyce AR, Karp PD, Broadbelt LJ, Hatzimanikatis V, Palsson BØ (2007) A genome-scale metabolic reconstruction for *Escherichia coli* K-12 MG1655 that accounts for 1260 ORFs and thermodynamic information. Mol Syst Biol 3:121
2. Sohn SB, Graf AB, Kim TY, Gasser B, Maurer M, Ferrer P, Mattanovich D, Lee SY (2010) Genome-scale metabolic model of methylotrophic yeast *Pichia pastoris* and its use for *in silico* analysis of heterologous protein production. Biotechnol J 5(7):705–715
3. Gianchandani EP, Joyce AR, Palsson BØ, Papin JA (2009) Functional states of the genome-scale *Escherichia coli* transcriptional regulatory system. PLoS Comput Biol 5(6):e1000403
4. Thiele I, Fleming RM, Bordbar A, Schellenberger J, Palsson BØ (2010) Functional characterization of alternate optimal solutions of *Escherichia coli*'s transcriptional and translational machinery. Biophys J 98(10):2072–2081
5. Park JH, Lee KH, Kim TY, Lee SY (2007) Metabolic engineering of *Escherichia coli* for the production of L-valine based on transcriptome analysis and *in silico* gene knockout simulation. Proc Natl Acad Sci USA 104(19):7797–7802
6. Becker J, Zelder O, Häfner S, Schröder H, Wittmann C (2011) From zero to hero–design-based systems metabolic engineering of *Corynebacterium glutamicum* for L-lysine production. Metab Eng 13(2):159–168
7. Cho BK, Zengler K, Qiu Y, Park YS, Knight EM, Barrett CL, Gao Y, Palsson BØ (2009) The transcription unit architecture of the *Escherichia coli* genome. Nat Biotechnol 27(11):1043–1049
8. Orth JD, Palsson BØ (2010) Systematizing the generation of missing metabolic knowledge. Biotechnol Bioeng 107(3):403–412
9. Bairoch A (2000) The ENZYME database in 2000. Nucleic Acids Res 28(1):304–305
10. Ashburner M, Ball CA, Blake JA, Botstein D, Butler H, Cherry JM, Davis AP, Dolinski K, Dwight SS, Eppig JT, Harris MA, Hill DP, Issel-Tarver L, Kasarskis A, Lewis S, Matese JC, Richardson JE, Ringwald M, Rubin GM, Sherlock G (2000) Gene ontology: tool for the unification of biology. The gene ontology consortium. Nat Genet 25(1):25–29
11. Kanehisa M, Goto S, Furumichi M, Tanabe M, Hirakawa M (2010) KEGG for representation and analysis of molecular networks involving diseases and drugs. Nucleic Acids Res 38(Database issue):D355–D360

12. Henry CS, DeJongh M, Best AA, Frybarger PM, Linsay B, Stevens RL (2010) High-throughput generation, optimization and analysis of genome-scale metabolic models. Nat Biotechnol 28(9):977–982
13. Caspi R, Altman T, Dale JM, Dreher K, Fulcher CA, Gilham F, Kaipa P, Karthikeyan AS, Kothari A, Krummenacker M, Latendresse M, Mueller LA, Paley S, Popescu L, Pujar A, Shearer AG, Zhang P, Karp PD (2010) The MetaCyc database of metabolic pathways and enzymes and the BioCyc collection of pathway/genome databases. Nucleic Acids Res 38(Database issue):D473–D479
14. Price ND, Papin JA, Schilling CH, Palsson BØ (2003) Genome-scale microbial *in silico* models: the constraints-based approach. Trends Biotechnol 21(4):162–169
15. Orth JD, Thiele I, Palsson BØ (2010) What is flux balance analysis? Nat Biotechnol 28(3):245–248
16. Palsson BØ (2006) Systems biology: properties of reconstructed networks. Cambridge University Press, New York
17. Becker SA, Feist AM, Mo ML, Hannum G, Palsson BØ, Herrgård MJ (2007) Quantitative prediction of cellular metabolism with constraint-based models: the COBRA toolbox. Nat Protoc 2(3):727–738
18. Reed JL, Patel TR, Chen KH, Joyce AR, Applebee MK, Herring CD, Bui OT, Knight EM, Fong SS, Palsson BØ (2006) Systems approach to refining genome annotation. Proc Natl Acad Sci USA 103(46):17480–17484
19. Segre D, Vitkup D, Church GM (2002) Analysis of optimality in natural and perturbed metabolic networks. Proc Natl Acad Sci USA 99(23):15112–15117
20. Klamt S, Stelling J (2003) Two approaches for metabolic pathway analysis? Trends Biotechnol 21(2):64–69
21. Dandekar T, Moldenhauer F, Bulik S, Bertram H, Schuster S (2003) A method for classifying metabolites in topological pathway analyses based on minimization of pathway number. Biosystems 70(3):255–270
22. Lewis NE, Cho BK, Knight EM, Palsson BØ (2009) Gene expression profiling and the use of genome-scale *in silico* models of *Escherichia coli* for analysis: providing context for content. J Bacteriol 191(11):3437–3444
23. Ibarra RU, Edwards JS, Palsson BO (2002) *Escherichia coli* K-12 undergoes adaptive evolution to achieve *in silico* predicted optimal growth. Nature 420(6912):186–189
24. Edwards JS, Palsson BØ (1999) Systems properties of the *Haemophilus influenzae* Rd metabolic genotype. J Biol Chem 274(25):17410–17416
25. Duarte NC, Becker SA, Jamshidi N, Thiele I, Mo ML, Vo TD, Srivas R, Palsson BØ (2007) Global reconstruction of the human metabolic network based on genomic and bibliomic data. Proc Natl Acad Sci USA 104(6):1777–1782
26. Ma H, Sorokin A, Mazein A, Selkov A, Selkov E, Demin O, Goryanin I (2007) The Edinburgh human metabolic network reconstruction and its functional analysis. Mol Syst Biol 3:135
27. Edwards JS, Palsson BØ (2000) The *Escherichia coli* MG1655 *in silico* metabolic genotype: its definition, characteristics, and capabilities. Proc Natl Acad Sci USA 97(10):5528–5533
28. Reed JL, Vo TD, Schilling CH, Palsson BØ (2003) An expanded genome-scale model of *Escherichia coli* K-12 (*i*JR904 GSM/GPR). Genome Biol 4(9):R54.1–R54.12
29. Alper H, Jin YS, Moxley JF, Stephanopoulos G (2005) Identifying gene targets for the metabolic engineering of lycopene biosynthesis in *Escherichia coli*. Metab Eng 7(3):155–164
30. Choi HS, Lee SY, Kim TY, Woo HM (2010) *In silico* identification of gene amplification targets for improvement of lycopene production. Appl Environ Microbiol 76(10):3097–3105
31. Jung YK, Kim TY, Park SJ, Lee SY (2010) Metabolic engineering of *Escherichia coli* for the production of polylactic acid and its copolymers. Biotechnol Bioeng 105(1):161–171
32. Burgard AP, Pharkya P, Maranas CD (2003) Optknock: a bilevel programming framework for identifying gene knockout strategies for microbial strain optimization. Biotechnol Bioeng 84(6):647–657

33. Patil KR, Rocha I, Forster J, Nielsen J (2005) Evolutionary programming as a platform for *in silico* metabolic engineering. BMC Bioinformatics 6:308
34. Pharkya P, Burgard AP, Maranas CD (2004) OptStrain: a computational framework for redesign of microbial production systems. Genome Res 14(11):2367–2376
35. Melzer G, Esfandabadi ME, Franco-Lara E, Wittmann C (2009) Flux design: *in silico* design of cell factories based on correlation of pathway fluxes to desired properties. BMC Syst Biol 3:120
36. Feist AM, Zielinski DC, Orth JD, Schellenberger J, Herrgård MJ, Palsson BØ (2010) Model-driven evaluation of the production potential for growth-coupled products of *Escherichia coli*. Metab Eng 12(3):173–186
37. Fong SS, Burgard AP, Herring CD, Knight EM, Blattner FR, Maranas CD, Palsson BØ (2005) *In silico* design and adaptive evolution of *Escherichia coli* for production of lactic acid. Biotechnol Bioeng 91(5):643–648
38. Krömer JO, Wittmann C, Schröder H, Heinzle E (2006) Metabolic pathway analysis for rational design of L-methionine production by *Escherichia coli* and *Corynebacterium glutamicum*. Metab Eng 8(4):353–369
39. Kim TY, Kim HU, Park JM, Song H, Kim JS, Lee SY (2007) Genome-scale analysis of *Mannheimia succiniciproducens* metabolism. Biotechnol Bioeng 97(4):657–671
40. Kjeldsen KR, Nielsen J (2009) *In silico* genome-scale reconstruction and validation of the *Corynebacterium glutamicum* metabolic network. Biotechnol Bioeng 102(2):583–597
41. Alam MT, Merlo ME, Hodgson DA, Wellington EM, Takano E, Breitling R (2010) Metabolic modeling and analysis of the metabolic switch in *Streptomyces coelicolor*. BMC Genomics 11:202
42. Sohn SB, Kim TY, Park JM, Lee SY (2010) *In silico* genome-scale metabolic analysis of *Pseudomonas putida* KT2440 for polyhydroxyalkanoate synthesis, degradation of aromatics and anaerobic survival. Biotechnol J 5(7):739–750
43. Lee J, Yun H, Feist AM, Palsson BØ, Lee SY (2008) Genome-scale reconstruction and *in silico* analysis of the *Clostridium acetobutylicum* ATCC 824 metabolic network. Appl Microbiol Biotechnol 80(5):849–862
44. Lee KY, Park JM, Kim TY, Yun H, Lee SY (2010) The genome-scale metabolic network analysis of *Zymomonas mobilis* ZM4 explains physiological features and suggests ethanol and succinic acid production strategies. Microb Cell Fact 9:94
45. Schilling CH, Covert MW, Famili I, Church GM, Edwards JS, Palsson BØ (2002) Genome-scale metabolic model of *Helicobacter pylori* 26695. J Bacteriol 184(16):4582–4593
46. Kim HU, Kim SY, Jeong H, Kim TY, Kim JJ, Choy HE, Yi KY, Rhee JH, Lee SY (2011) Integrative genome-scale metabolic analysis of *Vibrio vulnificus* for drug targeting and discovery. Mol Syst Biol 7:460
47. Oberhardt MA, Puchalka J, Fryer KE, Martins dos Santos VA, Papin JA (2008) Genome-scale metabolic network analysis of the opportunistic pathogen *Pseudomonas aeruginosa* PAO1. J Bacteriol 190(8):2790–2803
48. Kim HU, Kim TY, Lee SY (2010) Genome-scale metabolic network analysis and drug targeting of multi-drug resistant pathogen *Acinetobacter baumannii* AYE. Mol Biosyst 6(2):339–348
49. Kim TY, Kim HU, Lee SY (2010) Metabolite-centric approaches for the discovery of antibacterials using genome-scale metabolic networks. Metab Eng 12(2):105–111
50. Kitano H (2007) A robustness-based approach to systems-oriented drug design. Nat Rev Drug Discov 6(3):202–210
51. Kitano H (2007) Biological robustness in complex host-pathogen systems. Prog Drug Res 64(239):241–263
52. Mo ML, Palsson BØ, Herrgård MJ (2009) Connecting extracellular metabolomic measurements to intracellular flux states in yeast. BMC Syst Biol 3:37

53. Duarte NC, Herrgård MJ, Palsson BØ (2004) Reconstruction and validation of *Saccharomyces cerevisiae* iND750, a fully compartmentalized genome-scale metabolic model. Genome Res 14(7):1298–1309
54. Dobson PD, Smallbone K, Jameson D, Simeonidis E, Lanthaler K, Pir P, Lu C, Swainston N, Dunn WB, Fisher P, Hull D, Brown M, Oshota O, Stanford NJ, Kell DB, King RD, Oliver SG, Stevens RD, Mendes P (2010) Further developments towards a genome-scale metabolic model of yeast. BMC Syst Biol 4:145
55. Zomorrodi AR, Maranas CD (2010) Improving the *i*MM904 *S. cerevisiae* metabolic model using essentiality and synthetic lethality data. BMC Syst Biol 4:178
56. Herrgard MJ, Swainston N, Dobson P, Dunn WB, Arga KY, Arvas M, Bluthgen N, Borger S, Costenoble R, Heinemann M, Hucka M, Le Novere N, Li P, Liebermeister W, Mo ML, Oliveira AP, Petranovic D, Pettifer S, Simeonidis E, Smallbone K, Spasic I, Weichart D, Brent R, Broomhead DS, Westerhoff HV, Kirdar B, Penttila M, Klipp E, Palsson BØ, Sauer U, Oliver SG, Mendes P, Nielsen J, Kell DB (2008) A consensus yeast metabolic network reconstruction obtained from a community approach to systems biology. Nat Biotechnol 26 (10):1155–1160
57. David H, Ozcelik IS, Hofmann G, Nielsen J (2008) Analysis of *Aspergillus nidulans* metabolism at the genome-scale. BMC Genomics 9:163
58. Selvarasu S, Karimi IA, Ghim GH, Lee DY (2010) Genome-scale modeling and *in silico* analysis of mouse cell metabolic network. Mol Biosyst 6(1):142–151
59. de Oliveira Dal'Molin CG, Quek LE, Palfreyman RW, Brumbley SM, Nielsen LK (2010) AraGEM, a genome-scale reconstruction of the primary metabolic network in Arabidopsis. Plant Physiol 152(2):579–589
60. Lewis NE, Schramm G, Bordbar A, Schellenberger J, Andersen MP, Cheng JK, Patel N, Yee A, Lewis RA, Eils R, Konig R, Palsson BØ (2010) Large-scale *in silico* modeling of metabolic interactions between cell types in the human brain. Nat Biotechnol 28(12):1279–1285
61. Shlomi T, Cabili MN, Herrgard MJ, Palsson BØ, Ruppin E (2008) Network-based prediction of human tissue-specific metabolism. Nat Biotechnol 26(9):1003–1010
62. Pabinger S, Rader R, Agren R, Nielsen J, Trajanoski Z (2011) MEMOSys: bioinformatics platform for genome-scale metabolic models. BMC Syst Biol 5(1):20
63. Thiele I, Palsson BØ (2010) A protocol for generating a high-quality genome-scale metabolic reconstruction. Nat Protoc 5(1):93–121
64. Shen-Orr SS, Milo R, Mangan S, Alon U (2002) Network motifs in the transcriptional regulation network of *Escherichia coli*. Nat Genet 31(1):64–68
65. Thiele I, Jamshidi N, Fleming RM, Palsson BØ (2009) Genome-scale reconstruction of *Escherichia coli*'s transcriptional and translational machinery: a knowledge base, its mathematical formulation, and its functional characterization. PLoS Comput Biol 5(3):e1000312
66. Covert MW, Knight EM, Reed JL, Herrgard MJ, Palsson BØ (2004) Integrating high-throughput and computational data elucidates bacterial networks. Nature 429(6987):92–96
67. Klamt S, Saez-Rodriguez J, Lindquist JA, Simeoni L, Gilles ED (2006) A methodology for the structural and functional analysis of signaling and regulatory networks. BMC Bioinformatics 7:56
68. Barrett CL, Herring CD, Reed JL, Palsson BØ (2005) The global transcriptional regulatory network for metabolism in *Escherichia coli* exhibits few dominant functional states. Proc Natl Acad Sci USA 102(52):19103–19108
69. Gianchandani EP, Papin JA, Price ND, Joyce AR, Palsson BO (2006) Matrix formalism to describe functional states of transcriptional regulatory systems. PLoS Comput Biol 2(8):e101
70. Bonneau R (2008) Learning biological networks: from modules to dynamics. Nat Chem Biol 4(11):658–664
71. Workman CT, Mak HC, McCuine S, Tagne JB, Agarwal M, Ozier O, Begley TJ, Samson LD, Ideker T (2006) A systems approach to mapping DNA damage response pathways. Science (New York, NY) 312(5776):1054–1059

72. Shlomi T, Eisenberg Y, Sharan R, Ruppin E (2007) A genome-scale computational study of the interplay between transcriptional regulation and metabolism. Mol Syst Biol 3:101
73. Herrgard MJ, Lee BS, Portnoy V, Palsson BØ (2006) Integrated analysis of regulatory and metabolic networks reveals novel regulatory mechanisms in *Saccharomyces cerevisiae*. Genome Res 16(5):627–635
74. Covert MW, Xiao N, Chen TJ, Karr JR (2008) Integrating metabolic, transcriptional regulatory and signal transduction models in *Escherichia coli*. Bioinformatics 24(18):2044–2050
75. Lee JM, Gianchandani EP, Eddy JA, Papin JA (2008) Dynamic analysis of integrated signaling, metabolic, and regulatory networks. PLoS Comput Biol 4(5):e1000086
76. Cho BK, Charusanti P, Herrgard MJ, Palsson BØ (2007) Microbial regulatory and metabolic networks. Curr Opin Biotechnol 18(4):360–364
77. Cho BK, Barrett CL, Knight EM, Park YS, Palsson BØ (2008) Genome-scale reconstruction of the Lrp regulatory network in *Escherichia coli*. Proc Natl Acad Sci USA 105 (49):19462–19467
78. Gardner TS, di Bernardo D, Lorenz D, Collins JJ (2003) Inferring genetic networks and identifying compound mode of action via expression profiling. Science (New York, NY) 301 (5629):102–105
79. Calvo JM, Matthews RG (1994) The leucine-responsive regulatory protein, a global regulator of metabolism in *Escherichia coli*. Microbiol Rev 58(3):466–490
80. Newman EB, Lin R (1995) Leucine-responsive regulatory protein: a global regulator of gene expression in *E. coli*. Annu Rev Microbiol 49:747–775
81. Covert MW, Palsson BØ (2002) Transcriptional regulation in constraints-based metabolic models of *Escherichia coli*. J Biol Chem 277(31):28058–28064
82. Grigoriev A (2003) On the number of protein-protein interactions in the yeast proteome. Nucleic Acids Res 31(14):4157–4161
83. Jeong H, Tombor B, Albert R, Oltvai ZN, Barabasi AL (2000) The large-scale organization of metabolic networks. Nature 407(6804):651–654
84. Wagner A, Fell DA (2001) The small world inside large metabolic networks. Proc Biol Sci 268(1478):1803–1810
85. Venter JC, Adams MD, Myers EW, Li PW, Mural RJ, Sutton GG, Smith HO, Yandell M, Evans CA, Holt RA, Gocayne JD, Amanatides P, Ballew RM, Huson DH, Wortman JR, Zhang Q, Kodira CD, Zheng XH, Chen L, Skupski M, Subramanian G, Thomas PD, Zhang J, Gabor Miklos GL, Nelson C, Broder S, Clark AG, Nadeau J, McKusick VA, Zinder N, Levine AJ, Roberts RJ, Simon M, Slayman C, Hunkapiller M, Bolanos R, Delcher A, Dew I, Fasulo D, Flanigan M, Florea L, Halpern A, Hannenhalli S, Kravitz S, Levy S, Mobarry C, Reinert K, Remington K, Abu-Threideh J, Beasley E, Biddick K, Bonazzi V, Brandon R, Cargill M, Chandramouliswaran I, Charlab R, Chaturvedi K, Deng Z, Di Francesco V, Dunn P, Eilbeck K, Evangelista C, Gabrielian AE, Gan W, Ge W, Gong F, Gu Z, Guan P, Heiman TJ, Higgins ME, Ji RR, Ke Z, Ketchum KA, Lai Z, Lei Y, Li Z, Li J, Liang Y, Lin X, Lu F, Merkulov GV, Milshina N, Moore HM, Naik AK, Narayan VA, Neelam B, Nusskern D, Rusch DB, Salzberg S, Shao W, Shue B, Sun J, Wang Z, Wang A, Wang X, Wang J, Wei M, Wides R, Xiao C, Yan C, Yao A, Ye J, Zhan M, Zhang W, Zhang H, Zhao Q, Zheng L, Zhong F, Zhong W, Zhu S, Zhao S, Gilbert D, Baumhueter S, Spier G, Carter C, Cravchik A, Woodage T, Ali F, An H, Awe A, Baldwin D, Baden H, Barnstead M, Barrow I, Beeson K, Busam D, Carver A, Center A, Cheng ML, Curry L, Danaher S, Davenport L, Desilets R, Dietz S, Dodson K, Doup L, Ferriera S, Garg N, Gluecksmann A, Hart B, Haynes J, Haynes C, Heiner C, Hladun S, Hostin D, Houck J, Howland T, Ibegwam C, Johnson J, Kalush F, Kline L, Koduru S, Love A, Mann F, May D, McCawley S, McIntosh T, McMullen I, Moy M, Moy L, Murphy B, Nelson K, Pfannkoch C, Pratts E, Puri V, Qureshi H, Reardon M, Rodriguez R, Rogers YH, Romblad D, Ruhfel B, Scott R, Sitter C, Smallwood M, Stewart E, Strong R, Suh E, Thomas R, Tint NN, Tse S, Vech C, Wang G, Wetter J, Williams S, Williams M, Windsor S, Winn-Deen E, Wolfe K, Zaveri J, Zaveri K, Abril JF, Guigo R, Campbell MJ, Sjolander KV, Karlak B, Kejariwal A, Mi H, Lazareva B, Hatton T, Narechania A, Diemer K,

Muruganujan A, Guo N, Sato S, Bafna V, Istrail S, Lippert R, Schwartz R, Walenz B, Yooseph S, Allen D, Basu A, Baxendale J, Blick L, Caminha M, Carnes-Stine J, Caulk P, Chiang YH, Coyne M, Dahlke C, Mays A, Dombroski M, Donnelly M, Ely D, Esparham S, Fosler C, Gire H, Glanowski S, Glasser K, Glodek A, Gorokhov M, Graham K, Gropman B, Harris M, Heil J, Henderson S, Hoover J, Jennings D, Jordan C, Jordan J, Kasha J, Kagan L, Kraft C, Levitsky A, Lewis M, Liu X, Lopez J, Ma D, Majoros W, McDaniel J, Murphy S, Newman M, Nguyen T, Nguyen N, Nodell M, Pan S, Peck J, Peterson M, Rowe W, Sanders R, Scott J, Simpson M, Smith T, Sprague A, Stockwell T, Turner R, Venter E, Wang M, Wen M, Wu D, Wu M, Xia A, Zandieh A, Zhu X (2001) The sequence of the human genome. Science 291(5507):1304–1351

86. Schoeberl B, Eichler-Jonsson C, Gilles ED, Muller G (2002) Computational modeling of the dynamics of the MAP kinase cascade activated by surface and internalized EGF receptors. Nat Biotechnol 20(4):370–375
87. Wiley HS, Shvartsman SY, Lauffenburger DA (2003) Computational modeling of the EGF-receptor system: a paradigm for systems biology. Trends Cell Biol 13(1):43–50
88. van Drogen F, Stucke VM, Jorritsma G, Peter M (2001) MAP kinase dynamics in response to pheromones in budding yeast. Nat Cell Biol 3(12):1051–1059
89. Blagoev B, Ong SE, Kratchmarova I, Mann M (2004) Temporal analysis of phosphotyrosine-dependent signaling networks by quantitative proteomics. Nat Biotechnol 22(9):1139–1145
90. Ho Y, Gruhler A, Heilbut A, Bader GD, Moore L, Adams SL, Millar A, Taylor P, Bennett K, Boutilier K, Yang L, Wolting C, Donaldson I, Schandorff S, Shewnarane J, Vo M, Taggart J, Goudreault M, Muskat B, Alfarano C, Dewar D, Lin Z, Michalickova K, Willems AR, Sassi H, Nielsen PA, Rasmussen KJ, Andersen JR, Johansen LE, Hansen LH, Jespersen H, Podtelejnikov A, Nielsen E, Crawford J, Poulsen V, Sorensen BD, Matthiesen J, Hendrickson RC, Gleeson F, Pawson T, Moran MF, Durocher D, Mann M, Hogue CW, Figeys D, Tyers M (2002) Systematic identification of protein complexes in *Saccharomyces cerevisiae* by mass spectrometry. Nature 415(6868):180–183
91. Ideker T, Thorsson V, Ranish JA, Christmas R, Buhler J, Eng JK, Bumgarner R, Goodlett DR, Aebersold R, Hood L (2001) Integrated genomic and proteomic analyses of a systematically perturbed metabolic network. Science 292(5518):929–934
92. Boutros M, Kiger AA, Armknecht S, Kerr K, Hild M, Koch B, Haas SA, Paro R, Perrimon N (2004) Genome-wide RNAi analysis of growth and viability in *Drosophila* cells. Science 303(5659):832–835
93. Zheng L, Liu J, Batalov S, Zhou D, Orth A, Ding S, Schultz PG (2004) An approach to genomewide screens of expressed small interfering RNAs in mammalian cells. Proc Natl Acad Sci USA 101(1):135–140
94. Nielsen UB, Cardone MH, Sinskey AJ, MacBeath G, Sorger PK (2003) Profiling receptor tyrosine kinase activation by using Ab microarrays. Proc Natl Acad Sci USA 100(16): 9330–9335
95. Meyer T, Teruel MN (2003) Fluorescence imaging of signaling networks. Trends Cell Biol 13(2):101–106
96. Hyduke DR, Palsson BØ (2010) Towards genome-scale signalling-network reconstructions. Nat Rev Genet 11(4):297–307
97. Li F, Thiele I, Jamshidi N, Palsson BØ (2009) Identification of potential pathway mediation targets in Toll-like receptor signaling. PLoS Comput Biol 5(2):e1000292
98. Madigan MT, Brock TD (2009) Brock biology of microorganisms, 12th edn. Pearson/Benjamin Cummings, San Francisco
99. Streif S, Oesterhelt D, Marwan W (2010) A predictive computational model of the kinetic mechanism of stimulus-induced transducer methylation and feedback regulation through CheY in archaeal phototaxis and chemotaxis. BMC Syst Biol 4:27
100. van Albada SB, Ten Wolde PR (2009) Differential affinity and catalytic activity of CheZ in *E. coli* chemotaxis. PLoS Comput Biol 5(5):e1000378

101. Zhang J, Xu Y, Shen J, Luo X, Chen J, Chen K, Zhu W, Jiang H (2005) Dynamic mechanism for the autophosphorylation of CheA histidine kinase: molecular dynamics simulations. J Am Chem Soc 127(33):11709–11719
102. Kremling A, Heermann R, Centler F, Jung K, Gilles ED (2004) Analysis of two-component signal transduction by mathematical modeling using the KdpD/KdpE system of *Escherichia coli*. Biosystems 78(1–3):23–37
103. Kremling A, Bettenbrock K, Gilles ED (2007) Analysis of global control of *Escherichia coli* carbohydrate uptake. BMC Syst Biol 1:42

Chapter 2
Kinetic Modeling of Metabolic Networks

Daniel C. Zielinski and Bernhard Ø. Palsson

Contents

Abstract In the last decade, genome-scale stoichiometric models have played an increasing role in understanding metabolism under steady state. In order to study metabolic response to perturbation at timescales before a steady state is reached, however, a more explicit kinetic model must be developed. While kinetic models of metabolism have been around for longer than their stoichiometric counterparts, progress towards practical and useful kinetics models of metabolism has been slower, due to the difficulty of specifying necessary parameters. However, the

D.C. Zielinski • B.Ø. Palsson (✉)
Department of Bioengineering, University of California, San Diego, CA, USA
e-mail: bpalsson@bioeng.ucsd.edu

C. Wittmann and S.Y. Lee (eds.), *Systems Metabolic Engineering*,
DOI 10.1007/978-94-007-4534-6_2, © Springer Science+Business Media Dordrecht 2012

increased ability to measure metabolomics and proteomics profiles in high throughput may soon make accurate kinetic models of metabolism a reality. In this chapter, we review theoretical concepts useful for developing kinetic models of metabolism, practical difficulties with constructing such models, and methods that have been developed in an effort to circumvent these difficulties.

Keywords Kinetic modeling • Large-scale • Stoichiometric models • Kinetic parameters • Metabolite profiling • Proteome profiling • Dynamics • Structural hierarchies • Temporal hierarchies • Spatial heterogeneity • Gradients • Nonlinearity • Law of mass action • Gibbs free energy • Equilibrium constants • Reaction mechanism • Haldane • Rate laws • Elasticity coefficient • Michaelis-Menten • Hill • Linear analysis • Matrix • Data fitting • Parameter sensitivity

2.1 Introduction

Life is a dynamic process, and, as such, describing the time evolution of biological phenomena through mathematical modeling is of great practical and theoretical interest. Beginning nearly a century ago with the work of Michaelis and Menten on enzyme catalysis [1] and Lotka and Volterra on population dynamics [2, 3], mathematical models have now been constructed with varying success and scope for nearly every process in biology, from neuron synapse firing [4] to protein folding [5] to circadian rhythms [6]. Metabolic networks have been a major area of interest for kinetic modeling as early as the 1940s [7], and the field has remained active since.

Mathematical modeling of metabolism yielded some of the first examples of an integrative systems approach to understanding biological function. While first efforts were directed primarily at uncovering the dependencies of enzyme reaction rate on substrates, effectors, and environmental factors, these models were soon integrated to describe the dynamic function of canonical pathways such as glycolysis [8]. Iterative model construction eventually culminated in the full dynamic description of the metabolism of the red blood cell, a simple enucleated cell, based upon the assembly of individually-created enzyme models into a network model [9–12]. The red blood cell model represented one of the first successes of quantitative systems biology and was the result of iterative modeling based upon decades of detailed research in enzymology.

The construction of a realistic cell scale kinetic model of metabolism, even for a small cell in a known environment, is not a simple task. Efforts at kinetic modeling are inevitably limited by available measurement techniques and computational capabilities [13]. Although additional advances in large-scale kinetic modeling have been made, data availability limitations have spurred the development of steady-state constraint-based modeling methods, discussed in Chap. 1, as an alternative that requires less data to enable biologically interesting studies [14]. Steady-state models operate under the assumption that metabolic reaction rates have achieved a steady state at the timescale of interest, which is often that of bacterial growth on the order of a few hours. Kinetic models explicitly account for unsteady dynamics by developing systems of equations to describe how reaction rates respond to system variables away from steady-state. A kinetic model and steady-state model of the same metabolic system under the same condition would make identical predictions for reaction rates at steady-state. However, a kinetic model would also be able to predict reaction rates and trajectories at any arbitrary state away from steady-state, assuming that the factors not explicitly accounted for in the model, for example enzyme concentrations or pH, do not change over the timescale of interest. Obtaining the additional value given by explicit kinetic models requires both an accurate mathematical definition of the functional relationship between reaction rates and system variables, normally given in the form of rate laws, as well as full specification of kinetic parameters for those rate laws. Generally, both the mathematical form of reaction rate laws and the *in vivo* rate law parameters are highly difficult to specify due to limited data. However, with advances in high-throughput measurement capabilities and computational techniques, it may soon be possible to develop genome-scale kinetic models that will address a different set of questions than those addressable by steady-state models [15].

Given the difficulty in constructing kinetic models at a large-scale, it is worth considering the relative value of fully kinetic modeling as compared to steady-state modeling. For bacteria at least, the true value of kinetic models may lie in better understanding how metabolism interacts with other networks. Metabolite levels influence a host of processes, such as mRNA activity through riboswitch binding and transcription factor activity through allosteric regulation. Enzyme levels and active states are controlled by translation and post-translational modification. The specifics of these interactions are largely hidden from view in steady-state models but can be made explicit in kinetic models. In a cyclical fashion, an explicit understanding of the timescales that dynamic events occur on then may allow simplification of models to reduce the amount of data necessary to specify a useful model. For example, the steady-state assumption that works well for predicting bacterial growth rates is based upon a fundamental understanding that over the timescale of serial passages of batch growth, the dynamics of transcription and translation have relaxed and settled such that an optimum steady-state flux distribution is achieved. Thus, understanding the dynamics may enable simplifying assumptions that allow a wide range of questions to be probed using metabolic models with available data.

One prospective use of kinetic models of metabolism lies in cell-based bioprocessing. For practical reasons, biochemical processes are most often performed in batches, which inherently subjects cells to varying conditions as the batch progresses. Changing nutrient availability, pH, and temperature as well as byproduct accumulation causes an adaptive response that can alter the behavior of the strain in an adverse manner. Traditionally in the biochemical engineering literature, unstructured phenomenological models have been applied to describe the dependence of growth rate various factors. In the advent of genome-scale metabolic network reconstructions, mechanistic models may be constructed instead that provide more detailed information. For example, understanding the dynamics of the metabolic network may enable us to determine the signals that are being sent to regulatory networks in the form of altered metabolite levels. Determining the dynamics of the metabolic response thus would enable a mechanistic understanding of the origin of the adaptive regulatory response to changing conditions, likely of great interest toward optimizing strain performance.

This chapter describes the path to developing meaningful dynamic models of metabolic reaction networks. As the number of prior modeling efforts exceeds what can be reasonably reviewed in this space, we focus on fundamental concepts and methods at the foundation of current efforts toward developing kinetic models. First, methods and considerations related to modeling individual biochemical reactions are presented. Then, the analysis of dynamic models of reaction networks in meaningful contexts is discussed. After laying this groundwork, we discuss modeling frameworks that may enable the rapid development of systems-level kinetic models of metabolism. Finally, we present a perspective on the potential for large-scale kinetic models to contribute to our understanding of metabolism.

2.2 Modeling Chemical Reaction Dynamics

Dynamic processes in biology are characterized by complex structural and temporal hierarchies that span many orders of magnitude [16]. The choice of how to reduce these complexities to define a mathematical model of a particular process is worthy of considerable attention. First, we discuss biological processes that interact dynamically with the metabolic network and the assumptions required to make modeling of such networks a manageable undertaking. Then, we highlight mathematical formulations of computational models of metabolism. The thermodynamic relationships underlying biochemical reactions are examined to provide the physical basis for subsequent models. Focusing on deterministic models, forms of rate laws and motivations for their use are presented, with special attention to modeling enzyme-catalyzed reactions. This understanding of the properties of individual reaction dynamics will pave the way to the study of the dynamics of metabolic reaction networks.

2.2.1 Defining the Scope of a Metabolic Model

One of the most important considerations when building a model is defining the boundaries of the process to be modeled. The scope of the model should be determined by the questions that are going to be asked using the model. All processes related to the system of interest should be considered, or rejected with justification, to prevent the model from being subject to *a priori* bias. In this section, we highlight the diversity of biological components and properties that can affect

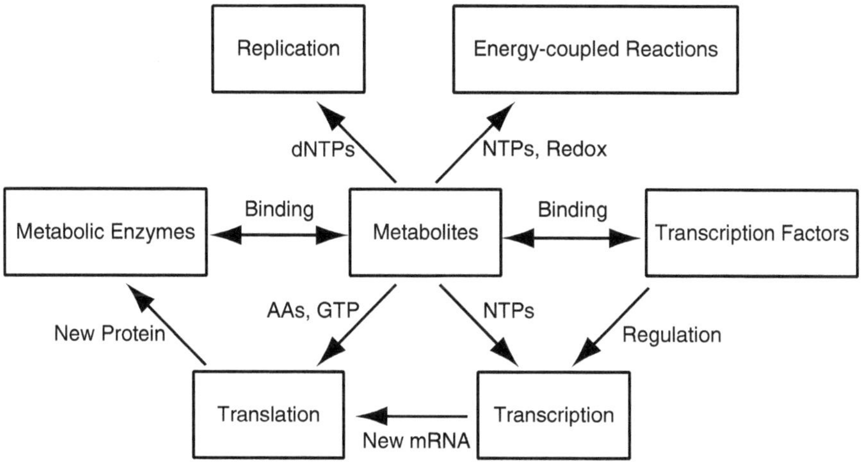

Fig. 2.1 Dynamic interactions of metabolism with various intracellular processes. The highly interconnected structure highlights the central importance of metabolism as well as the complexity associated with accurately modeling metabolic network dynamics

the dynamics of metabolic networks to emphasize the importance of a well-defined and justified model.

The cellular microenvironment is vastly different from a homogeneous solution in a test tube. One of the most immediately noticeable differences lies in the spatial heterogeneity present in the cell. The interior of the cell is separated by compartments that divide the cellular constituents and create chemical potential gradients that drive various transport and energy processes. Protein localization mechanisms create even further spatial inhomogeneity that can be defined qualitatively in terms of subcellular spaces within each compartment [17]. This inhomogeneity can result in variations in local pH [18], ionic strength [19], and metabolite concentration that complicate modeling efforts due to both the numerical difficulty of simulating spatial variation as well as the experimental difficulty of measuring small intracellular gradients. In addition to spatial considerations, the interior of a cell is also crowded [20]; protein concentration is so dense it is typically described as near crystalline, slowing diffusion as much as 1,000-fold in a size-dependent fashion and causing deviations from reaction kinetics predicted by simple models.

Almost every process in the cell has important interactions with the cellular metabolic network, through utilization, modulation, and sensing (Fig. 2.1). For example, transcription and translation depend on the presence of sufficient nucleotide and amino acid building blocks, and their availability can be rate limiting [21, 22]. Metabolites can act as allosteric effectors that modulate the activity of metabolite enzymes, signaling proteins, and transcription factors. Molecular motors, biosynthetic machinery, and transport processes are all dependent on the presence of adequate levels of high energy phosphate to make these reactions thermodynamically favorable through reaction coupling. Additionally, metabolic enzymes are subject to significant dynamic modulation by cellular regulatory systems. In addition to regulation of enzyme concentration through the transcription regulatory network, protein degradation levels are often tightly controlled through defined mechanisms [23]. Protein activity is modulated via allosteric modification by small molecules, covalent modification by signal transduction proteins, and sequestration via binding to certain proteins [24–26]. These processes often are occurring simultaneously during a system perturbation, and therefore careful thought must be put into which processes must be included to yield useful predictions.

A number of assumptions are commonly used to reduce the complexity of the kinetic model, explicitly or implicitly defining its scope. Biological systems are often assumed to be isothermal, isobaric, and at constant pH and ionic strength, except in studies that are explicitly intended to take reaction rate dependencies on such parameters into account [27]. These assumptions are important for the definition of thermodynamic parameters at a particular system state. For example, the Gibbs free energy that is commonly used when describing reaction thermodynamics is defined at isothermal and isobaric conditions. Additionally, while not required for thermodynamic convenience, volume is often assumed constant, thus assuming that there are no concentration changes due to volume fluctuations. This assumption may create significant error considering that the volumes of certain compartments in eukaryotic cells, such as the nucleus, are known to change significantly with time [28], and certainly bacterial volume changes along the progression of the cell cycle. Electric and osmotic potential effects also often are ignored, although these effects are critical in transport processes.

2.2.2 Dynamic Models of Metabolism

Dynamic models appear in various forms that can be categorized into distinct classifications. One such distinction is whether to model the system deterministically, where system variables are uniquely determined in time based on previous values, or stochastically, in which a random component exists and thus future states are described in the form of probabilities. Kinetic models of metabolism typically assume that the number of molecules involved in each reaction is large enough that stochastic effects can be averaged out to an expected value that accurately describes overall system dynamics. Stochastic models are popular for systems in which variables occur in very low copy number, such as in gene transcription [29].

Almost without exception, kinetic models of metabolism are accepted to be nonlinear in form due to multiplicative effects of concentrations on reaction rates for reaction containing more than one reactant or product. Nonlinearity creates complications in dynamic analysis that can be addressed in various ways, of which one common way is linearization, discussed in a later section. Another consequence of the nonlinear structure of individual reactions is that reaction networks are hypergraphs rather than graphs, complicating the application of many graph theory algorithms for the analysis of reaction networks [30].

Deterministic models can either be represented in terms of ordinary differential equations, in which the only dependent variable is time, or partial differential equations, in which dependent variables are both time and space. We only consider ordinary differential equations here, and thus assume spatial homogeneity of variables. While the spatial inhomogeneity of biological systems is well-known [17], the increase in the data required to define and specific parameters for such models precludes their widespread use.

Yet another distinction lies in whether changes in variables are described using functions that vary continuously over time, termed continuous models, or whether changes are defined in discrete steps from one state to the next, termed discrete models. Both methods have found use in modeling metabolic processes [31, 32], but here we focus solely upon continuous modeling.

2.2.3 The Law of Mass Action and Reaction Thermodynamics

In this section, several useful thermodynamic relationships associated with biochemical reactions are presented, focusing on the Law of Mass Action and the Gibbs free energy. These thermodynamic considerations are useful both practically in constraining the possible range of parameters as well as theoretically in defining physically meaningful dynamic models.

Every chemical reaction can be approximated as a series of elementary reactions, which are irreducible chemical events involving one or two compounds that proceed through a single transition state. Consider an elementary bi-molecular reaction that proceeds through a single transition state.

$$aA + bB \overset{K_{eq}}{\longleftrightarrow} pP + qQ \tag{2.1}$$

Collision theory states that for dilute, well-mixed solutions, the reaction velocity will be proportional to the product of reactant activities, which for metabolites in solution are the metabolite concentrations. For the above reaction, this yields an equation for reaction rate of the following form.

$$v = v^+ + v^- = k^+[A]^a[B]^b - k^-[P]^p[Q]^q \tag{2.2}$$

In this reaction equation, called a rate law, v is the reaction velocity, v^+ and v^- are the forward and reverse reaction rates, and the rate constants k^+ and k^- are proportionality constants for the forward and reverse reactions respectively. This equation is known as mass action kinetics (distinguished from the Law of Mass Action described below).

Chemical equilibrium is the state in which the forward and reverse rates of a reaction are equal and thus the reaction velocity is zero. The equilibrium constant K_{eq} is defined as the ratio of the activity of the products to the activity of the reactants. For reactions in an ideal solution, the activity, or effective concentration, of a compound is equal to its concentration, and the following relationship is obtained.

$$K_{eq} = \frac{[P]_{eq}^{p}[Q]_{eq}^{q}}{[A]_{eq}^{a}[B]_{eq}^{b}} \tag{2.3}$$

This relationship is known as the Law of Mass Action. The definition of the equilibrium constant is obtained in statistical mechanics from the principle of detailed balance, which in turn is derived from micro reversibility, of which the 2nd Law of Thermodynamics is also a consequence [33]. While mass action reaction kinetics, discussed above, only holds for elementary reactions in dilute, well-mixed solutions, the Law of Mass Action, defined in terms of activities, appears to hold for all cases. The value of the equilibrium constant is a function of temperature, ionic strength, and pH [27]. Fortunately, due to homeostatic processes and buffering, these variables most often can be assumed constant for biological systems. Thus, modeling efforts can be simplified by specifying a particular biological state at which equilibrium constants are independent of system variables.

For reaction mechanisms that proceed through multiple transition states, each individual step, if known, can be represented by a reaction that has its own forward and reverse rate constants and equilibrium constant. The intermediate equilibrium constants must then follow the so-called Haldane relationship such that the overall equilibrium constant of the reaction remains unchanged [34]. For example, consider an S_N1 reaction, in which the 'rate-determining' step is unimolecular, of the following scheme.

$$\text{Overall}: \quad A + 2B \overset{K_{eq}}{\longleftrightarrow} P + Q + R \tag{2.4}$$

$$1^{\text{st}}\ \text{Step}: \quad A \overset{K_{eq}^{1}}{\longleftrightarrow} A^{*} + Q \tag{2.5}$$

$$2^{\text{nd}}\ \text{Step}: \quad A^{*} + B \overset{K_{eq}^{2}}{\longleftrightarrow} P^{*} \tag{2.6}$$

$$3^{\text{rd}}\ \text{Step}: \quad P^{*} + B \overset{K_{eq}^{3}}{\longleftrightarrow} P + R \tag{2.7}$$

The following relationship must hold between the equilibrium constants, with terminology indicating superscripts rather than exponents.

$$K_{eq}^{1}K_{eq}^{2}K_{eq}^{3} = K_{eq} \tag{2.8}$$

Note that even the designation of the reaction as having a certain mechanism, such as S_N1, is an approximation based on relative rates of micro processes such as the dissociation of Q compared with the association of A and B. Only the equilibrium constant of the overall reaction defined in terms of the equilibrium concentrations can be seen as a universal relationship, for particular reaction conditions. Additional detail into the relationship between the reaction rates, equilibrium constants, and reactant and product concentrations can be gained by insightful modeling of the internal processes of the reaction, but knowledge of the associated assumptions is critical.

In contrast to chemical equilibrium, chemical steady-state is the state in which concentrations of reactants and products do not change but forward and reverse rates are not necessarily equal. Closed reaction systems only have a single steady-state, which is chemical equilibrium. Thus, an open system is required to maintain a steady-state through constant influx and efflux of material.

At any steady state, the distance of a reaction from equilibrium is given by the mass action ratio, once again assuming that the activities of the compounds are equal to their concentrations.

$$\Gamma = \frac{[P]^p [Q]^q}{[A]^a [B]^b} \tag{2.9}$$

Note that this definition is identical to that of the equilibrium constant, but the mass action ratio is not specified to be at equilibrium. The mass action ratio is useful in calculating the Gibbs free energy of reaction. The Gibbs free energy of reaction, defined at a reference temperature and pressure, is a useful quantity that can be given in terms of the equilibrium constant and mass action ratio.

$$\Delta G = -RT\ln(K_{eq}) + RT\ln(\Gamma) \tag{2.10}$$

In this equation, R is the universal gas constant and T is the temperature, which should correspond to the temperature at which the equilibrium constant is specified. The Second Law of Thermodynamics states that the Gibbs free energy for a reaction must be negative if the flux through the reaction has a positive steady-state value, and must be positive if the flux has a negative steady-state value [35].

$$v_{ss}\Delta G \leq 0 \tag{2.11}$$

Thus, the Gibbs free energy defines fundamental constraints on the relationship between the fluxes and concentrations away from equilibrium.

2.2.4 Phenomenological Rate Laws

The mass action reaction kinetics equation discussed above is one example of a rate law, which defines the reaction velocity in terms of parameters and variables. While the mass action rate law may provide an accurate approximation of dynamics for elementary reactions for certain conditions, mass action has notable shortcomings when used as a rate law for overall reactions in a biochemical system. First, the intracellular environment is known to be subject to significant macromolecular crowding, making the dilute, well-mixed assumption suspect [36]. It has been suggested that fractal kinetics, in which rate constants are time dependent for non-first order reactions, may be a better choice for modeling reaction kinetics in crowded conditions, at least when the reaction molecules are of same size as the crowding molecules, or in the case of restricted geometries [37]. Additionally, mass action kinetics using overall reaction stoichiometry ignores details of enzyme-catalyzed reaction mechanisms such as the allosteric and saturation effects that will be discussed in more detail later. While overall reactions can be represented as series of elementary reactions when mechanisms are known, this presents a challenge in both identifying such mechanisms as well as specifying necessary *in vivo* parameters. In order to describe the overall reaction kinetics using as few parameters as possible, several other rate laws have been proposed that seek to bypass the need for elementary reaction steps.

Two prominent examples of approximate rate laws are generalized mass action and linear-logarithmic rate laws. Generalized mass action, or power law, takes the same form as the mass action rate law but assumes arbitrary coefficients [38, 39].

$$v_j(x) = \alpha_j \prod_{i=1}^{m} x_i^{\beta_{ij}} \tag{2.12}$$

Where $\boldsymbol{\alpha}$ is the vector of rate constants and $\boldsymbol{\beta}$ is the vector of generalized (non-integer) kinetic orders. When used as an overall rate law, generalized mass action can provide a better approximation of reaction kinetics by a data fitting approach. When used as a rate law for elementary reactions, generalized mass action may give a better approximation of kinetics when the reaction space is topologically constrained. However, as neither the form of the rate law nor the values of the non-integer coefficients have a direct biochemical basis, the general applicability of this approach is difficult to assess.

Enzymatic reactions are known to exhibited saturation kinetics, meaning that the increase in reaction rate with addition of substrate becomes zero at a certain saturating amount of substrate. In order to capture the hyperbolic dependence of reaction rate on substrate concentration due to saturation using minimal parameters, linear-logarithmic rate laws have been proposed [40]. In lin-log rate laws, the dependency of the reaction rate on its reactants and products is given by the sum of the logarithms of the concentrations, scaled by elasticity coefficients (adapted from [41]).

$$v_j(x) = v_j^0 \left(\frac{E_T}{E_T^0}\right) \left(1 + \sum_i \eta_{\mathrm{ij}}^0 \ln\left(\frac{x_i}{x_i^0}\right)\right) \tag{2.13}$$

Where $\boldsymbol{E_T}$ is the vector of total enzyme concentrations, $\boldsymbol{\eta}$ is the elasticity coefficient vector, and the zero superscript indicates a reference state. The resulting equation is similar to log-transformed generalized mass action kinetics. The elasticity coefficients are local sensitivity parameters that define the responsiveness of the reaction rate to changes in each concentration at a particular reference state. Concentrations that have small elasticity coefficients indicate that the enzyme is effectively 'saturated' with respect to that concentration. As elasticity coefficients are approximated around particular reference states, lin-log models are condition-specific, and thus have many similarities to the mass action stoichiometric simulation (MASS) modeling approach discussed in a later section.

Although useful for dealing with large numbers of unknown parameters while still approximating biological function, approximate rate law models suffer from several drawbacks. First, as these models inherently ignore the underlying biochemistry of the reaction mechanism, it is difficult to interpret parameters biologically and set them directly using experimental data. Second, the original reasons for the development of such models, namely computational efficiency and limited data, are largely becoming obsolete due to technological and computational advances. The potential for accurate kinetic models based in data and reaction structure and interpretable in terms of underlying biochemistry may soon make phenomenological rate laws obsolete.

2.2.5 Enzyme Kinetics

The majority of metabolic reactions are catalyzed by enzymes that accelerate their reaction rates. Catalysts by nature cannot make a chemical reaction more thermodynamically favorable. Instead, catalysts lower the activation energy of the rate-limiting transition state of the reaction, and thus increase the frequency of collisions that have sufficient energy to pass through the high energy transition states responsible for the activation energy barrier for the reaction. This is accomplished by providing an alternate route of reaction that has a less unfavorable transition state, through various mechanisms such as bringing the reactants into proximity and destabilizing the reactants by inducing bond strain. A catalyst increases forward and reverse rates equally, leaving the equilibrium constant unchanged, as required by the 2nd Law of Thermodynamics. While the overall equilibrium constant for a reaction is a fixed quantity, the activation energy change provided by the enzyme is dependent upon the enzyme itself, and thus is susceptible to genetic change and evolutionary pressure [42]. Understanding the mechanistic link between amino acid composition of enzymes, reaction kinetics, and metabolic system dynamics would clarify the basis for many enzyme deficiency diseases.

In the description of enzyme-catalyzed reactions, it may be desirable to handle enzyme catalysis more rigorously to obtain an overall rate law using an assumed mechanism with certain approximations. One result of such efforts is the well-known Michaelis-Menten rate law [1]. The details and extensions of the Michealis-Menten approach can be found in any introductory biochemistry textbook [43], so here we present only key points related to the practical use of such rate laws.

The reaction that forms the basis of the Michaelis-Menten rate law is most often presented in the following form, although the form has since been extended to reversible reactions with various modifications.

$$S + E \leftrightarrow SE \rightarrow E + P \tag{2.14}$$

In this reaction, S is the substrate, P is the product, E is the free enzyme, and SE is the substrate-enzyme bound complex. The first step of the reaction is the reversible binding of the substrate to the enzyme, while the second step is catalytic and is assumed irreversible in this case. The Michaelis-Menten equation that results from this equation is as follows.

$$\frac{d[P]}{dt} = \frac{v_{MAX}[S]}{K_M + [S]} \tag{2.15}$$

This rate law thus establishes a relationship between the reaction rate and the substrate using two parameters. It can be seen by the form of the equation that the rate of reaction saturates to v_{MAX} as the concentration of S becomes large with respect to K_M. The popularity of the Michaelis-Menten rate law has led to a wealth of literature on Michaelis-Menten constants for various enzymes, organisms, and reaction conditions. The usefulness of this data in accurately describing *in vivo* dynamics is suspect, but Michaelis-Menten parameters are undeniably useful for certain applications, such as comparison with measured concentrations as an estimate of enzyme saturation [44].

The derivation of the Michaelis-Menten rate law from this reaction system still holds fundamental interest from a modeling perspective in its use of conserved moieties and time scale separation in model simplification. The first point of interest is the use of conserved moieties through the substitution of the total enzyme concentration, which can be experimentally measured, in place of free and bound enzyme concentrations, which are difficult to measure. These moieties appear in the left null space of the stoichiometric matrix, and the explicit inclusion of conserved enzyme concentrations in elementary reactions results in inherent saturation kinetics for an enzyme elementary reaction network. The second interesting step in the Michaelis-Menten derivation is the application of rudimentary time-scale decomposition in the form of a quasi-steady state assumption, where the binding step is assumed rapid with respect to the catalytic step. More rigorous time-scale decomposition is discussed in further sections, but the usefulness of the approach in the functional simplification of kinetic models can already be seen.

The Michaelis-Menten formalism can be extended for reaction mechanisms of arbitrary complexity, including allosteric regulation in various forms. While small molecule regulation of enzymes has traditionally been defined using the terms competitive, uncompetitive, and noncompetitive, these are in fact simply approximations useful to fit *in vitro* data with rate laws of specific form. More rigorously, algorithmic means to calculate an overall rate law for enzyme mechanisms of arbitrary form and regulation from elementary reaction networks exist in the form of the King-Altman method [45].

One commonly observed feature of enzyme kinetics arises from the fact that many enzymes exhibit cooperativity as a result of being multimeric, causing a sigmoidal rather than hyperbolic dependency of reaction rate on the cooperative compound concentration. Note that cooperativity is not restricted to enzymes, as perhaps the best-known example of cooperativity is the dependence of hemoglobin on oxygen. The net effect of cooperativity is that reaction rate, or binding, becomes much more sensitivity to changes in reactant concentration. Several models have been developed to explain the phenomenon. The most basic and easily applied is the Hill equation [46], which is similar in form to the Michaelis-Menten equation but in which components are raised to an empirically-determined power known as the Hill coefficient. A positive Hill coefficient indicates positive cooperation, such that binding of reactant increases the rate of reaction. Likewise a negative Hill coefficient indicates negative cooperation, such that the binding of a reactant reduces the rate of reaction. More detailed models that seek to describe the mechanistic basis for cooperation exist, the most accepted of which are the concerted model and the sequential model [47, 48].

Although overall rate law approximations hold advantages for their simplicity and experimental accessibility, enzyme kinetics are more fundamentally described using systems of elementary reactions representing a putative enzyme mechanism. Some of the most common mechanisms for bilinear reactions include the random bilinear-bilinear mechanism, the sequential bilinear-bilinear mechanism, and the ping pong mechanism [43]. Note that the sequential bilinear-bilinear mechanism is a special case of the random bilinear-bilinear in which the order of reaction steps is fixed. The advantage of using elementary reactions to describe enzyme kinetics is that every parameter in the model has a direct biochemical interpretation; for example, measurable enzyme-substrate dissociation constants are equilibrium constants for binding steps in an explicit mechanism. The corresponding challenge with implementing these methods is that the parameters are greater in number and difficult to measure, making construction of a validated model difficult.

2.3 Analyzing Metabolic Network Dynamics

The construction and analysis of dynamic models of metabolic networks requires a new set of tools and considerations than used in modeling individual reactions. In this section, we discuss the stoichiometric matrix **S** and its properties in the context of the dynamic mass balance equation. We discuss how aggregations of variables into pools and pathways can be defined based on either the structure of **S**, dynamic correlations, or biological criteria, thereby establishing a link between individual variables and systems-level properties. Finally, we present methods for the analysis of local dynamics around a reference state through the Jacobian matrix **J**. Numerical considerations with the development of large-scale kinetic models are discussed. This section lays the groundwork necessary for analyzing kinetic models of metabolism in a meaningful fashion. Other common topics in dynamic analysis, such as bifurcation analysis, are well established in the literature and are not discussed here in detail [49].

2.3.1 The Stoichiometric Matrix

While many of the original enzyme kinetics studies were conducted on single enzymes in isolation, a more realistic study requires the analysis of reactions in the context of the network in which they exist. Increasing the scale of the model from the individual reaction to a reaction network requires a new set of theoretical tools to facilitate model construction and analysis.

Assuming spatial homogeneity and constant volume, a system of reactions can be represented in terms of the dynamic mass balance equation.

$$\frac{dx}{dt} = \mathbf{S} \cdot \boldsymbol{v} \tag{2.16}$$

Where $\boldsymbol{x}$ is the metabolite concentrations, **S** is the matrix of reaction stoichiometries, and $\boldsymbol{v}$ is the vector of reaction rates or fluxes. The stoichiometric matrix, **S**, can contain lumped overall reactions or elementary reactions, depending on the detail on reaction mechanisms desired. The functional form of the rate laws, $\boldsymbol{v}$, is chosen based on considerations discussed above.

At steady-state, the changes in concentrations with time, d**x**/dt, are equal to 0. A non-equilibrium steady-state is only possible for open systems, which constantly intake and output material. The steady-state assumption plays an important role in the well-known flux balance analysis (FBA) problem [50]. The underlying assumption of a steady-state in FBA is valid only because the dynamics of interest, bacterial growth, occur on a time scale after the internal dynamics of the system have relaxed with respect to the environment. While this assumption has been shown to be valid for culture conditions [51], certain environmental communities may be subject to fluctuating nutrient conditions that keep the metabolic system away from steady-state. In kinetic modeling, by definition, we are interested in deciphering the dynamic fluctuations away from steady-state. We will see later that large-scale kinetic models can be built in a condition-specific fashion around biologically-relevant steady-states of the dynamic mass balance equations.

The stoichiometric matrix contains system-level biological properties that can be mathematically defined by analysis of the four fundamental subspaces of the matrix [42]. The right null space of **S**, here termed **N**, maps fluxes to a zero vector of concentration derivatives, and hence corresponds to linear combinations of fluxes that do not change the concentrations.

$$\mathbf{SN} = 0 \tag{2.17}$$

Each basis vector for the right null space can be thought of as a steady-state pathway, and any combination of weightings $\boldsymbol{\alpha}_N$ on such pathways defines a steady-state flux distribution $\boldsymbol{v}_{ss}$.

$$\mathbf{N}\alpha_N = \boldsymbol{v}_{ss} \tag{2.18}$$

The left null space of **S**, here termed **L**, contains weightings of concentrations that cannot be changed by any dynamics described in **S**, thus defining time-invariant pools of metabolites.

$$\mathbf{LS} = 0 \tag{2.19}$$

The invariant pools typically correspond to total concentration of cofactors, enzymes, or other moieties such as sulfur or phosphate when these are not exchanged with the environment. The concentration of metabolites in the network act as weightings on the basis vectors for the left null space that define the sizes of the invariant pools $\boldsymbol{x}_{pools}$.

$$\mathbf{L}\boldsymbol{x} = \boldsymbol{x}_{pools} \tag{2.20}$$

It is thus clear that the right and left null spaces of **S** have defined dynamic interpretations as the time-invariant flux and concentration properties of the network. The remaining subspaces contain the time-variant properties of the dynamic mass balance equations. The row space of **S** is orthogonal to the right null space and maps fluxes to non-zero concentration derivatives. The basis for the column space defines all non-steady state flux distributions. The column space of **S** is orthogonal to the left null space and defines all non-zero concentration derivatives. This understanding of the quantitative network information contained in **S** is useful at many stages when building and interpreting dynamic models of metabolic reaction networks.

2.3.2 Formation and Dynamics of Aggregate Variables

Here we discuss the use of phase portraits in studying the dynamic interactions of individual variables, as well as the usefulness of defining and analyzing physiologically-meaningful system variables. Dynamic phase portraits are a tool to study the dynamic interrelation of variables in a graphical manner. Dynamic phase portraits can be constructed by plotting trajectories formed by two variables against each other from an initial state to a resting state in a simulation of the dynamic mass balance equations. A positive slope of the trajectory indicates that the variables are correlated, or dynamic motion is in the same direction for both variables for the corresponding time scale. Similarly, a negative slope indicates that the variables are anti-correlated, indicating that motion occurs in opposite directions. A horizontal or vertical line in the phase portrait indicates that the motion of the variables is dynamically decoupled, as one variable moves while the other stays constant. The qualitative separation of dynamics at certain time scales results from order of magnitude differences in kinetic parameters and defines the time-scale decomposition of the network. Correlations can be used to define dynamically aggregating variables on certain time scales. A method to algorithmically calculate these pools is discussed later.

In addition to dynamic aggregate pools formed through correlation analysis, other pools can be formed based on biologically-relevant criteria. For example, a pool of total high energy phosphate can be formed by including metabolites that contribute phosphate to glycolysis. These pools are mathematically handled using the pooling matrix **P** that defines the relationship between the variables to be pooled, here assumed the concentrations $\boldsymbol{x}$, and the pool variables $\boldsymbol{p}$.

$$\boldsymbol{p} = \mathbf{P}\boldsymbol{x} \tag{2.21}$$

The dynamics of these pools then can be modeled using standard approaches simply by pre-multiplying the dynamic mass balance equation by the pooling matrix **P**.

$$\frac{d\boldsymbol{p}}{dt} = \mathbf{PS}\boldsymbol{v} \tag{2.22}$$

Defining aggregate variables in various terms enables the network characteristics to be studied rather than individual variables that are more difficult to interpret biologically. Thus, pooling methods attempt to address the issue of defining the physiologically-meaningful dynamics in the network.

2.3.3 Linear Analysis

Historically, dynamic simulations have been a primary method for studying systemic perturbation response. These methods are suited to study the nonlinear properties of small systems of equations far from steady state [52]. In dynamic simulation, the initial conditions must be carefully chosen such that a realistic process is being simulated. For instance, simulations involving the perturbation of a single internal concentration are not biologically-meaningful as concentrations cannot instantaneously vary independent from the rest of the system. Variations in external concentrations and load fluxes are more accurate approximations of biological phenomena and likely more yield meaningful and interpretable simulations. However, as the type of information that is obtained from simulations is always determined by the choice of simulation, these methods are subject to modeling bias, especially for large systems where the number of simulations for complete dynamic description becomes intractably large.

In contrast with dynamic simulation studies, linear (local) analysis studies the dynamic structure of metabolic systems in an unbiased fashion. Various nomenclatures for linear analysis exist, such as the well-known metabolic control analysis (MCA), but these methods share an underlying mathematical framework [53]. Linear analysis of metabolic networks [52] generally involves calculating or measuring the Jacobian matrix for the system of dynamic mass balance equations and decomposing the system response properties. The Jacobian matrix represents a product ($\mathbf{J_x} = \mathbf{SG}$) of network topology (through $\mathbf{S}$) and the reaction response times found in the gradient matrix $\mathbf{G}$, which is defined as the first order derivatives of the reaction fluxes with respect to the metabolite concentrations [15].

$$\mathbf{G} = \frac{\partial \boldsymbol{v}}{\partial \boldsymbol{x}} \tag{2.23}$$

Linear dynamic analysis reveals the underlying structure of the dynamic mass balance equation via a first-order Taylor series expansion of the dynamic mass balance equation around a reference steady-state concentration.

$$\frac{d\boldsymbol{x}}{dt} = \mathbf{S} \cdot \boldsymbol{v} \tag{2.24}$$

$$\boldsymbol{v}(\boldsymbol{x}) = \boldsymbol{v}(\boldsymbol{x}_0) + \frac{d\boldsymbol{v}}{d\boldsymbol{x}}|_{x_0} \cdot (\boldsymbol{x} - \boldsymbol{x}_0) + \frac{1}{2}\frac{d^2\boldsymbol{v}}{d\boldsymbol{x}^2}|_{x_0} \cdot (\boldsymbol{x} - \boldsymbol{x}_0)^2 + \cdots \tag{2.25}$$

$$\boldsymbol{v}(\boldsymbol{x}) = \boldsymbol{v}(\boldsymbol{x}_0) + \mathbf{G}|_{\boldsymbol{x}_0} \cdot (\boldsymbol{x} - \boldsymbol{x}_0) + \cdots \tag{2.26}$$

$$\frac{d\boldsymbol{x}}{dt} = \mathbf{S}\boldsymbol{v}(\boldsymbol{x}_0) + \mathbf{SG}|_{x_0} \cdot (\boldsymbol{x} - \boldsymbol{x}_0) + \cdots \tag{2.27}$$

$$\frac{d\boldsymbol{x}}{dt} = 0 + \mathbf{J}|_{\boldsymbol{x}_0} \cdot (\boldsymbol{x} - \boldsymbol{x}_0) + \cdots \approx \mathbf{J}|_{\boldsymbol{x}_0} \cdot \boldsymbol{x}_{ref} \tag{2.28}$$

This gives a first-order approximation of the non-linear dynamics in terms of the deviation variable $\boldsymbol{x}_{\boldsymbol{ref}} = (\boldsymbol{x} - \boldsymbol{x}_0)$. There are dual Jacobian matrices for every reaction system, depending on whether metabolite concentrations or reaction fluxes are chosen as the dependent variables [54], and these matrices share eigenvalues. A representation of the system dynamics in terms of fluxes is found taking the total derivative of $\boldsymbol{v}$ with respect to time and concentration.

$$\frac{d\boldsymbol{v}}{dt} = \frac{\partial \boldsymbol{v}}{\partial t} + \frac{\partial \boldsymbol{v}}{\partial \boldsymbol{x}}\frac{d\boldsymbol{x}}{dt} = 0 + \mathbf{GS}\boldsymbol{v}(\boldsymbol{x}_0) = \mathbf{J_v}\boldsymbol{v}(\boldsymbol{x}_0) \tag{2.29}$$

It is seen that the Jacobian matrix maps the concentration or flux variables onto their derivatives around a reference steady state, and these matrices are composed of the stoichiometric matrix and the gradient matrix. Subsequent modal analysis via an eigendecomposition of the Jacobian matrix decouples the system dynamics by defining modes (eigenrows) that move independently on characteristic time scales defined by the eigenvalues of the Jacobian matrix.

$$\mathbf{J_x} = \mathbf{M_x}\Lambda_\mathbf{x}\mathbf{M_x}^{-1} \quad \frac{d\boldsymbol{m}_x}{dt} = \Lambda_\mathbf{x}\boldsymbol{m}_x \quad \boldsymbol{m}_x = \mathbf{M}_\mathbf{x}^{-1}\boldsymbol{x} \tag{2.30a,b,c}$$

$$\mathbf{J_v} = \mathbf{M_v}\Lambda_\mathbf{v}\mathbf{M_v}^{-1} \quad \frac{d\boldsymbol{m}_v}{dt} = \Lambda_\mathbf{v}\boldsymbol{m}_v \quad \boldsymbol{m}_v = \mathbf{M}_\mathbf{v}^{-1}\boldsymbol{v} \tag{2.31a,b,c}$$

These modes are combinations of the original variables and their composition indicates which variables are moving at a particular time scale. The mapping between variables and modes is given by the eigenrows of the Jacobian matrix, termed the modal matrix. The time constant corresponding to each mode is the inverse of the eigenvalue for that mode and gives a measure of how quickly the mode moves in response to a perturbation.

$$\tau = \lambda^{-1} \tag{2.32}$$

Whether the mode is moving towards or away from the steady-state is determined by the sign of the corresponding eigenvector for that mode. Strictly negative eigenvalues indicate a stable steady-state. Stability considerations are discussed in more detail in a later section.

While the modes can be analyzed directly to decompose the dynamic motion of metabolites and variables, the modes can also be used to elucidate the time scale hierarchy of dynamic variable aggregation in the network [52]. Variables are considered to aggregate when their motion correlates at a certain time scale, and hence form a dynamic pool with components that changes in fixed ratios after the time scale of pooling. Variable correlation can be computed directly from the modal matrix by calculating the pair-wise angles between columns.

$$\Theta = \cos(\vartheta_{ij}) = \frac{\left(\mathbf{M}^{-1}\right)_i^T \cdot \left(\mathbf{M}^{-1}\right)_j}{\left\|\left(\mathbf{M}^{-1}\right)_i^T\right\| \left\|\left(\mathbf{M}^{-1}\right)_j\right\|} \tag{2.33}$$

Variables with angles close to one are considered correlated. Aggregation on progressively slow time scales can then be calculated by successively removing the row of the model matrix corresponding to the fastest mode, and recalculating the column angles. The resulting pooling structure across all time scales defines the topological relaxation of system dynamics. This reduction in the effective number of dynamic variables may be of interest for the creation of hybrid kinetic models, which seek to reduce the complexity of dynamic models by assuming certain sub-networks are internally at steady-state [55]. The calculation of the time-dependent pooling structure provides an algorithmic basis for the definition of such hybrid models.

The Jacobian matrix is decomposed into the stoichiometric matrix $\mathbf{S}$ and the gradient matrix $\mathbf{G}$. While the structural properties of $\mathbf{S}$ are well-known and have been discussed above, the properties of $\mathbf{G}$ are less familiar. The gradient matrix consists of derivatives of reaction rates with changes in concentrations, and thus consists of non-integer values that may vary on several orders of magnitude. Thus, the subspaces of $\mathbf{G}$ are less interesting than the elements of the matrix itself. The gradient matrix can be row-normalized or column-normalized to create decompositions that describe the kinetic properties of reactions or metabolites, respectively. The left normalization of $\mathbf{G}$ into a vector $\boldsymbol{\kappa}^{\mathrm{v}}$ and a matrix $\boldsymbol{\Gamma}^{\mathrm{v}}$ analyzes dynamic reaction properties.

$$\mathbf{G} = \boldsymbol{\kappa}^{\mathrm{v}}\boldsymbol{\Gamma}^{\mathrm{v}} \tag{2.34}$$

The matrix $\boldsymbol{\Gamma}^{\mathrm{v}}$ contains the directions and relative influences of the thermodynamic force on a reaction by each metabolite. The vector $\boldsymbol{\kappa}^{\mathrm{v}}$ contains the magnitude of dynamic response of each reaction. The right normalization of $\mathbf{G}$ into a matrix $\boldsymbol{\Gamma}^{\mathrm{x}}$ and a vector $\boldsymbol{\kappa}^{\mathrm{x}}$ analyzes dynamic metabolite properties.

$$\mathbf{G} = \boldsymbol{\Gamma}^{\mathrm{x}}\boldsymbol{\kappa}^{\mathrm{x}} \tag{2.35}$$

The matrix $\boldsymbol{\Gamma}^{\mathrm{x}}$ contains the scaled dynamic effects of each metabolite on the reaction in which it participates. The vector $\boldsymbol{\kappa}^{\mathrm{x}}$ contains the kinetic response tendencies for each metabolite. These decompositions can be used to analyze dynamic properties of reactions and metabolites in the context of the network.

The location of non-zero values in $\mathbf{G}$ is determined by the network structure described in $\mathbf{S}$, and thus $\mathbf{S}$ and $\mathbf{G}$ have complementary structure ($\mathbf{S} \sim -\mathbf{G}^{\mathrm{T}}$). As $\mathbf{G}$ has the same sparse structure as $\mathbf{S}^{\mathrm{T}}$, the Jacobian matrices are seen to be weighted adjacency matrices of $\mathbf{S}$, such that if $\mathbf{A}_{\mathbf{x}} = \mathbf{S}\mathbf{S}^{\mathrm{T}}$ and $\mathbf{A}_{\mathbf{v}} = \mathbf{S}^{\mathrm{T}}\mathbf{S}$, $\mathbf{J}_{\mathbf{x}} \sim \mathbf{A}_{\mathbf{x}}$ and $\mathbf{J}_{\mathbf{v}} \sim \mathbf{A}_{\mathbf{v}}$. This relationship likely plays an important role in the structure of modes and the dynamic aggregation of variables.

This section has discussed methods associated with the linear analysis of the dynamic mass balance equations. While these methods only strictly describe local dynamics for small perturbations around a steady-state, they provide a basis for understanding the dynamic structure of the network that is more scalable than dynamic simulation. Analysis of the properties of the gradient matrix $\mathbf{G}$ and modal matrix $\mathbf{M}^{-1}$ and calculation of the time-dependent aggregation of dynamic variables link the kinetics of individual components to the dynamic properties of the entire network through a structure-function relationship.

2.4 Towards Large-Scale Kinetic Models of Metabolism

Although the tools necessary to analyze the properties of kinetic models of metabolism exist, practical difficulties have limited the development of such models. There are two main issues that appear when attempting to scale the construction

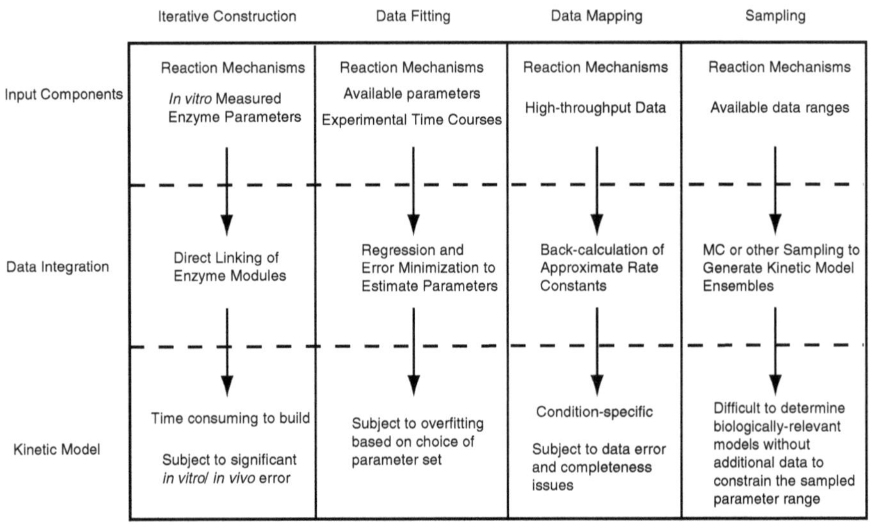

Fig. 2.2 Four approaches to developing large-scale kinetic models of metabolism. Each method suffers from practical concerns, so future modeling efforts will likely combine aspects from multiple approaches based on available data and size of the system

of kinetic modeling of metabolism to the systems level: the definition of the scope of the model, and specification of the model parameters. In this section, focus is given to the second question on how to address the need to specify model parameters given limited available data. Methods have been developed to address the data limitation issue that broadly can be classified into categories of iterative model construction, data fitting, data mapping, and parameter sampling (Fig. 2.2). We present a data mapping approach to the rapid construction of large-scale approximate kinetic models through the use of high-throughput metabolomic data. We then discuss the importance of sampling and parameter sensitivity analysis in addressing data completeness and data accuracy issues. This section is intended to serve as a roadmap to the practical construction of large-scale *in vivo* kinetic models.

2.4.1 Iterative Model Construction

Traditionally, network-scale kinetic models of metabolism have been constructed by measuring *in vitro* kinetic parameters of individual enzymes, and then linking enzyme modules into a coherent network model [12, 56]. This approach is the simplest and most direct method for the construction of network-scale models. The limitations of iterative model construction stem from the inaccuracies created by using *in vitro* kinetic parameters to create models that must simulate *in vivo* dynamics, as well as the often poorly defined culture conditions and enzyme concentrations in such enzyme kinetics studies that complicate systemic integration. The intracellular environment may be vastly different than the *in vitro* conditions and may vary in a condition-dependent manner. Efforts to define the global dynamic behavior of an enzyme are limited by the lack of knowledge of the *in vivo* state of the enzyme and inability to reproduce that state for *in vitro* experiments.

2.4.2 Data Fitting

Efforts have been made to deal with the discrepancy between *in vitro* and *in vivo* conditions by directly specifying kinetic parameters using *in vivo* dynamic measurements of specific concentration profiles after a system perturbation [56]. The challenge then becomes deconvoluting measurements to calculate individual kinetic parameters. These studies are often troubled by underdetermined systems, having fewer data measurements than necessary to specify the model. To deal with this issue, a variety of regression and error minimization methods can be used to determine the parameters that generate models that can best reproduce experimental simulations [57]. The underlying issue with these methods is that as the number of unspecified parameters grows, there may be many parameter sets that fit the limited available data equally well but may not perform equally well for arbitrary new experiments. This problem is known as over-fitting and creates a modeling bias where the model may appear to fit the data well but in reality has little predictive power. As a quote attributed to mathematician John von Newmann aptly claims: "With four parameters I can fit an elephant and with five I can make him wiggle his trunk" [58]. Unspecified parameters present a significant hurdle to generating large-scale kinetic models that are biologically predictive. Sampling methods may be used to avoid over-fitting, as the decision of picking a particular parameter set from the available range can be foregone in favor of sampling parameter sets from the entire range [59]. Such methods are discussed further in a subsequent section.

When available for well-defined culture conditions, initial rate data and progress curves for single enzyme experiments can also be used to define parameters for rate laws for individual enzymes. This process has traditionally been used with simple Michaelis-Menten equations using fitting approaches such as the Lineweaver-Burk plot, but more sophisticated mechanistic rate laws can also be specified using sufficient data and nonlinear fitting methods [60]. This approach appears to be less subject to overfitting than whole system time course fitting, due to the reduced scope of the model and relatively greater dimensionality of the corresponding data.

2.4.3 Construction of Kinetic Models via High-Throughput Data Mapping

With advances in high-throughput measurement of metabolite concentrations and reaction fluxes, as well as advances in estimation algorithms for thermodynamic properties, it may soon become possible to completely specify large-scale kinetic models of metabolism completely from data. One kinetic modeling approach that has been developed to address this oncoming onslaught of data is the mass action stoichiometric simulation (MASS) modeling approach [61].

The MASS approach is enabled by mapping *in vivo* data onto mass action equations based on network stoichiometry. Mass action reactions can be built from **S** in the following form:

$$v_i = k_i^+ \prod x_j^{\mathbf{S}_{ji}^n} - k_i^- \prod x_k^{\mathbf{S}_{ki}^p} \tag{2.36}$$

where v_{i} is the flux for reaction i, k_{i}^+ and k_{i}^- are the forward and reverse rate constants, $\mathbf{x}$ is the concentrations, and $\mathbf{S}^{\mathrm{n}}$ and $\mathbf{S}^{\mathrm{p}}$ are matrices of the same size as **S** containing the absolute values of the stoichiometry of reactants and products, respectively, in each reaction. Using the definition of the equilibrium constant $K_{eq,\mathrm{i}}$ in terms of the ratio of forward and reverse rate constants, the forward rate constant can be solved in terms of parameters for which data exists.

$$k_{\mathrm{i}}^+ = \frac{v_{\mathrm{i}}}{\left(\prod x_{\mathrm{j}}^{\mathbf{S}_{\mathrm{ji}}^{\mathrm{n}}} - \frac{\prod x_{\mathrm{k}}^{\mathbf{S}_{\mathrm{ki}}^{\mathrm{p}}}}{K_{eq,\mathrm{i}}}\right)} \tag{2.37}$$

The values of the steady-state fluxes, steady-state concentrations, and equilibrium constants thus enable the calculation of approximate reaction rate constants for mass action reactions. This rate constant is termed the pseudo-elementary rate constant (PERC) as it is not a true elementary rate constant but rather gives an approximation of the rate for the overall reaction based on the flux through the reaction, size of the involved concentration pools, and distance from equilibrium. Enzyme concentrations are subsumed into the PERC values, and thus the PERCs approximate the sum of regulatory and thermodynamic drivers for each reaction.

The MASS modeling formulation is an approximate approach to developing large-scale kinetic models rapidly using *in vivo* data. The reaction network stoichiometry serves as a platform for the functional integration of various data types in the form of a dynamic model. An important distinction of MASS models from many previous attempts at large-scale modeling is their condition-specific nature, similar to lin-log models. By using mass action rate laws based in metabolic reaction stoichiometry alone and incorporating condition-specific high-throughput data, the calculated PERCs only describe the dynamic properties at a particular system steady-state. The development of fully a comprehensive kinetic model of metabolism that explains the dynamics at all possible system states remains a significant theoretical and practical challenge [62]. The MASS approach seeks to use available data to create meaningful models that can be used to explain differential dynamic properties across multiple systems states for which *in vivo* measurements are available.

In the case when **S** contains specific enzyme mechanisms rather than overall reactions, the reactions in **S** become elementary reactions [61]. The incorporation of conserved enzyme moieties into **S** would recapitulate the saturation properties of enzymes that made Michaelis-Menten kinetics an attractive approximation in models thus far, but with no *a priori* quasi-steady state assumption required. This extension requires additional knowledge of the reaction mechanism, binding constants for reactants and allosteric effectors, the absolute concentration of total enzyme, and relative concentrations of the enzyme forms at steady-state. While binding constants and the concentration of total enzyme have been reported for many enzymes [63], reaction mechanisms are often disputed and the relative steady-state concentrations of enzyme forms are rarely experimentally known. However, careful use of parameter sampling along with historical enzyme kinetics studies may help to address these unknowns, as will be discussed, potentially enabling the generation of large-scale mechanistic kinetic models of metabolism based upon *in vivo* data.

High-throughput data-driven approaches are inherently susceptible to the issues regarding the completeness and accuracy of the data. Examining each the experimental origin of each data type exposes the source of these vulnerabilities. Steady-state reaction fluxes are typically estimated in one of two ways: experimental measurement using ^{13}C labeling experiments, or calculation through flux balance analysis [14, 64]. Isotope labeling experiments are subject to error associated with the measurement technique as well as in the model-based interpretation of mass distributions or NMR spectra in terms of reaction fluxes. Flux balance analysis calculations are dependent upon measured uptake rates and growth rates, as well as the assumed cellular objective function. Equilibrium constants can be experimentally measured at certain conditions or estimated using computational methods such as group contribution theory [27, 65]. Both methods are subject to error due to potentially unaccounted for factors in the *in vivo* thermodynamic state, although experimental measurement at certain conditions and subsequent transformation into *in vivo* conditions likely may give a more reliable estimate than current purely computational methods [27, 66]. While all measurements are subject to data error, metabolomics data suffers from the additional complication of data completeness. While *in vivo* metabolomics measurements are becoming increasingly high-throughput, the coverage of small molecule metabolomics is not yet at the genome scale [67]. Thus, many metabolites in the model may not have measurements at the condition of interest. Due to the many sources of error that may be present in data mapping methods, it may be desirable to couple such approaches with parameter sampling and parameter sensitivity approaches that assess the impact of error on the predictions made by condition-specific kinetic models.

2.4.4 Parameter Sensitivity and Random Sampling

Sampling methods bypass issues of possible data error by instead defining capabilities of the model within a certain range of parameters, thus effectively defining an ensemble of candidate metabolic models [68–70]. These methods can be implemented for any or all model parameters, and the utility of such methods is largely dependent upon the problem of interest.

A useful first approach to examining parameter effects is local parameter sensitivity analysis, varying each parameter in turn around some pre-determined state and observing the change in model function.

$$s_i = \frac{\partial o}{\partial p_i} \tag{2.38}$$

where s_i is the sensitivity of the output o to a change in parameter p_i. Parameter sensitivity can help determine the parameters for which it is most important to have accurate values. However, due to non-linearity present in most kinetic models, local parameter sensitivity may inaccurately predict changes in model function with large variation in a given parameter or variations in multiple parameters simultaneously.

In contrast to local sensitivity analysis, global sampling analysis can give insight into possible model properties over the entire spectrum of available parameters. Monte Carlo methods are a set of global sampling approaches that have seen significant adoption in metabolic systems biology [69, 71–73]. Such methods have had a large degree of success in characterizing linear convex spaces due to properties of these spaces that enable efficient uniform sampling. However, relevant parameter spaces in kinetic models are most often neither linear nor convex, and thus other approaches must be developed to sampling these spaces meaningfully. For example, the issue of non-linearity has been bypassed by instead sampling values of a scaled (linear, convex) Jacobian matrix for a network to determine the stability space across candidate steady-states [74]. Care must be taken with such methods that topological constraints such as $S \sim -G^T$, thermodynamic constraints placed by the 2nd Law, and individual variable constraints from data are not violated within the sampled space, to prevent the results from becoming biased by a potentially large number of physically meaningless sample points.

As mentioned previously, data mapping and sampling approaches are complementary. It is unlikely that it will be possible to simultaneously measure all variables necessary to completely define a large metabolic network at a high level of detail, placing limitations on purely data-driven modeling. Similarly, the large number of parameters makes it difficult to effectively define biologically relevant states using a pure parameter sampling approach. However, by using available data to constrain the parameter space, sampling attains additional utility and relevance, furthering enabling modeling of large-scale *in vivo* kinetics of metabolism.

2.4.5 Numerical Considerations

While the majority of setbacks associated with the construction of large-scale kinetic models are experimental in origin, there are also practical considerations of purely numerical origin. One critical aspect in building kinetic models is the stability of the steady-state around which the model is created, defined here as the qualitative behavior of the system after perturbation from a reference steady-state. The global stability of non-linear equations of arbitrary structure is a difficult problem. Research into Lyapunov functions has been useful for proving global stability of nonlinear equations of certain form [75]. However, no known Lyapunov function has been found to prove global stability for biochemical systems of realistic complexity, although Lyapunov functions of specified form, often quadratic, have been used to estimate the domain of attraction of locally stable steady-states of nonlinear dynamic systems.

In lieu of practical methods for determining global stability properties of the dynamic mass balance equations, local stability analysis can nonetheless provide useful information on the response to small perturbations around a particular state. Local stability is determined by the signs of the eigenvalues of the Jacobian matrix for the dynamic mass balance equations calculated at a particular steady state. A steady-state that is locally stable, corresponding to having eigenvalues with strictly negative real parts, returns to the same state for small perturbations. The presence of at least one eigenvalue with a positive real part results in an unstable steady-state that will cause the solution to move away from the steady-state with a perturbation. An unstable steady state can be thought of as a ball resting precariously at the top of a hill, such that a slight push will result in the ball settling upon a new resting state away from the initial steady state. Steady states with eigenvalues that have non-positive real parts but at least one real part equal to zero are considered semi-stable. If there is an imaginary component to the eigenvalues with zero real parts, the system will oscillate stably after perturbation. If there are zero eigenvalues with no imaginary part, certain perturbations can cause the system to reach a new steady-state within a defined manifold of solutions, typically as a result of conserved time-invariant quantities [49].

The eigenvalues of the system are roots of the characteristic equation of the Jacobian matrix and hence, by the well-known Abel-Ruffini theorem, cannot be analytically solved for systems larger than four roots. The Routh-Hurwitz stability criteria can be used to provide constraints on values of parameters required for stability or semi-stability of the steady-state [76]. These criteria are attractive as they provide analytical constraints on the parameters and thus define the stability space of the system. However, the Routh-Hurwitz criteria are immensely complex in practice and difficult to use for algorithmic purposes due to the requirement of determinant calculation to link stability criteria to the elements of the Jacobian matrix. Methods for effectively sampling the space defined by these constraints would be useful in determining stability properties of the available system state space.

Another entirely numerical complication resides in the potential for ill conditioning in dynamic systems. The condition number of a matrix, in this case the Jacobian for the system, is the ratio of the largest to the smallest singular value for the singular value decomposition of a matrix. The condition number gives a metric for the distortion of inputs that the matrix causes. The practical consideration is that the very small error in data or in solver variables can cause very large divergence in the result of calculations such as inversion and numerical integration. Ill conditioning creates many problems, one of which being that matrices with large condition numbers are considered effectively singular and any attempt to invert these matrices is subject to considerable error. From a simulation perspective, ill-conditioned systems are considered 'stiff,' and special care must be taken when solving stiff ODEs. Special solvers have been developed to deal with such systems, typically by dynamically altering step size based on the stiffness of the system. Ill-conditioned systems also pose challenges to optimization methods, as many rely on local gradients to search for optimal solutions and thus are vulnerable to scaling issues.

2.5 Closing: Practical Applications of Kinetic Models

The scope of experimentally-validated applications of kinetic models of metabolism has largely been limited to predicting time courses of substrate and product levels and local parameter sensitivities *in vitro*. While kinetic modeling already has been used for such applications as predicting possible effects of enzymopathies in the red blood cell [77, 78], the majority of *in vivo* applications of kinetic modeling are yet to come. Efforts at using metabolic control coefficients as a guide to metabolic engineering [79] are complicated by regulation at the transcription and translation level as well as the distal response to the overexpression of enzymes that are predicted to increase flux through desired pathways [80, 81]. Recent methods have effectively used knowledge of reaction thermodynamics to choose enzymes for synthetic pathways that result in higher product titer [82]; making use of dynamic network properties to improve titer is a rational next step, but significant practical hurdles may remain. It is possible that a shift in thinking is necessary to generate kinetic models of metabolism that are useful for industrial purposes [83]. For example, culture methods for engineered strains may 1 day incorporate dynamic nutrient levels to take advantage of metabolic properties that only exist transiently.

In searching for these practical applications, it may be useful to consider lessons from the comparative success of constraint-based modeling in applications relevant to metabolic engineering. There are two fundamental strengths of constraint-based modeling we observe that may have analogous strengths in kinetic models. First, bounds on metabolic capabilities are mathematically defined. For constraint-based models, the steady-state assumption and measured exchange rates define the space of feasible flux state; in kinetic models, bounds on dynamic response times are placed that may be related to the ability of the cell to respond to noise and shifting environmental conditions. Second, the predicted cell function is tied to a meaningful evolutionary selective pressure. In constraint-based modeling, the cell function is the predicted optimal growth rate, which is selected for when a growth pressure is placed on the population; however, corresponding selection pressures based in underlying network dynamics are not yet clear. While further elucidation of these concepts provides a practical motivation for developing kinetic models of metabolism, exploiting these concepts for metabolic engineering may require additional creativity and insight.

If efforts to use dynamic models of metabolism to aid in metabolic engineering efforts are to be successful, a greater understanding of the evolutionary origin of dynamic properties must be obtained. What fitness benefit is gained by certain kinetic properties over others? How does the metabolic network interact with signaling and transcription regulatory networks in a time-resolved fashion? How can the kinetic parameters be altered, through enzyme engineering, to elicit certain systemic dynamic behaviors that improve strain performance? The necessary tools are currently being developed, in the form of large-scale kinetic models fueled by high-throughput *in vivo* data sets, that may begin to provide the answers to these questions.

References

1. Michaelis L, Menten ML (1913) Die kinetik der invertinwirkung. Biochem Z 49 (333–369):352. doi:citeulike-article-id:5936552
2. Lotka AJ (1920) Analytical note on certain rhythmic relations in organic systems. Proc Natl Acad Sci USA 6:410–415
3. Volterra V (1927) Variazioni e fluttuazioni del numero d'individui in specie animali conviventi. Mem. R. Acad. Lincei 2:1–142
4. Hodgkin AL, Huxley AF (1952) A quantitative description of membrane current and its application to conduction and excitation in nerve. J Physiol Lond 117(4):500–544
5. Karplus M, Weaver DL (1976) Protein-folding dynamics. Nature 260(5550):404–406
6. Goldbeter A (2002) Computational approaches to cellular rhythms. Nature 420 (6912):238–245. doi:10.1038/Nature01259
7. Chance B (1943) The kinetics of the enzyme-substrate compound of peroxidase. J Biol Chem 151(2):553–577
8. Rapoport TA, Heinrich R, Rapoport SM (1976) The regulatory principles of glycolysis in erythrocytes in vivo and in vitro. A minimal comprehensive model describing steady states, quasi-steady states and time-dependent processes. Biochem J 154(2):449–469
9. Joshi A, Palsson BO (1990) Metabolic dynamics in the human red cell. Part IV – Data prediction and some model computations. J Theor Biol 142(1):69–85
10. Joshi A, Palsson BO (1990) Metabolic dynamics in the human red cell. Part III – Metabolic reaction rates. J Theor Biol 142(1):41–68
11. Joshi A, Palsson BO (1989) Metabolic dynamics in the human red cell. Part II – Interactions with the environment. J Theor Biol 141(4):529–545
12. Joshi A, Palsson BO (1989) Metabolic dynamics in the human red cell. Part I – A comprehensive kinetic model. J Theor Biol 141(4):515–528
13. Nicholson JK, Holmes E, Lindon JC, Wilson ID (2004) The challenges of modeling mammalian biocomplexity. Nat Biotechnol 22(10):1268–1274. doi:10.1038/Nbt1015

14. Price ND, Papin JA, Schilling CH, Palsson BO (2003) Genome-scale microbial in silico models: the constraints-based approach. Trends Biotechnol 21(4):162–169. doi: S0167779903000301 [pii]
15. Jamshidi N, Palsson BO (2008) Formulating genome-scale kinetic models in the post-genome era. Mol Syst Biol 4:171. doi:10.1038/msb.2008.8
16. Clauset A, Moore C, Newman ME (2008) Hierarchical structure and the prediction of missing links in networks. Nature 453(7191):98–101. doi:10.1038/nature06830
17. Horler RS, Butcher A, Papangelopoulos N, Ashton PD, Thomas GH (2009) EchoLOCATION: an in silico analysis of the subcellular locations of *Escherichia coli* proteins and comparison with experimentally derived locations. Bioinformatics 25(2):163–166. doi:10.1093/bioinformatics/btn596
18. Kurkdjian A, Guern J (1989) Intracellular Ph – measurement and importance in cell-activity. Annu Rev Plant Phys Plant Mol Biol 40:271–303
19. Dairaghi DJ, Oldham ER, Bacon KB, Schall TJ (1997) Communication – chemokine receptor CCR3 function is highly dependent on local pH and ionic strength. J Biol Chem 272 (45):28206–28209
20. Minton AP (2001) The influence of macromolecular crowding and macromolecular confinement on biochemical reactions in physiological media. J Biol Chem 276(14):10577–10580. doi:10.1074/jbc.R100005200
21. Gaal T, Bartlett MS, Ross W, Turnbough CL Jr, Gourse RL (1997) Transcription regulation by initiating NTP concentration: rRNA synthesis in bacteria. Science 278(5346):2092–2097
22. Elf J, Nilsson D, Tenson T, Ehrenberg M (2003) Selective charging of tRNA isoacceptors explains patterns of codon usage. Science 300(5626):1718–1722. doi:10.1126/science.1083811
23. Pogliano J, Lynch AS, Belin D, Lin EC, Beckwith J (1997) Regulation of *Escherichia coli* cell envelope proteins involved in protein folding and degradation by the Cpx two-component system. Genes Dev 11(9):1169–1182
24. Choudhary C, Kumar C, Gnad F, Nielsen ML, Rehman M, Walther TC, Olsen JV, Mann M (2009) Lysine acetylation targets protein complexes and co-regulates major cellular functions. Science 325(5942):834–840. doi:10.1126/science.1175371
25. Macek B, Mijakovic I, Olsen JV, Gnad F, Kumar C, Jensen PR, Mann M (2007) The serine/threonine/tyrosine phosphoproteome of the model bacterium *Bacillus subtilis*. Mol Cell Proteomics 6(4):697–707. doi:10.1074/mcp.M600464-MCP200
26. Low PS, Rathinavelu P, Harrison ML (1993) Regulation of glycolysis via reversible enzyme binding to the membrane protein, band 3. J Biol Chem 268(20):14627–14631
27. Alberty RA (2002) Thermodynamics of systems of biochemical reactions. J Theor Biol 215 (4):491–501. doi:10.1006/jtbi.2001.2516
28. Cavalier-Smith T (1978) Nuclear volume control by nucleoskeletal DNA, selection for cell volume and cell growth rate, and the solution of the DNA C-value paradox. J Cell Sci 34:247–278
29. Elowitz MB, Levine AJ, Siggia ED, Swain PS (2002) Stochastic gene expression in a single cell. Science 297(5584):1183–1186. doi:10.1126/science.1070919
30. Klamt S, Haus UU, Theis F (2009) Hypergraphs and cellular networks. PLoS Comput Biol 5 (5):e1000385. doi:10.1371/journal.pcbi.1000385
31. Mahadevan R, Edwards JS, Doyle FJ 3rd (2002) Dynamic flux balance analysis of diauxic growth in *Escherichia coli*. Biophys J 83(3):1331–1340. doi:10.1016/S0006-3495(02)73903-9
32. Meric PA, Wise MJ (1999) Quantitative, scalable discrete-event simulation of metabolic pathways. Proc Int Conf Intell Syst Mol Biol 187–194
33. Mahan BH (1975) Microscopic reversibility and detailed balance – analysis. J Chem Educ 52 (5):299–302
34. Haldane JBS (1930) Enzymes, Monographs on biochemistry. Longmans/Green, London/New York
35. Qian H, Beard DA, Liang SD (2003) Stoichiometric network theory for nonequilibrium biochemical systems. Eur J Biochem 270(3):415–421. doi:3357 [pii]

36. Zhou HX, Rivas G, Minton AP (2008) Macromolecular crowding and confinement: biochemical, biophysical, and potential physiological consequences. Annu Rev Biophys 37:375–397. doi:10.1146/annurev.biophys.37.032807.125817
37. Schnell S, Turner TE (2004) Reaction kinetics in intracellular environments with macromolecular crowding: simulations and rate laws. Prog Biophys Mol Biol 85(2–3):235–260. doi:10.1016/j.pbiomolbio.2004.01.012
38. Tucker W, Kutalik Z, Moulton V (2007) Estimating parameters for generalized mass action models using constraint propagation. Math Biosci 208(2):607–620. doi:DOI 10.1016/j.mbs.2006.11.009
39. Dreger A, Kronfeld M, Ziller MJ, Supper J, Planatscher H, Magnus JB, Oldiges M, Kohlbacher O, Zell A (2009) Modeling metabolic networks in *C. glutamicum*: a comparison of rate laws in combination with various parameter optimization strategies. BMC Syst Biol 3. doi:10.1186/1752-0509-3-5
40. Smallbone K, Simeonidis E, Swainston N, Mendes P (2010) Towards a genome-scale kinetic model of cellular metabolism. BMC Syst Biol 4:6. doi:10.1186/1752-0509-4-6
41. Steuer R, Junker BH (2009) Computational models of metabolism: stability and regulation in metabolic networks. Adv Chem Phys 142:105–251
42. Palsson B (2006) Systems biology: properties of reconstructed networks. Cambridge University Press, Cambridge/New York
43. Lehninger AL, Nelson DL, Cox MM (2008) Lehninger principles of biochemistry, 5th edn. W.H. Freeman, New York
44. Bennett BD, Kimball EH, Gao M, Osterhout R, Van Dien SJ, Rabinowitz JD (2009) Absolute metabolite concentrations and implied enzyme active site occupancy in *Escherichia coli*. Nat Chem Biol 5(8):593–599. doi:10.1038/nchembio.186
45. Cornish-Bowden A (1977) An automatic method for deriving steady-state rate equations. Biochem J 165(1):55–59
46. Hill AV (1910) The possible effects of the aggregation of the molecules of haemoglobin on its oxygen dissociation. J Physiol. doi:citeulike-article-id:440501
47. Monod J, Wyman J, Changeux JP (1965) On the nature of allosteric transitions: a plausible model. J Mol Biol 12:88–118
48. Changeux JP (1964) Allosteric interactions interpreted in terms of quaternary structure. Brookhaven Symp Biol 17:232–249
49. Strogatz SH (1994) Nonlinear dynamics and Chaos: with applications to physics, biology, chemistry, and engineering, Studies in nonlinearity. Addison-Wesley Pub, Reading
50. Orth JD, Thiele I, Palsson BO (2010) What is flux balance analysis? Nat Biotechnol 28(3):245–248. doi:10.1038/nbt.1614
51. Edwards JS, Ibarra RU, Palsson BO (2001) In silico predictions of *Escherichia coli* metabolic capabilities are consistent with experimental data. Nat Biotechnol 19(2):125–130. doi:10.1038/84379
52. Jamshidi N, Palsson BO (2008) Top-down analysis of temporal hierarchy in biochemical reaction networks. PLoS Comput Biol 4(9):e1000177. doi:10.1371/journal.pcbi.1000177
53. Heuett WJ, Beard DA, Qian H (2008) Linear analysis near a steady-state of biochemical networks: control analysis, correlation metrics and circuit theory. BMC Syst Biol 2:44. doi:10.1186/1752-0509-2-44
54. Jamshidi N, Palsson BO (2009) Flux-concentration duality in dynamic nonequilibrium biological networks. Biophys J 97(5):L11–L13. doi:10.1016/j.bpj.2009.06.049
55. Yugi K, Nakayama Y, Kinoshita A, Tomita M (2005) Hybrid dynamic/static method for large-scale simulation of metabolism. Theor Biol Med Model 2:42. doi:10.1186/1742-4682-2-42
56. Chassagnole C, Noisommit-Rizzi N, Schmid JW, Mauch K, Reuss M (2002) Dynamic modeling of the central carbon metabolism of *Escherichia coli*. Biotechnol Bioeng 79(1):53–73

57. Mannervik B (1982) Regression-analysis, experimental error, and statistical criteria in the design and analysis of experiments for discrimination between rival kinetic-models. Methods Enzymol 87:370–390
58. Dyson F (2004) A meeting with enrico fermi – how one intuitive physicist rescued a team from fruitless research. Nature 427(6972):297. doi:10.1038/427297a
59. Resendis-Antonio O (2009) Filling kinetic gaps: dynamic modeling of metabolism where detailed kinetic information is lacking. PLoS One 4(3). doi:10.1371/Journal.Pone.0004967
60. Beard DA, Vinnakota KC, Wu F (2008) Detailed enzyme kinetics in terms of biochemical species: study of citrate synthase. PLoS One 3(3):e1825
61. Jamshidi N, Palsson BO (2010) Mass action stoichiometric simulation models: incorporating kinetics and regulation into stoichiometric models. Biophys J 98(2):175–185. doi:10.1016/j.bpj.2009.09.064
62. Tomita M, Hashimoto K, Takahashi K, Shimizu TS, Matsuzaki Y, Miyoshi F, Saito K, Tanida S, Yugi K, Venter JC, Hutchison CA 3rd (1999) E-CELL: software environment for whole-cell simulation. Bioinformatics 15(1):72–84. doi:btc007 [pii]
63. Taniguchi Y, Choi PJ, Li GW, Chen H, Babu M, Hearn J, Emili A, Xie XS (2010) Quantifying *E. coli* proteome and transcriptome with single-molecule sensitivity in single cells. Science 329(5991):533–538. doi:10.1126/science.1188308
64. Zamboni N, Fendt SM, Ruhl M, Sauer U (2009) (13)C-based metabolic flux analysis. Nat Protoc 4(6):878–892. doi:10.1038/nprot.2009.58
65. Jankowski MD, Henry CS, Broadbelt LJ, Hatzimanikatis V (2008) Group contribution method for thermodynamic analysis of complex metabolic networks. Biophys J 95(3):1487–1499. doi:10.1529/biophysj.107.124784
66. Fleming RM, Thiele I, Nasheuer HP (2009) Quantitative assignment of reaction directionality in constraint-based models of metabolism: application to *Escherichia coli*. Biophys Chem 145 (2–3):47–56. doi:10.1016/j.bpc.2009.08.007
67. Bennett BD, Yuan J, Kimball EH, Rabinowitz JD (2008) Absolute quantitation of intracellular metabolite concentrations by an isotope ratio-based approach. Nat Protoc 3(8):1299–1311. doi:10.1038/nprot.2008.107
68. Tran LM, Rizk ML, Liao JC (2008) Ensemble modeling of metabolic networks. Biophys J 95 (12):5606–5617. doi:10.1529/biophysj.108.135442
69. Schellenberger J, Palsson BO (2009) Use of randomized sampling for analysis of metabolic networks. J Biol Chem 284(9):5457–5461. doi:10.1074/jbc.R800048200
70. Famili I, Mahadevan R, Palsson BO (2005) k-Cone analysis: determining all candidate values for kinetic parameters on a network scale. Biophys J 88(3):1616–1625. doi:10.1529/biophysj.104.050385
71. Thiele I, Price ND, Vo TD, Palsson BO (2005) Candidate metabolic network states in human mitochondria. Impact of diabetes, ischemia, and diet. J Biol Chem 280(12):11683–11695. doi:10.1074/jbc.M409072200
72. Barrett CL, Price ND, Palsson BO (2006) Network-level analysis of metabolic regulation in the human red blood cell using random sampling and singular value decomposition. BMC Bioinformatics 7. doi:10.1186/1471-2105-7-132
73. Barrett CL, Herrgard MJ, Palsson B (2009) Decomposing complex reaction networks using random sampling, principal component analysis and basis rotation. BMC Syst Biol 3:30. doi:10.1186/1752-0509-3-30
74. Steuer R, Gross T, Selbig J, Blasius B (2006) Structural kinetic modeling of metabolic networks. Proc Natl Acad Sci USA 103(32):11868–11873. doi:10.1073/pnas.0600013103
75. Clarke BL (1988) Stoichiometric network analysis. Cell Biophys 12:237–253
76. Routh EJ (1877) Adams prize essay: a treatise on the stability of a given state of motion, particularly steady motion. Macmillan and co, London
77. Holzhutter HG, Jacobasch G, Bisdorff A (1985) Mathematical modelling of metabolic pathways affected by an enzyme deficiency. A mathematical model of glycolysis in normal and pyruvate-kinase-deficient red blood cells. Eur J Biochem 149(1):101–111

78. Jamshidi N, Wiback SJ, Palsson BO (2002) In silico model-driven assessment of the effects of single nucleotide Polymorphisms (SNPs) on human red blood cell metabolism. Genome Res 12(11):1687–1692. doi:10.1101/Gr.329302
79. Visser D, Schmid JW, Mauch K, Reuss M, Heijnen JJ (2004) Optimal re-design of primary metabolism in *Escherichia coli* using linlog kinetics. Metab Eng 6(4):378–390. doi:10.1016/j.ymben.2004.07.001
80. Schuster S (1999) Use and limitations of modular metabolic control analysis in medicine and biotechnology. Metab Eng 1(3):232–242. doi:10.1006/mben.1999.0119
81. Fell DA (1992) Metabolic control analysis – a survey of its theoretical and experimental development. Biochem J 286:313–330
82. Bond-Watts BB, Bellerose RJ, Chang MCY (2011) Enzyme mechanism as a kinetic control element for designing synthetic biofuel pathways. Nat Chem Biol 7(4):222–227. doi: http://www.nature.com/nchembio/journal/v7/n4/abs/nchembio.537.html#supplementary-information
83. Andrianantoandro E, Basu S, Karig DK, Weiss R (2006) Synthetic biology: new engineering rules for an emerging discipline. Mol Syst Biol. doi:10.1038/Msb4100073

Chapter 3
Design of Superior Cell Factories Based on Systems Wide Omics Analysis

Katsunori Yoshikawa, Chikara Furusawa, Takashi Hirasawa, and Hiroshi Shimizu

Contents

Abstract The bioproduction industry is expanding towards sustainable production of energy, chemicals and materials requesting for superior, high-productivity cell factories. Recent advances in measurement technologies enable comprehensive analysis of cellular components, so-called "omics" analysis, which is expected to accelerate the construction of superior cell factories. As example, transcriptome analysis is widely used for genome-wide screening of candidate genes that may be manipulated to improve productivity. However, the massive amounts of data produced by this method, requests for smart approaches to narrow the selection of promising candidate genes as targets for higher productivity. In this chapter, we review several studies that demonstrate successful breeding based on omics data, and discuss how we can design experiments and screen for target genes to be manipulated for the development of superior cell factories.

K. Yoshikawa • C. Furusawa • T. Hirasawa • H. Shimizu (✉)
Department of Bioinformatic Engineering, Graduate School of Information Science and Technology, Osaka University, 1-5 Yamadaoka, Suita, Osaka 565-0871, Japan
e-mail: shimizu@ist.osaka-u.ac.jp

C. Wittmann and S.Y. Lee (eds.), *Systems Metabolic Engineering*,
DOI 10.1007/978-94-007-4534-6_3,

Keywords Systems wide omics analysis • Cell factory • Genome • Transcriptome • Metabolome • Proteome • Fluxome • Phenome • Multi-omics analyses • Fossil resources • Biofuel • Building block of chemicals • Breeding • Stress tolerance • Metabolic engineering • Systems metabolic engineering • DNA microarray • *In silico* simulation • Genome-scale metabolic model • Flux balance analysis • Metabolic flux analysis • Evolutionary engineering • Adaptive evolution • Next-generation sequencing technologies • R programming language

3.1 Introduction

Since ancient times, microorganisms have been used to manufacture useful products such as foods, medicines, and chemicals. Problems of global warming and depletion of fossil resources are now major drivers for the bioproduction of various products such as biofuels or chemicals from renewable raw materials replacing petroleum resources [40, 77]. Relevant examples, as will be discussed in the industrially relevant chapters of the book include ethanol [54, 80], higher alcohols [4, 138], lactate [57, 123], succinate [94, 126], or other building block of chemicals [38, 62, 130]. Breeding of superior cell factories to achieve industrial-scale production of these materials with high production yield and rate has been realized by genetic modification of metabolic pathways, known as metabolic engineering [73, 120]. Other approaches have aimed at improved stress tolerance of production strains which is also important for superior cell factories, since in the production process, cells are usually exposed to stresses such as high osmotic pressure, high concentrations of inhibiting products, high temperature, and oxidative conditions, all of which reduce productivity [6, 34].

The recent development of molecular biology techniques and systems wide omics technologies now enable the comprehensive analysis of intracellular components and their interactions including the genome, transcriptome, proteome, metabolome, and fluxome. These methods have advanced our understanding of cellular mechanisms [55, 58, 145] and expanded metabolic engineering to systems metabolic engineering [14, 61, 68, 97]. Based on the knowledge gained from omics data, molecular breeding to improve productivity is expected. In this chapter, we focus on the breeding of cells using transcriptome data, which is one of the most widely used omics data. The first experiment using a DNA microarray was performed to analyze genome-wide expression changes during the fermentation shift in *Saccharomyces cerevisiae* [22]. The use of DNA microarrays then spread rapidly (Fig. 3.1) and revealed novel insights into cellular mechanisms [29, 72]. DNA microarray technology is a powerful tool to analyze expression data for thousands of genes under various conditions and time-course series at a time; gene candidates are then identified to be manipulated to improve a desired phenotype.

Based on this technology, several studies have reported improved target phenotypes, including high productivity of target products and stress tolerance, as summarized in Table 3.1. For example, when comparing two strains of

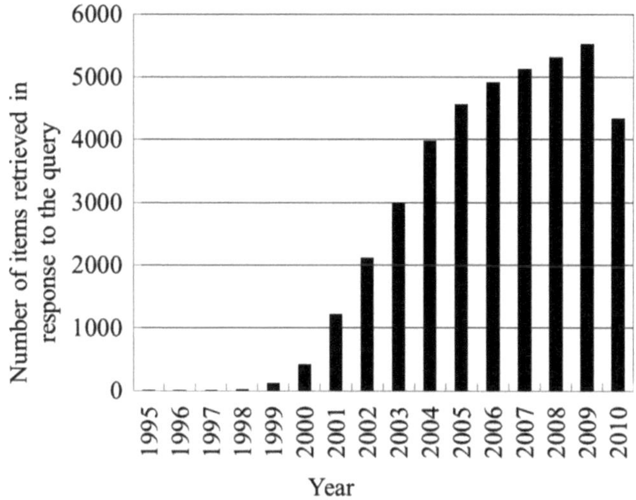

Fig. 3.1 Number of DNA microarray studies listed in PubMed (http://www.ncbi.nlm.nih.gov/pubmed). The number of search results from querying "DNA microarray gene expression" in each year is displayed

Table 3.1 Examples of transcriptome-based strain improvement

Target phenotype	Experimental design using transcriptome data	References
Lovastatin production	Obtained lovastatin production strains by molecular engineering and analyzed their transcriptome and metabolome data in *Aspergillus terreus*	Askenazi et al. [3]
Ethanol tolerance	Obtained an ethanol tolerant strain by adaptive evolution and compared transcriptome data in *E. coli*	Gonzalez et al. [35]
Galactose utilization	Compared transcriptome data of different galactose utilization strains and focused on galactose metabolism in yeast	Bro et al. [16]
Threonine production	Compared transcriptome data of different threonine production strains and integrated with *in silico* simulation in yeast	Lee et al. [76]
Osmotic stress tolerance	Compared transcriptome data among different osmotic tolerant strains in yeast	Hirasawa et al. [47]
Osmotic stress tolerance	Integrative analysis with transcription factor binding sites in yeast.	Pandey et al. [96]
Lysine production	Compared transcriptome data among strains showing different lysine productivity in *Corynebacterium glutamicum*	Sindelar and Wendisch [117]
Xylose utilization	Compared transcriptome data of different xylose-utilization strains	Bengtsson et al. [11]
Sulfur Dioxide production	Compared transcriptome data of different sulfur dioxide-producing strains and integrated with metabolome data in yeast	Yoshida et al. [140]

(continued)

Table 3.1 (continued)

Target phenotype	Experimental design using transcriptome data	References
Ethanol stress tolerance	Compared transcriptome data in different ethanol-tolerant strains and integrated phenotype data of single-deletion collection in yeast	Hirasawa et al. [47]
Lactate production	Comparison of different lactate-producing strains in yeast	Ookubo et al. [95]
Riboflavin production	Compared transcriptome data of different riboflavin-producing strains in *Bacillus subtilis*	Shi et al. [113]
Ethanol tolerance	Obtained ethanol-tolerant strains by adaptive evolution and compared transcriptome data in *E. coli*	Horinouchi et al. [51]
Arabinose utilization	Compared transcriptome data of different arabinose utilization strains and integrated with metabolome, fluxome and genome data in yeast.	Wisselink et al. [134]
Galactose utilization	Obtained increased galactose utilization strains by adaptive evolution and compared transcriptome and metabolome data of evolved stains in yeast	Hong et al. [50]
Diaminopentane production	Compared transcriptome data of different diaminopentane-producing strains in *C. glutamicum*	Kind et al. [63]

differing phenotypes, differentially expressed genes are generally thought to be responsible for the phenotypic difference. Then, the effects of manipulating these candidate genes are evaluated to achieve the preferred phenotype. In many cases, however, hundreds of candidate genes exhibit different expression levels between samples and many of these genes are not related to the target phenotype. Selection methods to extract target phenotype-related candidate genes from transcriptome data are needed. In this chapter, we review several reports on the construction of superior cells based on transcriptome data, and discuss experimental design and selection methods for effective identification of genes related to a given target phenotype. This is complemented by other omics studies which have proven valuable towards superior cell factories.

3.2 Transcriptome Analysis for Breeding of Superior Cell Factories

3.2.1 Transcriptome Analysis

Gene expression levels were first quantified by northern analysis [1]. Reverse transcription polymerase chain reaction (RT-PCR) is also widely used to quantify gene expression. These methods provide expression data for a small number of genes at a time. DNA microarrays [81, 107, 111], serial analysis of gene expression (SAGE) [125] and whole transcriptome shotgun sequencing (RNA-seq) enable us to analyze transcriptome at genome scale and obtain a comprehensive insight into the system of interest. In this section, we mainly describe the DNA microarray technology, which is still the most popular method for transcriptome analysis.

The principle of transcriptome analysis using DNA microarrays is hybridization of two DNA strands by hydrogen bonding between specific nucleotide pairs. DNA microarrays are constructed on glass and silicon chips. Nucleotide sequences (probes) complementary to the target sequence are spotted or directly synthesized [26, 92] on the chips. Full-length complementary DNA sequences (cDNAs) were used as probes in early DNA microarrays [107, 110], but have been replaced later by straightforward short oligonucleotide probes (25–60 bases) that complement specific parts of a target gene to avoid cross-hybridization with non-target sequences. RNA samples are reverse transcribed into fluorescence-labeled cDNA and hybridized to the probes. Gene expression is then measured as the fluorescence intensity of the labeled targets as they hybridize with the probe.

Another method for transcriptome analysis is SAGE [116], which is a sequence-based technique. A short nucleotide sequence tag is isolated from a specific position in each transcript, and these tags are linked to construct a longer chain. These chains are sequenced and expression levels are quantified based on the frequency of the tag sequence corresponding to each transcript. Although sequence-based methods provide more accurate data than DNA microarrays, such methods are not widely used since high-throughput DNA sequencing is required. Next-generation sequencing technologies have been developed [83, 87] to provide enough performance for sequence-based methods. Several methods using next-generation sequencing such as RNA-seq [89] and Super SAGE [82] provide more accurate and quantitative data and are likely to replace DNA microarrays [112].

The analysis of DNA microarray data first requires normalization, because the raw data often contain systematic errors caused by fluctuations in hybridization efficiency, labeling efficiency, or instrument noise. Removal and normalization of such errors is necessary to obtain reliable results for comparison with other microarray data. Several methods have been developed to solve such problems [13, 30, 52, 78]. Since DNA microarrays provide expression data for thousands of genes, bioinformatics algorithms are typically used to extract biologically meaningful information. Clustering is one of the most commonly used methods in this regard [23, 121]. Genes are grouped according to the similarity of their expression patterns (co-expression), and typical expression patterns are extracted from the transcriptome data. The biological significance of the expression patterns is extracted from the clustering results by analyzing the enrichment of gene functions in each cluster [104]. There are many other methods and software products for analyzing transcriptome data, such as principal component analysis (PCA; [79]), ontological analysis [59], visualization of the transcriptome data in the context of metabolic pathways [69, 84], inference of regulatory interaction [122], analysis based on a condition-specific entropy reduction [105], and network analysis [12, 49, 128, 136]. An open-source software package for DNA microarray analysis including data import, normalization, statistical analysis, etc., is available in Bioconductor [32], written in the R programming language (http://www.r-project.org/). Analysis of DNA microarray data using Bioconductor is well reviewed in Zhang et al. [144]. Thus, numerous algorithms and software for DNA microarray data analysis are available to help extracting the most biologically meaningful information.

DNA microarrays provide comprehensive transcriptome data, which enable us to screen candidate genes that may be manipulated to achieve a desired phenotype. Table 3.1 shows experimental designs and methods that have been successfully employed for such microarray-based strain improvements. Several studies have used transcriptome comparisons between strains of differing phenotypes to extract candidate genes for generating a target phenotype [11, 16, 63, 95, 117]. Transcriptome data have also been integrated with other information such as metabolome data, fluxome data or *in silico* simulation [3, 50, 76, 100, 134, 139]. In the following section, we review these studies and discuss how we can identify genes and cellular functions related to a given target phenotype can be identified from transcriptome data and used to create superior cell factories.

3.2.2 Comparative Transcriptome Analysis

Table 3.1 lists several studies that compared transcriptome data between strains exhibiting different stress tolerance and productivity phenotypes. Then, differently expressed genes are selected as candidates assuming that their differential expression levels reflect the phenotypic differences. However, even after extraction of the differently expressed genes, in many cases a large number of genes remain to be analyzed. The, experimental evaluation of all such candidate genes is difficult and often not feasible so that further screening is required to identify higher-potential candidates.

A common method of candidate gene selection is based on biological knowledge. This comprises genes whose functions are known to be associated with the target phenotype or the target metabolic pathway. Bro et al. [16] performed comparative transcriptome analysis between two mutant strains with improved galactose utilization and a reference strain of *S. cerevisiae*. Focusing on the expression of genes related to galactose utilization, only *PGM2*, encoding phosphoglucomutase, was significantly upregulated in both mutants. It was thus selected as the candidate gene responsible for the increased galactose utilization phenotype of the mutants. The effect of *PGM2* overexpression was investigated, and it was found that *PGM2* overexpression indeed led to increased galactose utilization. Ookubo et al. [95] used a similar method to improve lactate production in yeast, comparing transcriptome data between a lactate-producing strain carrying a gene encoding human L-lactate dehydrogenase (LDH) and a reference strain with the empty vector. Significant expression differences were found for 670 out of almost 5,800 analyzed genes. These genes were screened for their relation to lactic acid metabolism, and *CYB2*, encoding L-lactate dehydrogenase, was found to be strongly up-regulated in the lactate producing strain. The Cyb2p protein is required for L-lactate anabolism in yeast; thus, experimental deletion of *CYB2* prevented L-lactate anabolism in the lactate-producing strain. Lactate production in the *CYB2*-deleted strain could then be significantly increased at low pH in *S. cerevisiae* (pH 3.5). Kind et al. [63] also identified the genes whose overexpression enhanced 1,5-diaminopentane production by comparative transcriptome profiles between engineered diaminopentane producing-strain and its reference strain. 35 out of 3000 genes were significantly up-regulated in diaminopentane producing-strain. Among them, *cg2893* gene encoding permease was most up-regulated, and regarded as a candidate for potential diaminopentane transporter. Indeed deletion of *cg2893* exhibited 90% reduction of diaminopentane secretion and overexpression of *cg2893* showed 20% increased yield.

The above studies analyzed gene expression levels in strains that possess different phenotypes. Hirasawa et al. [46] analyzed gene expression dynamics in response to osmotic stress to generate an osmotic stress-tolerant strain of *S. cerevisiae*. Osmotic stress was created by the addition of NaCl to the culture medium. Temporal changes in gene expression after NaCl addition was then analyzed by microarrays in two strains that exhibited different osmotic stress tolerance. They focused on the genes that were upregulated by osmotic stress in both strains and more highly expressed in the tolerant strain. This gene class included *GPD1*, *ENA1*, and *CUP1*. *GPD1* and *ENA1* were known to be required for osmotic stress tolerance, while the relationship between *CUP1* and osmotic stress tolerance was unknown. The authors assumed that these more highly expressed genes mediated the osmotic tolerance of the tolerant strain. In fact, single gene overexpression of each of these genes improved the osmotic tolerance of the less-tolerant strain. Other studies that have focused on genes in the context of a target phenotype have successfully identified those genes which can enhance the desired phenotype [76, 113].

These studies suggest that focusing on genes based on biological knowledge such as gene function and metabolic pathway is an effective way to extract higher-potential candidate genes from transcriptome data. However, this method may miss other important genes whose relationship to the target phenotype are unknown. Other methods are required to select higher potential candidate genes without biological methods. Sindelar and Wendisch [117] and Bengtsson et al. [11] successfully identified genes that enhance lysine productivity in *Corynebacterium glutamicum* and xylose utilization in *S. cerevisiae* without biological knowledge.

In many cases, comparison of transcriptome data between a single superior strain and its reference strain yields hundreds of differentially expressed genes, making it difficult to identify specifically relevant candidates. To limit and identify reliable candidates, some studies have analyzed transcriptome data from several strains that express the target phenotype to different degrees and extracted those candidates which commonly exhibit higher or lower expression levels in the superior phenotype strains in comparison to the reference strains. Bengtsson et al. [11] obtained 13 candidate genes through this type of selection method and found that 5 of these candidates improved xylose utilization by single deletion or overexpression. These genes included novel genes for improvement of the target phenotype. Thus the comparison of transcriptome data between multiple strains possessing different phenotypes is an effective way to identify those genes that are responsible for the target phenotype without prior biological knowledge.

3.2.3 Relationship Between Transcriptome and Phenome Data

Generally, genes that are differentially expressed under certain conditions or in strains with different phenotypes are selected as candidates to be manipulated to achieve target phenotypes. For example, when cells are exposed to a stress condition, induced genes are thought to be responsible for response and adaptation to the corresponding stress, and overexpression of these genes is one strategy by which to enhance stress tolerance. Several studies have demonstrated successful breeding based on a similar screening scheme [46]. However, genes identified by this method are not always related to the target phenotype, and in many cases deletion and overexpression of these genes do not affect the phenotype. This means that it is still not clear how expression differences between different conditions and strains relate to phenotypic changes caused by specific gene manipulation. Genome-wide collections of single gene deletion and over expression strains in *Escherichia coli* [8, 132] and *S. cerevisiae* [31, 64, 119] now allow us to perform comprehensive analyses of the phenotype changes caused by gene deletion or overexpression (phenome), and evaluate the efficiency of selecting candidate genes from transcriptome data by comparison with phenome data.

Genes related to a target phenotype are identified from among the genes that exhibit different expression levels or dynamics in strains of different phenotypes [11, 16, 46, 47, 95, 117]. Comparative transcriptome analysis between strains of different phenotypes is an effective way to identify candidate genes. For example, Hirasawa et al. [48] used microarrays to quantify gene expression profiles in 2 lactate-producing *S. cerevisiae* strains and a non-lactate producing strain and compared these profile differences with lactate productivity of single-gene deletion strains. About 400 genes were commonly up-regulated or down-regulated in the 2 lactate-producing strains in comparison to the non-lactate-producing strain, suggesting that these genes might be involved in lactate production. In order to verify this hypothesis, the human *LDH* gene was introduced to the single gene deletion collection of yeast, and lactate production was evaluated. The genes whose deletion increased or decreased lactate productivity were significantly enriched in the collection of genes whose expression levels were commonly up-regulated or down-regulated in the lactate-producing strains. Thus, extraction of genes whose expression levels differ between strains of different phenotypes is an effective way to identify candidate genes to be manipulated to achieve a desired phenotype.

The relationship between transcriptome and phenome was analyzed from another perspective by Yoshikawa et al. [141], who investigated whether the genes responsible for a given phenotype can be extracted from transcriptome data of a single strain under different environmental conditions. First, they used DNA microarrays to determine the expression profiles of a yeast strain before and after ethanol stress. Second, growth rate changes produced by all possible single-gene deletions were quantified within a yeast gene deletion library. Then, these two collections of omics data, i.e., the transcriptome and phenome, were compared to analyze whether genes whose expression levels were changed by ethanol stress

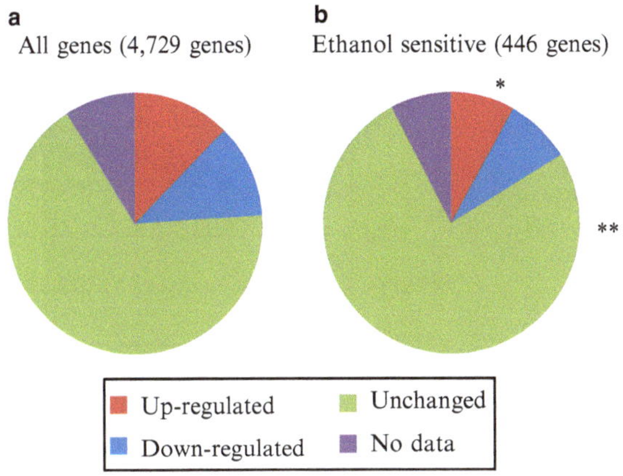

Fig. 3.2 Relationship between transcriptome and phenome data. Transcriptome data according to gene expression dynamics in response to high-concentration ethanol stress in yeast was obtained by Hirasawa et al. [47]; genes were categorized as "Up-regulated," "Down-regulated," and "Unchanged". Ethanol sensitivity of 4,729 single gene deletion strains in yeast was analyzed by Yoshikawa et al. [141]. Proportions of the transcriptome data gene categories were represented in all 4,729 genes (**a**), and in the 446 genes for which the deletion strains exhibited ethanol sensitivity (**b**). * indicates that the proportions were significantly low (randomized test, $P < 0.05$), and ** indicates that the proportions were significantly high ($P < 0.01$) in 446 ethanol-sensitive strains in comparison to the 4,729 deletion strains

were involved in the mechanism of ethanol stress tolerance. Yeast genes were categorized into three classes, i.e., up-regulated, down-regulated, and unchanged by ethanol stress, and genes were identified that, when deleted, yielded a growth defect under ethanol stress (ethanol-sensitive). The proportion of genes whose expression was changed after ethanol addition and which, when deleted, yielded ethanol sensitivity was not significantly high (Fig. 3.2). Other studies have demonstrated no correlation between gene expression changes and phenotype changes yielded by gene deletion during growth under stress conditions [33, 129]. These results suggest that selection of candidates based on the transcriptome dynamics of a single strain, focusing only on genes whose expression level changes with the environment is not an effective method to screen for target genes, and comparison of transcriptome data between strains of different phenotypes is important to identify those genes that are responsible for a given target phenotype.

3.2.4 Transcriptomics in Evolutionary Engineering

As described in previous sections, comparative transcriptome analysis between strains with different phenotypes is an effective way to identify candidate genes that may be manipulated for selective breeding. However, such strains are not

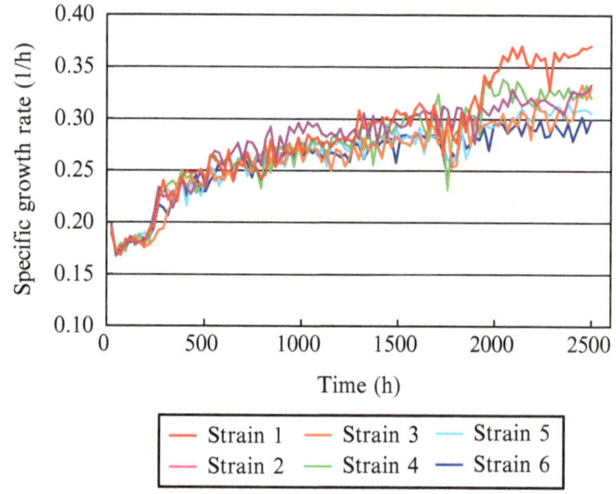

Fig. 3.3 Specific growth changes during adaptive evolution. Six parallel series of batch culture serial transfers every 24 h under 5% ethanol conditions. *Lines* represent the time-course of specifc growth rate in each serially transferred batch culture

always available for comparison, and sometimes it is difficult to apply the same strategy for breeding cells based on transcriptome data.

To obtain strains with a desired phenotype, adaptive evolution experiments have been introduced [106]. Adaptive evolution experiments are based on inherent cell mechanisms such as adaptability to environmental changes. For example, when cells are exposed to a severe stress condition in which growth rate significantly decreases, long-term cultivation by continuous culture or serial transfers of batch culture, cell growth rate may be gradually increases by iterations of selection and phenotypic changes by mutation or adaptation. This method has been applied to obtain industrial strains for their qualities of stress tolerance [5, 21, 27, 36, 71] and enhanced substrate utilization [28, 37, 50, 124, 134]. Several studies identified the genes and cellular functions responsible for evolved phenotypes by comparing the transcriptome data of the evolved strains with that of their ancestors, and further improved the target phenotype based on these findings [35, 51, 134].

In one such study, Horinouchi et al. [51] obtained ethanol-tolerant strains by adaptive evolution and successfully determined the cellular mechanisms responsible for ethanol tolerance in *E. coli* from transcriptome data. They cultured *E. coli* for 1,000 generations (2,500 h) under 5% ethanol stress by serial transfers of batch cultures and obtained six parallel-evolved ethanol-tolerant strains that exhibited twofold higher growth rates in comparison to the ancestor strain (Fig. 3.3). Subsequently, the transcriptome data between these evolved strains and the ancestor strain with and without ethanol stress were compared. Candidate genes that were relevant to the ethanol tolerance of evolved strains were those that were commonly up-regulated in the evolved strains. These genes were functionally enriched for those related to the biosynthesis of amino acids such as tryptophan, histidine, and branched-chain amino acids. The relationship between these amino acids and ethanol tolerance were analyzed by addition of isoleucine, tryptophan, and histidine to the culture medium. Growth of the ancestor strain was enhanced by addition of these amino acids to the culture medium under ethanol stress. This suggested that activation of amino acid biosynthesis increases growth under ethanol stress in *E. coli*.

Adaptive evolution is an effective method for improving a desired phenotype. Comparative analysis of transcriptome data between evolved strains and their ancestor strain could be used to identify the genes and cellular functions responsible for the improved phenotype of the evolved strains. This method, when applied in a parallel evolution experiment, has the advantage of yielding multiple strains exhibiting different degrees of the target phenotype, maximizing the potential of the identified candidate genes. Addition of other omics data and whole-genome sequences will provide more detailed insights into the cause of phenotype changes brought about by adaptive evolution, and will facilitate further improvement of the desired phenotype in the evolved strain.

3.3 Multi-omics Analyses for Breeding of Superior Cell Factories

3.3.1 Proteome, Metabolome, Fluxome Anlaysis

Recent developments in analytical technologies have enabled the comprehensive analysis of various cellular components, such as the proteome [7, 102, 103], metabolome [118], and genome [42, 83]. Moreover, improvements in information technology has enabled the in experimental quantification of the fluxome via ^{13}C labeling data [120, 135, 144] as well the silico simulation of metabolic flux profiles using genome-scale metabolic models and flux-balance analysis [24, 25, 43, 108].

These omics analyses have successfully identified the key cellular mechanisms for bioproduction, and contributed to breed superior cell factories. For example, proteome analysis could identify and quantify the intracellular protein concentrations. Gene expression level does not always reflect the protein expression level because of translational efficiency and degradation rate of mRNA, thus the proteome data can provide more accurate information on the amount of enzymes controlling cellular metabolism than the transcriptome data. Proteome analysis was applied for many bioproduction studies to understand the metabolic state in terms of protein expressions, e.g. industrial ethanol production process in yeast [18], glutamate production in *Corynebacterium glutamicum* [60], succinate production in *Mannheimia succiniciproducens* [75], and penicillin production in *Penicillium chrysogenum* [56].

Metabolome data are used to characterize intracellular amounts of metabolite, and we can identify the rate-limiting step of a metabolic reaction by accumulation or depletion of the corresponding metabolite linked to a specific reaction. Hasunuma et al. [39] identified rate-limiting step in ethanol production from xylose under the presence of week acids, such as acetate and formate, which are released from pretreatment of lignocelluloses and inhibits ethanol production. Metabolome analysis revealed the significant accumulation of metabolites involved in the non-oxidative pentose phosphate pathway (PPP) by the addition of acetate. This result suggested that the flux of the non-oxidative PPP was decreased by acetate and this pathway was a rate limiting step for ethanol production in the presence of acetate. Indeed overexpression of genes involved in PPP, *TAL1* or *TKL1* increased ethanol production in the presence of acetate and formate. Klimacek et al. [66] analyzed rate-limiting step of xylose fermentation in yeast based on metabolome data. Metabolome data was obtained from the genetic engineered xylose fermenting strains and their wild type strain. Thermodynamic analysis using metabolome data suggested that the activities of pentose phosphate pathway and intracellular pool of fluctose-6-phosphate are potential limiting step for xylose fermentation. Metabolome analysis also well captured metabolic state in other studies, e.g. various nutrient limitation condition in yeast [15], and genetic and temperature perturbation on fission yeast [101].

Fluxome analysis could quantify and simulate metabolic flux distribution on metabolic network. Metabolic flux analysis (MFA) is a powerful technique to quantify the *in vivo* carbon flux distribution on the metabolic network using isotopomer labeled substrate, mainly ^{13}C-labeled carbon source [120, 135]. MFA well captured the metabolic shift by the metabolic modification, and different metabolic phase. For example, metabolic shift by genetic modification in lysine producing *C. glutamicum* were identified by [9, 10]. Shirai et al. [115] revealed the increased flux of pyruvate decarboxylase in anaplerotic pathways plays an important role in the production phase of glutamate in *C. glutamicum*, by comparing flux distributions between growth phase and the production phase. ^{13}C-MFA was performed on around hundred of transcription regulator mutants and revealed the condition-specific transcriptional control in *E. coli* [41]. Basically the metabolic flux analysis requires long-time metabolic steady-state until ^{13}C isotopomer enrichments in metabolites reaches stationary. Recently non-stationary MFA based on the information of transient isotopomer labeling in metabolites was developed [91]. This method reduces the duration of the metabolic steady-state for ^{13}C-labelling experiment, and could expand MFA in application for the non-stable industrial production process like batch and fed batch fermentation in which the metabolic state was transient in short time [131].

Flux balance analysis (FBA) is another method for fluxome. FBA with *in silico* genome-scale metabolic model could simulate the genome-scale flux distribution with assumption of a steady state of metabolic reaction and optimization of an objective function, e.g. maximization of cell growth [24, 43, 108]. This *in silico* metabolic simulation enable to estimate the change in metabolic flux by genetic modification or culture conditions, and also design the optimal metabolic network [17, 137]. This method was further expanded to explore the novel metabolic pathway by introduction of heterologous pathway to produce the target product including non-native chemicals in host organisms [19, 45, 143]. Dozens of genome-scale metabolic models have been reconstructed in various species, such as *E. coli* [24],

S. cerevisiae [85], *C. glutamicum* [65, 114], *Bacillus subtilis* [44, 93], and photosynthetic bacteria *Synechocystis* [67, 86, 142]. *In silico* genetic modification using these genome-scale metabolic model have identified candidate genes whose deletion or amplification could improve the target production, such as succinic acid [74], valine [98], threonine [76], sesquiterpene [2], lycopene [20] and 3-hydroxypropanoate [45].

3.3.2 Integrative Multi-omics Analysis

Towards unraveling of the rather complex biological systems, it appears straightforward to integrate different approaches for a more complete picture on the cell and a rational basis for systems metabolic engineering. Most of multi-omics studies so far focus on industrially relevant microorganisms, identifying biotechnology applications as major driver. As example, the integration of the transcriptome with the fluxome unravelled the distribution of control between transcriptional and post-transcriptional regulation in the central carbon metabolism of the biotechnological working horses *B. subtilis* [109] and *C. glutamicum* [70]. In *E. coli* the combined analysis of transcriptome, metabolome and fluxome confirmed the existence of hidden reactions in the central metabolism of [90]. One of the most massive data sets was recently generated for a systematic analysis of *E. coli* cells to genetic and environmental perturbation [55] indicating only few and local responses to gene deletions, whereas the metabolism responded globally to change in the growth rate. In addition, integrative analysis of multi-omics data has provided more detailed information regarding cellular mechanisms in other organisms [36, 53, 88, 99].

Several studies with integrative analyses using multi-omics data and *in silico* simulation identified genes and cellular functions which improved the target phenotype ([3, 50, 76, 96, 127, 134]; Yoshida et al. [140]).

As example of multi-omics studies, Hong et al. [50] identified mutations conferring enhancement of galactose utilization by omics analysis including whole genome-sequencing on evolutionary engineered yeast strain which showed increased growth rate on galactose. Transcriptome and metabolome analysis of three parallel evolved strains and their ancestor strain showed different characteristics at transcriptome and metabolome levels in each evolved strain. Among them, they extracted common properties in three evolved strain, such as increase accumulation of intracellular glycogen and trehalose, and up-regulation of genes associated with these metabolisms and galactose metabolism. These results suggested that the improvement of galactose utilization could be explained by enhancement of glycogen and trehalose metabolism. To reveal these phenotypic changes, whole-genome resequence analysis of the evolved strains was performed, and no mutations were identified in genes associated with galactose metabolism and also in the glycogen and trehalose metabolism. Mutations were identified in genes associated with Ras/PKA pathway in all the evolved strains, which regulate glycogen and trehalose metabolism. This fact indicated that the metabolome and transcriptome shifts and enhancement of growth on galactose in the evolved strains were caused by mutation in regulatory system of these metabolism. Indeed, introduction of this mutation on the ancestor strain enhance the galactose uptake. From this study, they represented the two key points for success identification of genotype of the evolved strain; (1) combination of phenotypic analysis including of transcriptome and metabolome, and (2) comparison of these data of parallel evolved strains to identify the conserved mutation that resulted in the same improved phenotype.

Wisselink et al. [134] performed transcriptome, metabolome and fluxome analysis on arabinose fermenting strain which was genetic and evolutionary engineered [133] to understand the cellular mechanisms for fermentation of arabinose in yeast. Transcriptome analysis identified the up-regulation of genes associated with galactose metabolism including *GAL2* which encodes galactose permease inhibited growth. *GAL2* deletion completely inhibited growth on arabinose, thus galactose transporter was confirmed as essential for arabinose transport. Moreover, metabolic flux analysis and metabolome analysis indicated that higher flux in non-oxidative part of pentose phosphate pathway (PPP) and higher accumulation of metabolites related to PPP in evolved strain growing on arabinose in comparison with that on glucose. This suggested that the PPP was a key metabolic pathway for arabinose fermentation. Among expression data associated with PPP, the upregulation of *TKL2* and *YGR043C* encoding "minor" isoenzymes of transaldolase and transketolase were identified. Then, it was confirmed that these "minor" isoenzymes were key enzymes for arabinose fermentation, based on the fact that the deletion of these genes decreased the growth rate in the arabinose fermentation process.

Yoshida et al. [140] performed integrative analysis of transcriptome and metabolome data to breed a lager brewing yeast with low H_2S and high SO_2, which are related to beer flavor. They analyzed data from bottom-fermenting yeast *Saccharomyces pastorianus* and baker's yeast *S. cerevisiae* which produce different amounts of SO_2 and H_2S, and analyzed the differences in gene expression and metabolite amounts between these two strains, specifically in the context of SO_2 and H_2S metabolism. Metabolome data indicated that depletion of O-acetylhomoserine with which H_2S is metabolized to homocysteine in brewing yeast was the rate-limiting step to reduce H_2S production. Combining these results with the transcriptome data, they successfully constructed the desired strains.

Lee et al. [76] performed *in silico* simulations with a genome-scale metabolic model. Integrating transcriptome data, this study predicted the optimized expression of the candidate genes to increase threonine production in *E. coli*. It appeared straightforward to focus on the genes related to threonine biosynthesis and central metabolism. This identified *ppc*, encoding phosphoenolpyruvate carboxylase (PPC) that supplies the threonine precursor oxaloacetate. Surprisingly, *ppc* deletion and overexpression, decreased threonine production, indicating that optimal expression of *ppc* is required to increase threonine flux. *In silico* simulations using a genome-scale metabolic model of *E. coli* predicted a moderate over expression to maximize PPC flux for threonine production. Indeed a moderate overexpression of *ppc* yielded increased threonine production.

Many other studies also reported successful breeding and identification of key metabolic properties for bioproduction by integrative multi-omics analysis, e.g. transcriptome data and metabolite profile identified genetic and physiological control of lovastatin production in *Aspergillus terreus* [3], transcriptome data and transcription factor binding sites information identified importance of amino acids for osmotic stress tolerance in yeast [96], transcriptome and metabolome data identified the effect of protease supplementation on metabolism in beer fermentation [100]. Integrative analyses provide valuable information for molecular breeding which we cannot extract from transcriptome data alone. The rapid development of analytical and computational technologies will allow us to analyze multi-omics data sets with ease, and implement the integrative analysis of these data to clarify the molecular mechanisms of superior strains and accelerate molecular breeding.

3.4 Summary

In this chapter, we reviewed how to design superior cell factories based on omics data, mainly transcriptome data (summarized in Fig. 3.4). Although availability of DNA microarrays provides genome-wide transcriptome data easily and rapidly, the massiveness of the transcriptome data makes it difficult to identify genes that are specifically related to the target phenotype. Many successful studies and comparative analyses between transcriptome and phenome data suggest that the identification of differentially expressed genes by comparing transcriptome data between strains of different phenotypes is an effective way to identify genes related to a specific phenotype [48, 141]. When the number of screened genes remains too large, further selection based on biological knowledge or comparative analysis using several strains of different phenotypes can be used to identify those genes that confer the target phenotype. The use of several strains exhibiting different degrees of the target phenotype can be helpful. Adaptive evolution experiments have been successfully applied to obtain desired strains and transcriptome analysis of these evolved strains are available for identification of the genes responsible for the phenotype of an evolved stain. Moreover, recent developments of analytical technologies enable us to utilize other omics data such as fluxome, proteome or metabolome data, and these data also provide useful information for breeding of cell factories from different point of view. Integrative multi-omics analyses will provide new insights into cellular mechanisms and provide useful information for the effective design of cell factories.

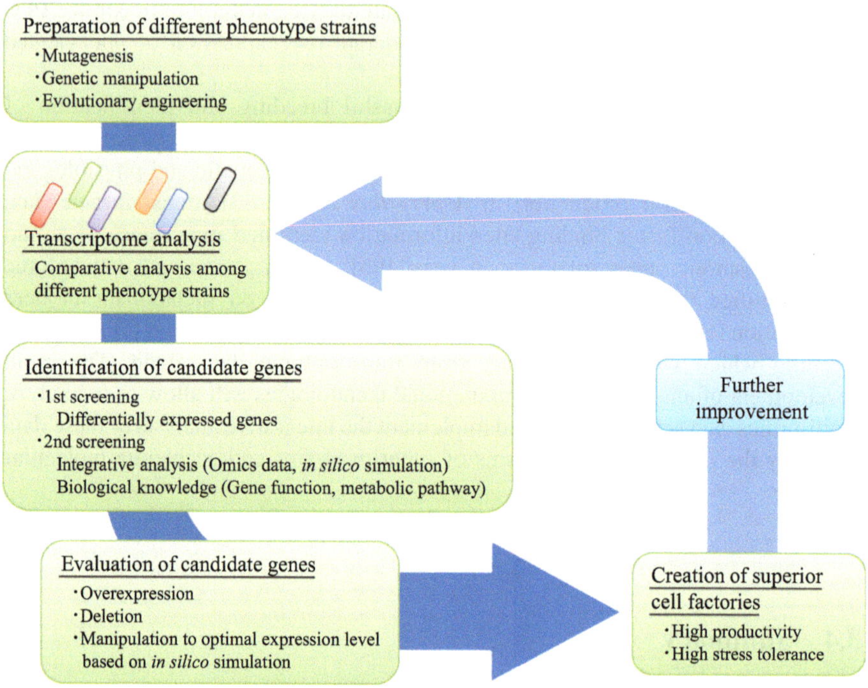

Fig. 3.4 Summary of experimental design to obtain superior cell factories based on transcriptome data

References

1. Alwine JC, Kemp DJ, Stark GR (1977) Method for detection of specific RNAs in agarose gels by transfer to diazobenzyloxymethyl-paper and hybridization with DNA probes. Proc Natl Acad Sci USA 74(12):5350–5354
2. Asadollahi MA, Maury J, Patil KR, Schalk M, Clark A, Nielsen J (2009) Enhancing sesquiterpene production in *Saccharomyces cerevisiae* through *in silico* driven metabolic engineering. Metab Eng 11(6):328–334. doi:10.1016/j.ymben.2009.07.001
3. Askenazi M, Driggers EM, Holtzman DA, Norman TC, Iverson S, Zimmer DP, Boers ME, Blomquist PR, Martinez EJ, Monreal AW, Feibelman TP, Mayorga ME, Maxon ME, Sykes K, Tobin JV, Cordero E, Salama SR, Trueheart J, Royer JC, Madden KT (2003) Integrating transcriptional and metabolite profiles to direct the engineering of lovastatin-producing fungal strains. Nat Biotechnol 21(2):150–156. doi:10.1038/nbt781
4. Atsumi S, Hanai T, Liao JC (2008) Non-fermentative pathways for synthesis of branched-chain higher alcohols as biofuels. Nature 451(7174):86–89. doi:10.1038/nature06450
5. Atsumi S, Wu TY, Machado IM, Huang WC, Chen PY, Pellegrini M, Liao JC (2010) Evolution, genomic analysis, and reconstruction of isobutanol tolerance in *Escherichia coli*. Mol Syst Biol 6:449. doi:10.1038/msb.2010.98
6. Attfield PV (1997) Stress tolerance: the key to effective strains of industrial baker's yeast. Nat Biotechnol 15(13):1351–1357. doi:10.1038/nbt1297-1351

7. Auerbach D, Thaminy S, Hottiger MO, Stagljar I (2002) The post-genomic era of interactive proteomics: facts and perspectives. Proteomics 2(6):611–623. doi:10.1002/1615-9861(200206)2:6<611::AID-PROT611>3.0.CO;2-Y
8. Baba T, Ara T, Hasegawa M, Takai Y, Okumura Y, Baba M, Datsenko KA, Tomita M, Wanner BL, Mori H (2006) Construction of *Escherichia coli* K-12 in-frame, single-gene knockout mutants: the Keio collection. Mol Syst Biol 2:2006.0008. doi:10.1038/msb4100050
9. Becker J, Klopprogge C, Herold A, Zelder O, Bolten CJ, Wittmann C (2007) Metabolic flux engineering of L-lysine production in Corynebacterium glutamicum–over expression and modification of G6P dehydrogenase. J Biotechnol 132(2):99–109. doi:10.1016/j.jbiotec.2007.05.026
10. Becker J, Klopprogge C, Wittmann C (2008) Metabolic responses to pyruvate kinase deletion in lysine producing *Corynebacterium glutamicum*. Microb Cell Fact 7:8. doi:10.1186/1475-2859-7-8
11. Bengtsson O, Jeppsson M, Sonderegger M, Parachin NS, Sauer U, Hahn-Hagerdal B, Gorwa-Grauslund MF (2008) Identification of common traits in improved xylose-growing *Saccharomyces cerevisiae* for inverse metabolic engineering. Yeast 25(11):835–847. doi:10.1002/yea.1638
12. Bhan A, Galas DJ, Dewey TG (2002) A duplication growth model of gene expression networks. Bioinformatics 18(11):1486–1493
13. Bilban M, Buehler LK, Head S, Desoye G, Quaranta V (2002) Normalizing DNA microarray data. Curr Issues Mol Biol 4(2):57–64
14. Blazeck J, Alper H (2010) Systems metabolic engineering: genome-scale models and beyond. Biotechnol J 5(7):647–659. doi:10.1002/biot.200900247
15. Boer VM, Crutchfield CA, Bradley PH, Botstein D, Rabinowitz JD (2010) Growth-limiting intracellular metabolites in yeast growing under diverse nutrient limitations. Mol Biol Cell 21(1):198–211. doi:10.1091/mbc.E09-07-0597
16. Bro C, Knudsen S, Regenberg B, Olsson L, Nielsen J (2005) Improvement of galactose uptake in *Saccharomyces cerevisiae* through overexpression of phosphoglucomutase: example of transcript analysis as a tool in inverse metabolic engineering. Appl Environ Microbiol 71(11):6465–6472. doi:10.1128/AEM.71.11.6465-6472.2005
17. Burgard AP, Pharkya P, Maranas CD (2003) Optknock: a bilevel programming framework for identifying gene knockout strategies for microbial strain optimization. Biotechnol Bioeng 84(6):647–657. doi:10.1002/bit.10803
18. Cheng JS, Qiao B, Yuan YJ (2008) Comparative proteome analysis of robust *Saccharomyces cerevisiae* insights into industrial continuous and batch fermentation. Appl Microbiol Biotechnol 81(2):327–338. doi:10.1007/s00253-008-1733-6
19. Cho A, Yun H, Park JH, Lee SY, Park S (2010) Prediction of novel synthetic pathways for the production of desired chemicals. BMC Syst Biol 4:35. doi:10.1186/1752-0509-4-35
20. Choi HS, Lee SY, Kim TY, Woo HM (2010) *In silico* identification of gene amplification targets for improvement of lycopene production. Appl Environ Microbiol 76(10):3097–3105. doi:10.1128/AEM.00115-10
21. Cullum AJ, Bennett AF, Lenski RE (2001) Evolutionary adaptation to temperature. IX. Preadaptation to novel stressful environments of *Escherichia coli* adapted to high temperature. Evolution 55(11):2194–2202
22. DeRisi JL, Iyer VR, Brown PO (1997) Exploring the metabolic and genetic control of gene expression on a genomic scale. Science 278(5338):680–686
23. Eisen MB, Spellman PT, Brown PO, Botstein D (1998) Cluster analysis and display of genome-wide expression patterns. Proc Natl Acad Sci USA 95(25):14863–14868
24. Feist AM, Henry CS, Reed JL, Krummenacker M, Joyce AR, Karp PD, Broadbelt LJ, Hatzimanikatis V, Palsson BO (2007) A genome-scale metabolic reconstruction for *Escherichia coli* K-12 MG1655 that accounts for 1260 ORFs and thermodynamic information. Mol Syst Biol 3:121. doi:10.1038/msb4100155

25. Feist AM, Herrgard MJ, Thiele I, Reed JL, Palsson BO (2009) Reconstruction of biochemical networks in microorganisms. Nat Rev Microbiol 7(2):129–143. doi:10.1038/nrmicro1949
26. Fodor SP, Read JL, Pirrung MC, Stryer L, Lu AT, Solas D (1991) Light-directed, spatially addressable parallel chemical synthesis. Science 251(4995):767–773
27. Fong SS, Burgard AP, Herring CD, Knight EM, Blattner FR, Maranas CD, Palsson BO (2005) *In silico* design and adaptive evolution of *Escherichia coli* for production of lactic acid. Biotechnol Bioeng 91(5):643–648. doi:10.1002/bit.20542
28. Fong SS, Joyce AR, Palsson BO (2005) Parallel adaptive evolution cultures of *Escherichia coli* lead to convergent growth phenotypes with different gene expression states. Genome Res 15(10):1365–1372. doi:10.1101/gr.3832305
29. Furusawa C, Kaneko K (2003) Zipf's law in gene expression. Phys Rev Lett 90(8):088102
30. Furusawa C, Ono N, Suzuki S, Agata T, Shimizu H, Yomo T (2009) Model-based analysis of non-specific binding for background correction of high-density oligonucleotide microarrays. Bioinformatics 25(1):36–41. doi:10.1093/bioinformatics/btn570
31. Gelperin DM, White MA, Wilkinson ML, Kon Y, Kung LA, Wise KJ, Lopez-Hoyo N, Jiang L, Piccirillo S, Yu H, Gerstein M, Dumont ME, Phizicky EM, Snyder M, Grayhack EJ (2005) Biochemical and genetic analysis of the yeast proteome with a movable ORF collection. Genes Dev 19(23):2816–2826. doi:10.1101/gad.1362105
32. Gentleman RC, Carey VJ, Bates DM, Bolstad B, Dettling M, Dudoit S, Ellis B, Gautier L, Ge Y, Gentry J, Hornik K, Hothorn T, Huber W, Iacus S, Irizarry R, Leisch F, Li C, Maechler M, Rossini AJ, Sawitzki G, Smith C, Smyth G, Tierney L, Yang JY, Zhang J (2004) Bioconductor: open software development for computational biology and bioinformatics. Genome Biol 5(10):R80. doi:10.1186/gb-2004-5-10-r80
33. Giaever G, Chu AM, Ni L, Connelly C, Riles L, Veronneau S, Dow S, Lucau-Danila A, Anderson K, Andre B, Arkin AP, Astromoff A, El-Bakkoury M, Bangham R, Benito R, Brachat S, Campanaro S, Curtiss M, Davis K, Deutschbauer A, Entian KD, Flaherty P, Foury F, Garfinkel DJ, Gerstein M, Gotte D, Guldener U, Hegemann JH, Hempel S, Herman Z, Jaramillo DF, Kelly DE, Kelly SL, Kotter P, LaBonte D, Lamb DC, Lan N, Liang H, Liao H, Liu L, Luo C, Lussier M, Mao R, Menard P, Ooi SL, Revuelta JL, Roberts CJ, Rose M, Ross-Macdonald P, Scherens B, Schimmack G, Shafer B, Shoemaker DD, Sookhai-Mahadeo S, Storms RK, Strathern JN, Valle G, Voet M, Volckaert G, Wang CY, Ward TR, Wilhelmy J, Winzeler EA, Yang Y, Yen G, Youngman E, Yu K, Bussey H, Boeke JD, Snyder M, Philippsen P, Davis RW, Johnston M (2002) Functional profiling of the *Saccharomyces cerevisiae* genome. Nature 418(6896):387–391. doi:10.1038/nature00935
34. Gibson BR, Lawrence SJ, Leclaire JP, Powell CD, Smart KA (2007) Yeast responses to stresses associated with industrial brewery handling. FEMS Microbiol Rev 31(5):535–569. doi:10.1111/j.1574-6976.2007.00076.x
35. Gonzalez R, Tao H, Purvis JE, York SW, Shanmugam KT, Ingram LO (2003) Gene array-based identification of changes that contribute to ethanol tolerance in ethanologenic *Escherichia coli*: comparison of KO11 (parent) to LY01 (resistant mutant). Biotechnol Prog 19(2):612–623. doi:10.1021/bp025658q
36. Goodarzi H, Bennett BD, Amini S, Reaves ML, Hottes AK, Rabinowitz JD, Tavazoie S (2010) Regulatory and metabolic rewiring during laboratory evolution of ethanol tolerance in *E. coli*. Mol Syst Biol 6:378. doi:10.1038/msb.2010.33
37. Guimaraes PM, Francois J, Parrou JL, Teixeira JA, Domingues L (2008) Adaptive evolution of a lactose-consuming *Saccharomyces cerevisiae* recombinant. Appl Environ Microbiol 74 (6):1748–1756. doi:10.1128/AEM.00186-08
38. Hanai T, Atsumi S, Liao JC (2007) Engineered synthetic pathway for isopropanol production in *Escherichia coli*. Appl Environ Microbiol 73(24):7814–7818. doi:10.1128/AEM.01140-07
39. Hasunuma T, Sanda T, Yamada R, Yoshimura K, Ishii J, Kondo A (2011) Metabolic pathway engineering based on metabolomics confers acetic and formic acid tolerance to a recombinant xylose-fermenting strain of *Saccharomyces cerevisiae*. Microb Cell Fact 10(1):2. doi:10.1186/1475-2859-10-2

40. Hatti-Kaul R, Tornvall U, Gustafsson L, Borjesson P (2007) Industrial biotechnology for the production of bio-based chemicals—a cradle-to-grave perspective. Trends Biotechnol 25 (3):119–124. doi:10.1016/j.tibtech.2007.01.001
41. Haverkorn van Rijsewijk BR, Nanchen A, Nallet S, Kleijn RJ, Sauer U (2011) Large-scale 13C-flux analysis reveals distinct transcriptional control of respiratory and fermentative metabolism in *Escherichia coli*. Mol Syst Biol 7:477. doi:10.1038/msb.2011.9
42. Hawkins RD, Hon GC, Ren B (2010) Next-generation genomics: an integrative approach. Nat Rev Genet 11(7):476–486. doi:10.1038/nrg2795
43. Heinemann M, Kummel A, Ruinatscha R, Panke S (2005) *In silico* genome-scale reconstruction and validation of the *Staphylococcus aureus* metabolic network. Biotechnol Bioeng 92 (7):850–864. doi:10.1002/bit.20663
44. Henry CS, Zinner JF, Cohoon MP, Stevens RL (2009) iBsu1103: a new genome-scale metabolic model of *Bacillus subtilis* based on SEED annotations. Genome Biol 10(6):R69. doi:10.1186/gb-2009-10-6-r69
45. Henry CS, Broadbelt LJ, Hatzimanikatis V (2010) Discovery and analysis of novel metabolic pathways for the biosynthesis of industrial chemicals: 3-hydroxypropanoate. Biotechnol Bioeng 106(3):462–473. doi:10.1002/bit.22673
46. Hirasawa T, Nakakura Y, Yoshikawa K, Ashitani K, Nagahisa K, Furusawa C, Katakura Y, Shimizu H, Shioya S (2006) Comparative analysis of transcriptional responses to saline stress in the laboratory and brewing strains of *Saccharomyces cerevisiae* with DNA microarray. Appl Microbiol Biotechnol 70(3):346–357. doi:10.1007/s00253-005-0192-6
47. Hirasawa T, Yoshikawa K, Nakakura Y, Nagahisa K, Furusawa C, Katakura Y, Shimizu H, Shioya S (2007) Identification of target genes conferring ethanol stress tolerance to *Saccharomyces cerevisiae* based on DNA microarray data analysis. J Biotechnol 131(1):34–44. doi:10.1016/j.jbiotec.2007.05.010
48. Hirasawa T, Ookubo A, Yoshikawa K, Nagahisa K, Furusawa C, Sawai H, Shimizu H (2009) Investigating the effectiveness of DNA microarray analysis for identifying the genes involved in *l*-lactate production by *Saccharomyces cerevisiae*. Appl Microbiol Biotechnol 84 (6):1149–1159. doi:10.1007/s00253-009-2209-z
49. Holter NS, Maritan A, Cieplak M, Fedoroff NV, Banavar JR (2001) Dynamic modeling of gene expression data. Proc Natl Acad Sci USA 98(4):1693–1698. doi:10.1073/pnas.98.4.1693
50. Hong KK, Vongsangnak W, Vemuri GN, Nielsen J (2011) Unravelling evolutionary strategies of yeast for improving galactose utilization through integrated systems level analysis. Proc Natl Acad Sci USA 108(29):12179–12184. doi:10.1073/pnas.1103219108
51. Horinouchi T, Tamaoka K, Furusawa C, Ono N, Suzuki S, Hirasawa T, Yomo T, Shimizu H (2010) Transcriptome analysis of parallel-evolved *Escherichia coli* strains under ethanol stress. BMC Genomics 11:579. doi:10.1186/1471-2164-11-579
52. Huber W, von Heydebreck A, Sultmann H, Poustka A, Vingron M (2002) Variance stabilization applied to microarray data calibration and to the quantification of differential expression. Bioinformatics 18(Suppl 1):S96–S104
53. Ideker T, Thorsson V, Ranish JA, Christmas R, Buhler J, Eng JK, Bumgarner R, Goodlett DR, Aebersold R, Hood L (2001) Integrated genomic and proteomic analyses of a systematically perturbed metabolic network. Science 292(5518):929–934. doi:10.1126/science.292.5518.929
54. Ingram LO, Aldrich HC, Borges AC, Causey TB, Martinez A, Morales F, Saleh A, Underwood SA, Yomano LP, York SW, Zaldivar J, Zhou S (1999) Enteric bacterial catalysts for fuel ethanol production. Biotechnol Prog 15(5):855–866. doi:10.1021/bp9901062
55. Ishii N, Nakahigashi K, Baba T, Robert M, Soga T, Kanai A, Hirasawa T, Naba M, Hirai K, Hoque A, Ho PY, Kakazu Y, Sugawara K, Igarashi S, Harada S, Masuda T, Sugiyama N, Togashi T, Hasegawa M, Takai Y, Yugi K, Arakawa K, Iwata N, Toya Y, Nakayama Y, Nishioka T, Shimizu K, Mori H, Tomita M (2007) Multiple high-throughput analyses

monitor the response of *E. coli* to perturbations. Science 316(5824):593–597. doi:10.1126/science.1132067
56. Jami MS, Barreiro C, Garcia-Estrada C, Martin JF (2010) Proteome analysis of the penicillin producer *Penicillium chrysogenum*: characterization of protein changes during the industrial strain improvement. Mol Cell Proteomics 9(6):1182–1198. doi:10.1074/mcp.M900327-MCP200
57. John RP, Nampoothiri KM, Pandey A (2007) Fermentative production of lactic acid from biomass: an overview on process developments and future perspectives. Appl Microbiol Biotechnol 74(3):524–534. doi:10.1007/s00253-006-0779-6
58. Jozefczuk S, Klie S, Catchpole G, Szymanski J, Cuadros-Inostroza A, Steinhauser D, Selbig J, Willmitzer L (2010) Metabolomic and transcriptomic stress response of *Escherichia coli*. Mol Syst Biol 6:364. doi:10.1038/msb.2010.18
59. Khatri P, Draghici S (2005) Ontological analysis of gene expression data: current tools, limitations, and open problems. Bioinformatics 21(18):3587–3595. doi:10.1093/bioinformatics/bti565
60. Kim J, Fukuda H, Hirasawa T, Nagahisa K, Nagai K, Wachi M, Shimizu H (2010) Requirement of de novo synthesis of the OdhI protein in penicillin-induced glutamate production by *Corynebacterium glutamicum*. Appl Microbiol Biotechnol 86(3):911–920. doi:10.1007/s00253-009-2360-6
61. Kim TY, Kim HU, Lee SY (2010) Data integration and analysis of biological networks. Curr Opin Biotechnol 21(1):78–84. doi:10.1016/j.copbio.2010.01.003
62. Kind S, Wittmann C (2011) Bio-based production of the platform chemical 1,5-diaminopentane. Appl Microbiol Biotechnol 91(5):1287–1296. doi:10.1007/s00253-011-3457-2
63. Kind S, Kreye S, Wittmann C (2011) Metabolic engineering of cellular transport for overproduction of the platform chemical 1,5-diaminopentane in *Corynebacterium glutamicum*. Metab Eng 13(5):617–627. doi:10.1016/j.ymben.2011.07.006
64. Kitagawa M, Ara T, Arifuzzaman M, Ioka-Nakamichi T, Inamoto E, Toyonaga H, Mori H (2005) Complete set of ORF clones of *Escherichia coli* ASKA library (a complete set of *E. coli* K-12 ORF archive): unique resources for biological research. DNA Res 12(5):291–299. doi:10.1093/dnares/dsi012
65. Kjeldsen KR, Nielsen J (2009) *In silico* genome-scale reconstruction and validation of the *Corynebacterium glutamicum* metabolic network. Biotechnol Bioeng 102(2):583–597. doi:10.1002/bit.22067
66. Klimacek M, Krahulec S, Sauer U, Nidetzky B (2010) Limitations in xylose-fermenting *Saccharomyces cerevisiae*, made evident through comprehensive metabolite profiling and thermodynamic analysis. Appl Environ Microbiol 76(22):7566–7574. doi:10.1128/AEM.01787-10
67. Knoop H, Zilliges Y, Lockau W, Steuer R (2010) The metabolic network of Synechocystis sp. PCC 6803: systemic properties of autotrophic growth. Plant Physiol 154(1):410–422. doi:10.1104/pp.110.157198
68. Kohlstedt M, Becker J, Wittmann C (2010) Metabolic fluxes and beyond-systems biology understanding and engineering of microbial metabolism. Appl Microbiol Biotechnol 88(5):1065–1075. doi:10.1007/s00253-010-2854-2
69. Kono N, Arakawa K, Ogawa R, Kido N, Oshita K, Ikegami K, Tamaki S, Tomita M (2009) Pathway projector: web-based zoomable pathway browser using KEGG atlas and Google Maps API. PLoS One 4(11):e7710. doi:10.1371/journal.pone.0007710
70. Krömer JO, Heinzle E, Schröder H, Wittmann C (2006) Accumulation of homolanthionine and activation of a novel pathway for isoleucine biosynthesis in *Corynebacterium glutamicum* McbR deletion strains. J Bacteriol 188(2):609–618. doi:10.1128/JB.188.2.609-618.2006

71. Kwon YD, Kim S, Lee SY, Kim P (2011) Long-term continuous adaptation of *Escherichia coli* to high succinate stress and transcriptome analysis of the tolerant strain. J Biosci Bioeng 111(1):26–30. doi:10.1016/j.jbiosc.2010.08.007
72. Laub MT, McAdams HH, Feldblyum T, Fraser CM, Shapiro L (2000) Global analysis of the genetic network controlling a bacterial cell cycle. Science 290(5499):2144–2148
73. Lee SY, Papoutsakis ET (1999) Metabolic engineering. Marcel Dekker, New York
74. Lee SJ, Lee DY, Kim TY, Kim BH, Lee J, Lee SY (2005) Metabolic engineering of *Escherichia coli* for enhanced production of succinic acid, based on genome comparison and *in silico* gene knockout simulation. Appl Environ Microbiol 71(12):7880–7887. doi:10.1128/AEM.71.12.7880-7887.2005
75. Lee JW, Lee SY, Song H, Yoo JS (2006) The proteome of *Mannheimia succiniciproducens*, a capnophilic rumen bacterium. Proteomics 6(12):3550–3566. doi:10.1002/pmic.200500837
76. Lee KH, Park JH, Kim TY, Kim HU, Lee SY (2007) Systems metabolic engineering of *Escherichia coli* for L-threonine production. Mol Syst Biol 3:149. doi:10.1038/msb4100196
77. Lee JW, Kim HU, Choi S, Yi J, Lee SY (2011) Microbial production of building block chemicals and polymers. Curr Opin Biotechnol. doi:10.1016/j.copbio.2011.02.011
78. Li C, Hung Wong W (2001) Model-based analysis of oligonucleotide arrays: model validation, design issues and standard error application. Genome Biol 2(8):RESEARCH0032
79. Li GZ, Bu HL, Yang MQ, Zeng XQ, Yang JY (2008) Selecting subsets of newly extracted features from PCA and PLS in microarray data analysis. BMC Genomics 9(Suppl 2):S24. doi:10.1186/1471-2164-9-S2-S24
80. Lin Y, Tanaka S (2006) Ethanol fermentation from biomass resources: current state and prospects. Appl Microbiol Biotechnol 69(6):627–642. doi:10.1007/s00253-005-0229-x
81. Lockhart DJ, Dong H, Byrne MC, Follettie MT, Gallo MV, Chee MS, Mittmann M, Wang C, Kobayashi M, Horton H, Brown EL (1996) Expression monitoring by hybridization to high-density oligonucleotide arrays. Nat Biotechnol 14(13):1675–1680. doi:10.1038/nbt1296-1675
82. Matsumura H, Yoshida K, Luo S, Kimura E, Fujibe T, Albertyn Z, Barrero RA, Kruger DH, Kahl G, Schroth GP, Terauchi R (2010) High-throughput SuperSAGE for digital gene expression analysis of multiple samples using next generation sequencing. PLoS One 5(8): e12010. doi:10.1371/journal.pone.0012010
83. Metzker ML (2010) Sequencing technologies—the next generation. Nat Rev Genet 11 (1):31–46. doi:10.1038/nrg2626
84. Mlecnik B, Scheideler M, Hackl H, Hartler J, Sanchez-Cabo F, Trajanoski Z (2005) PathwayExplorer: web service for visualizing high-throughput expression data on biological pathways. Nucleic Acids Res 33(Web Server issue):W633–W637. doi:10.1093/nar/gki391
85. Mo ML, Palsson BO, Herrgard MJ (2009) Connecting extracellular metabolomic measurements to intracellular flux states in yeast. BMC Syst Biol 3:37. doi:10.1186/1752-0509-3-37
86. Montagud A, Navarro E, Fernandez de Cordoba P, Urchueguia JF, Patil KR (2010) Reconstruction and analysis of genome-scale metabolic model of a photosynthetic bacterium. BMC Syst Biol 4:156. doi:10.1186/1752-0509-4-156
87. Morozova O, Marra MA (2008) Applications of next-generation sequencing technologies in functional genomics. Genomics 92(5):255–264. doi:10.1016/j.ygeno.2008.07.001
88. Moxley JF, Jewett MC, Antoniewicz MR, Villas-Boas SG, Alper H, Wheeler RT, Tong L, Hinnebusch AG, Ideker T, Nielsen J, Stephanopoulos G (2009) Linking high-resolution metabolic flux phenotypes and transcriptional regulation in yeast modulated by the global regulator Gcn4p. Proc Natl Acad Sci USA 106(16):6477–6482. doi:10.1073/pnas.0811091106
89. Nagalakshmi U, Wang Z, Waern K, Shou C, Raha D, Gerstein M, Snyder M (2008) The transcriptional landscape of the yeast genome defined by RNA sequencing. Science 320 (5881):1344–1349. doi:10.1126/science.1158441

90. Nakahigashi K, Toya Y, Ishii N, Soga T, Hasegawa M, Watanabe H, Takai Y, Honma M, Mori H, Tomita M (2009) Systematic phenome analysis of *Escherichia coli* multiple-knockout mutants reveals hidden reactions in central carbon metabolism. Mol Syst Biol 5:306
91. Noh K, Gronke K, Luo B, Takors R, Oldiges M, Wiechert W (2007) Metabolic flux analysis at ultra short time scale: isotopically non-stationary ^{13}C labeling experiments. J Biotechnol 129(2):249–267. doi:10.1016/j.jbiotec.2006.11.015
92. Nuwaysir EF, Huang W, Albert TJ, Singh J, Nuwaysir K, Pitas A, Richmond T, Gorski T, Berg JP, Ballin J, McCormick M, Norton J, Pollock T, Sumwalt T, Butcher L, Porter D, Molla M, Hall C, Blattner F, Sussman MR, Wallace RL, Cerrina F, Green RD (2002) Gene expression analysis using oligonucleotide arrays produced by maskless photolithography. Genome Res 12(11):1749–1755. doi:10.1101/gr.362402
93. Oh YK, Palsson BO, Park SM, Schilling CH, Mahadevan R (2007) Genome-scale reconstruction of metabolic network in *Bacillus subtilis* based on high-throughput phenotyping and gene essentiality data. J Biol Chem 282(39):28791–28799. doi:10.1074/jbc.M703759200
94. Okino S, Noburyu R, Suda M, Jojima T, Inui M, Yukawa H (2008) An efficient succinic acid production process in a metabolically engineered *Corynebacterium glutamicum* strain. Appl Microbiol Biotechnol 81(3):459–464. doi:10.1007/s00253-008-1668-y
95. Ookubo A, Hirasawa T, Yoshikawa K, Nagahisa K, Furusawa C, Shimizu H (2008) Improvement of *L*-lactate production by *CYB2* gene disruption in a recombinant *Saccharomyces cerevisiae* strain under low pH condition. Biosci Biotechnol Biochem 72(11):3063–3066
96. Pandey G, Yoshikawa K, Hirasawa T, Nagahisa K, Katakura Y, Furusawa C, Shimizu H, Shioya S (2007) Extracting the hidden features in saline osmotic tolerance in *Saccharomyces cerevisiae* from DNA microarray data using the self-organizing map: biosynthesis of amino acids. Appl Microbiol Biotechnol 75(2):415–426. doi:10.1007/s00253-007-0837-8
97. Park JH, Lee SY (2008) Towards systems metabolic engineering of microorganisms for amino acid production. Curr Opin Biotechnol 19(5):454–460. doi:10.1016/j.copbio.2008.08.007
98. Park JH, Lee KH, Kim TY, Lee SY (2007) Metabolic engineering of *Escherichia coli* for the production of L-valine based on transcriptome analysis and *in silico* gene knockout simulation. Proc Natl Acad Sci USA 104(19):7797–7802. doi:10.1073/pnas.0702609104
99. Patil KR, Nielsen J (2005) Uncovering transcriptional regulation of metabolism by using metabolic network topology. Proc Natl Acad Sci USA 102(8):2685–2689. doi:10.1073/pnas.0406811102
100. Piddocke MP, Fazio A, Vongsangnak W, Wong ML, Heldt-Hansen HP, Workman C, Nielsen J, Olsson L (2011) Revealing the beneficial effect of protease supplementation to high gravity beer fermentations using "-omics" techniques. Microb Cell Fact 10:27. doi:10.1186/1475-2859-10-27
101. Pluskal T, Nakamura T, Villar-Briones A, Yanagida M (2010) Metabolic profiling of the fission yeast S. pombe: quantification of compounds under different temperatures and genetic perturbation. Mol Biosyst 6(1):182–198. doi:10.1039/b908784b
102. Rabilloud T, Vaezzadeh AR, Potier N, Lelong C, Leize-Wagner E, Chevallet M (2009) Power and limitations of electrophoretic separations in proteomics strategies. Mass Spectrom Rev 28(5):816–843. doi:10.1002/mas.20204
103. Rabilloud T, Chevallet M, Luche S, Lelong C (2010) Two-dimensional gel electrophoresis in proteomics: past, present and future. J Proteomics 73(11):2064–2077. doi:10.1016/j.jprot.2010.05.016
104. Rivals I, Personnaz L, Taing L, Potier MC (2007) Enrichment or depletion of a GO category within a class of genes: which test? Bioinformatics 23(4):401–407. doi:10.1093/bioinformatics/btl633
105. Sangurdekar DP, Srienc F, Khodursky AB (2006) A classification based framework for quantitative description of large-scale microarray data. Genome Biol 7(4):R32. doi:10.1186/gb-2006-7-4-r32

106. Sauer U (2001) Evolutionary engineering of industrially important microbial phenotypes. Adv Biochem Eng Biotechnol 73:129–169
107. Schena M, Shalon D, Davis RW, Brown PO (1995) Quantitative monitoring of gene expression patterns with a complementary DNA microarray. Science 270(5235):467–470
108. Schilling CH, Schuster S, Palsson BO, Heinrich R (1999) Metabolic pathway analysis: basic concepts and scientific applications in the post-genomic era. Biotechnol Prog 15(3):296–303. doi:10.1021/bp990048k
109. Schilling O, Frick O, Herzberg C, Ehrenreich A, Heinzle E, Wittmann C, Stülke J (2007) Transcriptional and metabolic responses of Bacillus subtilis to the availability of organic acids: transcription regulation is important but not sufficient to account for metabolic adaptation. Appl Environ Microbiol 73(2):499–507. doi:10.1128/AEM.02084-06
110. Schmitt AP, McEntee K (1996) Msn2p, a zinc finger DNA-binding protein, is the transcriptional activator of the multistress response in *Saccharomyces cerevisiae*. Proc Natl Acad Sci USA 93(12):5777–5782
111. Shalon D, Smith SJ, Brown PO (1996) A DNA microarray system for analyzing complex DNA samples using two-color fluorescent probe hybridization. Genome Res 6(7):639–645
112. Shendure J (2008) The beginning of the end for microarrays? Nat Methods 5(7):585–587. doi:10.1038/nmeth0708-585
113. Shi S, Chen T, Zhang Z, Chen X, Zhao X (2009) Transcriptome analysis guided metabolic engineering of *Bacillus subtilis* for riboflavin production. Metab Eng 11(4–5):243–252. doi:10.1016/j.ymben.2009.05.002
114. Shinfuku Y, Sorpitiporn N, Sono M, Furusawa C, Hirasawa T, Shimizu H (2009) Development and experimental verification of a genome-scale metabolic model for *Corynebacterium glutamicum*. Microb Cell Fact 8:43. doi:10.1186/1475-2859-8-43
115. Shirai T, Fujimura K, Furusawa C, Nagahisa K, Shioya S, Shimizu H (2007) Study on roles of anaplerotic pathways in glutamate overproduction of *Corynebacterium glutamicum* by metabolic flux analysis. Microb Cell Fact 6:19. doi:10.1186/1475-2859-6-19
116. Shiraki T, Kondo S, Katayama S, Waki K, Kasukawa T, Kawaji H, Kodzius R, Watahiki A, Nakamura M, Arakawa T, Fukuda S, Sasaki D, Podhajska A, Harbers M, Kawai J, Carninci P, Hayashizaki Y (2003) Cap analysis gene expression for high-throughput analysis of transcriptional starting point and identification of promoter usage. Proc Natl Acad Sci USA 100 (26):15776–15781. doi:10.1073/pnas.2136655100
117. Sindelar G, Wendisch VF (2007) Improving lysine production by *Corynebacterium glutamicum* through DNA microarray-based identification of novel target genes. Appl Microbiol Biotechnol 76(3):677–689. doi:10.1007/s00253-007-0916-x
118. Soga T, Ohashi Y, Ueno Y, Naraoka H, Tomita M, Nishioka T (2003) Quantitative metabolome analysis using capillary electrophoresis mass spectrometry. J Proteome Res 2 (5):488–494
119. Sopko R, Huang D, Preston N, Chua G, Papp B, Kafadar K, Snyder M, Oliver SG, Cyert M, Hughes TR, Boone C, Andrews B (2006) Mapping pathways and phenotypes by systematic gene overexpression. Mol Cell 21(3):319–330. doi:10.1016/j.molcel.2005.12.011
120. Stephanopoulos G, Aristidou A, Nielsen J (1998) Metabolic engineering. Academic, San Diego
121. Tamayo P, Slonim D, Mesirov J, Zhu Q, Kitareewan S, Dmitrovsky E, Lander ES, Golub TR (1999) Interpreting patterns of gene expression with self-organizing maps: methods and application to hematopoietic differentiation. Proc Natl Acad Sci USA 96(6):2907–2912
122. Thomas R, Paredes CJ, Mehrotra S, Hatzimanikatis V, Papoutsakis ET (2007) A model-based optimization framework for the inference of regulatory interactions using time-course DNA microarray expression data. BMC Bioinformatics 8:228. doi:10.1186/1471-2105-8-228
123. Tokuhiro K, Ishida N, Nagamori E, Saitoh S, Onishi T, Kondo A, Takahashi H (2009) Double mutation of the *PDC1* and *ADH1* genes improves lactate production in the yeast *Saccharomyces cerevisiae* expressing the bovine lactate dehydrogenase gene. Appl Microbiol Biotechnol 82(5):883–890. doi:10.1007/s00253-008-1831-5

124. van Maris AJ, Geertman JM, Vermeulen A, Groothuizen MK, Winkler AA, Piper MD, van Dijken JP, Pronk JT (2004) Directed evolution of pyruvate decarboxylase-negative *Saccharomyces cerevisiae*, yielding a C2-independent, glucose-tolerant, and pyruvate-hyperproducing yeast. Appl Environ Microbiol 70(1):159–166
125. Velculescu VE, Zhang L, Vogelstein B, Kinzler KW (1995) Serial analysis of gene expression. Science 270(5235):484–487
126. Vemuri GN, Eiteman MA, Altman E (2002) Succinate production in dual-phase *Escherichia coli* fermentations depends on the time of transition from aerobic to anaerobic conditions. J Ind Microbiol Biotechnol 28(6):325–332. doi:10.1038/sj/jim/7000250
127. Vongsangnak W, Hansen K, Nielsen J (2011) Integrated analysis of the global transcriptional response to alpha-amylase over-production by *Aspergillus oryzae*. Biotechnol Bioeng 108 (5):1130–1139. doi:10.1002/bit.23033
128. Wang Y, Joshi T, Zhang XS, Xu D, Chen L (2006) Inferring gene regulatory networks from multiple microarray datasets. Bioinformatics 22(19):2413–2420. doi:10.1093/bioinformatics/btl396
129. Warringer J, Ericson E, Fernandez L, Nerman O, Blomberg A (2003) High-resolution yeast phenomics resolves different physiological features in the saline response. Proc Natl Acad Sci USA 100(26):15724–15729. doi:10.1073/pnas.2435976100
130. Whited GM, Feher FJ, Benko DA, Cervin MA, Chotani GK, McAuliffe JC, LaDuca RJ, Ben-Shoshan EA, Sanford KJ (2010) Development of a gas-phase bioprocess for isoprene-monomer production using metabolic pathway engineering. Ind Biotechnol 6(3):152–163
131. Wiechert W, Noh K (2005) From stationary to instationary metabolic flux analysis. Adv Biochem Eng Biotechnol 92:145–172
132. Winzeler EA, Shoemaker DD, Astromoff A, Liang H, Anderson K, Andre B, Bangham R, Benito R, Boeke JD, Bussey H, Chu AM, Connelly C, Davis K, Dietrich F, Dow SW, El Bakkoury M, Foury F, Friend SH, Gentalen E, Giaever G, Hegemann JH, Jones T, Laub M, Liao H, Liebundguth N, Lockhart DJ, Lucau-Danila A, Lussier M, M'Rabet N, Menard P, Mittmann M, Pai C, Rebischung C, Revuelta JL, Riles L, Roberts CJ, Ross-MacDonald P, Scherens B, Snyder M, Sookhai-Mahadeo S, Storms RK, Veronneau S, Voet M, Volckaert G, Ward TR, Wysocki R, Yen GS, Yu K, Zimmermann K, Philippsen P, Johnston M, Davis RW (1999) Functional characterization of the *S. cerevisiae* genome by gene deletion and parallel analysis. Science 285(5429):901–906. doi:10.1126/science.285.5429.901
133. Wisselink HW, Toirkens MJ, del Rosario Franco Berriel M, Winkler AA, van Dijken JP, Pronk JT, van Maris AJ (2007) Engineering of *Saccharomyces cerevisiae* for efficient anaerobic alcoholic fermentation of L-arabinose. Appl Environ Microbiol 73 (15):4881–4891. doi:10.1128/AEM.00177-07
134. Wisselink HW, Cipollina C, Oud B, Crimi B, Heijnen JJ, Pronk JT, van Maris AJ (2010) Metabolome, transcriptome and metabolic flux analysis of arabinose fermentation by engineered *Saccharomyces cerevisiae*. Metab Eng 12(6):537–551. doi:10.1016/j.ymben.2010.08.003
135. Wittmann C (2007) Fluxome analysis using GC-MS. Microb Cell Fact 6:6. doi:10.1186/1475-2859-6-6
136. Wu X, Dewey TG (2006) From microarray to biological networks: analysis of gene expression profiles. Methods Mol Biol 316:35–48
137. Yang L, Cluett WR, Mahadevan R (2011) EMILiO: a fast algorithm for genome-scale strain design. Metab Eng 13(3):272–281. doi:10.1016/j.ymben.2011.03.002
138. Yim H, Haselbeck R, Niu W, Pujol-Baxley C, Burgard A, Boldt J, Khandurina J, Trawick JD, Osterhout RE, Stephen R, Estadilla J, Teisan S, Schreyer HB, Andrae S, Yang TH, Lee SY, Burk MJ, Van Dien S (2011) Metabolic engineering of *Escherichia coli* for direct production of 1,4-butanediol. Nat Chem Biol 7(7):445–452. doi:10.1038/nchembio.580
139. Yoshida S, Hashimoto K, Shimada E, Ishiguro T, Minato T, Mizutani S, Yoshimoto H, Tashiro K, Kuhara S, Kobayashi O (2007) Identification of bottom-fermenting yeast genes expressed during lager beer fermentation. Yeast 24(7):599–606. doi:10.1002/yea.1494

140. Yoshida S, Imoto J, Minato T, Oouchi R, Sugihara M, Imai T, Ishiguro T, Mizutani S, Tomita M, Soga T, Yoshimoto H (2008) Development of bottom-fermenting saccharomyces strains that produce high SO_2 levels, using integrated metabolome and transcriptome analysis. Appl Environ Microbiol 74(9):2787–2796. doi:10.1128/AEM.01781-07
141. Yoshikawa K, Tanaka T, Furusawa C, Nagahisa K, Hirasawa T, Shimizu H (2009) Comprehensive phenotypic analysis for identification of genes affecting growth under ethanol stress in *Saccharomyces cerevisiae*. FEMS Yeast Res 9(1):32–44. doi:10.1111/j.1567-1364.2008.00456.x
142. Yoshikawa K, Kojima Y, Nakajima T, Furusawa C, Hirasawa T, Shimizu H (2011) Reconstruction and verification of a genome-scale metabolic model for *Synechocystis* sp. PCC6803. Appl Microbiol Biotechnol 92(2):347–358. doi:10.1007/s00253-011-3559-x
143. Yousofshahi M, Lee K, Hassoun S (2011) Probabilistic pathway construction. Metab Eng 13 (4):435–444. doi:10.1016/j.ymben.2011.01.006
144. Zamboni N, Fendt SM, Rühl M, Sauer U (2009) ^{13}C-based metabolic flux analysis. Nat Protoc 4(6):878–892. doi:10.1038/nprot.2009.58
145. Zhang Y, Szustakowski J, Schinke M (2009) Bioinformatics analysis of microarray data. Methods Mol Biol 573:259–284. doi:10.1007/978-1-60761-247-6_15
146. Zhang W, Li F, Nie L (2010) Integrating multiple 'omics' analysis for microbial biology: application and methodologies. Microbiology 156(Pt 2):287–301. doi:10.1099/mic.0.034793-0

Chapter 4
Technologies for Biosystems Engineering

Sonja Billerbeck, Sven Dietz, Gaspar Morgado, and Sven Panke

Contents

Abstract The rising knowledge for a variety of model organisms about chromosome compositions, gene regulation and molecular interactions which influence cell development and biological systems behavior drives an engineering effort to design and construct ever more complex novel molecular or cellular functions and behaviors. As molecular functions are encoded on DNA level, engineering of new and complex systems starts with the engineering of its encoding DNA. Although methods for genetic engineering are available since decades, their focus was the

---

S. Billerbeck • S. Dietz • G. Morgado • S. Panke (✉)
Department of Biosystems Science and Engineering, ETH Zurich,
Mattenstrasse 26, 4058 Basel, Switzerland
e-mail: sven.panke@bsse.ethz.ch

C. Wittmann and S.Y. Lee (eds.), *Systems Metabolic Engineering*,
DOI 10.1007/978-94-007-4534-6_4,

modification of rather small systems, such as cloning or modifying single genes. Therefore, the engineering of complex biological systems on DNA level, involving multiple genes up to multiple pathways or even entire genomes, requires revisiting the general usefulness of available methods and their potential for scalability. Here we review available methods and their applicability for biosystems engineering approaches as well as recent technological advances which expand the toolbox for understanding and engineering complex biological systems with novel traits.

Keywords Biosystems engineering • *De novo* gene synthesis • Gene and genome assembly • Parts • Streamlined genomes

4.1 Introduction

The level of ambition in manipulating biological systems has drastically increased over the last 10 years [1] from adding single functions to implementing complex phenotypes analogous to functionalities known from electrical engineering -such as counters [2], band pass filters [3], switches [4], and edge detectors [5] – or entire novel pathways [6]. This increase in scope brings with it the requirement to engineer a much larger set of genetic functions, either on the DNA fragment that one wants to integrate or in the host that is supposed to receive the recombinant DNA fragment. Such engineering might include adapting the overall expression or the relative protein concentrations within a metabolic or signaling pathway by tailoring promoter or ribosome binding strengths, mRNA or protein stability, the orchestrating of multiple regulatory responses to an environmental signal, or the removal of entire sections of a host to provide a clean background for engineering.

Such changes need to be quickly and efficiently programmed, even though in many cases it is unclear what to program exactly – at the current level of understanding, we are typically lacking even the most basic information to achieve a rational design of a biological systems property, such as a novel pathway for a microbial host. Consequently, in most cases the programming will contain a substantial component of exploration, during which multiple variants are tested until the best one is identified, either by evolutionary methods, by combinatorially testing suites of pre-defined parts [7], or by semi-rationally recombining natural parts [8].

In this chapter we will discuss a core of engineering approaches that is the contemporary basis to implement such large scale changes, with a focus on DNA-based methods. We will concentrate on the methods rather than on the applications, as those are discussed in the subsequent chapters of this book in depth. We will further mostly limit the discussion to approaches developed for *E. coli*. We argue that this is justified because here the technological development is farthest advanced in *E. coli*, even if other the most relevant hosts covered in the book such as *Corynebacterium glutamicum, Clostridium acetobutylicum* or eukaryotes such as yeast or filamentous fungi possess distinct advantages. However, extending this discussion to several model hosts is beyond the scope of this text. We will, however, include these hosts wherever exceptional technological developments require this.

We will start with the underlying technology of biosystems engineering, gene synthesis, and then proceed to discuss assembly of biosystems at DNA-level, sources of parts, tools to fine-tune system composition, and finally manipulations at genome level.

4.2 Fundamentals of Gene Synthesis

The technology that enables and drives our ambition to construct ever larger biological systems is the permanently improving ability to write long stretches of DNA *de novo*, i.e. without the requirement of a template (Table 4.1). This capability rests on techniques to chemically synthesize short DNA-oligonucleotides [20], which are later assembled to entire genes. This overall process from the definition of a DNA sequence to the assembly of an entire, effectively chemically synthesized gene is called gene synthesis.

The basic process of *de novo* oligonucleotide synthesis is based on the successive chemical addition of nucleotides to a growing oligomeric nucleotide chain, and has, in contrast to other DNA synthesis methods such as the polymerase chain reaction (PCR), no requirement for a template. This property allows a (nearly) unconstrained programming of the nucleotide sequence. The chemical synthesis is, however, still limited to chain lengths between 70 and 150 bp. Beyond such lengths, the error rates become unacceptably high (see below). Therefore, the oligonucleotides need to be assembled to longer sequences, usually exploiting biochemical *in vitro* steps or cellular assembly. This increase in length happens typically in steps, from oligonucleotides (100 bp) to double-stranded DNA fragments (appr. 700 bp) and further along several intermediate steps up to full genomes (10^6 bp [19]).

The synthesis of oligonucleotides is not a novel technology [11] and has been exploited already for a long time in the field of PCR. One prominent variant of the currently applied standard solid phase-based chemical phosphoramidite protocol is depicted in Fig. 4.1. Briefly, a first nucleotide with its 3′-OH function protected by a DMT-group is coupled to polystyrene beads as the solid phase. (Note that this also means that chemical DNA synthesis proceeds in 3′- to 5′-direction, unlike biochemical DNA-synthesis.) Next, the DMT-group is removed by acid treatment, generating a free 5′-OH-group. Then, the phosporamidite of choice (see Fig. 4.1a) is added and coupled to the previously terminal nucleotide, leaving a novel phosphite linkage. As the 5′-OH of the added nucleotide (+1 in Fig. 4.1b) is still protected, only one nucleotide can be added to the growing chain. The few 5′OH-groups that

Table 4.1 Milestones in DNA synthesis and assembly

Synthesized or assembled DNA	Length (bp)	Year	References	Comments
Ala-tRNA gene (yeast)	77	1970	[9]	None
Somatostatin gene	56	1977	[10]	None
Tyr-tRNA gene (*E. coli*)	207	1979	[11]	None
Poliovirus genome	7,558	2002	[12]	Overlapping fragments of 400–600 bp assembled from 69 bp oligonucleotides, fragments confirmed by sequencing and then combined to roughly 2.5 kbp fragments
Genome of Φ X174 bacteriophage	5,386	2003	[13]	Gel purification of oligonucleotides (to eliminate oligonucleotides of wrong length), PCA and PCR assembly, final test by transforming *E. coli* and selecting for functional phage, then sequencing to identify one correct genome
Polyketide synthase gene cluster	31,656	2004	[14]	1,600 oligonucleotides assembled to synthons of 500 bp and then in fragments of 5 kbp.
Genes encoding proteins the 30S ribosome subunit of *E. coli*	14,600	2004	[15]	Light-directed oligonucleotide synthesis on a chip, additional error correction by hybridization on control chip, then assembly in steps
Genome of the 1918 influenza virus	13,500	2005	[16, 17]	Separate genes assembled from oligonucleotides and later combined
Genome of *M. genitalium* JCVI-1.0	582,970	2008	[18]	Sequence-verified cassettes of 5–7 kbp were assembled in three sets of steps by *in vitro* recombination to 25 kbp, ~72 kbp and ~144 kbp, all stored in *E. coli* BACs. Final genome assembly in *S. cerevisiae* by homologous recombination.
Genome of *M. mycoides* JCVI-syn1.0	1,077,947	2010	[19]	1,078 sequence-verified cassettes of 1,080 bp (pieces) were assembled in three sets of steps to 10 kb, 100 kb and then to the full genome by homologous recombination in *S. cerevisiae*.

do not react need to be capped in an extra-reaction, so that they cannot continue to take part in the synthesis process and generate oligonucleotides with deletions (not shown in Fig. 4.1). Finally, the phosphite linkage is oxidized to a phosphodiester linkage with iodine. After repeating this sequence of steps for the required number of times, the oligonucleotide is finally cleaved off the column and treated with ammonium hydroxide at high temperature to remove all remaining protecting groups.

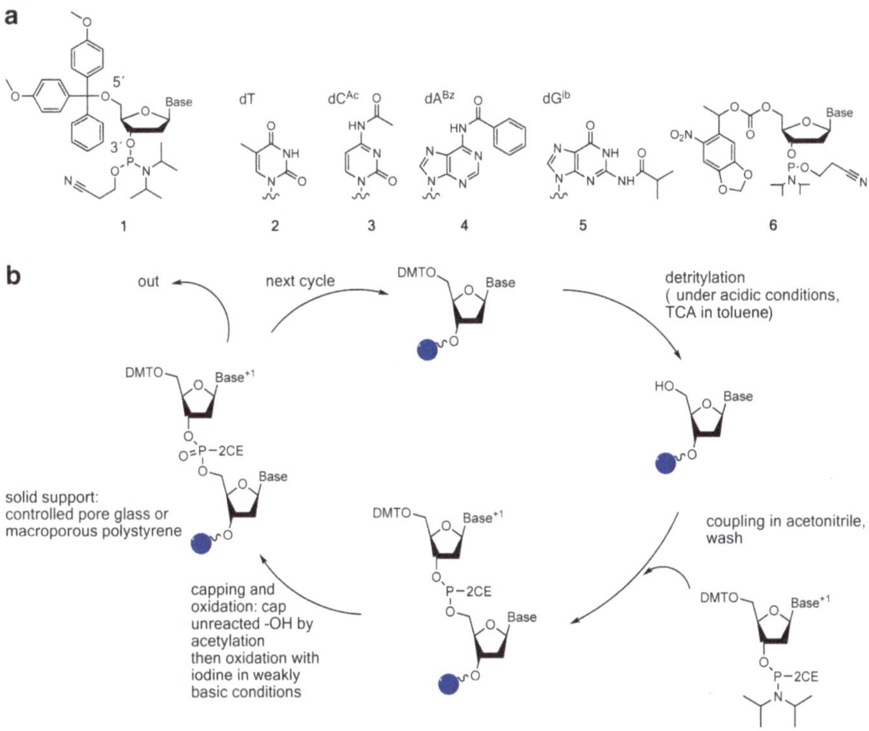

Fig. 4.1 Oligonucleotide synthesis. (**a**) Typical building blocks in oligonucleotide synthesis. *1*: nucleotide protected at its 5′-hydroxyl by the 4,4′-dimethoxytrityl-(DMT)-group and with an N,N-diisoproplylphosporamidite group at its 3′-hydroxyl (where part of the phosphoramidite group itself is protected by a 2-cyano-ethyl-group (CE-group)). For bases see *2* – *5*. Note that the sensitive functional groups of theses bases are protected as well. *6*: Nucleotide whose 5′-OH-group is protected by the UV-light (instead of acid)-sensitive protection group 5′-O-(α-methyl-6-nitropiperonyloxycarbonyl)-protecting group (NPPOC). (**b**) One cycle of extension in oligonucleotide synthesis. The blue sphere represents the attachment to the solid-phase at the 3′-hydroxyl-end of the growing oligonucleotide chain

An alternative to the classical scheme as shown in Fig 4.1a is to direct the deprotection of the 5′-OH-group of the terminal nucleotide by UV-light instead of a pH-shift. One typical reagent used in such schemes is compound six in Fig. 4.1a. This has the advantage that the deprotection can be applied in a spatially highly controlled fashion (achieved e.g. via micro-mirrors in photolithography), which has been exploited in the miniaturization of chemical oligonucleotide synthesis, e.g. in the design of DNA chips. A number of very promising recent gene-synthesis projects have exploited this strategy to enable efficient synthesis of the oligonucleotides for gene synthesis on microarrays [15, 21]. The most obvious drawback of this strategy is that it delivers less material for the subsequent steps [21], which requires additional PRC-amplification. Also, the existing protocols tend to be more error-prone than the more large-scale strategies exemplified above.

Next to light-directed deprotection, further attempts to miniaturize oligonucleotide synthesis to microarray format included inkjet printing [22], which is also implemented in the widely distributed Agilent DNA-synthesizers, and electrochemical arrays [23]. Recently, inkjet printing onto a novel, polymer-based structured microarray allowed even to combine oligonucleotide synthesis and on-chip gene synthesis [24]. Furthermore, a reduction of the amount of material required for DNA synthesis was achieved in microfluidic DNA synthesis, which led to multiple adaptations of existing chemical protocols to the requirements of the involved materials and actuators [25].

The common feature of all these methods is that they produce errors in the DNA sequence with a certain, non-negligible frequency of around 1 in 500 bp. Independent of the applied synthetic scheme, these errors have several sources. Most importantly, the coupling efficiency (the extent to which all available growing chains are extended in the following round), is never 100%. A coupling efficiency of 99% means that only 60% of all growing chains are still correct after 50 steps. This has considerable implications for the scale at which the synthesis needs to be started. Furthermore, also the capping does not proceed with total efficiency. As a result, a significant fraction of the molecules on the solid phase has a shorter length than originally intended. Next to such deletions, also chemical modifications play a role: for example, the phosphoramidites that are used for chain extension are not completely pure and oligonucleotide syntheses are prone to depurination (an adenine or a guanine is hydrolysed off the sugar-phosphate backbone), leaving a free hydroxyl group. Even though optimization of the synthesis process might lead to improved error rates [26], the errors introduced during the synthesis are one of the main limitations in contemporary gene synthesis. The error rate limits the length of the oligonucleotides that can be reliably used for gene synthesis and makes a final sequencing step inevitable in order to confirm the correctness of the sequence, which is one of the major bottlenecks in terms of overall gene synthesis rate and costs. Curiously, with only a few exceptions [27] this has not yet led to major efforts to further improve the chemical synthesis, but a number of strategies have been considered to identify and remove incorrect sequences from a pool.

Physical methods for the removal of erroneous oligonucleotides focus on the exploitation of size differences and the disturbances in the hybridization of error-containing complementary oligonucleotides. Polyacrylamide gel electrophoresis is for example easily sensitive enough to separate oligonucleotides of no more than one nucleotide difference. Therefore, subjecting oligonucleotides to a PAGE-purification can substantially reduce the number of erroneous oligonucleotides at the start of the experiment [13]. The same can be achieved by preparative HPLC, which is a standard technology if high-quality oligonucleotides are required. Alternatively, hybridization under stringent conditions can be used to identify mismatches. Perfect complementarity between two DNA strands leads to a maximum number of possible hydrogen bonds between the two DNA strands and thus a higher temperature is required to separate the molecules again ("melting temperature") as would be required for mismatched double-stranded DNA-fragments. This has been applied to reduce the error rate with light-directed, chip-based oligonucleotide synthesis [15]. The result of the latter strategy was actually quite promising: the authors reported on average 1 error in a sequence of 1,400 bp. Though this is still far too high for any large scale synthesis approach, it is by a factor of 3 better than standard solid phase approaches.

Alternatively, biochemical methods can be applied, e.g. based on enzymes that can detect DNA structures which typically occur when erroneous DNA molecules hybridize. For example, endonuclease VII of phage T4 can identify apurinic sites in DNA and then restricts both strands of the DNA molecule close to the lesion site [28]. This leads to shorter fragments which can again be separated from the correct oligonucleotides (see above) or are simply lost in subsequent cloning steps. Alternatively, *E. coli*'s MutHLS proteins can be used to detect mismatches and insertions/deletions in double stranded DNA. If such errors are present, fragments are cleaved at GATC sites [29]. Alternatively, deletions (and insertions) as main source of error can be identified by fusing the assembled gene to a reporter gene (such as the gene for a fluorescent protein) whose functional expression is only possible, if the assembled gene is inserted in-frame. Then, genes with deletions can be easily eliminated from the pool of genes because the reporter gene will not be correctly expressed. The latter approach was sufficient to reduce the error rate by an order of magnitude [30].

Another promising development is the exploitation of recursive DNA-synthesis strategies [31]. In this approach, DNA fragments are assembled from imperfect oligonucleotides in several PCR-steps and finally cloned and sequenced. Based on the determined sequence, the minimal set of error-free DNA fragments obtained during the PCR-steps which together could give rise to the desired long DNA-fragment (gene) is selected and combined by PCR. This way, DNA-fragments with errors are detected early and only error-free fragments are used in the following steps.

This concept of "working only with molecules of the correct sequence" was taken even one step further: oligonucleotides from light-directed synthesis (and with a high error rate) were removed from the solid substrate on which they had been synthesized and then fed into the regular protocol for next-generation sequencing (NGS). As oligonucleotides are short, the limited reach of NGS methods is no disadvantage. However, the capacity of NGS methods to treat copious numbers of oligonucleotides provided an excellent opportunity to identify those molecules of the correct sequence by direct sequencing. Correct oligonucleotides were recovered in a semi-automated fashion from the sequencing system and served to assemble genes [32]. This way, error rates as low as 1 error in 21 kb of DNA could be achieved – still much worse than template-driven natural DNA-synthesis, but already a factor of 40 better than the early typical error rates of 1 error in 500 bp.

Obviously a crucial step on the way from synthetic oligonucleotides to synthetic genes is the assembly of the oligonucleotides. There are several protocols to achieve this [33]. A frequently selected method to obtain DNA fragments of between 500 and 700 bp is polymerase chain assembly (PCA), in which a set of overlapping oligonucleotides is mixed in a reaction vessel and then subjected to a PCR reaction, which leads to successive elongation of the oligonucleotides until the entire desired fragment is produced at some point. This fragment can then be amplified by a second PCR-reaction based on an excess of primers homologous only to the end of the fragment [34]. Other methods pre-structure the extension process [35] or rely on the accuracy of the ligation reaction [13].

Using a selection from the methods discussed above, fragments of roughly the size of a gene can be relatively easily assembled and used directly in subsequent cloning schemes, while other methods are used to assemble these "synthetic genes" into even longer segments and ultimately genomes (see below).

4.3 From Genes to the Assembly of Genomes

The cloning and assembly of recombinant DNA by restriction enzymes was one of the most influential achievements of molecular biology in the last century. However, although still widely and successfully used for standard cloning of single genes, restriction enzyme-based DNA technologies become inefficient when it comes to the assembly of multiple fragments or in high throughput cloning approaches. Even though costs for synthesis of long DNA stretches are falling, rationally designing circuits with predictable output or readily engineering pathways with optimal production levels remains challenging. Construction followed by evaluation and another round of construction thus still reflects the reality in most synthetic biology or biotechnology projects. Therefore, fast implementation of designed biological circuits or pathways for rapid output evaluation is essential and requires efficient, reliable and scalable DNA assembly methods.

These methods should enable the ordered and parallel assembly of multiple parts to allow for the defined implementation of circuits and pathways as well as for combinatorial assembly of different parts to be able to create e.g. pathway libraries or fusion protein libraries [36] (Fig. 4.2). Especially the sequence specificity of restriction enzyme action becomes problematic as it restricts the required freedom in DNA sequence design for tailor-made pathway and circuit constructions. Seamless cloning is most notably required when open reading frames are combined with regulatory elements like promoters or ribosome binding sites (RBSs), as scar sequences can influence the expression level [37] and are therefore not desirable in constructing circuits with predictable output.

Efficiency and scalability of cloning methods becomes especially important in the combinatorial assembly of pathways for metabolic engineering when it is desirable to modify or diversify genes and their regulatory elements simultaneously in order to achieve the full potential of better combinations [38, 39].

To overcome the before mentioned limitations of restriction enzyme-based cloning, a variety of methods for DNA assembly have been established which prove to be robust enough to support the assembly of very large and/or very small genetic elements for the assembly of DNA fragments to different levels

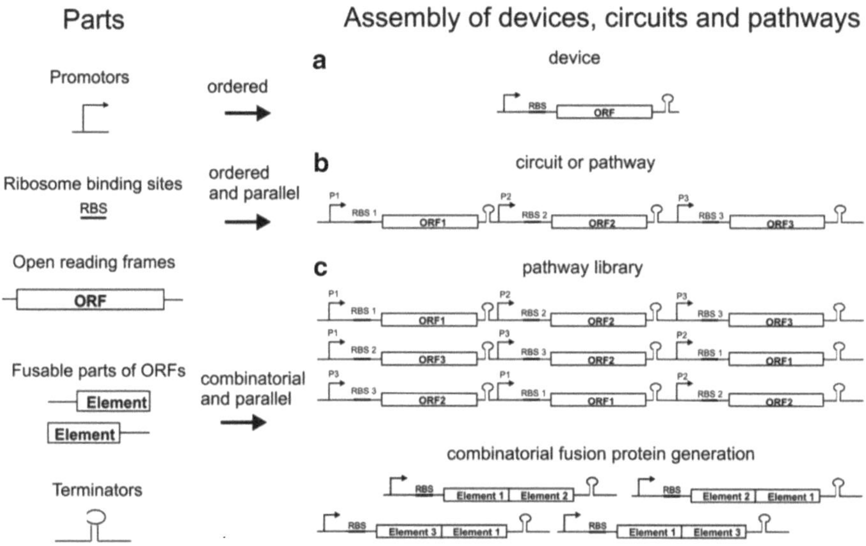

Fig. 4.2 Desired assembly strategies for the implementation and engineering of devices, circuits and pathways. (**a**) Devices are composed of parts: e.g. a promoter followed by a RBS and an open reading frame (ORF) or ORF fusion and need to be assembled in a defined order to generate a certain output as a promoter encoded downstream of an ORF would not give expression of a protein. (**b** and **c**) Circuits or metabolic pathways are composed of multiple devices which can be assembled in a defined ordered or in a combinatorial manner to generate libraries of different device orders. Multiple combinations of fusion protein can be efficiently assembled when parallel assembly makes the implementation of multiple ordered pathways and pathway libraries more efficient

of functional complexity: devices assembled from parts, circuits and pathways assembled from devices up to the assembly of full genomes. Conveniently, the corresponding methods are divided into *in vitro* and *in vivo* approaches.

4.3.1 In Vitro *Approaches*

Methods for the sequence-independent assembly of a limited set of PCR-based DNA-fragments were already introduced two decades ago when overlap extension PCR was first employed to generate a scarless fusion protein [40]. In this approach, PCR products from genes or fragments of interest are generated in separate reactions using primers which generate complementary sequences at the end of the first fragment and the beginning of the following fragment. In a second step, these PCR products are mixed, denatured, and re-annealed such that the strands having complementary sequences at their 3′-ends overlap and act as primers for each other. Extension by DNA polymerase then fuses both sequences to one molecule. A similar approach is used in DNA shuffling [41], where fragments of homologous sequences can be fused into one hybrid molecule in the presence of DNA polymerase due to priming by annealing of homologous sequences and recombination by template switches. Although overlap extension PCR and DNA shuffling are still widely applied for the generation of fusion proteins or small devices and protein engineering by directed evolution, the first is restricted to the assembly of smaller DNA stretches as it relies on amplification by PCR, which is limited by polymerase processivity and fidelity, and the second to the shuffling of only highly homologous DNA stretches.

Nevertheless a recent technique called "Circular Polymerase Extension Cloning" [42, 43] makes use of overlap extension to fuse a vector and several inserts by a single PCR cycle showing that the method can be extended to the generation of complete vector-embedded pathways as well as be used for the one-step assembly and cloning of multi-part combinatorial libraries.

Alternative methods like "Chain Reaction Cloning" (CRC) [44], type IIs restriction enzyme-based cloning, USER (uracil-specific excision reagent) fusion [45] and "Sequence and Ligation Independent Cloning" (SLIC) [46] could also overcome the limitations of conventional sequence-dependent cloning mechanisms in the assembly of multiple DNA fragments to longer stretches in one step.

In CRC several blunt ended PCR products can be stitched together and simultaneously be assembled into one vector as blunt-end ligation efficiency is raised in a product-driven manner due to the addition of bridging oligonucleotides to the reaction mix. These are homologous to the ends of the fragments which are supposed to be assembled and bring their termini close to each other after oligonucleotide hybridization. This facilitates the ligation reaction in the presence of a thermostable ligase at high temperatures. In the next round, the ligated strand bridges the ligation of the complementary strand to generate double stranded assemblies. With this approach, a 9.4 kb plasmid could be efficiently and specifically reassembled from 6 DNA fragments out of a pool of in total 12 available fragments [44].

An efficient method for parallel and combinatorial subcloning of DNA fragments or the construction of fusion proteins is type IIs restriction enzyme-based cloning. Type IIs restriction enzymes have the property to cut outside their recognition sequence. Proper design of the cleavage sites generates fragments with user-definable overhangs and allows the assembly of two or more fragments to a product lacking the original restriction site. In a first step, fragments are cloned into an entry vector where they can be released by type IIs restriction digest generating the user defined overhangs. The fragments can then simultaneously be subcloned into different host and expression vectors or assembled to longer fragments. As this assembly method does not use PCR-generated fragments for the final assembly, the assembled product sequence is not prone to accumulate point mutations, minimizing the requirement for quality control in the end. This method is therefore suitable for high throughput cloning, but it has also been used for efficient combinatorial cloning of a variety of fusion protein variants based on shuffling of a limited set of protein domains [47].

Sequence independent cloning of several PCR products into a vector can also be achieved by USER fusion [45] or SLIC [46]. Both methods rely on the generation of complementary single-stranded overhangs to mediate fusion of inserts and vector backbone. The USER protocol (see also the New England Biolabs website) employs PCR primers that contain a single desoxyuridine residue near their 5′ end. Due to the action of uracil DNA glycosylase (UDG) and the DNA glycosylase-lyase Endo VIII single stranded 3′-overhangs can be generated from these primers: UDG catalyses the excision of the uracil base and the lyase activity of Endo VIII breaks the phosphodiester backbone at the abasic site, generating a 8-nucleotide long 3′ single-stranded extension with the similarly treated destination vector. The overlap is sufficient to allow direct transformation of competent *E. coli* cells without prior ligation [48, 49]. In USER fusion cloning, this method was extended to simultaneously fuse and clone multiple PCR fragments in an ordered manner into a vector of interest by designing primers such that they complement the beginning of further PCR fragments. Even when shortening the complementary overhang to 7 bp, the method could be used to assemble 3 PCR products to a 3.9 kb insert into a vector with 90% success rate [45].

In SLIC single stranded overhangs are generated by exonuclease chew-back. Any destination vector can be readily used for SLIC-based assembly without adaptation. Fusion of complementary overhangs is achieved by *in vitro* homologous recombination and single-strand annealing. PCR fragments need to be designed such that they overlap with 30 bp homologous sequences with the next fragment or the vector backbone. The exonuclease activity of T4 DNA polymerase in the absence of dNTPs is used to chew back one strand of each fragment to leave ssDNA overhangs for homology searching. The destination vector is first linearized by restriction digest and then treated in the same way. The exonuclease activity can be stopped by adding dCTPs to the reaction. By mixing compatible fragments and vector backbone, recombination intermediates are created *in vitro*, either by addition of RecA or by RecA-independent single strand annealing. The recombination intermediates are then introduced into *E. coli* cells where the endogenous repair machinery allows finishing the recombination event to generate covalently closed double-stranded plasmids. SLIC was used for the assembly of up to 10 fragments between 200 and 900 bp simultaneously into one vector.

In an effort to assemble the synthetic *Mycoplasma genitalium JCVI-1.0* genome, an *in vitro* method was recently developed which could be used to reliably assemble DNA fragments to stretches of up to 72 kb in size from smaller cassettes [50]. The so-called isothermal or "Gibson assembly" is a one-pot one-temperature method which is similar to SLIC, except that it uses T5 exonuclease for overhang generation and a polymerase and a ligase to close gaps and seal the single stranded nicks in the end. Smart use of temperature (exonuclease loses its activity gradually at 50°C, allowing the polymerase and Taq ligase to overtake the system and perform the repairs) makes this method additionally attractive, but also limits its use to the assembly of DNA fragments that are at least 250 bp in length as there is a high chance that T5 exonuclease would otherwise digest a DNA fragment entirely before it can anneal and prime the polymerase for extension. Nevertheless, the fact that the isothermal assembly method does not require supplementation of further reagents during the process and only depends on a simple heating device and no other special equipment, allows for parallelization and automation.

A recent publication demonstrated that isothermal assembly cannot only be used for ordered assemblies but also to assemble promoters and gene cassettes in a combinatorial manner into a biochemical pathway to optimize utilization of acetate [51]. Unique 40 bp linker sequences at the end of each gene and the beginning of each promoter forced the assembly into a defined alternating promoter–gene order. 81% of colonies screened for acetate production contained the complete functional pathway and screening proved the library to be diverse. All methods described in this section are illustrated in Fig. 4.3.

4.3.2 In Vivo *Assembly of DNA Fragments*

Next to the different *in vitro* methods introduced above, DNA fragments can be seamlessly and efficiently assembled *in vivo* by homologous recombination using *Saccharomyces cerevisae* [52] or *Bacillus subtilis* [53] as host strains. The finding that linear DNA fragments can efficiently stimulate recombination in *S. cerevisiae* [54] invoked the use of this host for the manipulation of DNA *in vivo*. As about 15–30 bp homologous regions are sufficient to mediate homologous integration of DNA fragments into a linearized vector when co-transformed into yeast, cloning of PCR products *in vivo* became easy and efficient as homologous regions can be added via a primer. *In vivo* recombination is therefore not only used for plasmid construction but also to generate chimeric genes with precise fusion junctions [55], to generate diverse DNA libraries [56, 57] or to assemble metabolic pathways from multiple genes [58]. Especially the last approach, called "DNA assembler", is very useful for metabolic engineering and biotechnology and a good example how *in vitro* and *in vivo* assembly approaches can be combined to assemble large functional DNA sequences. To achieve pathway assembly, single expression cassettes were generated *in vitro* using overlap extension PCR [40] to fuse the desired open reading frames to promoter and terminator sequences. The 5′-end of the first gene expression cassette was designed to overlap with the destination vector while the 3′-end was designed to overlap with the second cassette. Successive cassettes were designed to overlap with flanking cassettes, and the 3′-end of the last cassette overlapped again with the destination vector. Overlaps were chosen to be 40 bp. When yeast cells were simultaneously transformed with the linearized vector and the linear expression cassettes, DNA assembler was able to assemble a 3-gene pathway (9 kb) in a single step with 100% efficiency (10 out of 10 clones carried the right construct), a 5-gene pathway (10.6 kb) with 80% efficiency, or an 8-gene pathway (19.6 kb) with 20% efficiency.

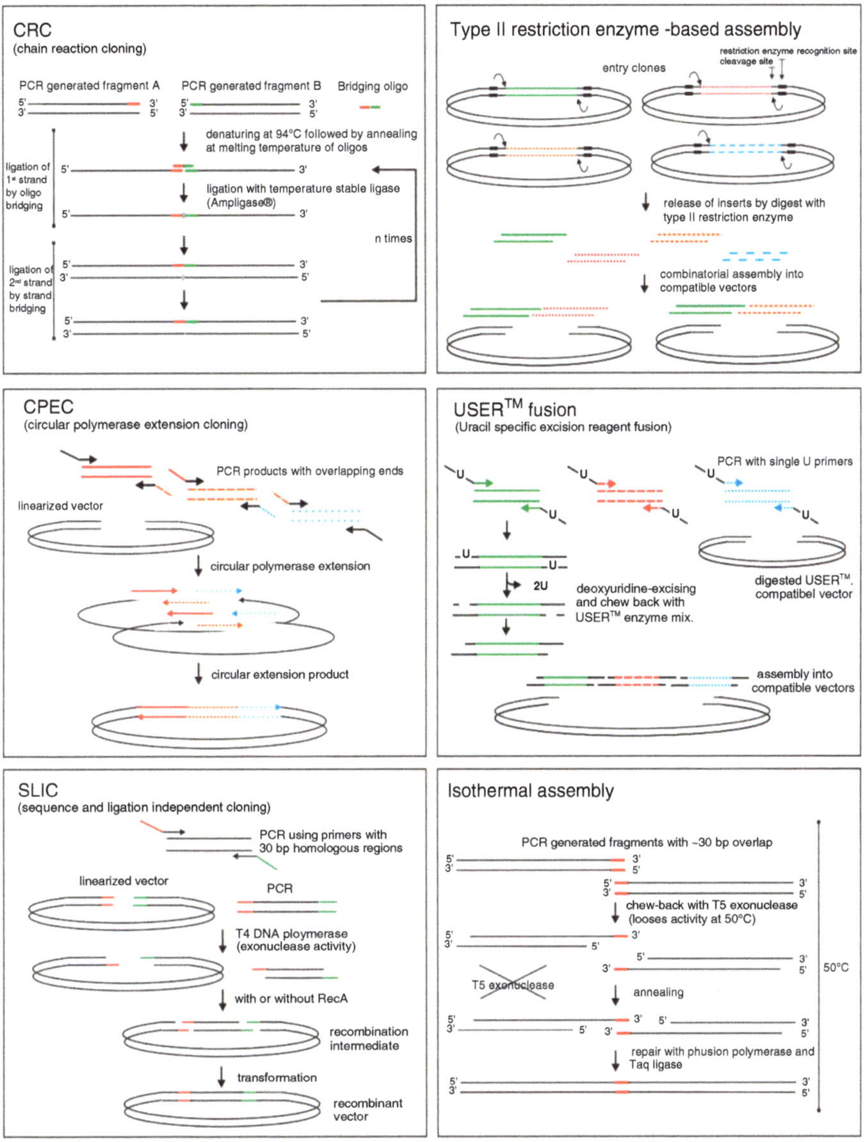

Fig. 4.3 *In vitro* methods for the seamless ordered or combinatorial assembly of multiple DNA fragments for the construction of longer DNA stretches

4.3.3 Assembly of Long DNA Stretches up to Entire Genomes

The methods discussed above raise the question of how far these assembly processes can be developed. To this end, it is interesting to note that these methods have been used to assemble entire genomes (see also Table 4.1): In an attempt to synthesize the 583 kbp of *M. genitalium* JCV-1.0 (a slightly modified variant of a wild-type *M. genitalium* strain), sequence verified pre-assembled cassettes of 5–7 kbp were assembled in three sets of steps by *in vitro* recombination to 25 kbp, ~72 kbp and ~144 kbp, and the intermediate steps were all stored in *E. coli* artificial chromosomes (BACs). The large cassettes were then assembled into a complete genome by homologous recombination in *S. cerevisiae* [18]. In an even more audacious attempt, the same laboratory also assembled 1,078 sequence-verified cassettes of 1,080 bp in only three sets of steps (10 kb and 100 kb intermediates and then the full genome) to the entire 1.1 Mbp genome of *M. mycoides* JCVI-syn1.0 (a slightly modified variant of a wild-type *M. mycoides* strain) by homologous recombination in *S. cerevisiae* [19].

Besides yeast, *B. subtilis* has been used as a vehicle for the assembly of a series of smaller genomes like the mouse mitochondrial genome (16.3 kbp) and the rice chloroplast genome (134.5 kbp) by homologous recombination. Even though the so-called domino method developed for *B. subtilis* recombination-based assemblies is very laborious when compared to the yeast assembly, it was used to assemble the highly repetitive rice chloroplast genome [59].

4.4 Tools for (Semi-) Rational and Combinatorial Biosystems Engineering

Synthetic biology's core goal is the rational design and construction of complex biological systems, and the methods to assemble the required large fragments of *de novo* synthesized genes are becoming available (see above). In parallel, tools emerge to program such large coding sequences. In this section, we do not discuss the development of complex synthetic phenotypes, but rather the tools that are available to source the functions required and to fine-tune the assembly of genes into a functioning system.

One aspect that deserves specific mentioning here is the scale of the undertaking. While it is standard practice today to identify, clone, and finally optimize the gene expression of a single recombinant gene, carrying out the entire trajectory is still a quite laborious undertaking. Doing this for an entire system – for example, for a multi-step metabolic pathway [7, 8] – is then a much more intense effort, as it is essential not only to identify multiple genes, organize them into operons and take care of the proper regulation of expression, but it is also important to fine-tune the relative amounts of system members, which is a highly involved activity if the system consists of, say, ten members. If we assume that all ten genes were in one operon and we would like to test five variants of promoter strength and gene expression level (adjusted via different RBSs), then we would have to screen through a search space of 5^{11} or more than 48 million possibilities. The task becomes correspondingly more involved if variants of specific genes or different principle designs are included. These numbers should make clear that it will not be sufficient to find a unique solution for one specific case, but that there is a need to systematize the process of implementing a novel system, which includes all aspects of the process, from the source of genes via the availability of suites of similar parts that differ only in one parameter to computational tools that aid in planning the system construction. In other words, it seems essential that biosystem construction becomes a true engineering effort that can take advantage of the vast experience in design that is available in classical engineering disciplines [60, 61].

4.4.1 Promoters, Ribosome Binding Sites, Terminators, and RNA Tools

The essential actuating elements in constructing a biological system are those that direct gene expression. Correspondingly, substantial effort has been devoted to provide suites of tools to fine-tune this process [37, 62, 63].

4.4.1.1 Promoters

Well-characterized open-loop "controllers" with predictable properties are necessary for rational development of advanced synthetic networks [61]. Hence, large libraries of e.g. promoters must be characterized in terms of performance, compatibility, reliability, output demand [64] and noise level [63, 65] in order to extract quantitative and qualitative conclusions that will ultimately allow the mathematical modeling of the promoter behavior, but for the time being allow the operator to rationally test hypotheses on how to improve system performance. Several authors reported the usefulness of alterations in promoter properties such as strength [62, 66], transfer function [63, 67], regulation [66, 68], and noise characteristics [63, 65].

Promoter engineering can be performed over the entire sequence targeting any of the seven main motifs of promoter sequence (or several of them): upstream region, UP element, -35 box, spacer region, Pribnow box, discriminator, initial transcribed region and optionally, operator sites [65, 69–78].

To obtain promoters with predictive properties that could be useful for rationally designed complex biological networks, the construction of large libraries is imperative. In this regard, the promoters employed as start point should have well-characterized parameters under static and dynamic performance [79], small noise level, high reliability over time and preferably be not the target of host regulation nor induce pleiotropic effects in the cell [80].

Likewise, constitutive promoters should always display constant expression levels, while inducible promoters should be tightly regulated, achieve pre-specified expression levels after induction, employ non-toxic and cheap inducer molecules or conditions which do not produce pleiotropic effects or lead to bistability, and the switch to the "on" state should be fast, full and smooth across the population [80, 81].

As most natural promoters do not fulfill the above requirements customized promoters have to be generated from large synthetic libraries. Several approaches were employed for design and construction of synthetic promoter libraries (SPLs). Firstly, saturation libraries, were constructed by degenerating the promoter sequence at a specific point (specific base pairs or specific motifs) or along the entire promoter sequence in order to obtain a sufficient library size allowing a wide range of promoter properties. This strategy allowed covering a broad range of promoter activities up to a 400-fold change in strength [67, 70, 73, 76, 82] and changes in promoter strength without changing the induction characteristics [66].

Secondly, rather than covering a specific design space exhaustively, mutagenesis libraries were produced by error-prone PCR targeting the entire promoter sequence, introducing only a limited number of mutations. Different authors reported promoter variants produced in this manner with customized promoter properties [62, 67, 83] including engineering of the induction characteristics [81].

Finally, the two previous approaches were combined in combinatorial promoter libraries [77, 78], which enabled complex responses to intracellular changes [84], such as the programming of gene expression for different logics [72] and the engineering of the noise level [63, 77]. A common approach to this end is the recombination at different promoter positions of multiple operator binding sites with diverse strengths and modulation patterns (repression, activation, or both).

An important alternative to pre-manufacturing promoter variants and exhaustively characterizing them for biosystem construction would be prediction of the properties from the DNA sequence. Over the last years, substantial efforts in promoter strength prediction from DNA sequence were reported [73, 77, 85–87]. Three kinds of models showed promising results: (i) models based on position weight matrices with separate scores for each promoter motif could predict within a certain limit the strength of several genomic promoters [86]; (ii) for combinatorial libraries, models based on the thermodynamics of transcription factor- and RNA polymerase-binding went a long way in explaining the variance in gene expression levels [77, 78] and regulatory logic functions [84]; and finally, estimates for binding parameters were also successfully obtained from correlating sequence conservation with protein-DNA binding strength [87].

An important development in the design of promoters for biosystems engineering is the attempt to construct promoters that are context-insensitive. Such promoters would come very close to the ideal design proposition, in which a promoter performs its assigned function irrespective of the specific situation, which is currently not the case [88]. First attempts in this direction have been made by including sequences up- and downstream of promoters and documenting their effect in terms of making promoter performance more predictable and less dependent of the specific context [89].

4.4.1.2 Ribosome Binding Sites

RBSs direct the translation of the mRNA into protein. As part of the mRNA, they are included in a number of regulatory mechanisms, which will be discussed further below. Here, the focus is on exploiting the role of the RBS in determining the strength of the translation, i.e. how many protein-syntheses will be initiated from a specific mRNA molecule per time. Even though a number of reports in the literature suggest the existence of generally strong or weak RBSs [4], in fact the efficiency of an RBS is a strongly influenced by the context in which it is placed (e.g. [90]). This fact can be compensated for to a certain extent by modeling which takes into account the crucial elements of the process such as the secondary structure of the mRNA around the start codon and the ribosome-binding affinity [91]. Corresponding models based exclusively on thermodynamic considerations of RNA-folding [37] or on a combination of thermodynamic and kinetic considerations [91] have recently been published with the aim of explaining the effect of existing RBSs on relative protein translation or allowing deriving a suitable RBS for a given gene sequence [37, 92]. Of course, these tools are not (yet) exact, but rather give a certain probability with which a specific RBS will lead to a certain relative translation initiation rate. Still, such tools will be very helpful to reduce the combinatorial search space for biosystems assembly.

4.4.1.3 Transcriptional Terminators

Embedding of regulatory sequences into genomes will require some extra design constraints to insulate such sequences from the context in which they are placed. For example, interference between neighboring operons can occur by RNA polymerase read-through due to inefficient transcription termination, leading to occlusions, collisions or "sitting-duck" interferences [93, 94]. To prevent – or exploit – such incidences, suites of terminators of known effectiveness would be very useful. Here, we focus on Rho-independent terminators, which have almost exclusively been applied in biosystems engineering.

Terminators are RNA-elements composed of stable GC-rich stem loops flanked by A/U-tracts and are responsible for transcription elongation attenuation. Engineering of terminator efficiency was achieved by synthetic libraries employed for mutagenesis, the introduction of tandem repetition, and combinatorics. Clearly, termination efficiency is a sequence-dependent function rather than an additive property, thus various degrees of leakiness need to be taken into account [95, 96].

The most preferred terminator regions for mutagenesis were the hairpin [97, 98], the termination site [94] and the poly A/U-tracts [99], either separately or all together [100]. Alternatively, tandem repeats of short or long terminators were introduced into mono- or polycistronic gene constructs [95, 96]. Finally, A-tracts, stem-loops, U-tracts, 3′-tails (the "termination site" region, [101]) and "stuffer regions" (terminator-unrelated DNA [94]) were combinatorially recombined to produce differentially efficient terminators. Even though these three strategies delivered terminator variants that achieved up ~95% efficiency, most of them had lower efficiency than the wild-type starting sequence. Effectively, the availability of a diverse set of reliably efficient terminators is still an issue, as it is with RBSs. To this end, terminator efficiency can in principle be modeled according to similar principles as applied to RBS-efficiency models [102, 103].

4.4.1.4 RNA-Tools

RBSs and terminators are functional elements that exercise their function on the level of mRNA. In fact, it is debatable whether it is justified to discuss them as single functional entities as above, as they are in fact very context-sensitive. Correspondingly, they are also used as parts of larger regulatory RNA-based structures. In general, the use of RNA-tools to fine tune gene expression in synthetic biosystems has become an extremely productive area of research over the last 10 years [104] based on the ability of RNAs to form diverse secondary structures that influence stability and can be influenced by binding partners and that allow conditional sequestration of essential signals (such as the RBS). Consequently, they have been used to fine-tune the enzyme composition of bacterial pathways [38], to implement cellular switches [105] and logic operators [106] in mammalian cells and yeast [107], and to construct small molecule- [108] or protein- [109] responsive and tunable [110] gene regulators in yeast and mammalian cells. A particular interesting feature is the potential for modularity, which – at least in principle – opens up the road to generally applicable engineering RNA-tools [110, 111].

4.4.2 Design Software and Registries

As already pointed out, designing at biosystems-scale is a very laborious effort, which can be substantially facilitated by computational tools. The requirements for support range from the selection of oligonucleotides of comparable properties (such as melting points, see above) in the design of synthetic genes to the question of which specifications are required to implement a specific complex circuit-behavior. A variety of tools have been developed to support design questions at the appropriate levels, and summaries of these tools are available [112, 113].

A crucial point is to source the genetic parts that are required to build a system to specifications. While the methods to modify important parameters such as promoter strength, RBS strength, etc. emerge (see above), it is a separate question to have all these parts at hand once they are needed, and with the required documentation relevant for design. Even though DNA-synthesis has become very efficient, this does not mean – examples to the contrary notwithstanding [19] – that large biosystems are regularly assembled entirely from *de novo* synthesized DNA; in particular not if a large component of combinatorial optimization is involved. In fact, biosystems assembly is such a crucial aspect that entire infrastructures are dedicated to this topic, as the recently founded "Biofab", a multi-institutional facility to professionally design and build biosystems (www.biofab.org).

An additional point involves the format in which such parts become available. Clearly, on large scale it would be helpful if certain formatting rules were adhered to which would allow using these parts in an automated fashion [64]. These concerns – part availability, documentation of parts, and availability in the proper format according to useful standards – are the central ideas behind the "parts registries" that are emerging at various institutions to fuel biosystems assembly (http://partsregistry.org/Main_Page, http://www.bioss.uni-freiburg.de/cms/index.php?id=308). A central point will be to fill these registries with useful parts and documentation, which is an unsolved issue in the face of frequently unclear intellectual property issues connected to the genes and regulatory elements deposited in the registry.

4.5 Engineering and Streamlining Genomes

Biosystems are designed for *in vitro* and *in vivo* applications. Even though *in vitro* applications that are increasingly emerging can be conceptually very instructive [8, 114–116] the majority of projects in synthetic biology deals with constructing biosystems in living cells, which can due to their complexity interfere with the designed phenotype in various ways. Therefore, manipulation of the future host's genome and its conversion into a "chassis" (ideally, a cellular background that is suitable to harbor multiple designs while hardly interfering with the design) is essential. This might include deletion or replacement of host genes, or, in a somewhat more radical strategy, the elimination of apparently unnecessary genetic material from the host's genome. Such "streamlined" genomes might have important beneficial properties in biotechnology, such as reduced genetic drift [117]. An important hope in streamlining host cells is that a reduced genome is equivalent to a reduced network of interactions and thus to higher predictability. However, this concept still needs to be experimentally supported [118].

4.5.1 Engineering the Chromosome of E. coli*: Deletions, Insertions and Modifications*

Complete knowledge of the genome sequence of *E. coli* strain K-12 has opened the road for a variety of chromosomal engineering approaches. For instance gene functions can be addressed by precise deletion of the corresponding reading frame, gene regulatory properties can be modified by exchanging promoter sequences and unnatural functions like reporter genes or foreign metabolic pathways or gene circuits can be chromosomally inserted. Engineered variants of a protein can be expressed at wild type levels from their natural genome location [119] and chromosomal insertions are maintained in a stable way without the requirement for selection.

For *E. coli* different methods have been developed to implement these engineering approaches, being either specifically suitable for deletions or applicable for genome modifications in general. They also differ in whether or not they allow for seamless genome engineering or leave a scar behind.

In the following, we limit the discussion mostly to methods that allow controlling the specific locations of engineering. However, it should be made clear that over the last years also a number of methods have emerged that allow efficient genome-wide engineering if an efficient phenotypic selection is available. Specifically, methods that allow modulating multiple genomic functions by changing just one gene – such as global transcription machinery engineering [120] – or methods that allow the repeated facilitated integration of short mutagenic oligonucleotides into the genome at sites predefined by the oligonucleotides sequence [121] should be considered.

4.5.1.1 Methods for General Genome Modifications

The plasmid-mediated Rec recombination-dependent "in-out" method [122–125] and the linearized fragment-mediated λ Red recombination-dependent "gene-gorging" were primarily developed for the seamless introduction of modifications into the chromosome. Nevertheless both can also be used to insert new sequences or delete sequences at a chromosomal target site. In the "in-out" method the desired modification or insertion, flanked by 300–500 bp-long regions homologous to the target site within the chromosome, needs to be encoded on a temperature sensitive vector carrying a *sacB* gene and introduced into the cell by transformation. A Rec-mediated single homologous crossover results in co-integration of the whole plasmid into the genome at the designated target site with the plasmid vector between the original wild type sequence and the modified sequence ("in" step). This event can be selected for at the non-permissive temperature, as the antibiotic resistance gene of the plasmid can only be maintained by integration on the chromosome. The resulting co-integrate is resolved by a second single crossover ("out" step), which can be selected for by the absence of the *sacB* gene, whose gene product leads to toxic products produced from sucrose. At least theoretically, in 50% of the cases in and out crossovers span the mutant site, and the desired mutation is transferred to the genome. Otherwise, the wild type sequence is restored. In case the modification does not mediate a selectable or screenable phenotype, colony PCR-based screening for the mutations is required. Based on this system, a more efficient strategy was proposed [125], where the previously spontaneous occurrence of the second cross over event is forced by an *I-SceI* nuclease-induced double strand break. The *I-SceI* nuclease recognizes a specific 18 bp site encoded on the plasmid which is integrated into the chromosome during the "in" step. The cell is then transformed with a helper plasmid constitutively expressing the *I-SceI* nuclease to induce the double strand break and stimulate recombination. Using this strategy, the frequency of the second cross over event was effectively enhanced, showing 100% cross over when expressing high levels of nuclease, abolishing the need for SacB counter selection on sucrose. Still, realistically speaking, when introducing a modification which affects viability of the cells the ratio between wild type and desired mutant will shift making the screening for mutants a laborious task.

Gene gorging has the same applications as the "in-out" method but was simplified to involve only one recombination event and works without a selection step. Like in the 'in-out' method, the mutant allele, flanked by homologous regions, is introduced into the cell on a plasmid which additionally encodes for an *I-SceI* nuclease cleavage site. Thus, the plasmid can be linearized *in vivo* by *I-SceI* nuclease co-expression. Recombination with the chromosomal locus is mediated by the efficient λ Red system rather than the *E. coli* inherent Rec system. It has been shown that the λ Red system promotes a greatly enhanced recombination frequency compared to the *E. coli* Rec system when using linearized DNA fragments [126]. Since every cell receives a linear donor fragment, the fraction of cells carrying the modifications incorporated into the genome was determined to be at least 1%. This makes it possible to identify a colony carrying the desired genotype by colony PCR. Nevertheless, the method brings along the same problem as the in-out method mentioned above. In case the introduced mutation has a negative effect on cell viability the efficiency of incorporation might decrease making it laborious to screen for such rare events.

So far there is no method available which would allow for the seamless modification of the genome and additionally tightly select in all steps for the desired modification

4.5.1.2 Methods for Site-Specific In-Frame Deletions

The PCR-based one-step deletion method introduced by Wanner and Datsenko [127] is frequently used as it is easily scalable and was used to create the "KEIO collection" – a collection of in-frame single gene deletions of all non-essential genes in *E. coli* K-12 [128]. In this method PCR is used to amplify a resistance marker with primers encoding 50–60 bp long regions which are homologous to the sites in the chromosome flanking the desired deletion. Cells expressing the efficient λ Red recombination system from a temperature sensitive helper plasmid are transformed with the PCR product and λ red-based recombination is mediated via the homologous regions. Successful recombination events can be selected for with the resistance marker. The resistance marker template comes with two FLP recombinase recognition sites (FRT sites) and the resistance marker can easily be removed in a second step by expression of the corresponding recombinase gene from a temperature-sensitive helper plasmid. This latter step leaves behind a 250 bp scar, but the elimination of the resistance is important for the creation of multiple knock outs. Of course, the resistance marker can be replaced by alternative suitable markers, such as genes for fluorescent proteins [119, 129]. To overcome the problem of the remaining scar, the FLP recombinase sites flanking the inserted marker were later replaced by *I-SceI* nuclease cleavage sites. Like in the improved "in-out" method, expression of the nuclease from a helper plasmid stimulates a double strand break, and thus, induces recombination between the flanking part of the desired deletion, and a copy of this part encoded on the PCR product [130]. Although generation of the PCR product used for the deletion needs prior design and cloning steps, the method allows for site-specific chromosome surgeries and was actually used to create a minimized genome of *E. coli* by sequential large scale deletions of non-essential chromosomal stretches (see below).

Table 4.2 Genome reduction efforts

Strain	Extent of genome reduction	Effects on strain	Experimental approach	References
E. coli MG1655	Up to 15.3% (MDS43, 708 kbp)	Genetically stable cloning vehicle, high electroporation efficiency, comparable protein overexpression, improved threonine production	Recursive targeted deletion of rationally selected genes	[132]
E. coli W3110	22.2% (1,030 kbp)	Extended growth, increased final cell density, improved threonine production	Recursive targeted deletion of rationally selected genes	[131]
E. coli MG1655	5% (200 kbp)	Not mentioned	Recursive transposon-mediated arbitrary deletions in the geneome	[133]
E. coli MG1655	7% (313 kbp)	Normal growth under standard laboratory conditions	Recursive transposon-mediated arbitrary deletions in the geneome	[134]
E. coli	29.7% (1,377 kbp)	Changes in cell length and width, changes in nucleoid structure, extended generation times	Recursive targeted deletion of rationally selected genes	[135]
B. subtilis 168	25% (1,000 kb)	Slightly reduced growth rate, superior protein secretion	Targeted deletion of rationally selected genes in non-vital segments >10 kb	[136, 137]
B. subtilis 168	7.7% (322 kbp)	No major changes in growth rate, metabolic flux patterns, natural competence, sporulation, extracellular proteome, protein secretion. Reduced strain has higher motility	Recursive targeted deletion of rationally selected genes	[138]

4.5.2 Streamlining Microbial Hosts

A number of efforts are currently under way in order to streamline organisms with the objective of reducing cellular complexity and providing a "cell factory" with more predictable and controllable behavior, ultimately working towards a more deterministic approach to metabolic engineering [131] (Table 4.2). Several target organisms are under study, including obligate parasites like *M. genitalium* (580 kbp), laboratory model bacteria such as *E. coli* (4.6 Mbp) or *B. subtilis*, and *S. cerevisiae* (6 Mbp).

Recently, a streamlined version of *E. coli* with a 15% smaller genome was presented [132]. By comparing multiple, fully sequenced *E. coli* genomes the authors had extracted a "core genome" and successively deleted pathogenicity islands, cryptic sequences, biosynthetic pathways that are not required under standard culture conditions and other genes by targeted large-scale gene disruption (see above). Despite having fewer genes than common laboratory strains, the cells with reduced genome grew to slightly higher density in certain media, tolerated a greater variety of plasmids and had higher electroporation efficiency. In subsequent investigations, the strain was found to be useful in protein overproduction [139], had an improved stability [140, 141], and over-produced threonine more efficiently than a parent strain [142]. These results can be seen as evidence that reducing the complexity of the network at least does not have to reduce the performance of a host strain, but rather can provide beneficial properties.

Next to the effects discussed above, it should be pointed out that strains with a reduced genome might also be very useful from a more conceptual point of view: In a recent contribution, Trinh et al. demonstrated that constraining the available metabolic network of an *E. coli* cell led to increased ethanol production by eliminating alternative pathways [143]. Of course, this work is only one example of longstanding efforts in metabolic engineering to increase yields by optimizing metabolic networks. While the authors in the mentioned study needed only eight knock-outs to achieve the required constraining, the argument can be made in principle as well: an *E. coli* strain whose genome has been reduced to its "core" without interfering with the biotechnological capacity under specific laboratory conditions would represent a much more predictable basis or "chassis" [144, 145] for metabolic and other engineering endeavors. Rather than adapting the wild-type strain each and every time to a specific task, one can argue that in the future, the sequence of events should be reversed: given the ease, with which large scale *de novo* DNA synthesis projects can currently be carried out [19], a strain engineering project should start with a basic chassis strain and then implement the required functionalities into a background of considerably reduced complexity [118]. While such a "core" *E. coli* strain will still be rather complex, its overall complexity should be substantially reduced and the room for unexpected reactions considerably constrained.

In contrast, redundancy is considered to be an important part of cellular robustness [145], so constraining redundancy might have difficult-to-predict consequences on the biotechnological performance of reduced strains. However, it is important to note that one important feature of robustness is that it ensures proper behavior in highly changing environments. Even though strains pass through a variety of environments on their way from a cell bank to the harvesting stage in a large-scale bacterial fermentation, this spectrum is only a fraction of what nature has actually prepared *E. coli* for. In other words, it is completely unclear to which extent robustness based on redundancy is required for strains that have to perform biotechnological tasks under the highly controlled conditions of the laboratory or a production process.

Of course, an intriguing question is how far genome streamlining could proceed at least in principle. In a project similar to one discussed above, up to 30% of

E. coli's genome were deleted, leading to clear effects on physiology [135]. However, as these reductions were achieved by deleting large chunks of DNA in one step, it is not clear whether a more careful deletion procedure removing only smaller parts of the genome and thus capable of identifying and preventing physiological problems would have had the same effect.

But even if such "minimal genome" strains might hold little attraction for high performance biotechnology, such strains would be highly instructive in order to delineate the mechanisms of cellular complexity and the requirements to evolve from a strain with minimal capabilities to a highly accomplished occupant of a complex ecological niche – which in turn might inform the quest for a strain that is optimally suited for the controlled conditions of a bioreactor.

This quest for the minimal genome can be carried out in two ways: bottom-up and top-down. In the bottom-up approach, the theoretical absolute minimal set of genes (e.g. [146, 147]) would be synthesized and transplanted and additional genes would be added until a viable cell can be obtained. The corresponding technology has been demonstrated with the recent genome synthesis projects for *M. genitalium* [18] and *M. mycoides* [19]. By implementing the entire experimental trajectory from chemical DNA synthesis via whole genome assembly to transplantation into cytoplasm (at least of cell wall-less parasitic bacterial cells), these projects have delivered the prerequisite for identifying and building a minimal genome bottom up.

The top-down approach, on the other hand, would start with an established laboratory organism like *E. coli* and incrementally remove genes until the organism is no longer viable under the specified culture conditions. The genome-reduced *E. coli* discussed above is a prominent example for the rational implementation of this approach, which has also been applied to *B. subtilis* [136, 137]. Alternatively, the gene material to delete could be identified by a random approach, in which "anchor sites" are produced by transposon mutagenesis and then the material between anchor sites is deleted. Of course, such an approach has the advantage that no prior knowledge is required on which genes to delete. However, so far these random methods have not been applied to delete large sections of the genome [133, 134], as the identification of a mutant with a "comparable" physiology to the parent is very laborious.

4.6 Summary

Clearly, the focus of biotechnology has shifted from simple, single-gene or oligogene phenotypes to more complex, system-level phenotypes. Manipulation at the system level is a different game on many levels. Technologically, a variety of promising tools are emerging to encode and implement multiple changes to a host strain's genome. Even though these tools are in most cases not new, but build upon the techniques of genetic engineering, the new focus on scaling requires many adaptations and variations of these methods. In particular, in our view, bioengineering needs to re-emphasize the second of its two foundations – the "engineering". (Semi-) Rationally manipulating cells at systems level is prohibitively difficult without basic efforts in computational design, standards [64], registries, and documentation, and probably also without achieving a minimum of abstract formalisms [148].

Conceptually, major hurdles still need to be overcome. Cellular systems are complex, and this complexity in all its facets interferes with predictability and true rational design. Recent advances in orthogonal systems [149–151] point to a promising direction, but tangible successes on the road to a generally useful chassis are still scarce. However, given the increasing pervasiveness of systems analysis methods on all relevant molecular levels, from DNA to metabolites, and the major advances in large scale gene synthesis, major advancements can be expected in the near future.

Acknowledgements The authors wish to acknowledge support from the EU (FP6 projects NANOMOT and EMERGENCE, FP7 project ST-FLOW), and from the ESF/SNF (EuroCORE project Nanocell). G.M. is a holder of Becas Chile-scholarship (granted by CONICYT-Government of Chile).

References

1. Carr PA, Church GM (2009) Genome engineering. Nat Biotechnol 27:1151–1162
2. Friedland AE, Lu TK, Wang X, Shi D, Church G, Collins JJ (2009) Synthetic gene networks that count. Science 324:1199–1202
3. Basu S, Gerchman Y, Collins CH, Arnold FH, Weiss R (2005) A synthetic multicellular system for programmed pattern formation. Nature 434(7037):1130–1134
4. Gardner TS, Cantor CR, Collins JJ (2000) Construction of a genetic toggle switch in *Escherichia coli*. Nature 403(6767):339–342
5. Tabor JJ, Salis HM, Simpson ZB, Chevalier AA, Levskaya A, Marcotte EM, Voigt CA, Ellington AD (2009) A synthetic genetic edge detection program. Cell 137:1272–1281
6. Ro DK, Paradise EM, Ouellet M, Fisher KJ, Newman KL, Ndungu JM, Ho KA, Eachus RA, Ham TS, Kirby J, Chang MCY, Withers ST, Shiba Y, Sarpong R, Keasling JD (2006) Production of the antimalarial drug precursor artemisinic acid in engineered yeast. Nature 440:940–943
7. Ajikumar PK, Xiao W-H, Tyo KEJ, Wang Y, Simeon F, Leonard E, Mucha O, Phon TH, Pfeifer B, Stephanopoulos G (2010) Isoprenoid pathway optimization for taxol precursor overproduction in *Escherichia coli*. Science 330:70–74
8. Bujara M, Schumperli M, Pellaux R, Heinemann M, Panke S (2011) Optimization of a blueprint for in vitro glycolysis by metabolic real-time analysis. Nat Chem Biol 7 (5):271–277. doi:nchembio.541 [pii] 10.1038/nchembio.541
9. Agarwal KL, Büchi H, Caruthers MH, Gupta NK, Khorana HG, Kleppe K, Kumar A, Ohtsuka E, RajBhandary EL, van de Dande JH, Sgaramella V, Weber H, Yamada T (1970) Total synthesis of the gene for an alanine transfer ribonucleic acid from yeast. Nature 227:27–34
10. Itakura K, Hirose T, Crea R, Riggs AD, Heyneker HL, Bolivar F, Boyer HW (1977) Expression in *Escherichia coli* of a chemically synthesized gene for the hormone somatostatin. Science 198:1056–1063
11. Khorana HG (1979) Total synthesis of a gene. Science 203:614–625
12. Cello J, Paul AV, Wimmer E (2002) Chemical synthesis of poliovirus cDNA: generation of infectious virus in the absence of natural template. Science 297:1016–1018

13. Smith HO, Hutchison CA 3rd, Pfannkoch C, Venter JC (2003) Generating a synthetic genome by whole genome assembly: ϕX174 bacteriophage from synthetic oligonucleotides. Proc Natl Acad Sci USA 100(26):15440–15445
14. Kodumal SJ, Patel KG, Reid R, Menzella HG, Welch M, Santi DV (2004) Total synthesis of long DNA sequences: synthesis of a contiguous 32-kb polyketide synthase gene cluster. Proc Natl Acad Sci USA 101(44):15573–15578
15. Tian J, Gong H, Sheng N, Zhou X, Gulari E, Gao X, Church G (2004) Accurate multiplex gene synthesis from programmable DNA microchips. Nature 432(7020):1050–1054
16. Kobasa D, Jones SM, Shinya K, Kash JC, Copps J, Ebihara H, Hatta Y, Kim JH, Halfmann P, Hatta M, Feldmann F, Alimonti JB, Fernando L, Li Y, Katze MG, Feldmann H, Kawaoka Y (2007) Aberrant innate immune response in lethal infection of macaques with the 1918 influenza virus. Nature 445(7125):319–323
17. Tumpey TM, Basler CF, Aguilar PV, Zeng H, Solórzano A, Swayne DE, Cox NJ, Katz JM, Taubenberger JK, Palese P, García-Sastre A (2005) Characterization of the reconstructed 1918 Spanish influenza pandemic virus. Science 310:77–80
18. Gibson DG, Benders G, Andrews-Pfannkoch C, Denisova EA, Baden-Tillson H, Zaveri J, Stockwell TB, Brownley A, Thomas DW, Algire MA, Merryman C, Young L, Noskov VN, Glass JI, Venter JC, Hutchison CA III, Smith HO (2008) Complete chemical synthesis, assembly, and cloning of a *Mcyoplasma genitalium* genome. Science 319:1215–1220. doi:10.1126/science.1151721
19. Gibson DG, Glass JI, Lartigue C, Noskov VN, Chuang R-Y, Algire MA, Benders GA, Montague MG, Ma L, Moodie MM, Merryman C, Vashee S, Krishnakumar R, Assad-Garcia N, Andrews-Pfannkoch C, Denisova EA, Young L, Qi Z-Q, Segall-Shapiro TH, Calvey CH, Parmar PP, Hutchison CA, Smith HO, Venter JC (2010) Creation of a bacterial cell controlled by a chemically synthesized genome. Science 329:52–56. doi:10.1126/science.1190719
20. Carlson R (2009) The changing economics of DNA synthesis. Nat Biotechnol 27:1091–1094
21. Kosuri S, Eroshenko N, LeProust EM, Super M, Way J, Li JB, Church GM (2010) Scalable gene synthesis by selective amplification of DNA pools from high-fidelity microchips. Nat Biotechnol 28:1295–1299
22. Lausted C, Dahl T, Warren C, King K, Smith K, Johnson M, Saleem R, Aitchison J, Hood L, Lasky SR (2004) POSaM: a fast, flexible, open-source, inkjet oligonucleotide synthesizer and microarrayer. Genome Biol 5:R58
23. Egeland RD, Southern EM (2005) Electrochemically directed synthesis of oligonucleotides for DNA microarray fabrication. Nucleic Acids Res 33:e125
24. Quan J, Saaem I, Tang N, Ma S, Negre N, Gong H, White KP, Tian J (2011) Parallel on-chip gene synthesis and application to optimization of protein expression. Nat Biotechnol 29 (5):449–452. doi:nbt.1847 [pii] 10.1038/nbt.1847
25. Tian J, Ma K, Saaem I (2009) Advancing high-throughput gene synthesis technology. Mol Biosyst 5:714–722
26. LeProust EM, Peck BJ, Spirin K, McCuen HB, Moore B, Namsaraev E, Caruthers MH (2010) Synthesis of high-quality libraries of long (150mer) oligonucleotides by a novel depurination controlled process. Nucleic Acids Res 38:2522–2540
27. Sierzchala AB, Dellinger DJ, Betley JR, Wyrzykiewicz TK, Yamada CM, Caruthers MH (2003) Solid-phase oligodeoxynucleotide synthesis: a two-step cycle using peroxy anion deprotection. J Am Chem Soc 125:13427–13441
28. Greger B, Kemper B (1998) An apyrimidinic site kinks DNA and triggers incision by endonuclease VII of phage T4. Nucleic Acids Res 26:4432–4438
29. Smith J, Modrich P (1997) Removal of polymerase-produced mutant sequences from PCR products. Proc Natl Acad Sci USA 94(13):6847–6850
30. Kim H, Han H, Shin D, Bang D (2010) A fluorescence selection method for accurate large-gene synthesis. Chembiochem 11:2448–2452. doi:10.1002/cbic.201000368

31. Linshiz G, Yehezkel TB, Kaplan S, Gronau I, Ravid S, Adar R, Shapiro E (2008) Recursive construction of perfect DNA molecules from imperfect oligonucleotides. Mol Syst Biol 4:191. doi:10.1038/msb.2008.26
32. Matzas M, Stähler PF, Kefer N, Siebelt N, Boisguérin V, Leonard JT, Keller A, Cord F, Stähler PH, Gharizadeh B, Babrzadeh F, Church GM (2010) High-fidelity gene synthesis by retrieval of sequence-verified DNA identified using high-throughput pyrosequencing. Nat Biotechnol 28:1291–1294
33. Xiong A-S, Peng R-H, Zhuang J, Gao F, Li Y, Cheng Z-M, Yao Q-H (2008) Chemical gene synthesis: strategies, softwares, error corrections, and applications. FEMS Microbiol Rev 32:522–540
34. Stemmer WP, Crameri A, Ha KD, Brennan TM, Heyneker HL (1995) Single-step assembly of a gene and entire plasmid from large numbers of oligodeoxyribonucleotides. Gene 164 (1):49–53
35. Gao X, Yo P, Keith A, Ragan TJ, Harris TK (2003) Thermodynamically balanced inside-out (TBIO) PCR-based gene synthesis: a novel method of primer design for high fidelity assembly of longer gene sequences. Nucleic Acids Res 31:e143
36. Ellis T, Adie T, Baldwin GS (2011) DNA assembly for synthetic biology: from parts to pathways and beyond. Integr Biol 3(2):109–118. doi:10.1039/c0ib00070a
37. Salis HM, Mirsky EA, Voigt CA (2009) Automated design of synthetic ribosome binding sites to control protein expression. Nat Biotechnol 27:946–950
38. Pfleger BF, Pitera DJ, Smolke CD, Keasling JD (2006) Combinatorial engineering of intergenic regions in operons tunes expression of multiple genes. Nat Biotechnol 24 (8):1027–1032
39. Smolke CD, Keasling JD (2002) Effect of gene location, mRNA secondary structures, and RNase sites on expression of two genes in an engineered operon. Biotechnol Bioeng 80 (7):762–776. doi:10.1002/bit.10434
40. Horton RM, Hunt HD, Ho SN, Pullen JK, Pease LR (1989) Engineering hybrid genes without the use of restriction enzymes: gene splicing by overlap extension. Gene 77(1):61–68. doi:0378-1119(89)90359-4 [pii]
41. Stemmer WP (1994) DNA shuffling by random fragmentation and reassembly: in vitro recombination for molecular evolution. Proc Natl Acad Sci USA 91(22):10747–10751
42. Quan J, Tian J (2009) Circular polymerase extension cloning of complex gene libraries and pathways. PLoS One 4(7):e6441. doi:10.1371/journal.pone.0006441
43. Quan J, Tian J (2011) Circular polymerase extension cloning for high-throughput cloning of complex and combinatorial DNA libraries. Nat Protoc 6(2):242–251. doi:nprot.2010.181 [pii] 10.1038/nprot.2010.181
44. Pachuk CJ, Samuel M, Zurawski JA, Snyder L, Phillips P, Satishchandran C (2000) Chain reaction cloning: a one-step method for directional ligation of multiple DNA fragments. Gene 243(1–2):19–25. doi:S0378-1119(99)00508-9 [pii]
45. Geu-Flores F, Nour-Eldin HH, Nielsen MT, Halkier BA (2007) USER fusion: a rapid and efficient method for simultaneous fusion and cloning of multiple PCR products. Nucleic Acids Res 35(7):e55. doi:gkm106 [pii] 10.1093/nar/gkm106
46. Li MZ, Elledge SJ (2007) Harnessing homologous recombination *in vitro* to generate recombinant DNA via SLIC. Nat Methods 4(3):251–256. doi:nmeth1010 [pii] 10.1038/nmeth1010
47. Peisajovich SG, Garbarino JE, Wei P, Lim WA (2010) Rapid diversification of cell signaling phenotypes by modular domain recombination. Science 328(5976):368–372. doi:328/5976/368 [pii] 10.1126/science.1182376
48. Nisson PE, Rashtchian A, Watkins PC (1991) Rapid and efficient cloning of Alu-PCR products using uracil DNA glycosylase. PCR Methods Appl 1(2):120–123
49. Nour-Eldin HH, Hansen BG, Norholm MH, Jensen JK, Halkier BA (2006) Advancing uracil-excision based cloning towards an ideal technique for cloning PCR fragments. Nucleic Acids Res 34(18):e122. doi:gkl635 [pii] 10.1093/nar/gkl635

50. Gibson DG, Young L, Chuang RY, Venter JC, Hutchison CA, Smith HO (2009) Enzymatic assembly of DNA molecules up to several hundred kilobases. Nat Methods 6(5):343–345. doi:nmeth.1318 [pii] 10.1038/nmeth.1318
51. Ramon A, Smith HO (2011) Single-step linker-based combinatorial assembly of promoter and gene cassettes for pathway engineering. Biotechnol Lett 33(3):549–555. doi:10.1007/s10529-010-0455-x
52. Larionov V, Kouprina N, Graves J, Chen XN, Korenberg JR, Resnick MA (1996) Specific cloning of human DNA as yeast artificial chromosomes by transformation-associated recombination. Proc Natl Acad Sci USA 93(1):491–496
53. Tsuge K, Matsui K, Itaya M (2003) One step assembly of multiple DNA fragments with a designed order and orientation in Bacillus subtilis plasmid. Nucleic Acids Res 31(21):e133
54. Orr-Weaver TL, Szostak JW, Rothstein RJ (1981) Yeast transformation: a model system for the study of recombination. Proc Natl Acad Sci USA 78(10):6354–6358
55. Oldenburg KR, Vo KT, Michaelis S, Paddon C (1997) Recombination-mediated PCR-directed plasmid construction *in vivo* in yeast. Nucleic Acids Res 25(2):451–452. doi:gka088 [pii]
56. Hua SB, Luo Y, Qiu M, Chan E, Zhou H, Zhu L (1998) Construction of a modular yeast two-hybrid cDNA library from human EST clones for the human genome protein linkage map. Gene 215(1):143–152. doi:S0378-1119(98)00258-3 [pii]
57. Schaerer-Brodbeck C, Barberis A (2004) Coupling homologous recombination with growth selection in yeast: a tool for construction of random DNA sequence libraries. Biotechniques 37(2):202–206
58. Shao Z, Zhao H (2009) DNA assembler, an *in vivo* genetic method for rapid construction of biochemical pathways. Nucleic Acids Res 37(2):e16. doi:gkn991 [pii] 10.1093/nar/gkn991
59. Itaya M, Fujita K, Kuroki A, Tsuge K (2008) Bottom-up genome assembly using the Bacillus subtilis genome vector. Nat Methods 5(1):41–43. doi:nmeth1143 [pii] 10.1038/nmeth1143
60. Endy D (2005) Foundations for engineering biology. Nature 438(7067):449–453
61. Clancy K, Voigt CA (2010) Programming cells: towards an automated 'Genetic Compiler'. Curr Opin Biotechnol 21(4):572–581. doi:S0958-1669(10)00124-2 [pii] 10.1016/j.copbio.2010.07.005
62. Alper H, Fischer C, Nevoigt E, Stephanopoulos G (2005) Tuning genetic control through promoter engineering. Proc Natl Acad Sci USA 102(36):12678–12683. doi:0504604102 [pii] 10.1073/pnas.0504604102
63. Murphy KF, Balazsi G, Collins JJ (2007) Combinatorial promoter design for engineering noisy gene expression. Proc Natl Acad Sci USA 104(31):12726–12731. doi:0608451104 [pii] 10.1073/pnas.0608451104
64. Canton B, Labno A, Endy D (2008) Refinement and standardization of synthetic biological parts and devices. Nat Biotechnol 26(7):787–793. doi:10.1038/nbt1413
65. Murphy KF, Adams RM, Wang X, Balazsi G, Collins JJ (2010) Tuning and controlling gene expression noise in synthetic gene networks. Nucleic Acids Res 38(8):2712–2726. doi:gkq091 [pii] 10.1093/nar/gkq091
66. Ellis T, Wang X, Collins JJ (2009) Diversity-based, model-guided construction of synthetic gene networks with predicted functions. Nat Biotechnol 27(5):465–471. doi:nbt.1536 [pii] 10.1038/nbt.1536
67. Nevoigt E, Kohnke J, Fischer CR, Alper H, Stahl U, Stephanopoulos G (2006) Engineering of promoter replacement cassettes for fine-tuning of gene expression in Saccharomyces cerevisiae. Appl Environ Microbiol 72(8):5266–5273. doi:72/8/5266 [pii] 10.1128/AEM.00530-06
68. Dubendorff JW, Studier FW (1991) Controlling basal expression in an inducible T7 expression system by blocking the target T7 promoter with *lac* repressor. J Mol Biol 219(1):45–59
69. Kobayashi M, Nagata K, Ishihama A (1990) Promoter selectivity of *Escherichia coli* RNA polymerase: effect of base substitutions in the promoter −35 region on promoter strength. Nucleic Acids Res 18(24):7367–7372

70. Jensen PR, Hammer K (1998) The sequence of spacers between the consensus sequences modulates the strength of prokaryotic promoters. Appl Environ Microbiol 64(1):82–87
71. Hsu LM, Cobb IM, Ozmore JR, Khoo M, Nahm G, Xia L, Bao Y, Ahn C (2006) Initial transcribed sequence mutations specifically affect promoter escape properties. Biochemistry 45(29):8841–8854. doi:10.1021/bi060247u
72. Cox RS 3rd, Surette MG, Elowitz MB (2007) Programming gene expression with combinatorial promoters. Mol Syst Biol 3:145. doi:msb4100187 [pii] 10.1038/msb4100187
73. De Mey M, Maertens J, Lequeux GJ, Soetaert WK, Vandamme EJ (2007) Construction and model-based analysis of a promoter library for *E. coli*: an indispensable tool for metabolic engineering. BMC Biotechnol 7:34. doi:1472-6750-7-34 [pii] 10.1186/1472-6750-7-34
74. Krueger M, Scholz O, Wisshak S, Hillen W (2007) Engineered Tet repressors with recognition specificity for the *tetO*-4C5G operator variant. Gene 404(1–2):93–100. doi:S0378-1119 (07)00462-3 [pii] 10.1016/j.gene.2007.09.002
75. Antunes LC, Ferreira RB, Lostroh CP, Greenberg EP (2008) A mutational analysis defines Vibrio fischeri LuxR binding sites. J Bacteriol 190(13):4392–4397. doi:JB.01443-07 [pii] 10.1128/JB.01443-07
76. Braatsch S, Helmark S, Kranz H, Koebmann B, Jensen PR (2008) *Escherichia coli* strains with promoter libraries constructed by Red/ET recombination pave the way for transcriptional fine-tuning. Biotechniques 45(3):335–337. doi:000112907 [pii] 10.2144/000112907
77. Gertz J, Siggia ED, Cohen BA (2009) Analysis of combinatorial cis-regulation in synthetic and genomic promoters. Nature 457(7226):215–218. doi:nature07521 [pii] 10.1038/nature07521
78. Mogno I, Vallania F, Mitra RD, Cohen BA (2010) TATA is a modular component of synthetic promoters. Genome Res 20(10):1391–1397. doi:gr.106732.110 [pii] 10.1101/gr.106732.110
79. Leveau JH, Lindow SE (2001) Predictive and interpretive simulation of green fluorescent protein expression in reporter bacteria. J Bacteriol 183(23):6752–6762. doi:10.1128/JB.183.23.6752-6762.2001
80. Tyo KE, Nevoigt E, Stephanopoulos G (2011) Directed evolution of promoters and tandem gene arrays for customizing RNA synthesis rates and regulation. Methods Enzymol 497:135–155. doi:B978-0-12-385075-1.00006-8 [pii] 10.1016/B978-0-12-385075-1.00006-8
81. Nevoigt E, Fischer C, Mucha O, Matthaus F, Stahl U, Stephanopoulos G (2007) Engineering promoter regulation. Biotechnol Bioeng 96(3):550–558. doi:10.1002/bit.21129
82. Imburgio D, Rong M, Ma K, McAllister WT (2000) Studies of promoter recognition and start site selection by T7 RNA polymerase using a comprehensive collection of promoter variants. Biochemistry 39(34):10419–10430. doi:bi000365w [pii]
83. Jensen K, Alper H, Fischer C, Stephanopoulos G (2006) Identifying functionally important mutations from phenotypically diverse sequence data. Appl Environ Microbiol 72 (5):3696–3701. doi:72/5/3696 [pii] 10.1128/AEM.72.5.3696-3701.2006
84. Buchler NE, Gerland U, Hwa T (2003) On schemes of combinatorial transcription logic. Proc Natl Acad Sci USA 100(9):5136–5141. doi:10.1073/pnas.0930314100 0930314100 [pii]
85. Liang G, Li Z (2007) Scores of generalized base properties for quantitative sequence-activity modelings for *E. coli* promoters based on support vector machine. J Mol Graph Model 26 (1):269–281. doi:doi:S1093-3263(06)00147-1 [pii] 10.1016/j.jmgm.2006.12.004
86. Rhodius VA, Mutalik VK (2010) Predicting strength and function for promoters of the *Escherichia coli* alternative sigma factor, σ^E. Proc Natl Acad Sci USA 107(7):2854–2859. doi:0915066107 [pii] 10.1073/pnas.0915066107
87. Hardiman T, Meinhold H, Hofmann J, Ewald JC, Siemann-Herzberg M, Reuss M (2010) Prediction of kinetic parameters from DNA-binding site sequences for modeling global transcription dynamics in *Escherichia coli*. Metab Eng 12(3):196–211. doi:S1096-7176(09) 00092-5 [pii] 10.1016/j.ymben.2009.10.006
88. Anand R, Rai N, Thattai M (2011) Promoter reliability in modular transcriptional networks. Methods Enzymol 497:31–49

89. Davis JH, Rubin AJ, Sauer RT (2011) Design, construction and characterization of a set of insulated bacterial promoters. Nucleic Acids Res 39:1131–1141
90. Goltermann L, Jensen MB, Bentin T (2011) Tuning protein expression using synonymous codon libraries targeted to the 50 mRNA coding region. Protein Eng Des Sel 24:123–129
91. Na D, Lee S, Lee D (2010) Mathematical modelling of translation initiation for the estimation of its efficiency to computationally design mRNA sequences with desired expression levels in prokaryotes. BMC Syst Biol 4:71
92. Na D, Lee D (2010) RBSDesigner: software for designing synthetic ribosome binding sites that yields a desired level of protein expression. Bioinformatics 26:2633–2634
93. Sneppen K, Dodd IB, Shearwin KE, Palmer AC, Schubert RA, Callen BP, Egan JB (2005) A mathematical model for transcriptional interference by RNA polymerase traffic in *Escherichia coli*. J Mol Biol 346(2):399–409. doi:S0022-2836(04)01556-6 [pii] 10.1016/j.jmb.2004.11.075
94. Macdonald LE, Durbin RK, Dunn JJ, McAllister WT (1994) Characterization of two types of termination signal for bacteriophage T7 RNA polymerase. J Mol Biol 238(2):145–158. doi: S0022-2836(84)71277-0 [pii] 10.1006/jmbi.1994.1277
95. Macdonald LE, Zhou Y, McAllister WT (1993) Termination and slippage by bacteriophage T7 RNA polymerase. J Bacteriol 232(4):1030–1047. doi:S0022-2836(83)71458-0 [pii] 10.1006/jmbi.1993.1458
96. Du L, Gao R, Forster AC (2009) Engineering multigene expression in vitro and in vivo with small terminators for T7 RNA polymerase. Biotechnol Bioeng 104(6):1189–1196. doi:10.1002/bit.22491
97. Cheng SW, Lynch EC, Leason KR, Court DL, Shapiro BA, Friedman DI (1991) Functional importance of sequence in the stem-loop of a transcription terminator. Science 254 (5035):1205–1207
98. Wilson KS, von Hippel PH (1995) Transcription termination at intrinsic terminators: the role of the RNA hairpin. Proc Natl Acad Sci USA 92(19):8793–8797
99. Lynn SP, Kasper LM, Gardner JF (1988) Contributions of RNA secondary structure and length of the thymidine tract to transcription termination at the thr operon attenuator. J Biol Chem 263(1):472–479
100. Christie GE, Farnham PJ, Platt T (1981) Synthetic sites for transcription termination and a functional comparison with tryptophan operon termination sites *in vitro*. Proc Natl Acad Sci USA 78(7):4180–4184
101. Reynolds R, Chamberlin MJ (1992) Parameters affecting transcription termination by *Escherichia coli* RNA. II. Construction and analysis of hybrid terminators. J Mol Biol 224 (1):53–63. doi:0022-2836(92)90575-5 [pii]
102. d'Aubenton Carafa Y, Brody E, Thermes C (1990) Prediction of rho-independent *Escherichia coli* transcription terminators. A statistical analysis of their RNA stem-loop structures. J Mol Biol 216(4):835–858
103. Larson MH, Greenleaf WJ, Landick R, Block SM (2008) Applied force reveals mechanistic and energetic details of transcription termination. Cell 132(6):971–982. doi:S0092-8674(08) 00127-X [pii] 10.1016/j.cell.2008.01.027
104. Isaacs FJ, Dwyer DJ, Collins JJ (2006) RNA synthetic biology. Nat Biotechnol 24:545–554
105. Deans TL, Cantor CR, Collins JJ (2007) A tunable genetic switch based on RNAi and repressor proteins for regulating gene expression in mammalian cells. Cell 130:363–372
106. Rinaudo K, Bleris L, Maddamsetti R, Subramanian S, Weiss R, Benenson Y (2007) A universal RNAi-based logic evaluator that operates in mammalian cells. Nat Biotechnol 25:795–801
107. Win MN, Smolke CD (2008) Higher-order cellular information processing with synthetic RNA devices. Science 322:456–460
108. Bayer TS, Smolke CD (2005) Programmable ligand-controlled riboregulators of eukaryotic gene expression. Nat Biotechnol 23(3):337–343

109. Culler SJ, Hoff KG, Smolke CD (2010) Reprogramming cellular behavior with RNA controllers responsive to endogenous proteins. Science 330:1251–1255
110. Babiskin AH, Smolke CD (2011) A synthetic library of RNA control modules for predictable tuning of gene expression in yeast. Mol Syst Biol 7:471. doi:10.1038/msb.2011.4
111. Win MN, Smolke CD (2007) A modular and extensible RNA-based gene-regulatory platform for engineering cellular function. Proc Natl Acad Sci USA 104(36):14283–14288. doi:10.1073/pnas.0703961104
112. MacDonald JT, Barnes C, Kitney RI, Freemont PS, Stan G-BV (2011) Computational design approaches and tools for synthetic biology. Integr Biol 3:97–108
113. Purnick PEM, Weiss R (2009) The second wave of synthetic biology: from modules to systems. Nat Rev Mol Cell Biol 10:410–422
114. Kim J, Winfree E (2011) Synthetic in vitro transcriptional oscillators. Mol Syst Biol 7:465
115. Montagne K, Plasson R, Sakai Y, Fujii T, Rondelez Y (2011) Programming an *in vitro* DNA oscillator using a molecular networking strategy. Mol Syst Biol 7:466
116. Noireaux V, Maeda YT, Libchaber A (2011) Development of an artificial cell, from self-organization to computation and self-reproduction. Proc Natl Acad Sci USA 108:3473–3480
117. Foley PL, Shuler ML (2010) Considerations for the design and construction of a synthetic platform cell for biotechnological applications. Biotechnol Bioeng 105:26–36
118. Dietz S, Panke S (2010) Microbial systems engineering: first successes and the way ahead. BioEssays 32(4):356–362
119. Reyes-Lamothe R, Sherratt DJ, Leake MC (2010) Stoichiometry and architecture of active DNA replication machinery in *Escherichia coli*. Science 328(5977):498–501. doi:328/5977/498 [pii] 10.1126/science.1185757
120. Alper H, Moxley J, Nevoigt E, Fink GR, Stephanopoulos G (2006) Engineering yeast transcription machinery for improved ethanol tolerance and production. Science 314 (5805):1565–1568. doi:314/5805/1565 [pii] 10.1126/science.1131969
121. Wang HH, Isaacs FJ, Carr PA, Sun ZZ, Xu G, Forest CR, Church GM (2009) Programming cells by multiplex genome engineering and accelerated evolution. Nature 460 (7257):894–898. doi:nature08187 [pii] 10.1038/nature08187
122. Hamilton CM, Aldea M, Washburn BK, Babitzke P, Kushner SR (1989) New method for generating deletions and gene replacements in *Escherichia coli*. J Bacteriol 171 (9):4617–4622
123. Blomfield IC, Vaughn V, Rest RF, Eisenstein BI (1991) Allelic exchange in *Escherichia coli* using the *Bacillus subtilis sacB* gene and a temperature-sensitive pSC101 replicon. Mol Microbiol 5(6):1447–1457
124. Link AJ, Phillips D, Church GM (1997) Methods for generating precise deletions and insertions in the genome of wild-type *Escherichia coli*: application to open reading frame characterization. J Bacteriol 179(20):6228–6237
125. Posfai G, Kolisnychenko V, Bereczki Z, Blattner FR (1999) Markerless gene replacement in *Escherichia coli* stimulated by a double-strand break in the chromosome. Nucleic Acids Res 27(22):4409–4415
126. Murphy KC (1998) Use of bacteriophage lambda recombination functions to promote gene replacement in *Escherichia coli*. J Bacteriol 180(8):2063–2071
127. Datsenko KA, Wanner BL (2000) One-step inactivation of chromosomal genes in *Escherichia coli* K-12 using PCR products. Proc Natl Acad Sci USA 97(12):6640–6645
128. Baba T, Ara T, Hasegawa M, Takai Y, Okumura Y, Baba M, Datsenko KA, Tomita M, Wanner BL, Mori H (2006) Construction of *Escherichia coli* K-12 in-frame, single-gene knockout mutants: the Keio collection. Mol Syst Biol 0008. doi:msb4100050 [pii] 10.1038/msb4100050
129. Reyes-Lamothe R, Possoz C, Danilova O, Sherratt DJ (2008) Independent positioning and action of *Escherichia coli* replisomes in live cells. Cell 133(1):90–102. doi:S0092-8674(08)00209-2 [pii] 10.1016/j.cell.2008.01.044

130. Kolisnychenko V, Plunkett G, Herring CD, Feher T, Posfai J, Blattner FR, Posfai G (2002) Engineering a reduced *Escherichia coli* genome. Genome Res 12(4):640–647. doi:10.1101/gr.217202
131. Mizoguchi H, Mori H, Fujio T (2007) *Escherichia coli* minimum genome factory. Biotechnol Appl Biochem 46(Pt 3):157–167. doi:BA20060107 [pii] 10.1042/BA20060107
132. Posfai G, Plunkett G, Feher T, Frisch D, Keil GM, Umenhoffer K, Kolisnychenko V, Stahl B, Sharma SS, de Arruda M, Burland V, Harcum SW, Blattner FR (2006) Emergent properties of reduced-genome *Escherichia coli*. Science 312(5776):1044–1046. doi:1126439 [pii] 10.1126/science.1126439
133. Goryshin IY, Naumann TA, Apodaca J, Reznikoff WS (2003) Chromosomal deletion formation system based on Tn5 double transposition: use for making minimal genomes and essential gene analysis. Genome Res 13(4):644–653
134. Yu BJ, Sung BH, Koob MD, Lee CH, Lee JH, Lee WS, Kim MS, Kim SC (2002) Minimization of the *Escherichia coli* genome using a Tn5-targeted Cre/*loxP* excision system. Nat Biotechnol 20:1018–1023
135. Hashimoto M, Ichimura T, Mizoguchi H, Tanaka K, Fujimitsu K, Keyamura K, Ote T, Yamakawa T, Yamazaki Y, Mori H, Katayama T, J-i K (2005) Cell size and nucleoid organization of engineered *Escherichia coli* cells with a reduced genome. Mol Microbiol 55(1):137–149
136. Ara K, Ozaki K, Nakamura K, Yamane K, Sekiguchi J, Ogasawara N (2007) *Bacillus* minimum genome factory: effective utilization of microbial genome information. Biotechnol Appl Biochem 46:169–178. doi:10.1042/BA20060111
137. Morimoto T, Kadoya R, Endo K, Sawada MTK, Liu S, Ozawa T, Kodama T, Kakeshita H, Kageyawa Y, Manabe K, Kanaya S, Ara K, Ozaki K, Ogasarawa N (2008) Enhanced recombinant protein productivity by genome reduction in *Bacillus subtilis*. DNA Res 15:73–81
138. Westers H, Dorenbos R, van Dijl JM, Kabel J, Flanagan T, Devine KM, Jude F, Seror SJ, Beekman AC, Darmon E, Eschevins C, de Jong A, Bron S, Kuipers OP, Albertini AM, Antelmann H, Hecker M, Zamboni N, Sauer U, Bruand C, Ehrlich DS, Alonso JC, Salas M, Quax WJ (2003) Genome engineering reveals large dispensable regions in *Bacillus subtilis*. Mol Biol Evol 20(12):2076–2090
139. Sharma SS, Blattner FR, Harcum SW (2007) Recombinant protein production in an *Escherichia coli* reduced genome strain. Metab Eng 9:133–141
140. Chakiath CS, Esposito D (2007) Improved recombinational stability of lentiviral expression vectors using reduced-genome *Escherichia coli*. Biotechniques 43(4):466, 468, 470. doi:000112585 [pii]
141. Umenhoffer K, Fehér T, Balikó G, Ayaydin F, Blattner JPFR, Pósfai G (2010) Reduced evolvability of *Escherichia coli* MDS42, an IS-less cellular chassis for molecular and synthetic biology applications. Microb Cell Fact 9:38
142. Lee JH, Sung BH, Kim MS, Blattner FR, Yoon BH, Kim JH, Kim SC (2009) Metabolic engineering of a reduced-genome strain of *Escherichia coli* for L-threonine production. Microb Cell Fact 8:2
143. Trinh CT, Unrean P, Srienc F (2008) Minimal *Escherichia coli* cell for the most efficient production of ethanol from hexoses and pentoses. Appl Environ Microbiol 74:3634–3643
144. Heinemann M, Panke S (2006) Synthetic biology – putting engineering into biology. Bioinformatics 22:2790–2799
145. Kitano H (2004) Biological robustness. Nat Rev Genet 5(11):826–837. doi:nrg1471 [pii] 10.1038/nrg1471
146. Gil R, Silva FJ, Pereto J, Moya A (2004) Determination of the core of a minimal bacterial gene set. Microbiol Mol Biol Rev 68(3):518–537
147. Glass JI, Assad-Garcia N, Alperovich N, Yooseph S, Lewis MR, Maruf M, Hutchison CA III, Smith HO, Venter JC (2006) Essential genes of a minimal bacterium. Proc Natl Acad Sci USA 103(2):425–430

148. Bayer TS (2010) Transforming biosynthesis into an information science. Nat Chem Biol 6:859–861
149. Dueber JE, Wu GC, Malmirchegini GR, Moon TS, Petzold CJ, Ullal AV, Prather KLJ, Keasling JD (2009) Synthetic protein scaffolds provide modular control over metabolic flux. Nat Biotechnol 27:753–759
150. Neumann H, Wang K, Davis L, Garcia-Alai M, Chin JW (2010) Encoding multiple unnatural amino acids via evolution of a quadruplet-decoding ribosome. Nature 464:441–444
151. Wörsdörfer B, Woycechowsky KJ, Hilvert D (2011) Directed evolution of a protein container. Science 331:589–592

Chapter 5
Systems Metabolic Engineering of *Escherichia coli* for Chemicals, Materials, Biofuels, and Pharmaceuticals

Dokyun Na, Jin Hwan Park, Yu-Sin Jang, Jeong Wook Lee, and Sang Yup Lee

Contents

D. Na • J.H. Park • Y.-S. Jang • J.W. Lee
Metabolic and Biomolecular Engineering National Research Laboratory, Department of Chemical and Biomolecular Engineering (BK21 Program), BioProcess Engineering Research Center, Korea Advanced Institute of Science and Technology (KAIST), 291 Daehak-ro, Yuseong–gu, Daejeon 305-701, South Korea

Center for Systems and Synthetic Biotechnology, Institute for the BioCentury, KAIST, 291 Daehak-ro, Yuseong–gu, Daejeon 305-701, South Korea
e-mail: dkna@kaist.ac.kr; jinhwan@kaist.ac.kr; nall55@kaist.ac.kr; l.jw@kaist.ac.kr

S.Y. Lee (✉)
Metabolic and Biomolecular Engineering National Research Laboratory, Department of Chemical and Biomolecular Engineering (BK21 Program), Center for Systems and Synthetic Biotechnology, Institute for the BioCentury, KAIST, Daejeon, Republic of Korea
e-mail: leesy@kaist.ac.kr

C. Wittmann and S.Y. Lee (eds.), *Systems Metabolic Engineering*,
DOI 10.1007/978-94-007-4534-6_5, © Springer Science+Business Media Dordrecht 2012

Abstract Microorganisms are small but efficient micro-factories for producing bulk and value-added products. Systems biology and synthetic biology advanced during the last decade are enabling us to unravel the underlying complicated intracellular and intercellular mechanisms, to comprehensively understand them as a whole, and to rationally reconstruct metabolic and gene regulatory networks for the optimized production of target materials including non-innate materials. Recent advances in these fields have opened a new way of metabolic engineering, termed systems metabolic engineering. In this article, we review the achievements in systems metabolic engineering for the production of chemical compounds and materials, biofuels, and pharmaceuticals in a widely used microbial platform factory, *Escherichia coli*.

Keywords *Escherichia coli* • value-added products • systems biology • synthetic biology • cell factory • network reconstruction • metabolic capability • spider silk protein • polyhydroxyalkanoate • polylactic acid • propanediol • diamine • putrescine • cadaverine • ethanol • butanol • isobutanol • biodiesel • hydrogen • farnesol • biofuel • amino acids • valine • threonine • isoprenoids • artemisinin • taxol • pharmaceutical • systems metabolic engineering • computational model • bio-materials • polymer • genome-scale model • transcriptional profiling • sustainable production • bio-industry

5.1 Introduction

Metabolic engineering based on the comprehensive understanding on complex biological systems supported by systems biology, which is termed systems metabolic engineering, has enabled us to perform engineering of microorganisms into microbial factories by optimally reconstructing metabolic and regulatory networks. Recent advances in synthetic biology are also allowing us to create novel biological systems, thereby empowering microorganisms to gain new metabolic capabilities, which otherwise might not be obtained (see Chaps. 1, 2, 3, and 4). Here we review recent developments in systems metabolic engineering for the production of chemicals, materials, biofuels, and pharmaceuticals in *E. coli*.

5.2 Chemicals and Materials

5.2.1 Spider Silk Protein

Due to the high strength equivalent to Kevlar, spider silk fibers are promising multifunctional materials for their use as parachute cords, protective clothing, composite materials in aircrafts, and so on [46, 73]. Particularly, the biocompatible property of silk fibers permits their use as sutures for repairing ligament, bone,

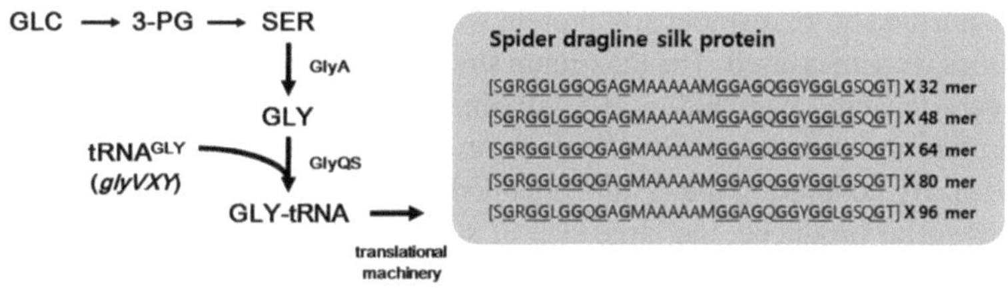

Fig. 5.1 Metabolic engineering of *E. coli* for spider dragline silk protein production. Comparative proteomic analysis revealed that host cells struggled to cope with stresses caused by the depletion of glycine and glycyl-tRNAs, which is highly frequent in the coding sequence of the recombinant spider protein. For an improved production titre, the *glyVXY* genes encoding glycyl-tRNAs and the *glyA* gene encoding an enzyme converting serine to glycine were amplified to increase the glycyl-tRNA pool, and eventually the titre of spider silk protein. Abbreviations are: *GLC* glucose, *3-PG* 3-phosphoglycerate, *SER* serine *GLY* glycine

and other tissues, and as biomaterial scaffolds for drug delivery system [46]. Despite diverse potential applications, the difficulty in mass production of spider silk protein has posed obstacles towards its commercialization. Recently, native-sized recombinant spider silk protein was successfully produced by employing metabolically engineered *E. coli* [73]. In this section, the detailed approach for the production of the exotic recombinant spider silk protein is described.

The basic monomeric motif of the dragline silk component protein of *Nephilia clavipes* was multimerized by iterated ligation of compatible but non-redigestable restriction enzyme sites to synthesize recombinant spider silk genes of 32–96 repeats encoding proteins sized 100.7–284.9 kDa. The molecular weight of 96-mer silk protein (284.9 kDa) is equivalent to that of the native silk protein [64]. The constructed recombinant genes with various lengths were expressed in *E. coli.* The expression level of the silk proteins decreased dramatically as its molecular weight increased. Comparative proteomic analysis showed that glycine biosynthesis enzymes and glycyl-tRNA synthetase were induced during the production of silk proteins, representing host cells were struggling to cope with stresses of glycyl-tRNA depletion derived from the highly biased codon frequency of the recombinant proteins (44.9% of the codons in the proteins are glycine). The *glyVXY* genes encoding tRNAs recognizing the frequently occurring glycine codons in the proteins, GGU and GGC, were amplified to supply sufficient number of tRNAs recognizing those codons (Fig. 5.1). The amplification increased titre of larger proteins (48- to 96-mer). Further amplification of the *glyVXY* genes in two copies increased the titre of the top largest proteins (80- and 96-mer), indicating that tRNA pool availability was one of limiting factors for highly codon-biased spider silk protein production. In addition to tRNAs, the glycine pool was increased by engineering the glycine pathway. The *glyA* gene was over-expressed to enhance glycine pool from serine, which increased the titre of the large spider proteins up to 35 folds. After fermentation and purification steps, 1.2 g/L of the largest silk protein (96-mer) with a purity of greater than 90% could be obtained.

The purified 96-mer spider silk protein was spun into a fiber, and its mechanical properties were measured. The maximum strength a fiber, that can take before being torn apart (tenacity), was 508 ± 108 MPa and its extensibility was 15 ± 5%, which are similar to those of native *N. clavipes* dragline silk (740–1,200 MPa and 18–27%). By comparing the results obtained with smaller recombinant spider silk proteins so far produced, the large size of silk protein was found to be crucial for better mechanical properties. The tenacity of the 96-mer protein was the highest ever reported for recombinant spider silk proteins.

In the above work, the systematic analysis of cellular physiology has allowed to tackle down difficulties in producing exotic but industrially attractive recombinant proteins having high biased amino acid repetition, high GC-content, and large size. By doing so, the large artificial spider silk fibers derived from the recombinant proteins could reproduce good mechanical properties of native spider dragline silk fibers. This strategy can also be applied to other promising biomaterials such as elastin, collagen, byssus, and resilin.

5.2.2 *Polyhydroxyalkanoate*

During the last several decades, the demand for biodegradable materials has been gradually increasing. More recently, production of various polymers from renewable resources using microorganisms itself has been drawing more attention [43, 51]. Among the biodegradable materials, polyhydroxyalkanoates (PHAs) have been receiving much attention as good alternatives to petrochemical-derived polymers. PHAs are linear polyesters accumulated in many microorganisms as energy and carbon reserve materials under nutrient-limiting (e.g., N or P limited condition) but carbon-rich conditions [34]. The chain length and side-chain types-of PHAs are responsible for their diverse material properties. For example, short-chain-length (SCL) PHAs can be used as thermoplastics, while medium-chain-length (MCL) PHAs can be used as elastomers. The SCL and MCL PHA copolymers display physical properties similar to a low density polyethylene (Fig. 5.2).

The composition of PHAs monomer highly depends on the substrate specificity of the PHA synthase and on the metabolic characteristics of the host microorganism. Therefore, it is crucial to understand the characteristics of enzymes involved in the PHA production and the metabolic capability of host organisms in order to develop microbial factories for the production of diverse PHAs with novel properties. Recent advances in systems metabolic engineering allow us to understand cellular metabolism at global-scale and improve the performance of PHA producers as well. As an example, *in silico* simulation based on a metabolic network model was adopted to estimate the distribution of metabolic fluxes in *E. coli* producing poly(3-hydroxybutyrate), P(3HB), and predicted that the enhancement of the Entner-Doudoroff (ED) pathway leads to increased accumulation of P(3HB) [21]. To prove the role of the ED pathway during P(3HB) production, Eda (2-oxo-3-deoxy-6-phosphogluconate aldolase/2-keto-4-hydroxyglutarate

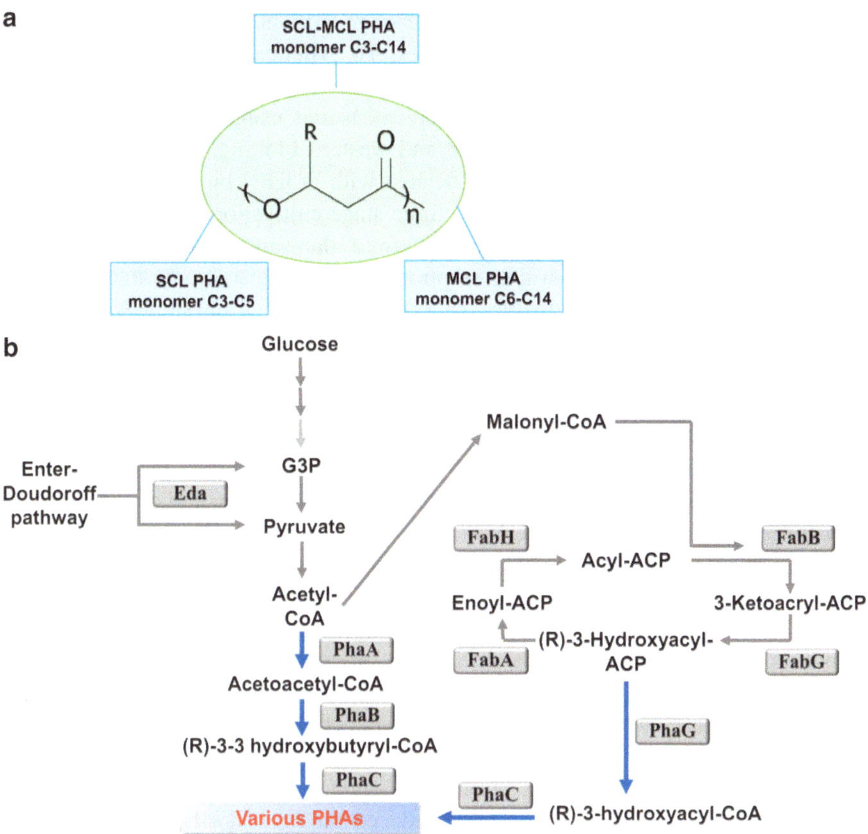

Fig. 5.2 Polyhydroxyalkanoates. (**a**) General structure and three types of PHAs. Short-chain-length (*SCL*) PHA composed of C3–C5 monomers, stiff crystalline material similar to the properties of polypropylene; Medium-chain-length (*MCL*) PHA composed of C6–C14 monomers, semicrystalline elastomer; SCL-MCL PHA composed of C3–C14 monomers, physical properties similar to those of low density polyethylene (*LDPE*). (**b**) Pathways for the production of various PHAs in *E. coli*. *Thick blue arrows* denote heterologous enzymes introduced into *E. coli* to produce PHAs. The enzymes are listed below along with the representative sources of the foreign enzymes: *FabA* β-hydroxyl-ACP dehydrase, *FabB* β-ketoacryl-ACP synthase/malonyl-ACP decarboxylase, *FabF* β-ketoacyl-ACP synthase, *FabG* β-ketoacyl-ACP reductase, *FabH* β-ketoacyl-CoA thiolase, *PhaA* β-keththiolase from *Cupriavidus necator*, *PhaB* acetoacetyl-CoA reductase from *C. necator*, *PhaC*, PHA synthase from type I, II, III, and IV, *PhaG* (R)-3-hydroxydecanoyl-CoA:ACP transacylase from *Pseudomonas putida*

aldolase), one of the enzymes in the ED pathway was knocked out and restored in an *E. coli* strain producing P(3HB). As a result, it was found that the ED pathway plays a crucial role in the P(3HB) production because it can contribute to increase the NADPH pool that is required during P(3HB) production.

Another organism, *Cupriavidus necator*, was also systematically analyzed for their performance on P(3HB) production [33]. The transcriptional analysis of genes 1 related to P(3HB) homeostasis in *C. necator* H16 (formerly, *Ralstonia eutropha* H16) was performed during three-stage cultivation: cell growth, P(3HB) biosynthesis, and P(3HB) utilization stages. To understand how these genes are synchronized and how they break down and assemble P(3HB) granules, intracellular mRNA levels and protein levels were measured using reverse transcriptase quantitative polymerase chain reaction and western blotting. This result showed three expression patterns of genes associated with P(3HB) biosynthesis, degradation, and granule formation during the three-stage cultivation. Consequently, five genes related to P(3HB) biosynthesis showed the same transcriptional trend, decreasing during ammonium consumption and keeping a steady state after the absence of ammonium. Meanwhile, the expression level of phasin (*phaP*), a gene encoding the granule-associated protein, was also coupled to P(3HB) concentration. The identified genes would be potential targets for rational metabolic engineering of P(3HB) production.

In addition to PHAs, polylactic acid (PLA) can be a good alternative to petrochemical-derived polymers. PLA has several desirable properties as a promising polymer such as biocompatibility, biodegradability and compostability. Because of these properties, PLA and its copolymers have various biomedical applications including dissolvable sutures, drug delivery system, and bone fracture internal fixation devices [40]. Currently, PLA is produced by chemical synthesis through ring-opening polymerization of lactide, which is derived from lactic acid produced by fermentation. For a sustainable production of PLA from renewable resources, there have been several attempts to systematically engineer microorganisms for one-step direct fermentative production instead of the current two-step processes. In order to produce PLA in microorganisms such as *E. coli*, propionate CoA transferase (Pct_{Cp}) originated from *Clostidium propionicum* and PHA synthase originated from *Pseudomonas* sp. were introduced into *E. coli* [74]. Directed, random, and saturation mutagenesis were carried out to improve enzyme specificities, because the wild-type of Pct_{Cp} and PHA synthase did not efficiently convert lactate to lactyl-CoA and did not accept lactyl-CoA as a substrate, respectively [74]. Furthermore, in order to increase the contents of the polymer, *in silico* genome-scale flux response analysis and gene knockout simulations were performed to identify metabolic genes to be manipulated [25]. The target metabolic pathways selected by *in silico* simulations were engineered to enhance the PLA production by increasing the precursor pool. Consequently, PLA homopolymer and P(3HB-co-LA) copolymer could be produced up to 11% (g/g) of dry cell weight (DCW) and 56% (g/g) of DCW. The lactate fraction in the copolymer could be controlled [74]. Systems metabolic engineering through the comprehensive understanding of genome-scale cellular metabolism allowed the engineering of the host strain to produce biopolymers [36].

5.2.3 Propanediol

1,2-Propanediol (1,2-PDO), also referred to as propylene glycol, is a three-carbon compound with the chemical formula $C_3H_8O_2$ or HO-CH_2-CHOH-CH_3. It has

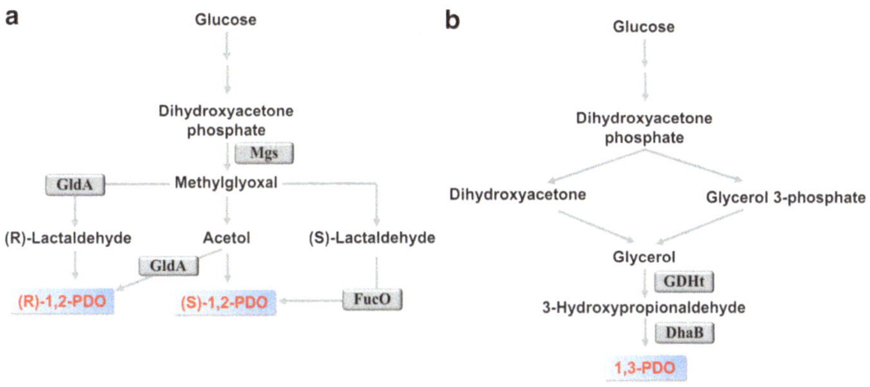

Fig. 5.3 Metabolic pathways for the production of propanediol from glucose or glycerol. (**a**) Biosynthetic pathways for 1,2-propanediol (1,2-PDO) from glucose and (**b**) biosynthetic pathways for 1,3-propanediol (1,3-PDO) from glucose and glycerol

attracted great interest from researchers in chemical industries due to its possible applications as a less toxic anti-freezer, a de-icer, a food additive for coloring, flavors in food industries, a carrier of oils, moisturizer in cosmetics, and precursors for pharmaceuticals [24]. 1,2-PDO has been produced *via* chemical synthesis involving the hydration of propylene oxide. However, this petrochemical process is not sustainable and environmentally favorable, since the precursor, propylene oxide, is a non-renewable and hazardous petrochemical derivative [5]. 1,3-Propanediol (1,3-PDO) is also a three-carbon diol with formula $C_3H_8O_2$ or HO-CH_2-CH_2-CH_2-OH. It is well-known as a building block of polymers such as poly-trimethylene terephthalate, and can also be used as a precursor for various industrially important materials including composites, adhesives, laminates, coatings, moldings, aliphatic polyesters, co-polyesters, and as an antifreeze and wood paint as well. With a variety of industrial applications, 1,3-PDO has been produced by the hydration of acrolein, or by the hydroformylation of ethylene oxide to afford 3-hydroxypropionaldehyde (3-HPA) but these methods are expensive and generates much waste water. These environmental and economic concerns urged researchers to design metabolically engineered microorganisms for the production of 1,2-PDO and 1,3-PDO.

There are several well-known pathways to produce 1,2-PDO in microorganisms. Among the possible pathways, a pathway *via* methylglyoxal has been used as a standard route for the production of 1,2-PDO (Fig. 5.3a). Wild type *E. coli* incapable of producing 1,2-PDO has been metabolically engineered by overexpressing methylglyoxal synthase (Mgs) and glycerol dehydrogenase (GldA). The engineered strain was able to convert glucose to 1,2-PDO *via* the activated pathway, although the titer was low at 0.2 g/L [4]. In another report, chromosomal genes of *E. coli* including *tpiA*, *edd*, *aldA*, *gloA*, *ldhA*, *lpd*, *pflAB*, and *adhE* were attenuated or inactivated to maintain a high methylglyoxal level [63]. In addition, the *gapA* gene was overexpressed to produce sufficient NADH for balancing redox potential.

The engineered strain was further evolved under selection pressure towards a state where a higher level of 1,2-PDO is obtained. The combined evolutionary and rational metabolic engineering allowed production of 3.7 g/L of 1,2-PDO from 18.0 g/L glucose.

Glycerol is a preferred carbon source to others because only two enzymatic steps are enough to convert glycerol to 1,3-PDO (Fig. 5.3b). Glycerol dehydratase (GDHt) catalyzes glycerol to 3-HPA, and 1,3-PDO oxidoreductase (DhaB), encoded by *dhaB*, then catalyzes 3-HPA to 1,3-PDO [44]. On the other hand, an attempt to produce 1,3-PDO in *E. coli* was carried out as well from a sugar substrate, mainly glucose [12]. The pathway for the 1,3-PDO production includes glycerol as an intermediate and, thus, the carbon flow moves from glucose to dihydroxyacetone phosphate *via* the glycolytic pathway, glycerol 3-phosphate (or dihydroxyacetone), glycerol, 3-hydroxypropionaldehyde, and finally to 1,3-PDO. This engineered *E. coli* strain includes the attenuation of phosphotransferase, ubiquitin-protein ligase, and glyceraldehyde-3-phosphate dehydrogenase in order to achieve a high level 1,3-PDO production. In addition, GDHt and DhaB were overexpressed to facilitate 1,3-PDO synthesis. This engineered *E. coli* was able to produce about 129 g/L of 1,3-PDO from 380 g/L glucose in 74 h.

5.2.4 Diamine

Diamine is a kind of polyamine having two amino groups. Diamines of aliphatic linear carbon chain backbone are called ethylene diamine (C2), diaminopropane (C3), putrescine (C4), cadaverine (C5), and hexamethylenediamine (C6) according to the carbon chain length. Among them, putrescine and cadaverine have been attracting much interest due to their potential use in polymer industries [27, 59]; for example, putrescine is used as a monomer of Nylon-4,6. Currently, industrial supply of putrescine highly relies on petrochemical synthesis by hydrogenation of succinonitrile derived by a reaction of hydrogen cyanide and acrylonitrile [58]. The European market size of putrescine was about 16 million Euro in 2007, and is expected to grow [60].

Putrescine is also found in various organisms as it plays roles in cell proliferation and cell growth [66]. Putrescine is produced through the pathways including either ornithine decarboxylase and/or arginine decarboxylase in various animals, plants, and microorganisms [59]. Recently, there have been several reports on microbial production of putrescine by employing metabolically engineered *E. coli* and *Corynebacterium glutamicum* [14, 53, 59].

After the first report on the production of 5.1 g/L of putrescine in *E. coli* [14], *E. coli* has been metabolically engineered using a system-wide approach for enhanced putrescine production. Also, its cultivation condition was optimized. As a result, the engineered *E. coli* strain was able to produce putrescine up to 24.2 g/L with a volumetric productivity of 0.75 g/L/h in a minimal medium [53]. To achieve this performance, they strategically engineered *E. coli* by combining local pathway engineering and systems metabolic engineering. The putrescine-producing base strain of *E. coli* was constructed by disrupting major competing pathways and degrading pathways, and by overexpressing the key enzyme ornithine decarboxylase. This strain yield 1.2 g/L putrescine with a growth rate of 0.43 h^{-1}. To further increase the putrescine yield, genome engineering was carried out to replace native promoters of those genes in the ornithine production pathway and putrescine/ornithine antiporter with the strong *tac* promoter. This engineered strain produced 1.6 g/L putrescine, which is 33% higher than that obtained with the base strain. The stress responsive sigma factor RpoS which controls a large number of *E. coli* genes was additionally blocked, to minimize expression of several genes which are not associated with putrescine production. This resulting strain yielded 24.2 g/L putrescine with a productivity of 0.75 g/L/h by the pH-stat fed-batch cultivation.

In addition to *E. coli*, putrescine production in *C. glutamicum* has more recently been reported. *C. glutamicum* is a well-known industrial microorganism not only for various amino acid production but also for production of ethanol, sugar alcohols, organic acids, vitamins, and biopolymers [59]. As this Gram-positive soil bacterium can tolerate up to 0.5 M putrescine without any significant growth retardation, it is also considered as a promising microbial host for putrescine production. As the putrescine biosynthesis pathway in *C. glutamicum* has not been characterized yet, the well-characterized genes involved in putrescine production in *E. coli* were heterologously introduced. First, an arginine overproducing *C. glutamicum* strain was engineered to overexpress arginine decarboxylase and agmatinase, but failed to covert arginine to putrescine efficiently. Alternatively, ornithine decarboxylase (SpeC) originated from *E. coli* was introduced into an arginine auxotrophic *C. glutamicum* strain, and the resulting strain produced 6 g/L of putrescine successfully but with a relatively low yield.

In summary, both *E. coli* and *C. glutamicum* could be successfully engineered by using the ornithine decarboxylase pathway rather than the arginine decarboxylase pathway (Fig. 5.4). The arginine decarboxylase pathway was not preferred since it intrinsically produces urea which is toxic to cells, and is negatively inhibited *via* a feedback mechanism by the accumulated arginine. Additionally, it is still difficult to increase the activities of arginine decarboxylase and agmatinase [53, 59]. Based on the recent developments in putrescine producing strains, fine-tuning of expression levels in ornithine decarboxylase pathway and optimizing the fluxes through systems metabolic engineering will further enhance putrescine production.

5.3 Biofuels

5.3.1 Ethanol

Currently, bioethanol is used as a transportation fuel throughout the world including USA and Brazil, two largest producing countries. Fuel ethanol is used

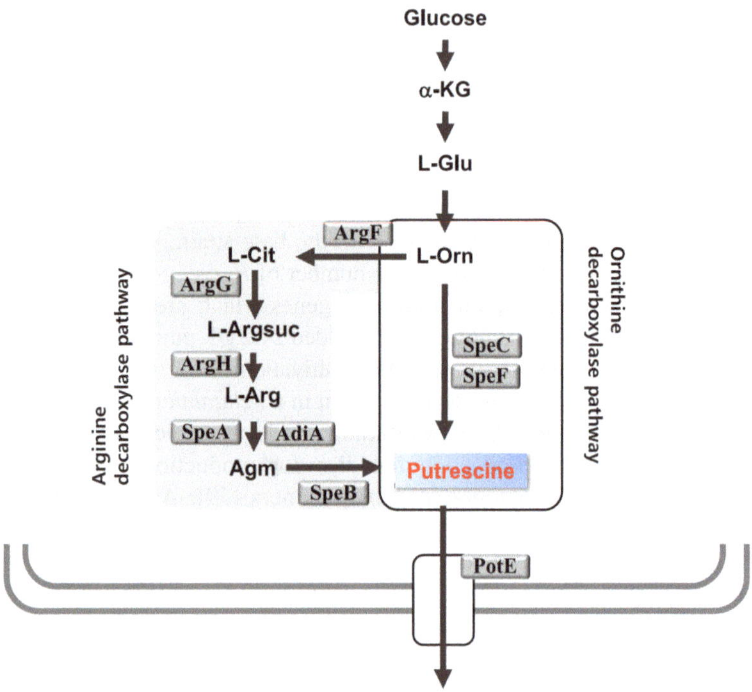

Fig. 5.4 Metabolic pathways for putrescine production. Biosynthetic pathways to putrescine synthesis *via* arginine decarboxylase pathway and ornithine decarboxylase pathway are shown. The ornithine decarboxylase pathway is preferred to the arginine decarboxylase pathway for the production of putrescine because the arginine decarboxylase pathway inevitably produces toxic urea and negatively inhibited by arginine before being converted to putrescine. Abbreviations are: *α-KG* α-ketoglutarate, *AdiA* degradative arginine decarboxylase, *AGM* Agmatine, *ArgF* ornithine carbamoyltransferase chain F-monomer, *ArgG* argininosuccinate synthase, *ArgH* argininosuccinate lyase, *GLC* glucose, *L-ARG* L-arginine, *L-ARGSUC* L-argininosuccinate, *L-CIT* L-citrulline, *L-GLU* L-glutamate, *L-ORN* L-ornithine, *SpeA* biosynthetic arginine decarboxylase, *SpeB* agmatinase, *SpeC*/*SpeF* biosynthetic ornithine decarboxylase

as a gasoline-ethanol blend in general gasoline engines, but ethanol can completely replace the petroleum fuels in flexible engines. The energy density of ethanol is lower than that of gasoline at about 66% due to partially oxidized characteristics of ethanol. Yeast, especially *Saccharomyces,* has been used as a main host strain for commercial ethanol production. *Zymommonas mobilis* also produces ethanol as a major product. Despite their inherent capability to produce ethanol, their incapability to utilize major compoenents of hemicellulose, such as xylose, as a carbon source prevents their use as a general platform host for ethanol production. On the other hand, *E. coli* can be a good host strain for ethanol production as it can utilize hexose and pentose sugars, but does not produce ethanol at high yield (Fig. 5.5). Thus, metabolic engineering has been performed to improve ethanol production by *E. coli*, impressively by L. Ingram's group.

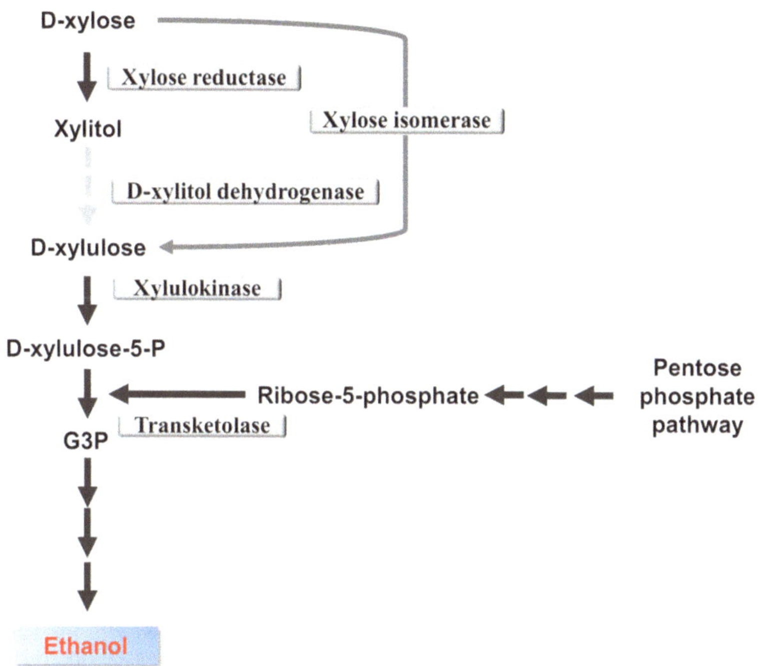

Fig. 5.5 Metabolic pathway converting xylose to ethanol. *S. cerevisiae* lacks in D-xylitol dehydrogenase (*dashed arrow*), and thus cannot utilize xylose as a carbon source. Metabolic bypassing through the pathway *via* xylitol using xylose isomerase enables *S. cerevisiae* to grow in xylose-containing medium and produce ethanol. On the other hand, *E. coli* can utilize xylose as carbon source, but its ethanol-producing pathway is not sufficiently efficient. In *E. coli*, the ethanol-producing pathway is a major target for metabolic engineering, while in *S. cerevisiae* the xylose-utilizing pathway is a major target

A synthetic pathway converting xylose to xylulose has been employed in *S. cerevisiae*, since the organism can utilize xylulose instead of xylose. The pathway was reconstructed using heterologous enzymes, xylose reductase (XYL1) and xylitol dehydrogenase (XYL2) from *Pichia stipitis* [29, 30, 55, 56, 67, 68]. However, the engineered strain produced low titre of ethanol (0.04 g/g xylose), and relatively high amounts of xylitol (0.47 g/g xylose) as a by-product from 50 g/L xylose [68]. It was later discovered that *S. cerevisiae* possesses an xylose reductase encoded by *GRE3* gene. Both xylose reductases are negatively inhibited by xylitol, and thus their activities were dramatically reduced by feedback mechanism [32]. To avoid xylitol-induced feedback inhibition, an alternative pathway from xylose to xylulose mediated by xylose isomerase was employed in *S. cerevisiae*. Additional knock-out of the *GRE3* gene and amplification of genes involved in the pentose phosphate pathway resulted in increased ethanol yield (0.43 g/g xylose) with the reduction of xylitol yield (0.003 g/g xylitol) from 20 g/L xylose. However, cell growth was retarded by 30% [32].

Zymomonas mobilis strain has also been engineered to express xylose isomerase, xylulokinase and transketolase, but was unable to grow on xylose as a sole carbon source. Additional expression of *E. coli* transaldolase encoded by the *talB* gene together with the genes encoding xylose isomerase, xylulokinase, and transketolase allowed *Z. mobilis* to grow on xylose [76]. The metabolically engineered *Z. mobilis* ZM4(pZB5) produced 65–70 g/L ethanol from the mixtures of 75 g/L glucose and 75 g/L xylose [23].

An ethanologic *E. coli* KO11 strain was constructed from *E. coli* W through the integration of *Z. mobilis* pyruvate decarboxylase (*pdc*) and alcohol dehydrogenase (*adhB*) genes into its chromosome, and the disruption of the *frd* gene responsible for succinic acid production. The resulting strain produced 54.4 g/L and 41.6 g/L in media containing 10% glucose and 8% xylose, respectively [45]. The *E. coli adhE, ackA,* and *ldhA* genes were further disrupted together with the *mgsA* gene encoding the enzyme that inhibits sugar metabolism to drive metabolic fluxes towards ethanol formation. This engineered ethanologic *E. coli* strain successfully produced more than 40 g/L ethanol when fermented in media supplemented with mixture of diverse sugars: glucose, xylose, arabinose, mannose, and galactose (2% each sugar) [75]. On the other hand, the high-yield ethanol producer, *E. coli* SZ420, was constructed by eliminating the competing fermentation pathways corresponding to *frdBC, ldh, ackA, folA-pfl* genes and by expressing the pyruvate dehydrogenase complex (*aceEF-lpd,* a typical aerobically-expressed operon) under anaerobic condition. The engineered strain could achieve a 90% ethanol yield from xylose [80]. More recently, an adaptive evolution of *E. coli* KC01 strain (*ldhA pflB ackA frdBC pdhR::pflBp6-aceEF-lpd*) has been performed to achieve higher ethanol yield, and consequently succeeded in 94% ethanol yield from xylose [70, 71].

Sensitivity of *E. coli* to ethanol and furfural is one of the hurdles in producing ethanol with high efficiency, especially from cellulosic biomass. Furfural is produced as a by-product during the acid hydrolysis of cellulosic biomass. A furfural-resistant mutant of ethanologenic *E. coli* EMFR9 was isolated, and comparative transcriptome study between the mutant strain and its parental strain was performed to identify genetic differences [41]. The silenced response of the *yqhD* and *dkgA* genes encoding for enzymes having activities of NADPH-dependent furfural reductase led to increased furfural tolerance. Meanwhile, a genetic mutant of *E. coli* tolerant to ethanol has also been isolated, and their distinctive genetic alterations were identified through transcriptomic analysis [17]. The tolerant strain showed silence of the *fnr* gene, increased expression of the *gcv* gene improving glycine metabolism, increased expression of the *betIBA* genes involved in betaine production, and increased expression of the *sdaC* and *sdaB* genes involved in serine uptake and deamination. These results provided hints in designing culture conditions for better ethanol tolerance, and consequently, supplementing glycine, betaine, and serine enhanced ethanol tolerance. The identified genes and their roles would be promising targets for developing a more efficient ethanol producer. It is expected that combined gene regulatory engineering and metabolic engineering will be performed to tackle multiple problems of substrate utilization, flux optimization, and substrate utilization, which will lead to the development of engineered super ethanol producer.

5.3.2 Butanol and Isobutanol

Butanol possesses similar characteristics to gasoline to be used directly in conventional gasoline engines without modification and substitution of existing equipment. Butanol is superior to ethanol as a fuel additive in many regards: higher energy content, lower volatility, less hygroscopy (thus, butanol does not pick up water), and lower corrosiveness. For example, a gasoline car with an unmodified engine was filled with 100% butanol and successfully ran almost 10,000 miles across the USA in 2005. This result clearly demonstrates that biobutanol would be an attractive alternative liquid fuel. Thus, the development of a sustainable platform for the efficient production of butanol is in great demand and of absolute necessity.

In nature, several strains from the genus *Clostridium* can produce butanol (Fig. 5.6a), and can utilize a variety of substrates including pentoses, hexoses, and cellulose. Recently, *E. coli* was metabolically engineered to produce butanol or isobutanol.

The biphasic fermentation is one of characteristics of butanol-producing *Clostridia*. During the first phase (acidogenic phase), weak acids including acetate and butyrate are produced as major products, while ATPs are generated. During the second phase (solventogenic phase), acids are reassimilated and used as precursors for the production of solvents. During the solventogenic phase, by-products including acetone and ethanol are also produced along with butanol. To characterize strains lacking acids forming pathway, the *buk* (encoding butyrate kinase in the butyrate forming pathway) and *pta* (encoding phosphotransacetylase in the acetate forming pathway) genes were inactivated. The *buk*-disrupted *C. acetobutylicum* PJC4BK strain showed increased butanol production by 10% compared with the parental *C. acetobutylicum* ATCC 824 strain [19]. After the optimization of culture condition, the PJC4BK strain was able to produce higher concentration of butanol (16.7 g/L) with some byproducts (4.4 g/L acetone and 2.6 g/L ethanol) [20]. However, the *pta* disruption in *C. acetobutylicum* did not result in redirection of carbon flow from acid formation to solvent formation. In the butanol production industry, butanol selectivity to total solvent has been considered important to reduce the recovery cost. For the sake of the increased butanol selectivity, *C. acetobutylicum* M5, a strain lacking the pSOL1 megaplasmid containing the genes responsible for acetone, butanol, and ethanol production, was metabolically engineered by introducing the *adhE1* gene (aldehyde/alcohol dehydrogenase) and the *ctfAB* gene (CoA transferase). The engineered strain was able to produce 11.4 g/L butanol with a high selectivity to total solvents (0.84), which was much better than that (10.9 g/L butanol with a selectivity of 0.57) obtained with the wild-type ATCC 824 strain [38]. Despite the high butanol selectivity of the engineered strain, it also produced high concentration of acetate. An *in silico* flux balance analysis was performed to investigate the tendency of metabolic fluxes in a qualitative way. The results revealed that the high accumulation of acetate was due to the cell's effort to compensate ATP production. The systematic analysis gives a clue for further engineering of *C. acetobutylicum* for enhanced butanol production.

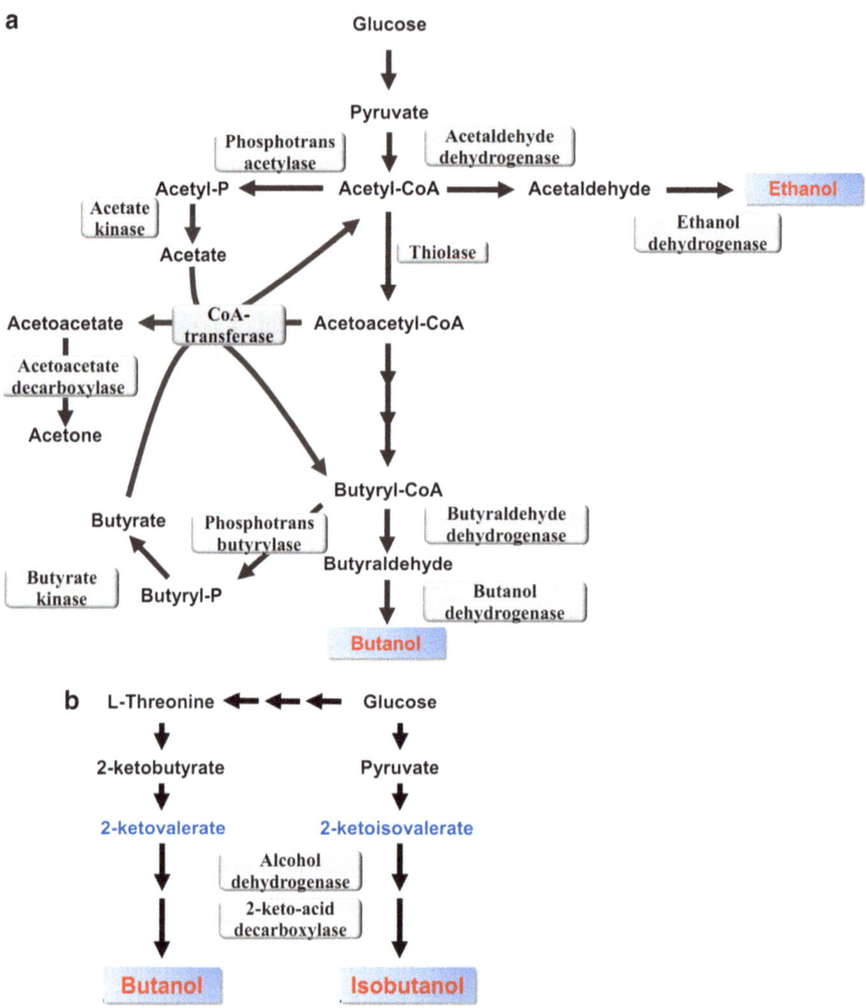

Fig. 5.6 Biosynthetic pathway for butanol synthesis in *C. acetobutylicum* and a metabolically engineered pathway in *E. coli* for butanol production. (**a**) A metabolic pathway for the production of acetone, butanol and ethanol from glucose in *C. acetobutylicum* is shown. (**b**) A synthetic pathway producing isobutanol and butanol from glucose in *E. coli* is shown

The difficulties in genetic engineering of *Clostridia* have forced to develop butanol producing microbial factories based on well-studied model organisms, such as *E. coli*. The whole butanol biosynthetic pathway genes responsible for butanol formation in *C. acetobutylicum*, were introduced into *E. coli*, which allowed production of small amounts (139 mg/L) of butanol. Butanol production by *E. coli* could then be increased up to 1.2 g/L by employing a different alcohol dehydrogenase gene (*adhE1* or *adhE*) instead of *adhE2* [6]. Other than the *Clostridial* pathway for butanol production, the 2-keto acid pathway was also used for the production of butanol and isobutanol in *E. coli* by overexpressing the 2-keto acid decarboxylase and alcohol dehydrogenase (Fig. 5.6b) [6]. To increase the flux toward 2-ketovalerate, the *ilvA-leuABCD* genes were overexpressed. Threonine dehydratase encoded by *ilvA* catalyzes the reaction of L-threonine to 2-ketobutyrate, while the enzymes encoded by the *leuABCD* genes catalyze the conversion of 2-ketobutyrate to 2-ketovalerate. To improve butanol production further, the *ilvD* gene was additionally disrupted. This resulted in the production of about 9 mM butanol in the medium supplemented with 8 g/L L-threonine. To produce isobutanol, the *ilvIHCD* genes were overexpressed, which enhanced 2-ketoisovalerate biosynthesis, and the alcohol producing pathway (2-keto acid decarboxylases and alcohol dehydrogenases) was introduced. Furthermore, in order to further increase isobutanol production, the *adhE*, *ldhA*, *frdAB*, *fnr*, *pta*, and *pflB* genes were knocked out while the *alsS* gene from *Bacillus subtilis* was introduced. This systematic approach led to the production of 22 g/L isobutanol [6].

More recently, two different synthetic pathways producing butanol were also constructed in *E. coli* by assembling heterologous genes. One of the synthetic pathways was composed of the *phaA* (*R. eutropha*), *hbd*, *crt*, and *adhE2* (*C. acetobutylicum*), and *ter* encoding trans-enoyl-CoA reductase (*Treponema denticola*) [11]. The *E. coli* strain containing the pathway was further engineered to provide reducing equivalents by overexpressing the native *aceEF-lpd* operon. Consequently, a maximum of 4.7 g/L of butanol could be produced by the metabolically engineered *E. coli* in the 250-mL flask culture sealed with parafilm [11]. Another pathway was composed of the *hbd*, *crt* and *adhE2* (*C. acetobutylicum*), *ter* (*T. denticola*), and *atoB* encoding thiolase (*E. coli*) [61]. In the *E. coli* containing the second pathway, *ldh*, *adh*, *frd*, and *pta* genes were knocked out, while *fdh* gene originated from *Candida boidinii* was overexpressed [61]. Anaerobic culture of the engineered *E. coli* strain resulted in 15–30 g/L of butanol from fed-batch culture [61].

Clearly, butanol producing strain needs to be optimized further so that the concentration, productivity, and yield of butanol together with solvent tolerance can be increased.

5.3.3 Biodiesel

Another potential alternative biofuel is biodiesel produced from vegetable oil or animal fat by trans-esterification of tri-acylglycerides [15], and biodiesel has quite similar chemical structures with fossil-derived diesel. Owing to structural similarity with petroleum-derived diesels, biodiesel is gaining increasing attraction as an alternative biofuel. Biodiesel can reduce emission of carbon mono-oxide and sulfur during combustion, and maintain environmental carbon balance by recycling carbon dioxide in case of producing from biomass. In Europe, commercial use of biodiesel has greatly increased and is further increasing [77].

Biodiesel consists mostly of fatty acid methyl esters (FAMEs), propyl esters, and ethyl esters. Biodiesel can be produced by engineering of the fatty acid β-oxidation

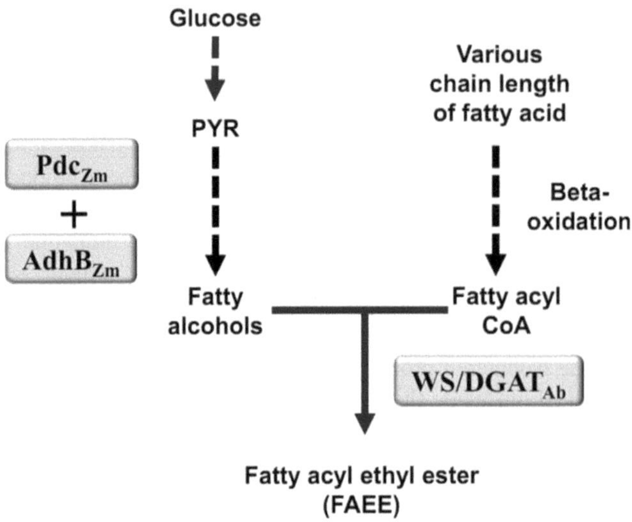

Fig. 5.7 Metabolically engineered pathway for biodiesel production. Biodiesel is mainly composed of fatty acid alcohol esters, and metabolic reconstruction of the pathways towards its precursors (fatty alcohol and fatty acyl CoA), and finally towards fatty acyl ethyl esters has been attracting much attention. *PYR* pyruvate, *Pdc* pyruvate decarboxylase, *Adh* alcohol dehydrogenase, *WS/DGAT* acyltransferase

pathway of microorganisms (Fig. 5.7). So far, biodiesel was mainly produced by *in vitro* transesterification of triacylglycerol to fatty acid alkyl esters (alkali or acid-catalyzed transesterification) of fatty acids, vegetable oil or animal fat with short chain alcohols. However, the high sensitivity of the transesterification process to the contents of free fatty acid leading to unfavorable saponification is the main cause of low biodiesel production yield and requires complicated separation steps. Moreover, the process has several difficulties in waste treatment, as well as corrosion of the equipment. Therefore, environment-friendly biocatalytic process using lipase has been gaining great interest. More than 80% (mol/mol) of ethyl ester was produced from palmitic acid and industrial palm fatty acid distillate (PFAD) by using cross-linked protein coated microcrystalline lipase from *Thermomyces lanuginosus* in the presence of ethanol [54]. The microcrystalline lipase has high stability and no significant loss in production yield, and therefore is an economical biocatalyst for biodiesel production from various biomass feedstock with high contents of free fatty acids.

Different from *in vitro* biocatalytic esterification using lipase, biodiesel production through metabolic engineering has been also reported. *E. coli* has been metabolically engineered to produce biodiesel by the heterologous expression of pyruvate decarboxylase and alcohol dehydrogenase originated from *Z. mobilis*, and acyltransferase originated from *Acinetobacter baylyi* strain ADP1 (Fig. 5.7). The engineered bacteria trans-esterified ethanol with fatty acyl-CoA, and produced 1.3 g/L of biodiesel by fed-batch cultivation [26].

E. coli has been metabolically engineered to utilize plant biomass as carbon source to produce biodiesel [65]. The inherent metabolic pathway of the host strain was engineered to accumulate fatty acyl CoA, and then the ethanol pathway with WS/DGAT was also introduced to convert cellular fatty acyl CoA to fatty acyl ethyl ester (FAEEs). This engineered *E. coli* strain produced 11.6 mg/L of biodiesel from 2% xylan.

Biodiesel is a promising alternative and advanced fuel for the global fossil fuel crisis, because of its relatively lower carbon dioxide emission compared to petroleum diesel and the possibility of production from biomass. Therefore, metabolic engineering of microorganisms has been considered as one of key technologies to cope with world energy crisis.

5.3.4 Other Biofuels

Recently, hydrogen has been considered as an alternative fuel due to its high energy density and sustainability. Hydrogen is produced by photosynthetic microorganisms from water using light as energy source, but at rather low efficiency. In photosynthetic algae and cyanobacteria either hydrogenase or nitrogenase catalyzes hydrogen production [8, 16]. *Chlamydomonas reinhardtii* Stm6 isolated by random screening with the goal of increasing H^+ and e^- supply to the hydrogenase is able to produce hydrogen at five times higher yield than its parent strain [31]. Metabolic engineering studies on the Smt6 strain were performed to understand its internal cellular activities and thereby to engineer to produce more hydrogen through enhanced water photolysis and light capture efficiency. The metabolically engineered *C. reinhardtii* Smt6 strain resulted in an increased growth rate by 20–30% at high light (800 mE), and followed by 1.5 times higher hydrogen production rate compared to its parent strain [7].

Isoprenoid is an organic compound composed of two or more building units of hydrocarbons. As isoprenoids directly converted to cyclic alkenes or branched-chain hydrocarbons by terpene synthase or alcohols by pyrophosphatase, isoprenoid is also considered as an important precursor for biofuel synthesis. Isoprenoids are synthesized from isoprenyl pyrophosphate and dimethylallyl pyrophosphate via several enzymatic steps, which are synthesized through mevalonate pathway and 1-deoxy-D-xylulose 5-phosphate pathway, respectively [28, 42, 50, 78]. The chemical binding of isoprenyl pyrophosphate and dimethylallyl pyrophosphate produces geranyl pyrophosphate of C10, followed by farnesyl pyrophosphate of C15, and geranylgeranyl pyrophosphate of C20 by the reaction of the prenyltransferases. Terpene synthases convert isoprenoids to cyclic alkenes or branched chains, while pyrophosphatase hydrolyzes isoprenoids to alcohols. *E. coli* has been engineered to enhance the production of farnesol, which is considered as a fuel alcohol [70, 71]. Introduction of the foreign mevalonate pathway and overexpression of native farnesyl pyrophosphate synthase in *E. coli* resulted in the production of 135.5 mg/L farnesol. Recently, a variety of metabolic engineering has been attempted to overproduce isoprenoids in different organisms, which have been reviewed in the recent review papers [28, 42]. These isoprenoids are very important source for biofuel production, because the metabolic products, including monoterpenes and diterpenes (cyclic alkenes), sesquiterpenes (branched alkanes) and farnesol, geraniol, and isopentenol (alcohols) have been proposed as gasoline substitutes.

5.4 Pharmaceuticals

5.4.1 Amino Acids

Systems metabolic engineering has been successfully applied to the development of strains for amino acids production [48]. Amino acids are important cellular components and are also key precursors to valuable chemicals and materials. The world market size for amino acids is 4.5 billon dollars in 2004. Among them, L-valine is particularly outstanding in the growing market as it is used as a component of pharmaceuticals and cosmetics as well as animal feed additives. Currently, more than 500 tons of L-valine is produced annually by fermentation [22]. L-threonine is one of the three major amino acids being produced by microbial fermentation. All amino acid producing strains have been developed by repeated random mutation and selection processes. Due to the unknown metabolic and regulatory characteristics of these randomly mutated strains, it is very difficult to further improve their performance even though various omics techniques and computational tools are now available. Thus, a new strategy for the development of genetically well-defined strains, which can overcome the disadvantages of the classical method of strain development by random mutagenesis, is urgently needed. In this section, examples of systematically engineering microorganisms for enhanced production of useful amino acids taking L-valine and L-threonine are presented.

5.4.1.1 L-Valine

An *E. coli* strain overproducing L-valine was constructed by genome engineering combined with transcriptome profiling and gene knockout simulation of an *in silico* genome-scale metabolic network [47]. The systems metabolic engineering strategy for the stepwise improvement of L-valine producing strain is described in Fig. 5.8. First, all known regulations that negatively affect L-valine production were removed by site-specific genome engineering. Feedback inhibition in an enzyme encoded by *ilvH* was removed by changing two nucelotides in the chromosome. Transcriptional attenuation regulations in *ilvGMEDA* and *ilvBN* operons were also removed by replacing the native attenuator leader region with the strong *tac* promoter in the chromosome. Next, pathways that compete with L-valine biosynthesis were blocked by knocking out the *ilvA*, *leuA*, and *panB* genes. Combined transcriptome profiling

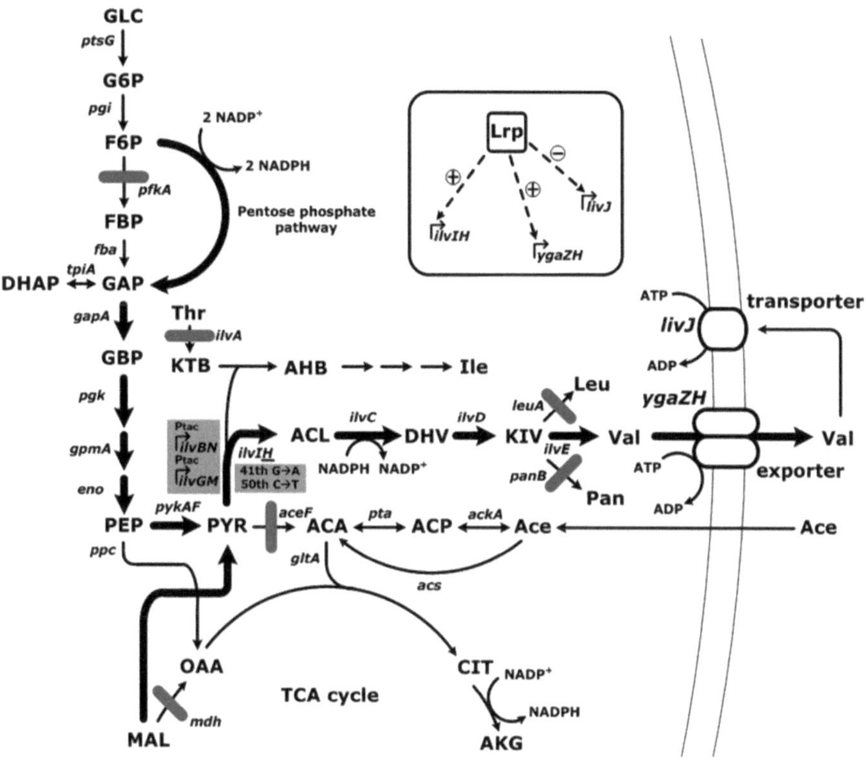

Fig. 5.8 A representative example of systems metabolic engineering for L-valine production in *E. coli* (Figure modified from [49]). The shaded boxes represent genetic modifications, and the grey bars indicate genes that were deleted. The thick arrows indicate increased flux by directly overexpressing the corresponding genes or by knocking out the target genes. The inlet box of Lrp represents the regulatory circuit of Lrp. The dashed lines with arrow head indicate the increased activity of Lrp by directly overexpressing *lrp* gene. The plus (+) and minus (−) symbols indicate activation and repression of gene expression by Lrp, respectively. Abbreviations are: *GLC* glucose, *G6P* glucose 6-phosphate, *F6P* fructose 6-phosphate, *FBP* fructose 1, 6-bisphosphate, *GAP* glyceraldehyde 3-phosphate, *DHAP* dihydroxyacetonephosphate, *GBP* 1,3-bisphosphoglycerate, *PEP* phosphoenolpyruvate, *PYR* pyruvate, *ACA* acetyl-CoA, *ACP* acetyl-phosphate, *Ace* acetate, *ACL* 2-acetolactate, *DHV* 2,3-dihydroxyisovalerate, *KIV* 2-ketoisovalerate, *Val* L-valine, *Leu* L-leucine, *Pan* pantothenate, *Thr* L-threonine, *KTB* 2-ketobutyrate, *AHB* 2-aceto-2-hydroxybutyrate, *Ile* L-isoleucine, *CIT* citrate, *AKG* α-ketoglutarate, *MAL* malate, *OAA* oxaloacetate

and *in silico* gene knockout simulation were conducted for further strain improvement. L-valine biosynthetic *ilvCED* genes, global regulator Lrp, and L-valine exporter YgaZH were amplified, and found to be beneficial for the enhanced production of L-valine. Furthermore, the effect of Lrp and YgaZH was synergistic due to the novel function of Lrp as an activator of *ygaZH* expression. Finally, triple knockout mutations (Δ*aceF* Δ*mdh* Δ*pfkA*) suggested by *in silico* gene knockout simulation were introduced into the strain, resulting in a more than twofold increase in L-valine production. The final engineered strain was able to produce L-valine with an impressively high yield of 0.38 g L-valine per g glucose. These results suggest that an industrially competitive strain can be successfully constructed by systems metabolic engineering based on combined genome-scale metabolic engineering, transcriptome profiling, and systems-level *in silico* gene knockout simulation.

Most recently, fed-batch fermentation process of this engineered *E. coli* strain was developed for the production of high level L-valine based on the systems-level *in silico* flux response analysis [49]. Before developing the fed-batch culture of the strain having high L-valine yield, further increase in L-valine production was obtained by using a mutated *ilvN* encoding a feedback-resistant acetohydroxy acid synthase (AHAS) I. The result of *in silico* flux response analysis revealed that ATP plays an important role in L-valine production. Furthermore, it was in good accordance with the experimental proof that ATP is one limiting factor for L-valine production. For the cultivation of the high-yield L-valine production strain, in which the *aceF* gene (encoding pyruvate dehydrogenase) was deleted, acetic acid should be fed to maintain reasonable growth. During fed-batch fermentation of the high-yield L-valine production strain, an optimal level of external acetic acid feeding was examined by *in silico* flux response analysis for the efficient production of L-valine. By adjusting the acetic acid feeding strategy, the L-valine concentration obtained was dramatically increased up to 32.3 g/L, which is the highest concentration ever reported for *E. coli*. This study showed that fed-batch fermentation process of *E. coli* could be optimized by systems-level *in silico* flux response analysis.

5.4.1.2 L-Threonine

Understanding the metabolic characteristics of strains developed through classical random mutagenesis and selection can often provide good information for further strain improvement. The transcriptome, proteome, and nucleotide sequences between a parental *E. coli* W3110 strain and a classically constructed L-threonine producer strain of *E. coli* TF5015 were compared to understand regulatory mechanisms in L-threonine biosynthesis and the physiological changes in the classical L-threonine production strain [35]. The comparison revealed that genes involved in the glyoxylate shunt, TCA cycle, and amino acid biosynthesis, were significantly upregulated, whereas ribosomal protein genes were downregulated. The most notable finding is that two significant mutations in *thrA* (encoding aspartate kinase I) and *ilvA* (threonine dehydratase) genes were identified essential for overproduction of L-threonine. It is obvious that these mutations can be also beneficial for the enhanced production of L-threonine in rationally engineered strain.

Another good example of systematically engineering bacterial strains for the enhanced production of L-threonine was performed based on genome-scale metabolic engineering combined with transcriptome analysis and *in silico* flux response analysis [37]. The detailed strategy for strain development is described in Fig. 5.9. First, a L-threonine production base strain was constructed by removing

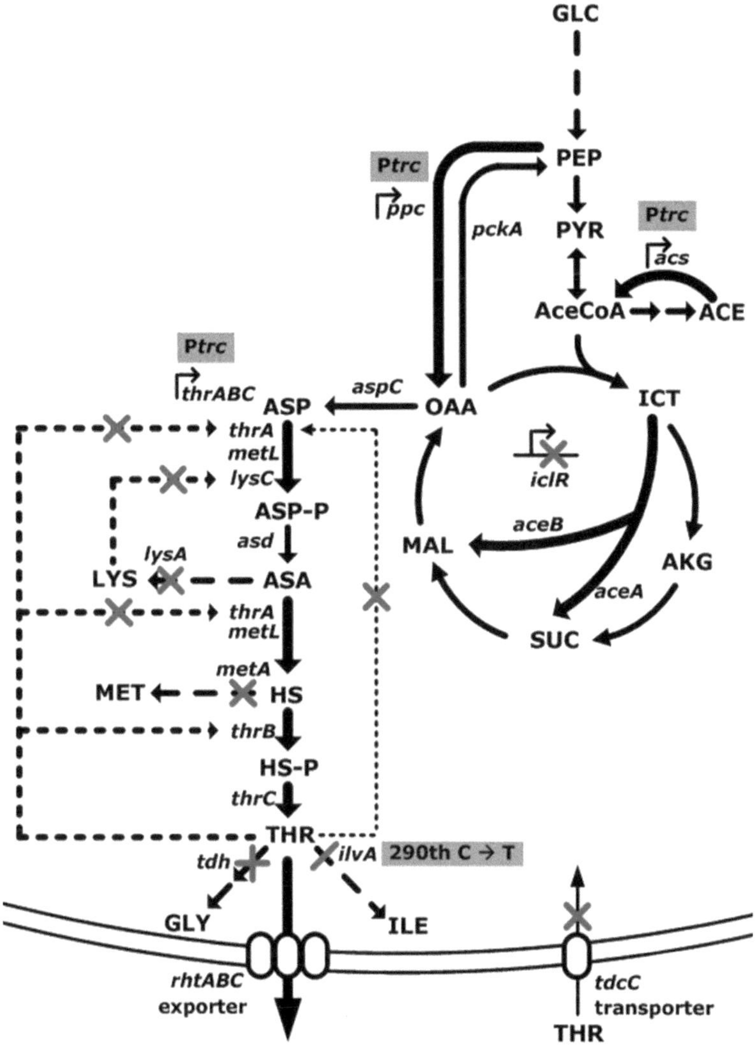

Fig. 5.9 A representative example of systems metabolic engineering for L-threonine production in *E. coli*. *Dotted lines* denote feedback inhibition. The *thin dotted line* denotes transcriptional attenuation regulation. The *gray* symbol X indicates that the corresponding genes are knocked out or their inhibition/repression is removed. The *shaded boxes* denote targeted mutations introduced into the chromosome. The *thick arrows* indicate an increased flux or activity by amplifying corresponding genes through plasmid-based overexpression or chromosomal promoter replacement. The *gray bar* indicates that a point mutation was introduced into the *ilvA* gene to decrease the activity of threonine dehydratase. Genes are given in italics. *GLC* glucose, *PEP* phosphoenolpyruvate, *PYR* pyruvate, *AceCoA* acetyl-CoA, *ACE* acetic acid, *ICT* isocitrate, *AKG* α-ketoglutarate, *SUC* succinate, *MAL* malate, *OAA* oxaloacetate, *ASP* L-aspartate, *ASP-P* Aspartyl phosphate, *ASA* aspartate semialdehyde, *HS* homoserine, *HS-P* homoserine phosphate, *THR* L-threonine, *ILE* L-isoleucine, *LYS* L-lysine, *MET* L-methionine, *GLY* glycine

all negative regulations including feedback inhibition and transcriptional attenuation regulation. Then a competing pathway that negatively affects L-threonine production was also removed. Second, more target genes were identified by transcriptome profiling, and their desirable expression levels were examined by *in silico* flux response analysis. The two anaplerotic enzymes, phosphoenolpyruvate carboxylase (PPC) and isocitrate lyase (ICL) were selected as targets for overexpression from transcriptome data, and their desirable expression levels were decided by *in silico* flux response analysis. The L-threonine production increased by 51.4% with optimized co-amplification of PPC and ICL. Further improvement was obtained through transporter and exporter engineering, and fed-batch fermentation process was also optimized to reduce acetate production with the aid of *in silico* flux response analysis. The final engineered *E. coli* strain was able to produce L-threonine with a high yield of 0.393 g L-threonine per g glucose, and 82.4 g/L L-threonine by fed-batch culture. The values obtained in this work are almost comparable to those typically obtained from industrial strains. This is the first successful example that fine-control of gene expression levels identified by transcriptome profiling based on the *in silico* genome-scale flux response analysis is crucial for enhanced production yield.

More recently, a reduced-genome strain of *E. coli* (MDS42), constructed through sequential deletion of nonessential genes, has also been used for L-threonine production [39]. The MDS42 strain, a reduced-genome strain of *E. coli* [52], showing improved genetic stability and robust metabolic performance, was engineered to produce L-threonine. First, the *thrA*BC* genes encoding feedback-resistant aspartate kinase I, homoserine kinase and threonine synthase, respectively, were overexpressed. The pathway that negatively affected L-threonine formation was removed by deleting the *tdh* (encoding L-threonine dehydrogenase) gene. Then, *tdcC* and *sstT* genes encoding L-threonine transporters were deleted to prevent re-uptaking extracellular L-threonine into the cell. Finally, a mutant L-threonine exporter (*rhtA23*) was introduced into the MDS42. The resulting MDS-205 strain was able to produce 40.1 g/L L-threonine, which is 83% higher than that obtained from a wild-type *E. coli* strain MG1655 in which the same L-threonine-specific modifications were introduced (MG-105). Increased L-threonine production in MDS42 probably resulted from a decrease in the metabolic burden due to the genome reduction. The result obtained in this report clearly demonstrates that a reduced-genome strain of *E. coli* can be a good platform for the efficient production of L-threonine and other useful bioproducts.

5.4.2 Isoprenoids

Isoprenoids form a large family of natural products that play a wide variety of roles in plant and animal physiology, and contribute to the biological synthesis of diverse key metabolites. Many isoprenoids have great values because of its diverse use as flavors, fragrances, food colorants, antioxidants, steroids, natural polymers or drugs. Isoprenoids are extracted from plants that are main producers, but the method is neither economical nor eco-friendly due to the low extraction yield from plants. Moreover, the complex structure of isoprenoids also obscures from their *de novo* chemical synthesis. Alternatively, microbial production of isoprenoids has been considered to be a promising method that may produce a large amount of commercial isoprenoids in an efficient and sustainable way.

Isoprenoids are classified into groups according to the number of carbons contained within: monoterpenes (C10), sesquiterpenes (C15), diterpenes (C20), and triterpenes (C30). All isoprenoids are synthesized from two common precursors, isopentenyl pyrophosphate (IPP, also known as isopentenyl diphosphate) and dimethylallyl pyrophosphate (DMAPP, also known as dimethylallyl diphosphate). These compounds are mainly derived through methylerythritol phosphate (MEP) pathways in eubacteria.

Systems metabolic engineering in combination with classical random mutagenesis has been employed to develop bacterial strains for isoprenoid production. Among diverse isoprenoid superfamily compounds, the approach has been applied mostly to carotenoid groups due to their colorimetric properties, which permits efficient screening by color without other additional assay processes. One of good examples of microbial isoprenoids production using systems-level metabolic engineering is the strain development of lycopene over-producing *E. coli* (Fig. 5.10) [3]. Lycopene is a bright red carotenoid pigment responsible for the red color of tomatoes, carrots, and so on, and has been considered as a potent anticancer agent for prevention of several types of cancers. System-level metabolic pathway analysis, specifically, a genome-wide stoichiometric flux balance analysis (FBA), has identified many target genes located in the MEP pathway and even those outside of the pathway for lycopene overproduction while maintaining an acceptable growth rate (Fig. 5.10). The identified target genes were *gdhA*, *gpmA*/*gpmB*, *aceE*, *fdhE*, and *talB*. The effects of single knock-outs were computationally evaluated, and consequently *gdhA* was predicted to produce more increased lycopene than other target genes. Based on the *gdhA* knocked out strain, *in silico* simulation was further performed to identify the next target gene. Sequential virtual process finally predicted that the best multiple knockout targets are *gdhA*, *aceE*, and *fdhF*. As predicted, the real implementation in *E. coli* enhanced lycopene production by 40% over that of its parental strain.

A random approach has been also developed to enhance the lycopene production [2]. The protein sequence of a sigma factor (σ^{70}), mediating transcription by recruiting RNA polymerase, was randomly mutated to induce chromosomal gene transcription randomly. Various sigma factor mutants enhancing lycopene production were screened, and interestingly a single step of this sigma factor-based engineering outperformed that of single gene knock-out. It would be because that the screened sigma factor mutants controlled unidentified target genes globally.

There have been great advances in tools for developing improved microbial strains for isoprenoid production, but expression of P450 in an active form, that catalyzes the reaction of initial isoprenoid products, is one of the challenges. Most bacteria including *E. coli* do not possess any native P450 and therefore

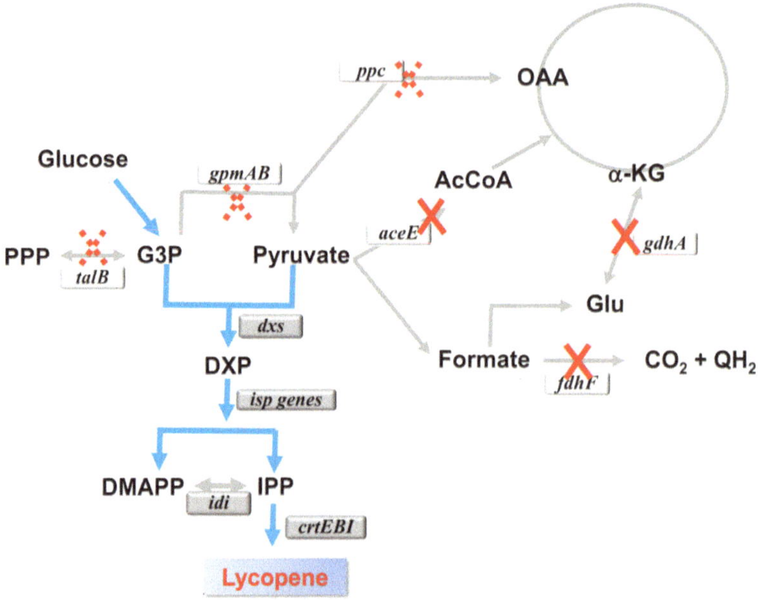

Fig. 5.10 Lycopene production pathway and virtually identified knockout targets in *E. coli*. Lycopene synthesis in *E. coli* begins with the condensation reaction of glycolytic intermediates, glyceraldehyde 3-P (G3P) and pyruvate, and continues through MEP pathway. For lycopene production in *E. coli*, usually *crtEBI* operon was heterologously expressed and *idi*, *ispFD*, and *dxs* are overexpressed. The virtually identified knockout targets are shown as *X marks*. Among the targets, the knockout combination of *gdhA*, *aceE*, and *fdhF* (*solid X marks*) resulted in the best yield. Abbreviations are: *OAA* oxaloacetate, *AcCoA* acetyl coenzyme A, *αKG* alpha-ketoglutarate, *Glu* glutamate, *DXP* 2-C-Methyl-D-erythritol 4-phosphate, *IPP* isopentenyl pyrophosphate, *DMAPP* dimethyl allyl pyrophosphate

expression of P450 in *E. coli* is one of the critical issues. Bacterial P450 as a substitute of plant P450, or chimeras of bacterial and plant P450 have been studied as an alternative to express soluble enzymes with activity in microbial hosts such as *E. coli* [9, 62]. However, development of methods for the expression of initial isoprenoids-modifying enzymes in microbial hosts remains an important task for the complete biosynthesis of isoprenoids in microbial cell.

5.4.3 *Artemisinin*

Recently, synthetic biology is making it possible to reconstruct synthetic metabolic pathways that do not exist in nature. Thus, novel pathways producing non-innate metabolites that cannot be produced in wild-type strains can be designed and constructed. Such approach has been successfully applied to produce therapeutic compounds in *E. coli*. Artemisinin is an effective drug against multi-drug-resistant and cerebral malaria-causing strains of *Plasmodium falciparum*. Annually 100 million people are infected with malaria and 1 million of them die of the disease. Despite of the efficacy of artemisinin, the relatively low yield (0.01–0.8%) of artemisinin in *Artemisia annua* L (sweet wormwood) limits its massive production and wide application to control malaria. In order to stabilize the supply of artemisinin, alternative methods have been developed to produce artemisin intermediates: artemisinic acid in *S. cerevisiae* and amorpha-4,11-diene in *E. coli* (Fig. 5.11) [57, 69].

S. cerevisiae is inherently incapable of producing artemisinic acid, and therefore was engineered by employing a synthetic metabolic pathway to convert an intracellular metabolite, farnesyl pyrophosphate (FPP), to artemisinic acid (Fig. 5.11) [57]. The synthetic pathway was constructed by incorporating amorphadiene synthase (ADS) and cytochrome P450 (CYP71AV1/CPR). ADS cloned from *A. annua* converts FPP to amorpha-4-11-diene, and cytochrome P450 cloned from *A. annua* as well converts amorpha-4,11-diene to artemisinic acid *via* three enzymatic steps. NADPH:cytochrome P450 oxidoreductase (CPR) was also cloned from *A. annua* to ensure the function of CYP71AV1. In addition to the synthetic pathway, the internal metabolic pathway of *S. cerevisiae* was also engineered to enhance the intracellular level of FPP, a precursor for amorpha-4,11-diene. To increase FPP level, genes involved in the mevalonate pathway were cloned and amplified: truncated form of 3-hydroxy-3-methyl-glutaryl-coenzyme A reductase (tHMGR), FPP synthase (*erg20*), and *upc2-1*. Theupc2-1 is a regulator of the mevalonate pathway. Thereby, its amplification eventually increases the expression of its target genes (*erg13*, *erg12*, and *erg8*) and as a result indirectly enhances the flux towards FPP. On the other hand, flux towards a competing pathway consuming FPP was reduced by down-regulating *erg9* promoter. The engineered *S. cerevisiae* produced a high titre up to 100 mg/L of artemisinic acid.

E. coli (W3110) is also incapable of producing artemisinin precursors, but it was engineered to produce another artemisinin precursor (amorpha-4,11-diene) through the reconstruction of the mevalonate pathway and the optimization of cultivation conditions [69]. The *erg12*, *erg8*, and *erg19* were cloned from *S. cerevisiae* to reconstruct the mevalonate pathway, and ADS was cloned from *A. annua* to convert FPP to amorpha-4,11-diene. As the native enzymes of *S. cerevisiae* in the mevalonate pathway, HMG-CoA synthase (HMGS, *erg13*) and HMG-CoA reductase (tHMGR), were reported to possess low activity and thereby to be pathway bottlenecks, the two enzymes were replaced with more active ones from *Staphylococcus aureus*. The engineered strain was cultivated in a glucose-restricted fed-batch with ammonia maintained between 30 and 60 mM. After 20 h cultivation, the cells were induced by IPTG to activate the reconstructed synthetic pathway. The process achieved up to 27.4 g/L amorpha-4,11-diene. Through the use of the synthetic metabolic pathways, *S. cerevisiae* and *E. coli* could be engineered to produce non-innate artemisinic acid and amorphadiene, respectively.

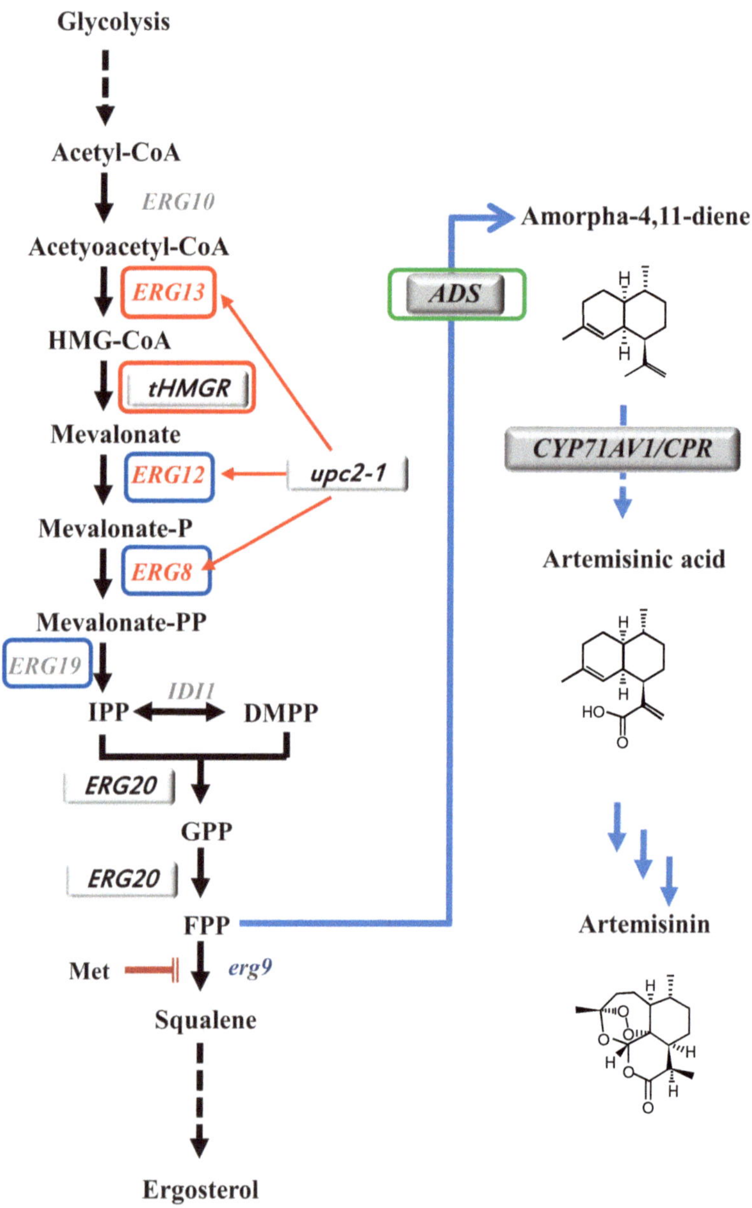

Fig. 5.11 Engineered artemisinic acid biosynthetic pathway. The synthetic artemisinin biosynthetic pathways reconstructed in *S. cerevisiae* and *E. coli* are shown. In the mevalonate pathway, genetic manipulations performed in *S. cerevisiae* are shown. The farnesyl pyrophosphate (FPP) pool was increased through engineering of the FPP biosynthetic pathway and then FPP was converted to the precursor of artemisinin, artemisinic acid, by the synthetic artemisinic acid biosynthetic pathway containing amorphadiene synthase, a novel cytochrome P450, and its redox partner from *Artemisia annua*. Truncated HMGR (tHMGR) and *erg20* were amplified

5.4.4 Taxol

Taxol (paclitaxel) is another example that employed synthetic pathway reconstruction strategy. Taxol was firstly extracted from the bark of Pacific yew tree (*Taxus brevifolia*) in the effort for screening new anti-cancer drugs from numerous samples of plants [18, 72]. Taxol is a mitotic inhibitor interfering with the breakdown of microtubules during cell division and used as a potent anti-cancer drug to treat lung, breast, head and neck cancer. Until 1993, almost all taxol was extracted from Pacific yew trees, and the amount of taxol derived from 2 to 4 fully grown trees was sufficient only for one patient. Although chemical synthesis was attempted, due to the complex structure of taxol the synthesis required 35–51 steps and produced only small amount of taxol with a yield of 0.4%.

In the last decade, to avoid the destructive and costly processes various taxol-producing plant cells were cultured *in vitro* with optimized culture conditions such as osmotic pressure, medium compositions, bioprocessing processes, cultivation methods, feeding carbon sources [79]. In a report, an industrially interesting level of taxol (512 mg/L) was obtained from the culture of *T. chinensis* [13].

A semi-synthetic production of a taxol precursor (taxadiene) has been attempted in plant cells (*Arabidopsis*) [10]. The first step for taxol production was to synthesize a precursor, taxadiene, from geranylgeranyl diphosphate (GGPP) by utilizing taxadiene synthase. GGPP is mostly supplied from the MEP pathway that synthesizes prenyl diphosphate precursors for the production of isoprenoids (Fig. 5.12). The constitutively produced taxadiene synthase consumes GGPP to produce taxadiene. Then the taxadiene was further processed 12 enzymatic reactions to become a taxol. The later process required the cytochrome P450 monooxygenase hydroxylating taxadiene to the next intermediate of the taxol pathway. The imbalance of GGPP supply due to taxadiene synthase resulted in the imbalance of endogenous isoprenoids that are key molecules for plant growth and development. As the expression of taxadiene synthase resulted in dramatic growth retardation, to avoid such interference the synthase gene was redesigned to be expressed under the presence of inducer molecules, a synthetic glucocorticoid dexamethasone (DEX). Transgenic *Arabidopsis* showed a normal growth rate, and treated leaves with DEX produced taxadiene up to 600 ng/g of DCW.

Fig. 5.11 (continued) while *erg8*, *erg12*, and *erg13* were indirectly amplified by the incorporation of a semi-dominant mutant of *upc2-1*. The *erg9* was down-regulated by replacing its promoter. The pathway intermediates IPP, DMAPP and GPP denotes isopentenyl pyrophosphate, dimethyl allyl pyrophosphate and geranyl pyrophosphate, respectively. The synthetic pathway containing ADS, CYP71AV1 and CPR was employed from *A. annua*. In the metabolic engineering of *E. coli*, to reconstruct the mevalonate pathway *erg12*, *erg8* and *erg19* were cloned from *S. cerevisiae*, and HMG-CoA synthase (HMGS, *erg13*) and HMG-CoA reductase (HMGR) were cloned from *S. aureus*. As the Erg13 and tHMGR in *S. cerevisiae* have a low catalytic activity, more active ones were cloned from *Staphylococcus aureus* instead. To convert FPP to amorpha-4,11-diene ADS was cloned from *A. annua*

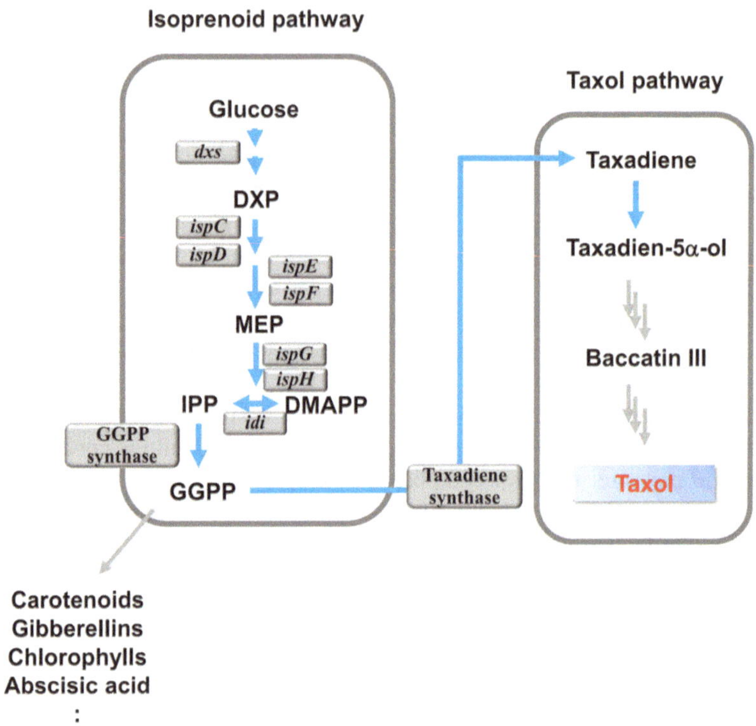

Fig. 5.12 Biosynthetic pathway of taxol. The biosynthetic pathway of taxol is shown. The isoprenoid pathway converts glucose to geranylgeranyl diphosphate (GGPP) which is further converted to diverse isoprenoids. In plants, GGPP is a precursor of several growth-related hormones such as gibberellins, abscisic acid, and etc. GGPP is also converted to a taxol precursor (taxadiene) by taxadiene synthase. Taxadiene can be further converted to taxol *via* several enzymatic or chemical reactions. Abbreviations are: *DXP* deoxy-xylulose-5-phosphate, *MEP* methylerythritol 4-phosphate, *IPP* isopentenyl diphosphate, *DMAPP* dimethylallyl diphosphate, *GGPP* geranylgeranyl diphosphate

Recently, taxadiene has been produced in *E. coli*, with a yield of 1 g/L (~15,000-fold increased yield) through metabolic engineering [1]. In the approach, a metabolic pathway of taxol was reconstructed in *E. coli*. The pathway comprises of the *E. coli*'s native isoprenoid pathway and a heterologous terpenoid pathway. As pathway components are highly interconnected, the outcome of pathway engineering frequently displays a complex non-linearity due to the possibility of the accumulation of toxic intermediates, the existence of competing pathways or metabolites inhibiting the production of a target compound. Therefore, combinatorial approaches examining all search spaces are required to find the parameter set for the maximal production yield. Due to the unavailability of high-throughput screening of taxadiene and its intermediates, the taxol pathway was partitioned into two parts (the native isoprenoid pathway and the heterologous terpenoid pathway) and each partitioned pathway was examined. The gene expression of the enzymes in each part was varied by modifying promoter strength and gene copy number

(Fig. 5.12), and simultaneous examination of the combinations was conducted to search for the best balanced pathway fluxes and to avoid any bottle-necks as well. As a result, it was discovered that high taxadiene accumulation can be obtained from modifications within a narrow range of gene expression levels of enzymes involved in the pathways, and consequently up to 1 g/L taxadiene was obtained from fed-batch cultivation.

Taxol was approved by the FDA for the treatment of several cancers and is the best-selling anti-cancer drug to date. The current metabolic engineering studies have focused on the production of a taxol precursor, taxadiene, in a model organism. The accumulated knowledge on the components in the taxol-producing pathway and advances in synthetic biology tools would potentially facilitate to engineer the whole taxol pathway in *E. coli* for industrial scale production.

5.5 Summary

The current global crises such as climate change and fossil fuel depletion have increased the demand for sustainable production of biofuels and biomaterials. Microorganisms have gained increasing attraction as a platform for the production of bulk and valuable chemicals from renewable sources. *E. coli* provides a perfect platform for metabolic engineering owing to highly advanced genetic engineering technologies for this microorganism, vast amount of accumulated knowledge on its physiology, and developed computational models and tools. Recent great leap in synthetic and systems biology opens a new path to metabolic engineering of *E. coli*: rational reconstruction of metabolic pathways and gene regulatory networks for the production of innate and non-innate chemical compounds (termed systems metabolic engineering). Systems metabolic engineering approach has succeeded in producing many biofuels, biomaterials, and pharmaceuticals in *E. coli* and promises transition from conventional chemical industry to bio-industry.

Our work described in this chapter was supported by the Advanced Biomass R&D Center of Korea (ABC-2010-0029799) through the Global Frontier Project of Ministry of Education, Science and Technology.

References

1. Ajikumar P, Xiao W, Tyo K, Wang Y, Simeon F, Leonard E, Mucha O, Phon T, Pfeifer B, Stephanopoulos G (2010) Isoprenoid pathway optimization for taxol precursor overproduction in *Escherichia coli*. Science 330(6000):70–74
2. Alper H, Stephanopoulos G (2007) Global transcription machinery engineering: a new approach for improving cellular phenotype. Metab Eng 9(3):258–267
3. Alper H, Miyaoku K, Stephanopoulos G (2005) Construction of lycopene-overproducing *E. coli* strains by combining systematic and combinatorial gene knockout targets. Nat Biotechnol 23(5):612–616

4. Altaras NE, Cameron DC (1999) Metabolic engineering of a 1,2-propanediol pathway in *Escherichia coli*. Appl Environ Microbiol 65(3):1180–1185
5. Andersen F (1994) Final report on the safety assessment of propylene glycol and polypropylene glycols. J Am Coll Toxicol 13:437–491
6. Atsumi S, Cann A, Connor M, Shen C, Smith K, Brynildsen M, Chou K, Hanai T, Liao J (2008) Metabolic engineering of *Escherichia coli* for 1-butanol production. Metab Eng 10(6):305–311
7. Beckmann J, Lehr F, Finazzi G, Hankamer B, Posten C, Wobbe L, Kruse O (2009) Improvement of light to biomass conversion by de-regulation of light-harvesting protein translation in *Chlamydomonas reinhardtii*. J Biotechnol 142(1):70
8. Benemann JR, Weare NM (1974) Hydrogen evolution by nitrogen-fixing *Anabaena cylindrica* cultures. Science 184(4133):174
9. Bernhardt R (2006) Cytochromes P450 as versatile biocatalysts. J Biotechnol 124(1):128–145
10. Besumbes O, Sauret Gueto S, Phillips M, Imperial S, Rodriguez Concepcion M, Boronat A (2004) Metabolic engineering of isoprenoid biosynthesis in Arabidopsis for the production of taxadiene, the first committed precursor of taxol. Biotechnol Bioeng 88(2):168
11. Bond-Watts BB, Bellerose RJ, Chang MCY (2011) Enzyme mechanism as a kinetic control element for designing synthetic biofuel pathways. Nat Chem Biol 7:222–227
12. Cervin MA, Soucaille P, Valle F (2008) Process for the biological production of 1,3-propanediol with high yield. United States Patent WO/2004/033646
13. Choi H, Kim S, Son J, Hong S, Lee H, Lee H (2000) Enhancement of paclitaxel production by temperature shift in suspension culture of Taxus chinensis. Enzyme Microb Technol 27(8):593–598
14. Eppelmann K, Nossin PMM, Kremer SM, Wubbolts MG (2006) Biochemical synthesis of 1,4-butanediamine. US Patent 2009/0275093
15. Fortman JL, Chhabra S, Mukhopadhyay A, Chou H, Lee TS, Steen E, Keasling JD (2008) Biofuel alternatives to ethanol: pumping the microbial well. Trends Biotechnol 26(7):375–381
16. Gaffron H, Rubin J (1942) Fermentative and photochemical production of hydrogen in algae. J Gen Physiol 26(2):219
17. Gonzalez R, Tao H, Purvis J, York S, Shanmugam K, Ingram L (2003) Gene array based identification of changes that contribute to ethanol tolerance in ethanologenic *Escherichia coli*: comparison of KO11 (parent) to LY01 (resistant mutant). Biotechnol Prog 19(2):612–623
18. Goodman J, Walsh V (2001) The story of taxol: nature and politics in the pursuit of an anti-cancer drug. Cambridge University Press, New York
19. Green E, Bennett G (1996) Inactivation of an aldehyde/alcohol dehydrogenase gene from *Clostridium acetobutylicum* ATCC 824. Appl Biochem Biotechnol 57(1):213–221
20. Harris L, Desai R, Welker N, Papoutsakis E (2000) Characterization of recombinant strains of the *Clostridium acetobutylicum* butyrate kinase inactivation mutant: need for new phenomenological models for solventogenesis and butanol inhibition? Biotechnol Bioeng 67(1):1–11
21. Hong SH, Park SJ, Moon SY, Park JP, Lee SY (2003) *In silico* prediction and validation of the importance of the Entner-Doudoroff pathway in poly(3-hydroxybutyrate) production by metabolically engineered *Escherichia coli*. Biotechnol Bioeng 83(7):854–863
22. Ikeda M (2003) Amino acid production processes. Adv Biochem Eng Biotechnol 79:1–35
23. Joachimsthal E, Haggett K, Rogers P (1999) Evaluation of recombinant strains of *Zymomonas mobilis* for ethanol production from glucose/xylose media. Appl Biochem Biotechnol 77(1):147–157
24. Jung JY, Choi ES, Oh MK (2008) Enhanced production of 1,2-propanediol by *tpil* deletion in *Saccharomyces cerevisiae*. J Microbiol Biotechnol 18(11):1797–1802
25. Jung YK, Kim TY, Park SJ, Lee SY (2010) Metabolic engineering of *Escherichia coli* for the production of polylactic acid and its copolymers. Biotechnol Bioeng 105(1):161–171
26. Kalscheuer R, Stolting T, Steinbuchel A (2006) Microdiesel: *Escherichia coli* engineered for fuel production. Microbiology 152(9):2529

27. Kind S, Wittmann S (2011) Bio-based production of the platform chemical 1,5-diaminopentane. Appl Microbiol Biotechnol 91(5):1287
28. Kirby J, Keasling JD (2008) Metabolic engineering of microorganisms for isoprenoid production. Nat Prod Rep 25(4):656
29. Kotter P, Ciriacy M (1993) Xylose fermentation by *Saccharomyces cerevisiae*. Appl Microbiol Biotechnol 38(6):776–783
30. Kotter P, Amore R, Hollenberg C, Ciriacy M (1990) Isolation and characterization of the *Pichia stipitis* xylitol dehydrogenase gene, XYL2, and construction of a xylose-utilizing *Saccharomyces cerevisiae* transformant. Curr Genet 18(6):493–500
31. Kruse O, Rupprecht J, Bader KP, Thomas-Hall S, Schenk PM, Finazzi G, Hankamer B (2005) Improved photobiological H_2 production in engineered green algal cells. J Biol Chem 280(40):34170
32. Kuyper M, Hartog M, Toirkens M, Almering M, Winkler A, Dijken J, Pronk J (2005) Metabolic engineering of a xylose isomerase expressing *Saccharomyces cerevisiae* strain for rapid anaerobic xylose fermentation. FEMS Yeast Res 5(4–5):399–409
33. Lawrence AG, Schoenheit J, He AM, Tian JM, Liu PH, Stubbe JA, Sinskey AJ (2005) Transcriptional analysis of *Ralstonia eutropha* genes related to poly-(R)-3-hydroxybutyrate homeostasis during batch fermentation. Appl Microbiol Biotechnol 68(5):663–672
34. Lee SY (1996) Bacterial polyhydroxyalkanoates. Biotechnol Bioeng 49(1):1–14
35. Lee JH, Lee DE, Lee BU, Kim HS (2003) Global analyses of transcriptomes and proteomes of a parent strain and an L-threonine-overproducing mutant strain. J Bacteriol 185(18):5442
36. Lee SY, Lee DY, Kim TY (2005) Systems biotechnology for strain improvement. Trends Biotechnol 23(7):349–358
37. Lee KH, Park JH, Kim TY, Kim HU, Lee SY (2007) Systems metabolic engineering of *Escherichia coli* for L-threonine production. Mol Syst Biol 3:149
38. Lee JY, Jang YS, Lee J, Papoutsakis ET, Lee SY (2009) Metabolic engineering of *Clostridium acetobutylicum* M5 for highly selective butanol production. Biotechnol J 4(10):1432–1440
39. Lee J, Sung B, Kim M, Blattner F, Yoon B, Kim J, Kim S (2009) Metabolic engineering of a reduced-genome strain of *Escherichia coli* for L-threonine production. Microb Cell Fact 8(1):2
40. Mehta R, Kumar V, Bhunia H, Upadhyay SN (2005) Synthesis of poly(lactic acid): a review. J Macromol Sci-Polym Rev C45(4):325–349
41. Miller EN, Jarboe LR, Yomano LP, York SW, Shanmugam KT, Ingram LO (2009) Silencing of NADPH-dependent oxidoreductases (*yqhD* and *dkgA*) in furfural-resistant ethanologenic *Escherichia coli*. Appl Environ Microbiol 75(13):4315–4323
42. Muntendam R, Melillo E, Ryden A, Kayser O (2009) Perspectives and limits of engineering the isoprenoid metabolism in heterologous hosts. Appl Microbiol Biotechnol 84(6):1003
43. Nair LS, Laurencin CT (2007) Biodegradable polymers as biomaterials. Prog Polym Sci 32(8–9):762–798
44. Nakamura CE, Whited GM (2003) Metabolic engineering for the microbial production of 1,3-propanediol. Curr Opin Biotechnol 14(5):454–459
45. Ohta K, Beall D, Mejia J, Shanmugam K, Ingram L (1991) Genetic improvement of *Escherichia coli* for ethanol production: chromosomal integration of *Zymomonas mobilis* genes encoding pyruvate decarboxylase and alcohol dehydrogenase II. Appl Environ Microbiol 57(4):893
46. Omenetto F, Kaplan D (2010) New opportunities for an ancient material. Science 329(5991):528
47. Park JH, Lee KH, Kim TY, Lee SY (2007) Metabolic engineering of *Escherichia coli* for the production of L-valine based on transcriptome analysis and *in silico* gene knockout simulation. Proc Natl Acad Sci USA 104(19):7797–7802
48. Park JH, Lee SY, Kim TY, Kim HU (2008) Application of systems biology for bioprocess development. Trends Biotechnol 26(8):404–412
49. Park JH, Kim TY, Lee KH, Lee SY (2011) Fed batch culture of *Escherichia coli* for L valine production based on *in silico* flux response analysis. Biotechnol Bioeng 108(4):934–946

50. Peralta-Yahya PP, Keasling JD (2010) Advanced biofuel production in microbes. Biotechnol J 5(2):147
51. Philip S, Keshavarz T, Roy I (2007) Polyhydroxyalkanoates: biodegradable polymers with a range of applications. J Chem Technol Biotechnol 82(3):233–247
52. Posfai G, Plunkett G III, Feher T, Frisch D, Keil GM, Umenhoffer K, Kolisnychenko V, Stahl B, Sharma SS, de Arruda M, Burland V, Harcum SW, Blattner FR (2006) Emergent properties of reduced-genome *Escherichia coli*. Science 312(5776):1044–1046
53. Qian Z-G, Xia X-X, Lee SY (2009) Metabolic engineering of *Escherichia coli* for the production of putrescine: a four carbon diamine. Biotechnol Bioeng 104(4):651–662
54. Raita M, Laothanachareon T, Champreda V, Laosiripojana N (2011) Biocatalytic esterification of palm oil fatty acids for biodiesel production using glycine-based cross-linked protein coated microcrystalline lipase. J Mol Catal B Enzym 73(1–4):74–79. doi:10.1016/j.molcatb.2011.07.020
55. Rizzi M, Erlemann P, Bui-Thanh N, Dellweg H (1988) Xylose fermentation by yeasts. Appl Microbiol Biotechnol 29(2):148–154
56. Rizzi M, Harwart K, Erlemann P, Bui-Thanh N, Dellweg H (1989) Purification and properties of the NAD + −xylitol-dehydrogenase from the yeast *Pichia stipitis*. J Ferment Bioeng 67(1):20–24
57. Ro D-K, Paradise EM, Ouellet M, Fisher KJ, Newman KL, Ndungu JM, Ho KA, Eachus RA, Ham TS, Kirby J, Chang MCY, Withers ST, Shiba Y, Sarpong R, Keasling JD (2006) Production of the antimalarial drug precursor artemisinic acid in engineered yeast. Nature 440(7086):940–943
58. Sanders J, Scott E, Weusthuis R, Mooibroek H (2007) Bio-refinery as the bio-inspired process to bulk chemicals. Macromol Biosci 7(2):105–117
59. Schneider J, Wendisch VF (2010) Putrescine production by engineered *Corynebacterium glutamicum*. Appl Microbiol Biotechnol 88(4):859–868
60. Scott E, Peter F, Sanders J (2007) Biomass in the manufacture of industrial products – the use of proteins and amino acids. Appl Microbiol Biotechnol 75(4):751–762
61. Shen CR, Lan EI, Dekishima Y, Baez A, Cho KM, Liao JC (2011) High titer anaerobic 1-butanol synthesis in *Escherichia coli* enabled by driving forces. Appl Environ Microbiol. doi:10.1128/AEM.03034-10
62. Shimoji M, Yin H, Higgins L, Jones JP (1998) Design of a novel P450: a functional bacterial-human cytochrome P450 chimera. Biochemistry 37(25):8848–8852
63. Soucaille P, Voelker F, Figge R (2008) New microorganisms for the production of 1,2-propanediol obtained by a combination of evolution and rational design. WO2008/116852
64. Sponner A, Schlott B, Vollrath F, Unger E, Grosse F, Weisshart K (2005) Characterization of the protein components of *Nephila clavipes* dragline silk. Biochemistry 44(12):4727–4736
65. Steen EJ, Kang Y, Bokinsky G, Hu Z, Schirmer A, McClure A, Cardayre SB, Keasling JD (2009) Microbial production of fatty-acid-derived fuels and chemicals from plant biomass. Nature 463:559
66. Tabor C, Tabor H (1985) Polyamines in microorganisms. Microbiol Mol Biol Rev 49(1):81
67. Tantirungkij M, Nakashima N, Seki T, Yoshida T (1993) Construction of xylose-assimilating *Saccharomyces cerevisiae*. J Ferment Bioeng 75(2):83–88
68. Toivari M, Aristidou A, Ruohonen L, Penttila M (2001) Conversion of xylose to ethanol by recombinant *Saccharomyces cerevisiae*: importance of xylulokinase (XKS1) and oxygen availability. Metab Eng 3(3):236–249
69. Tsuruta H, Paddon CJ, Eng D, Lenihan JR, Horning T, Anthony LC, Regentin R, Keasling JD, Renninger NS, Newman JD (2009) High-level production of amorpha-4,11-diene, a precursor of the antimalarial agent artemisinin, in *Escherichia coli*. PLoS One 4(2):e4489
70. Wang Y, Manow R, Finan C, Wang J, Garza E, Zhou S (2010) Adaptive evolution of nontransgenic *Escherichia coli* KC01 for improved ethanol tolerance and homoethanol fermentation from xylose. J Ind Microbiol Biotechnol. doi:10.1007/s10295-010-0920-5
71. Wang C, Yoon SH, Shah AA, Chung YR, Kim JY, Choi ES, Keasling JD, Kim SW (2010) Farnesol production from *Escherichia coli* by harnessing the exogenous mevalonate pathway. Biotechnol Bioeng 107(3):421

72. Wani M, Taylor H, Wall M, Coggon P, McPhail A (1971) Plant antitumor agents. VI. Isolation and structure of taxol, a novel antileukemic and antitumor agent from Taxus brevifolia. J Am Chem Soc 93(9):2325–2327
73. Xia XX, Qian ZG, Ki CS, Park YH, Kaplan DL, Lee SY (2010) Native-sized recombinant spider silk protein produced in metabolically engineered *Escherichia coli* results in a strong fiber. Proc Natl Acad Sci USA 107(32):14059–14063
74. Yang TH, Kim TW, Kang HO, Lee S-H, Lee EJ, Lim S-C, Oh SO, Song A-J, Park SJ, Lee SY (2010) Biosynthesis of polylactic acid and its copolymers using evolved propionate CoA transferase and PHA synthase. Biotechnol Bioeng 105(1):150–160
75. Yomano L, York S, Shanmugam K, Ingram L (2009) Deletion of methylglyoxal synthase gene (*mgsA*) increased sugar co-metabolism in ethanol-producing *Escherichia coli*. Biotechnol Lett 31(9):1389–1398
76. Zhang M, Eddy C, Deanda K, Finkelstein M, Picataggio S (1995) Metabolic engineering of a pentose metabolism pathway in ethanologenic *Zymomonas mobilis*. Science 267(5195):240
77. Zhang Y, Dube MA, McLean DD, Kates M (2003) Biodiesel production from waste cooking oil: 2. Economic assessment and sensitivity analysis. Bioresour Technol 90(3):229–240
78. Zhang F, Rodriguez S, Keasling JD (2011) Metabolic engineering of microbial pathways for advanced biofuels production. Curr Opin Biotechnol 22:1
79. Zhong J-J (2002) Plant cell culture for production of paclitaxel and other taxanes. J Biosci Bioeng 94(6):591–599
80. Zhou S, Iverson A, Grayburn W (2008) Engineering a native homoethanol pathway in *Escherichia coli* B for ethanol production. Biotechnol Lett 30(2):335–342

Chapter 6
Systems Metabolic Engineering of *Corynebacterium glutamicum* for Biobased Production of Chemicals, Materials and Fuels

Judith Becker, Stefanie Kind, and Christoph Wittmann

Contents

Abstract Systems metabolic engineering integrates systems wide understanding of biological systems with targeted genetic modification towards optimum production performance. Supported by novel powerful tools and technologies from systems biology, strategies for industrial strain engineering evolve more and

J. Becker • S. Kind • C. Wittmann (✉)
Institute of Biochemical Engineering, Technische Universität Braunschweig,
Gaußstrasse 17, 38106 Braunschweig, Germany
e-mail: c.wittmann@tu-braunschweig.de

C. Wittmann and S.Y. Lee (eds.), *Systems Metabolic Engineering*,
DOI 10.1007/978-94-007-4534-6_6,

more from trial and error into knowledge based rational development. For the soil bacterium *Corynebacterium glutamicum*, a working horse in industrial biotechnology since more than 50 years, this opens a broad avenue to create and shape a versatile cell factory with superior properties for many purposes. As reviewed in this chapter, applications of systems metabolic engineering to *C. glutamicum* deeply open the new era of sustainable bio-economy with various chemicals, materials and fuels obtained from renewable feed stocks. Additionally, a first example from lysine production gives a flavor on the next future level of strain engineering, i.e. synthetic metabolic engineering, enabling genome scale models and synthetic biology for a priori global strain design.

Keywords Metabolic engineering • *Corynebacterium glutamicum* • Chemicals • Bio-materials • Bio-fuels • Genetic engineering • Strain engineering • Industrial biotechnology • Fluxomics • Systems biology • Synthetic biology • Sustainable bio-economy • Renewable feedstocks • Model-based design • Hemicellulose • Xylose • Arabinose • Cellobiose • Glycerol • Silage • Starch • Whey • Amino acids • Lysine • Glutamate • Methionine • Valine • Serine • Tryptophan • Vitamins • Flavor • Fragrances • Diamines • Diaminopentane • Diaminobutane • Succinic acid • Propanediol, Ethanol • Isobutanol • Rational Strain design • Metabolic model • Multi omics • Sugar assimilation • Biosynthesis • Model prediction • Genome breeding • Cadaverine • Putrescine

6.1 Introduction

The soil bacterium *Corynebacterium glutamicum* was isolated about 60 years ago as natural glutamate producer [68, 142]. Immediately from its discovery research aimed at the development of strains with improved production properties for glutamate and other amino acids, rapidly identified as commercial products of this bacterium. Strain optimization was strongly driven by the steadily increasing demand and the high competition between the major suppliers [113, 149]. In the early years, strain optimization involved random mutagenesis and selection [97, 99]. This was inherently linked to the accumulation of detrimental side-mutations causing undesired growth deficiencies, weak stress tolerance, elevated nutritional requirements or suboptimal dead-end mutants resisting further improvement. Today we know about the high connectivity of the underlying metabolic routes with central switch points feeding several biosynthetic routes and complex and strict regulation networks which display a demanding labyrinth to be explored and understood prior to success. The high industrial relevance of *C. glutamicum* thus stimulated intense efforts to analyze and modify the underlying metabolic and regulatory networks in a targeted manner. The rapid development of sophisticated experimental and computational tools in systems biology has provided an excellent fundament to overcome the limitations of classical strain engineering and initiated the era of metabolic engineering and now systems metabolic engineering. The major achievements towards systems-level analysis of *C. glutamicum* include the complete decoding of the genome sequence [32, 47, 56, 141] which opened the door to genome scale stoichiometric modeling [70, 76, 89, 130]. The screening of the annotated genome sequence of *C. glutamicum* ATCC 13032 using bioinformatics approaches [20, 22] revealed the existence of 159 genes encoding transcription regulators, response regulators of two-component systems and sigma factor subunits of RNA-polymerase to coordinate the expression of the about 3,000 predicted protein-coding genes under varying environmental conditions [21]. Beyond the valuable predictions of metabolic capabilities possible by *in silico* models, the estimation of *in vivo* fluxes of central metabolic pathways (fluxomics) displayed a breakthrough towards understanding and engineering of the metabolism of *C. glutamicum* [72, 148]. Especially for bio-production, fluxomics has proven relevant due to the close connection between the central metabolic pathways, accessible by this approach, and many relevant product biosynthetic routes. Important studies to be mentioned are the comparative analysis of fluxes during growth, glutamate, and lysine production [85, 136], in different mutants of a lysine producing strain genealogy [151], during co-utilization of acetate and glucose [146], and during lysine production on different industrially relevant carbon sources [63, 152]. The integration of data from proteomics [24, 36, 133], transcriptomics [35, 75, 145] or metabolomics [19, 75] now enables multi omics studies that deliver key information for rational strain engineering on a truly global level [72]. This provides optimal data sets to elucidate regulatory mechanisms on a global cellular level, thereby making ambitious multidimensional optimization approaches possible which require a comprehensive understanding of the cell. Initial steps taken in this direction [75, 77, 133] have opened the door for novel systems-level design of *C. glutamicum* as efficient cell factory for the production of an array of bio-based goods. Beyond traditional products such as glutamate or lysine systems metabolic engineering has recently led to an enormous expansion of the product portfolio of *C. glutamicum*. Today a large number of products are successfully produced by engineered *C. glutamicum* strains. This includes different chemicals, materials and fuels towards the new era of bio-based economy as sustainable alternative to petrochemical routes. *Corynebacterium glutamicum* represents an excellent platform organism and systems metabolic engineering strongly drives this microorganism into a multi-functional production host as will be explained in more detail throughout this chapter.

6.2 Alternative Raw Materials

C. glutamicum can naturally grow on a variety of different substrates ensuring flexibility and competitiveness to permanently changing nutrient availability in its natural environment (Fig. 6.1). For industrial application this is considerably advantageous because typical fermentation media contain complex substrate mixtures [41, 149]. The major raw materials for industrial production are cane molasses, beet molasses or starch hydrolysates, which can be easily obtained and display well-established feed stocks for traditional bio-products in *C. glutamicum* [88]. Among the carbohydrates contained are glucose, sucrose and fructose which are all taken up

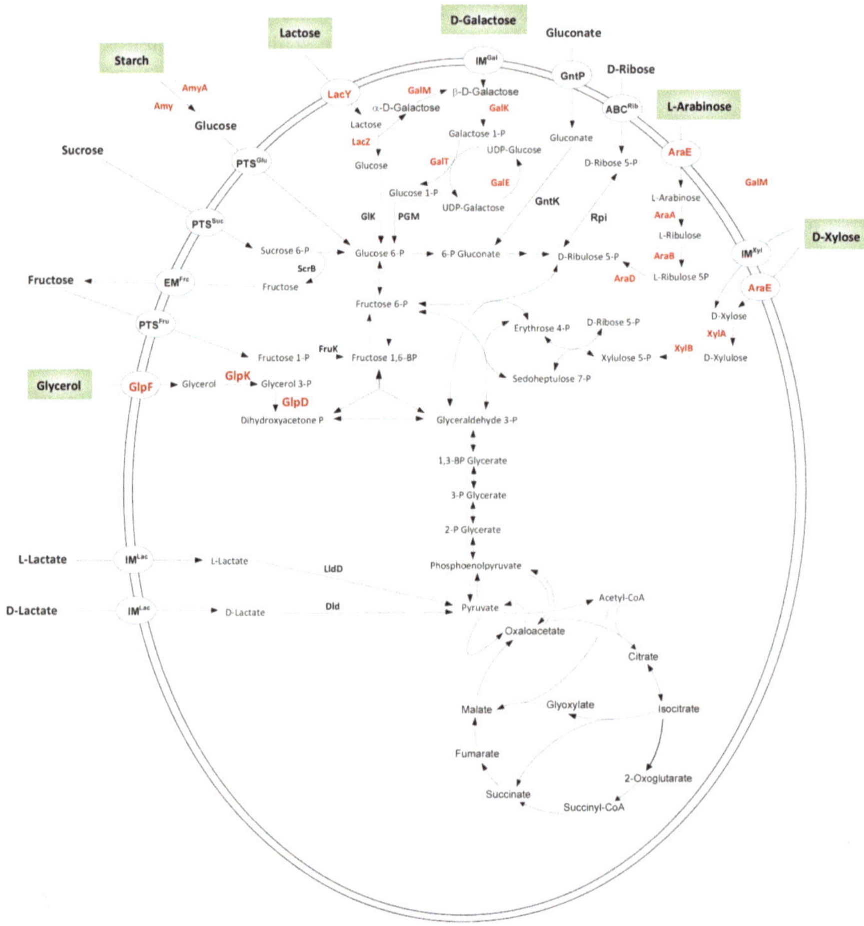

Fig. 6.1 Pathways for utilization of industrially relevant substrates in *C. glutamicum*. Non-natural substrates (glycerol, starch, lactose, galactose, L-arabinose and D-xylose), recently made available for *C. glutamicum* by metabolic engineering are highlighted by *green boxes*. The corresponding heterologous proteins expressed in *C. glutamicum* to mediate growth on the novel substrates are indicated in *red*. Endogenous proteins involved in channeling pathway intermediates into central metabolism are indicated in *black*. ABC^{rib} ATP-binding cassette transporter for D-ribose import, *AraA* arabinose isomerase, *AraB* ribulokinase, *AraE* L-arabinose transporter, *AraD* L-ribulose-5-phosphate 4-epimerase, *Dld* D-lactate dehydrogenase, EM^{fru} fructose export mechanism, *FruK* fructose 1-phosphate kinase, *GalK* galactose kinase, *GalE* UDP-glucose 4-epimerase, *GalM* aldose 1-epimerase, *GalT* UDP-glucose-hexose-1-phosphate uridylyltransferase, *Glk* glucokinase, *GlpF* aquaglyceroporin, *GlpK* glycerokinase, *GlpD* glycerol 3-phosphate dehydrogenase, *Gntk* gluconate kinase, *GntP* gluconate permease, IM^{gal} galactose importer, IM^{xyl} D-xylose importer, *LacY* lactose permease, *LacZ* b-galactosidase, *LldD* L-lactate dehydrogenase, *PGM* phosphoglucomutase, PTS^{fru} fructose-specific phosphotransferase system, PTS^{Glu} glucose-specific phosphotransferase system, PTS^{suc} sucrose-specific phosphotransferase system, *Rpi* ribose-5-phosphate epimerase, *ScrB* sucrose-6-phosphate hydrolase, *XylA* xylose isomerase, *XylB* xylulokinase

via phosphotransferase systems [27, 92, 105]. Their use for the production of low price bulk chemicals or fuels, however, is controversially discussed due to rising of raw material costs and competition with human nutrition. The extension of the substrate spectrum to other cheap and eco-efficient feedstocks is of substantial importance with regard to bio-economy applications, driving systems metabolic engineering of *C. glutamicum* to the utilization of non-natural carbon sources.

The entry point into metabolism thereby significantly influences the production characteristics of *C. glutamicum* [62] and thus the choice and suitability of carbon feedstocks for industrial production. The relative contribution of the individual pathways to growth, product formation and survival of *C. glutamicum* strongly depends on the carbon source. From systems wide ^{13}C metabolic flux analysis it became obvious that during growth on glucose and sucrose, glycolysis and PPP almost equally contribute to carbon conversion [77, 152], while growth on fructose strongly relies on high glycolytic fluxes [63] often accompanied with the formation of the overflow metabolites dihydroxyacetone and glycerol [28]. The different efforts to engineer the substrate utilization by *C. glutamicum* are reviewed below.

6.2.1 Hemicellulose Sugars

Interesting non-food substrates comprise cellulose and hemicellulose derived from lignocellulosic biomass which is the most abundant organic source on earth [2]. The pentoses xylose and arabinose display the major fraction of this biomass. To a smaller extent, hemicellulose additionally contains hexoses such as glucose, mannose or galactose. The major production strains as well as the best described type strain *C. glutamicum* ATCC 13032 are unable to grow on the major hemicellulose sugars. It was therefore quite important to broaden the substrate spectrum.

6.2.1.1 Xylose

Xylose metabolization is in general mediated by the activity of xylose isomerase (*xylA*) and xylulokinase (*xylB*) that convert xylose to the pentose phosphate pathway intermediate xylulose 5-phosphate (Fig. 6.1). *C. glutamicum* only possesses a rudiment xylulokinase activity and is thus unable to utilize this pentose [11, 58]. Implementation of the *E. coli* gene *xylA*, encoding xylose isomerase is sufficient to mediate xylose metabolization but the intrinsic xylulokinase activity of *C. glutamicum* is still limiting efficient growth [58]. This could then be improved by additional heterologous expression of the *xylB* gene from *E. coli* [58]. Under oxygen-limited conditions, the derived xylose-consuming strains efficiently produced and secreted organic acids, predominantly lactic and succinic acid, from xylose with yields of up to 54% (lactate) and 25% (succinate) [58]. More recently, metabolic engineering of *C. glutamicum* for the production of 1,5-diaminopentane as building block for bio-based materials, from xylose was reported [11]. With over expression of xylose isomerase (*xylA*) and xylulokinase (*xylB*) from *E. coli* in a diaminopentane producing background, the obtained mutant successfully produced 1,5-diaminopentane from xylose, glucose and mixtures of both sugars. For the first time, this enabled direct bio-production by *C. glutamicum* from crude hemicellulose hydrolysates [11]. In a two-step process involving enzymatic hydrolysis of oat spelt hemicellulose and fermentation on the obtained hydrolyzate, *C. glutamicum* exhibits an elevated carbon yield of 50 mmol c-mol^{-1} from the hemicellulose as compared to glucose (45 mmol c-mol^{-1}) or xylose (33 mmol c-mol^{-1}). During growth on mixtures of glucose and xylose the xylose-consuming *C. glutamicum* strains exhibited a clear preference for glucose, whereby xylose was only metabolized after glucose was depleted. This could be overcome by additional engineering of the xylose transport. Using the *araE* gene from *C. glutamicum* ATCC 31831, a natural arabinose consumer improved xylose uptake [121]. Despite its said role as arabinose transporter, AraE significantly supported growth on xylose. Additional xylose-specific transporter systems have not been identified so far in *C. glutamicum* and still display interesting targets for further optimization.

6.2.1.2 Arabinose

The ability to use the pentose arabinose as carbon source differs between different *C. glutamicum* isolates. The wild type strain *C. glutamicum* ATCC 13032, however, cannot utilize this sugar [13]. The engineering process for arabinose utilization by *C. glutamicum* requires at least three heterologous genes. The studies so far focused on the *E. coli araABD* gene cluster encoding for L-arabinose isomerase (*araA*), L-ribulokinase (*araB*), and L-ribulose-5-phosphate 4-epimerase (*araD*), respectively [59, 124] (Fig. 6.1). This gene cluster did not only mediate growth but additionally allowed production of lactic acid and succinic acid under oxygen-deprivation from arabinose [59] as well as the production of several amino acids under aerobic conditions [125]. First proof-of-concepts also demonstrated the production of 37 mM L-glutamate and 55 mM L-lysine from minimal medium with 500 mM arabinose as sole carbon source [124]. Unlike xylose there is obviously no strict separation between glucose and arabinose consumption when substrate mixtures are used [59]. Arabinose uptake is, at least in *C. glutamicum* ATCC 31831, partially mediated by a highly affine arabinose-inducible H^+ symporter encoded by *araE* [60]. There are, however, probably more transport systems involved in arabinose uptake also in other *C. glutamicum* species that have not been identified so far [13].

6.2.1.3 Cellobiose

In addition to monosaccharides such as xylose, arabinose or glucose hemicellulose hydrolyzates contain di-saccharides such as sucrose or cellobiose. The latter is highly abundant in cellulose hydrolyzates also representing an interesting fermentation feedstock. Cellobiose uptake and metabolization in *C. glutamicum* can be mediated by a modified PTS system that in addition to methyl beta-glucosides shows side activity for cellobiose (Fig. 6.1). This is related to a point mutation in the *bglF* gene encoding the beta-glucoside specific enzyme IIBCA component of the PTS system. Amino acid substitution V317A or V317M in the membrane-spanning IIC domain of BglF generated cellobiose-utilizing strains of *C. glutamicum* by modifying the selective filter of the PTS [74].

6.2.2 Glycerol

Glycerol is a major by-product from biodiesel production. Large amounts of crude glycerol are available as waste stream from the biodiesel industry making this polyalcohol an interesting raw material for bio-based production processes. The wild type *C. glutamicum* ATCC 13032 naturally cannot grow on glycerol. It was thus engineered to utilize this carbon source by plasmid-based heterologous expression the *E. coli* glycerol utilization genes [118] (Fig. 6.1). This involved facilitated uptake via aquaglyceroporin (GlpF) and two-step conversion into the glycolytic intermediate dihydroxyacetone 3-phosphate, encoded by glycerokinase (GlpK) and glycerol 3-phosphate dehydrogenase (GlpD). The two latter genes were sufficient to enable glycerol-dependent growth in *C. glutamicum*, but additional expression of the facilitator increased growth rate and biomass yield. Expression of all three genes, *glpF*, *glpK*, and *glpD* in the *C. glutamicum* wild type enabled glutamate production from glycerol with a product yield of 0.11 g g^{-1}. Introduction of the corresponding plasmid into the lysine producer *C. glutamicum* DSM1730 allowed production of lysine on glycerol as well as on mixtures of glycerol and glucose, although at lower yield as compared to glucose. At concentrations above 18 g L^{-1} glycerol, growth of *C. glutamicum* was perturbed. Increased tolerance to higher levels thus needs to be optimized further.

6.2.3 Silage

A widespread method for stabilization of agricultural raw materials is silage, having a 70% higher nutritional value then hay. Beyond the application for animal nutrition or production of biogas, silage appears as an interesting raw material also for higher value chemicals. The major constituents of silage juice are lactate and glucose. *C. glutamicum* can grow on lactate. Lactate utilization is mediated by L-lactate (*lldD*) and D-lactate dehydrogenase (*dld*) as well as putative permease for uptake (Fig. 6.1). Hereby, *lldD* is induced by its substrate, whereas *dld* is expressed constitutively at low level. To enable efficient conversion of the typically racemic lactate mixtures from silage raw materials, *dld* was overexpressed in the lysine producing mutant *C. glutamicum* $lysC^{fbr}$ [143]. The mutant strain grew well on lactate and also on mixtures with glucose. By additional amplification of pyruvate carboxylase (*pycA*) and malic enzyme (*malE*) the anaplerotic supply of the lysine precursor oxaloacetate could be optimized leading to increased lysine yield up to 0.15 C-mol C-mol^{-1}. For the application to real silage juice additional modifications of *C. glutamicum* appear useful, e.g. enabling utilization of xylose, which is also contained.

6.2.4 Starch

Starchy materials from corn, wheat, potato or cassava are widely available in nature as carbon sources for *C. glutamicum* and other microorganisms. Starch degradation involves glucoamylase and α-amylase [96]. However, *C. glutamicum* lacks these enzymes and cannot utilize starch directly, but requires preceding hydrolysis into free sugar. This could be overcome by surface display of α -amylase from *Streptococcus bovis* 148 (AmyA) [139] (Fig. 6.1). The resulting lysine producing strain was capable to directly utilize starch and produced up to 6 g L^{-1} of lysine from 50 g L^{-1} of starch. In a parallel study *C. glutamicum* was engineered to secrete α-amylase into the broth [126, 140]. Although high-molecular-weight degradation products remained, the recombinant strain, expressing the *amy* gene from *Streptomyces griseus* IMRU 3570, effectively used soluble starch. Complemented by heterologous expression of lysine decarboxylase from *E. coli* (*cadA*) starch utilizing *C. glutamicum* could be recently also applied for diaminopentane production [140].

6.2.5 Whey

Whey-based media contain lactose and galactose, which cannot by metabolized by *C. glutamicum* (Fig. 6.1). In this regard, *C. glutamicum* was engineered towards utilization of lactose and galactose as sole carbon sources by heterologous expression of the lactose and galactose operon from *Lactobacillus delbrueckii ssp. bulgaris* and *Lactococcus lactis ssp. cremoris* [5]. Recombinant *C. glutamicum* overexpressing lactococcal aldose-1-epimerase (*galM*) and the genes of the Leloir pathway, i.e. galactokinase (*galK*), UDP-glucose-1-P-uridylyltransferase (*galT*), and UDP-galactose-4-epimerase (*galE*) in association with lactose permease (*lacY*) and beta-galactosidase (*lacZ*) could utilize lactose and also galactose as sole carbon source and showed growth comparable to that on glucose indicating efficient utilization. Beyond this, lysine production on a whey-based medium was significantly improved upon implementation of the lactose and galactose utilization genes. When grown in whey-based medium, the engineered *C. glutamicum* strain produced lysine at concentrations of up to 2 g L^{-1}, which represented a tenfold increase over the results obtained with the lactose- and galactose-negative control, *C. glutamicum* 21253. Despite their increased catabolic flexibility, however, the modified Corynebacteria exhibited slower growth rates and plasmid instability. The overall approach was not optimized and lysine production thus not competitive to traditional production processes, but provides a good proof-of-concept awaiting further optimization.

6.3 Human and Animal Nutrition

6.3.1 Amino Acids

Amino acids belong to the most important products in industrial biotechnology (Table 6.1). They are produced worldwide by fermentation using microorganisms, with *C. glutamicum* and *E. coli* as major producers. Amino acid production with *C. glutamicum* has a long tradition. Since its discovery in the 1960s as natural glutamate producer, strains for production of other valuable amino acids such as lysine [69] or tyrosine [138] have been generated from early on. Today, the broad range of amino acids produced by *C. glutamicum* covers also threonine [29], valine [14], arginine [51], isoleucine [94], alanine [54] tryptophan and other aromatic amino acids [42], serine [112] or methionine [17, 106]. In wild type strains, the withdrawal of carbon precursors for amino acid biosynthesis is strictly regulated by feedback inhibition [55, 93, 98, 124, 138] and expression control [51, 98, 128]. One key step towards improved production thus implies deregulation of these mechanisms [42, 51, 55, 93, 129]. Moreover, the amino acid biosynthetic routes are closely linked to central metabolism, so that global flux redirection by systems wide pathway engineering is a crucial prerequisite to generate superior producers.

6.3.1.1 Lysine

The essential amino acid lysine belongs to the aspartate family. With a world market of more than 1,000,000 tons per year mainly for animal nutrition it displays a major industrial product. Since many decades, *C. glutamicum* is the dominating production organism so that we have substantial knowledge on the pathways of lysine biosynthesis in this organism (Fig. 6.2).

Lysine is synthesized from the carbon building blocks pyruvate and oxaloacetate and thus directly originates from central metabolism. In addition, four mol of NADPH are required per mol lysine as reducing power. In *C. glutamicum* there

Table 6.1 Current market size of amino acids produced by fermentation [41, 42, 71, 79]

Amino acid	Market size [t/a]
L-Glutamate	2,500,000
L-Lysine	1,500,000
L-Tryptophan	10,000
L-Glutamine	2,000
L-Arginine	1,200
L-Phenylalanine	8,000[a]
L-Valine	500[a]
L-Histidine	400[a]

[a]Not exclusively produced by *C. glutamicum* fermentation

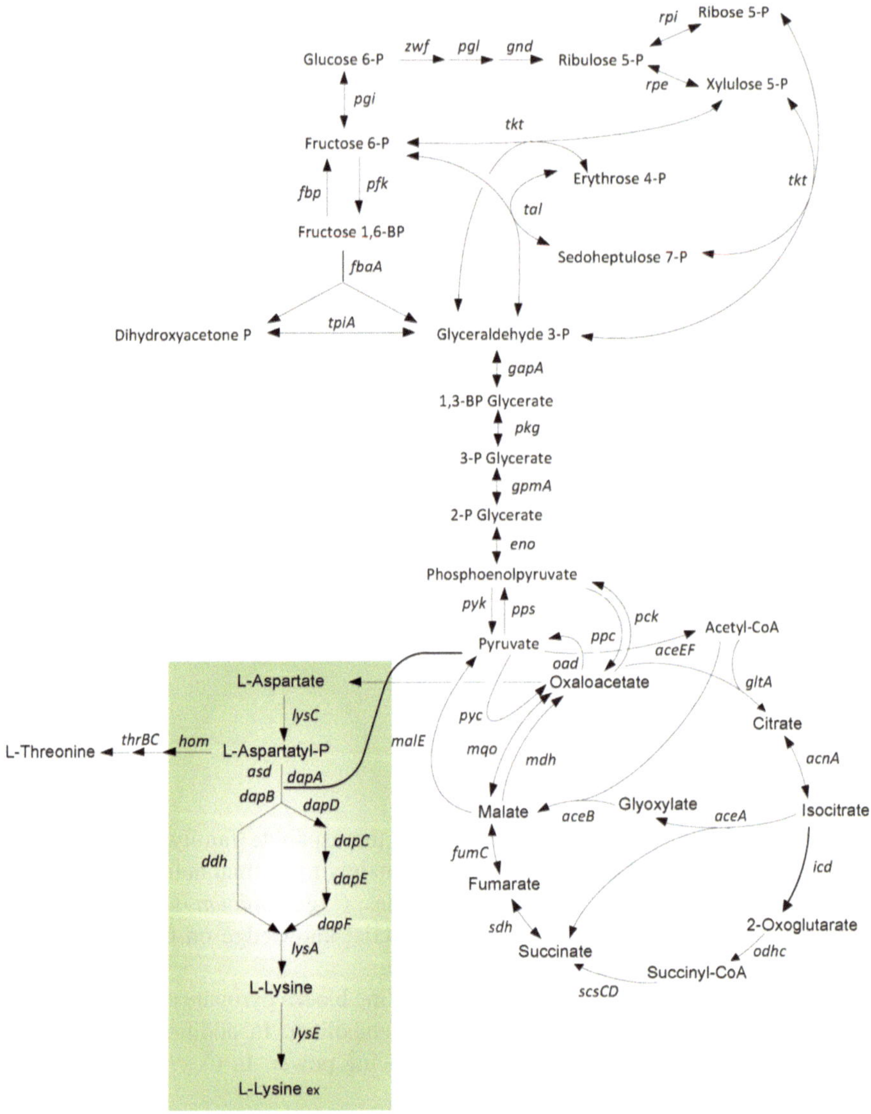

Fig. 6.2 Central metabolism of *Corynebacterium glutamicum* and biosynthetic lysine route via the parallel succinylase pathway and dehydrogenase pathway. Enzymes involved in lysine formation are aspartate kinase (*lysC*), aspartate semialdehyde dehydrogenase (*asd*), dihydrodipicolinate synthase (*dapA*), dihydrodipicolinate reductase (*dapB*) and subsequent split into (i) the succinylase-pathway comprising tetrahydrodipicolinate succinylase (*dapD*), succinyl-amino-ketopimelate transaminase (*dapC*), succinyl-diaminopimelate desuccinylase (*dapE*), diaminopimelate epimerase (*dapF*) and (ii) the dehydrogenase pathway with diaminopimelate dehydrogenase (*ddh*). The final common steps are catalyzed by diaminopimelate decarboxylase (*lysA*) and lysine permease (*lysE*)

exist two alternative biosynthetic routes for lysine formation that allow flexible adjustment to altering ammonium availability in the growth environment [23, 135]. The high relevance of lysine for animal nutrition and the steadily increasing demand has resulted in a footrace of the major suppliers for novel promising genetic targets. Key studies recruited systems wide metabolic flux analysis to elucidate the contribution of central metabolic pathways for lysine biosynthesis. As an example, the PPP was identified as major source for NADPH in *C. glutamicum* in a genealogy of lysine production mutants [151]. The flux into the PPP was found quite flexible depending on the carbon source [63, 150, 152] or the metabolic burden of the cell [85]. In addition the key role of oxaloacetate as precursor for lysine biosynthesis was unraveled together with the complex network of carboxylation and decarboxylation reactions linked to oxaloacetate supply and withdrawal [108]. The outcome of these fascinating insights stimulated a number of metabolic engineering studies which proved the relevance of specific reactions and pathways for lysine over production and provided strategies for targeted strain improvement. Strategies involved the pyruvate node to increase the anabolic net flux from glycolytic C_3 metabolites to C_4 metabolites of the TCA cycle. Beneficial for lysine production were over expression and modification of the major anaplerotic enzyme pyruvate carboxylase (*pycA*) [101, 110] as well as deletion of the counteracting enzyme PEP carboxykinase (*pepck*) [109]. Successful strategies towards improved supply of the reduction equivalent NADPH focused on increasing flux through the pentose phosphate pathway (PPP) as major NADPH source in *C. glutamicum*. Some studies hereby reported improved lysine production by directly modifying the NADPH generating enzymes glucose 6-phosphate dehydrogenase (*zwf*) [9] and 6-phosphogluconate dehydrogenase (*gnd*) [102]. The inactivation of the glycolytic flux by deletion of phosphoglucoisomerase (*pgi*) forces *C. glutamicum* to exclusively utilize the PPP during growth on glucose [86]. This generated high amounts of NADPH, although at perturbed growth behavior. Efficient flux redirection from glycolysis towards the PPP could be realized by targeted over expression of the gluconeogenic enzyme fructose 1,6-bisphosphatase [8]. This resulted in improved lysine yield on different sugars including glucose, fructose and sucrose which are the major constituents of industrial raw materials such as molasses or starch hydrolyzates. Beyond, promising targets comprised attenuation or deletion of competing reactions. One prominent example in this regard is down-regulation or complete deletion of the biosynthetic pathways towards threonine production [101, 127]. This strategy benefits from two positive effects on lysine production: a reduced withdrawal of the lysine precursor aspartate semialdehyde and a reduced inhibitory effect of threonine on the activity of the key enzyme aspartokinase [75]. In addition, the TCA cycle strongly competes with lysine production for carbon as demonstrated by genome scale modeling [89] or ^{13}C metabolic flux analysis [151]. Deletion of pyruvate dehydrogenase at the entry into the TCA cycle could redirect carbon towards lysine biosynthesis [15]. The interruption of aerobic energy formation, however, introduced growth deficiencies, enhanced nutrient demand and elevated by-product formation [15]. Further studies, hence, focused on targeted down-regulation by start codon exchange, a more gentle and moderate genetic

Table 6.2 Systems metabolic engineering of *C. glutamicum* towards improved lysine production using genome breeding identifying promising targets via comparative sequence analysis of the wild type and a classically derived lysine producer

Generation	Target gene	Modification	Yield [%]	Titer [g L^{-1}]	Productivity [g L^{-1} h^{-1}]
1	*lysC*	T311I	21[a]	55	1.8[a]
2	*hom*	V59A	29[a]	75	2.5[a]
3	*pyc*	P458S	31[a]	80	3.0
4	*gnd*	S361F	36	90	–
5	*mqo*	W224opal[b]	40	100	–

Description of the strains obtained by genome breeding is taken from [50, 101, 102]. The production characteristics were determined during fed batch fermentation in 5 L jar fermenters using a glucose based fermentation medium
[a]Estimated from reference
[b]Non-sense mutation

modification acting on the translational efficiency [10, 11]. Substitution of the common ATG start codon by the rare GTG variant in the genes encoding pyruvate dehydrogenase [11] and isocitrate dehydrogenase [10] had significant influence on the enzyme activity and resulted in an improved production performance by 17% and 40%, respectively. These basic metabolic engineering strategies increased the production performance of individual strains and in addition contributed significantly to the current knowledge on production-relevant pathways and regulatory mechanism in the cell. None of the above strains, however, reached a performance attractive for industrial production. Due to the complex underlying networks or metabolic reactions and regulatory interactions, systems metabolic engineering with global rather than local strain optimization was needed to create competitive production strains. An impressive pioneering study in this regard was performed in Japan, the home country of *C. glutamicum* [101]. Comparative sequence analysis of a classical lysine producer and the wild type identified single nucleotide exchanges which had been introduced into the genome during many rounds of random mutagenesis and selection. Stepwise implementation of a selected set of these point mutations in the background of the wild type strain (genome breeding) yielded efficient production strains with a clearly improved phenotype (Table 6.2). Aspartokinase (*lysC*) the key regulatory enzyme for lysine biosynthesis was released from feedback inhibition by lysine and threonine. The activity of the competing enzymes homoserine dehydrogenase (*hom*) and malate:quinone oxidoreductase (*mqo*) was impaired by leaky mutations. Improved kinetic properties reduced the sensitivity of pyruvate carboxylase (*pycA*) and 6-phosphogluconate dehydrogenase (*gnd*) towards inhibiting metabolites [50, 101, 102]. After five generations, the obtained mutant produced 100 g L^{-1} of lysine with a product yield of 40%. Although this did not reach the performance of classically derived strains with titers of 100–120 g L^{-1} and yields in the range of 40–55%, this work represents a key study for systems metabolic engineering in *C. glutamicum*. Beyond this, synthetic metabolic engineering has quite recently derived a wild type based rational lysine producer of *C. glutamicum* which carries only 12 mutations in the genome and accumulates 120 g L^{-1} of lysine within 30 h at a yield of 55% [12].

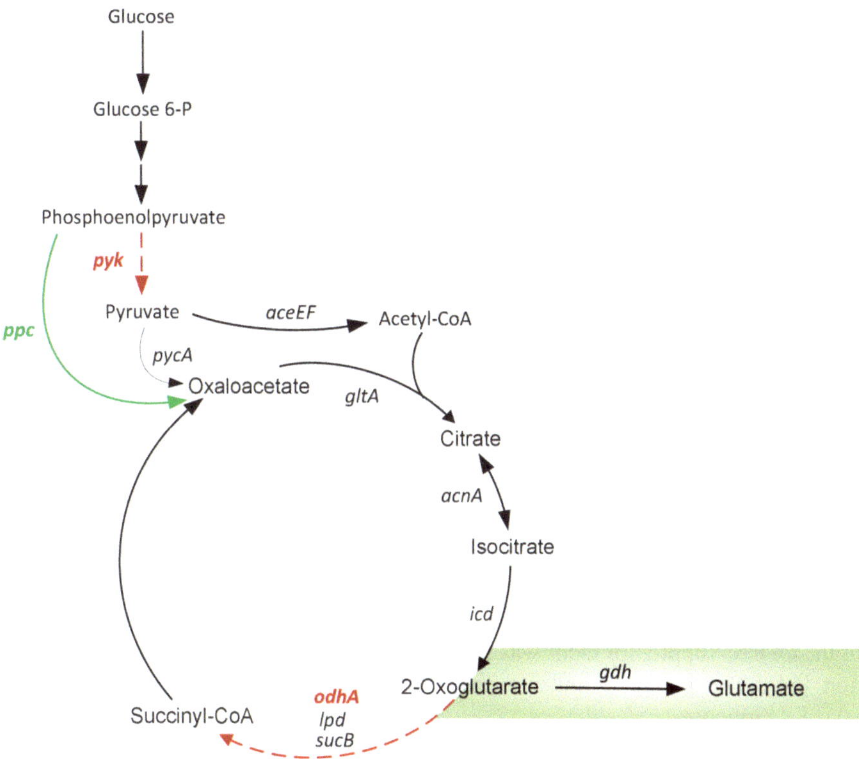

Fig. 6.3 Central pathways in *Corynebacterium glutamicum* for glutamate overproduction. Metabolic engineering strategies to direct carbon flux towards glutamate are highlighted in *red* (chromosomal deletion) and *green* (plasmid based overexpression). Genes: *aceEF* pyruvate dehydrogenase complex, *acnA* aconitase, *gdh* glutamate dehydrogenase, *gltA* citrate synthase, *icd* isocitrate dehydrogenase, *odhA*/*lpd*/*sucB* 2-oxoglutarate complex, *ppc* phosphoenolpyruvate carboxylase, *pycA* pyruvate carboxylase, *pyk* pyruvate kinase

In this work, the *in vivo* pathway activity of the wild type – obtained by ^{13}C metabolic flux analysis – and a theoretical flux distribution of an optimal lysine producer – obtained by *in silico* modeling – were used for global design of a metabolic blueprint of *C. glutamicum*. This novel strategy is described in more detail in Sect. 6.6.

6.3.1.2 Glutamate

With an annual production of more than 2.5 million tons, the synthesis of glutamate using *C. glutamicum* as biocatalyst is one of the major industrial fermentation processes. Since many decades this amino acid serves as important flavor enhancer in a variety of processed foods. From a metabolic perspective, glutamate overproduction is closely linked to the central metabolism (Fig. 6.3). Clearly, efficient production demands for substantial redirection of flux. For glutamate biosynthesis, significant amounts of the TCA cycle intermediate 2-oxoglutarate are consumed, demanding for efficient anaplerotic replenishment [110, 132]. Increased anaplerotic carboxylation in the wild type of *C. glutamicum* could be achieved by deletion of the central glycolytic enzyme pyruvate kinase (*pykA*) [123] as well as be overexpression of phosphoenolpyruvate carboxylase (*ppc*) [120, 122] resulting in improved glutamate production under biotin-limited conditions.

Focussing on the direct glutamate precursor 2-oxoglutarate, flux analysis of isocitrate dehydrogenase and 2-oxoglutarate dehydrogenase (ODHC) revealed that ICD activity does not significantly change throughout the fermentation, while that of the ODHC significantly decreases after the induction of glutamate production [131]. Beyond this, intensive studies of the enzymatic set around the pyruvate node revealed strong changes as response to glutamate production [34]. Major findings involved decreased activities of pyruvate and 2-oxoglutarate dehydrogenase. The effect on pyruvate carboxylase activity, however, strongly depended on the glutamate trigger. In the light of this finding, the major role in flux control at the key branch point of 2-oxoglutarate was attributed to ODHC. Glutamate production was successfully improved by deletion of *odhA*, encoding a subunit of the enzyme complex [3]. A similar effect was attained via *odhA* antisense RNA expression that leads to a decreased specific activity of ODHC [65]. Accordingly, overexpression of the ODHC suppressor OdhI was observed to be beneficial for glutamate production [64]. The obtained mutants produced about 20 g L^{-1} of glutamate from 80 g L^{-1} glucose.

6.3.1.3 Methionine

The sulfur containing amino acid methionine is widely used as supplement in animal nutrition with an estimated market of 2.2 billion US$. Despite the preeminence of biotechnological production processes in amino acid manufacturing, methionine is still produced as racemic mixture by chemical synthesis [144]. Related to the requirement of hazardous chemicals and cost intensive purification of the product, an biotechnological production of methionine is highly attractive [76]. Numerous elaborate attempts have, however, not yielded in generating efficient production strains so far [78, 91].

Still, there are first promising studies that revealed the feasibility of fermentative methionine production with engineered *C. glutamicum* [106]. These strains, which were derived from lysine producing strain MH20-22B and resistant to the methionine analogue ethionine, were additionally modified by deletion of the threonine pathway and release of homoserine dehydrogenase from feedback inhibition (Fig. 6.4). The finally achieved methionine titer of 2.9 g L^{-1} was, however, very low. This is likely related to the highly complex regulatory network that controls the energy-demanding biosynthesis of methionine involving transcriptional repression in combination with metabolic regulation [80, 116, 117]. One of the

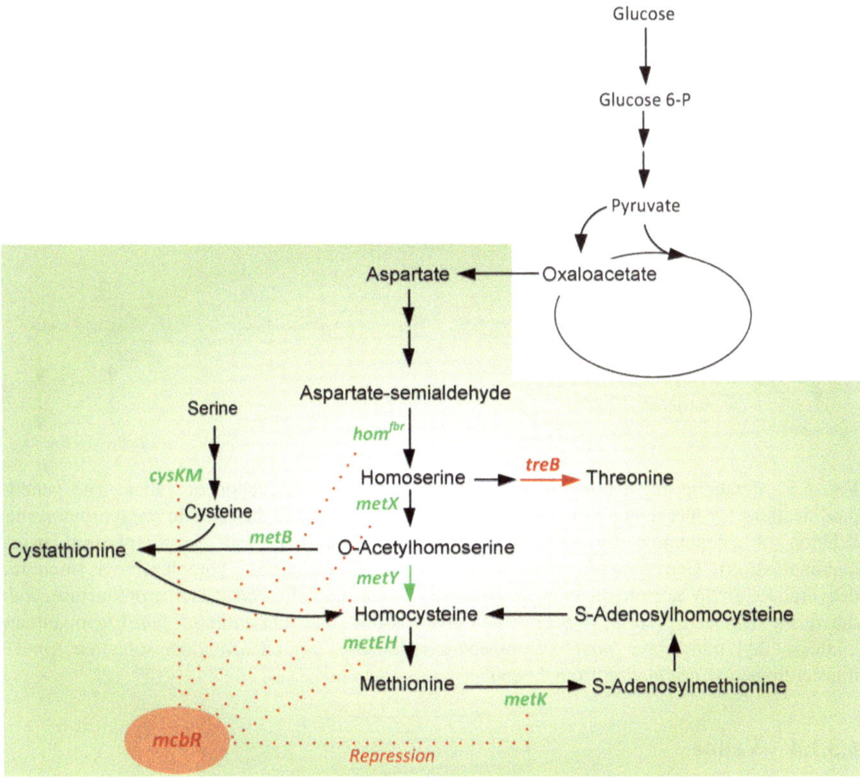

Fig. 6.4 Metabolic engineering of *C. glutamicum* towards methionine production. The genetic modifications are highlighted in *red* (genome-based deletion or leaky mutation) and *green* (plasmid-based over expression, release from feedback inhibition and repression). Genes: *cysKM* cysteine synthase, *hom* homoserine dehydrogenase, *metEH* methionine synthases, *metK* methionine adenosyltransferase, *metX* homoserine acetyltransferase, *metY* O-acetlyhomoserine sulfhydrolase

global regulators in *C. glutamicum*, the methionine and cysteine biosynthesis repressor (McbR) (Fig. 6.4), and the metabolic response to its deletion have been studied in detail [77, 116, 117] in order to collect an extended dataset that might give access to novel metabolic engineering strategies for rational strain development. Beyond this, *in silico* studies revealed metabolic drawbacks as well as process-relevant parameters including identification of methanethiol as optimal sulfur source for production [76]. Extended studies on the metabolism gave a detailed insight into sulfur assimilation from methanethiol and identified *metY* as responsible enzyme [17]. From measurement of the intracellular metabolite pool size it became obvious that over expression of *metY* was beneficial for methionine production but there still remain metabolic barriers that need to be dispelled to obtain competitive methionine production strains [17].

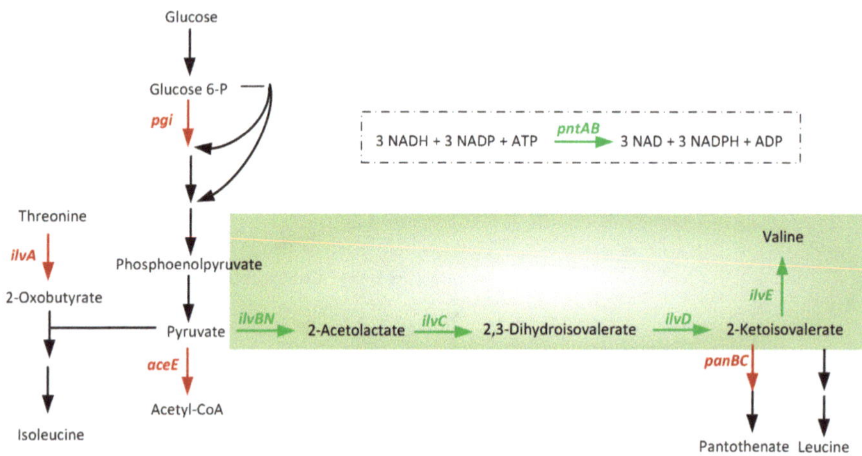

Fig. 6.5 Metabolic engineering of *C. glutamicum* for overproduction of valine. The genetic modifications for flux redirection towards valine production are highlighted in *red* (chromosomal deletion or attenuation by introduction of leaky mutation) and *green* (plasmid based overexpression). Genes: *aceE* subunit of pyruvate dehydrogenase complex, *ilvA* threonine dehydratase, *ilvBN* acetohydroxy acid synthase, *ilvC* acetohydroxyacid isomeroreductase, *ilvD* dihydroxy acid dehydratase, *ilvE* branched chain amino acid transaminase, *panB* ketopantoate hydroymethyl transferase, *panC* pantothenate synthetase, *pgi* phosphoglucoisomerase, *pntAB* transhydrogenase originating from *E. coli*

6.3.1.4 Valine

The pyruvate-derived amino acid valine belongs to the branched-chain amino acids and is mainly applied in the pharmaceutical and agricultural sectors. As displayed in Fig. 6.5, biosynthesis is closely connected to isoleucine, leucine and pantothenate synthesis by sharing the identical enzymatic set encoded by *ilvBN*, *ilvC*, *ilvD* and *ilvE* [147]. In consequence, these reaction as well as reactions that feed into isoleucine and leucine synthesis have to be well balanced to favor valine production. Related to this, over expression of IlvBNCD resulted in valine accumulation in *C. glutamicum* which was enhanced by deletion of *ilvA* initiating isoleucine synthesis from threonine [115]. Further improvement was achieved upon deletion of *panBC* involved in pantothenate production [115].

The positive effect on the production can both result from a reduced carbon withdrawal from the valine pathway and also from an increased availability of the building block pyruvate. Pantothenate serves as precursor for coenzyme A (CoA) formation, an important co-factor of pyruvate dehydrogenase. Limited supply of CoA was hence considered to result in flux re-routing from pyruvate dehydrogenase towards valine production [115]. This assumption was later supported by the construction of pyruvate dehydrogenase deficient valine producers that accumulate large amounts of pyruvate in addition to valine pointing at further limitations in the biosynthetic pathway [14]. Further studies focussed on genome-based modulation of the expression level of beneficial enzymes and competing pathways using

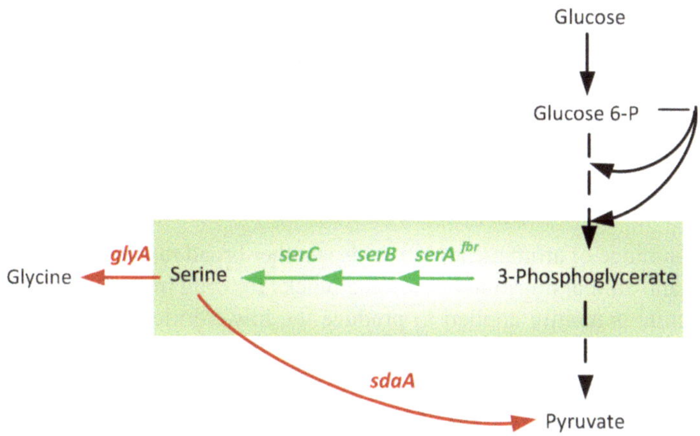

Fig. 6.6 Metabolic engineering of *C. glutamicum* towards generation of a serine producer. The genetic modifications strictly focus on the serine node and comprise genome-based deletion or down-regulation (leaky mutants) highlighted in *red* of serine-consuming reactions and plasmid-based amplification of the biosynthetic gene cluster for serine highlighted in *green*. Genes: *glyA* serine hydroxymethyltransferase, *sdaA* serine dehydratase, *serA* 3-phosphoglycerate dehydrogenase, *serB* phosphoserine aminotransferase, *serC* phosphoserine phosphatase

promoters of different strength to support optimal production flux. The resulting strain was able to produce 136 mM valine within 48 h [38]. In addition to engineering pathway flux and supply of carbon building blocks, valine production also benefited from optimized supply of NADPH by deletion of phosphoglucoisomerase (*pgi*) [6] or heterologous expression of the *E. coli* transhydrogenase *pntAB* [7].

6.3.1.5 Serine

The annual demand of serine of about 300 tons is mainly required for the pharmaceutical and cosmetic industry. For the biotechnological production of serine there are strongly deviating methods described including serine production from methanol and glycine with methanol-utilizing *Methylobacterium* sp NM43 [33] and serine production from glucose with *Brevibacterium flavum* [41]. The generation of serine-producing *C. glutamicum* strains was performed directly from the scratch by metabolic engineering thereby building a clear contrast to the traditional development of first lysine producers. Initial steps towards strain development included amplified expression of the biosynthetic reactions comprising 3-phosphoglycerate dehydrogenase (SerA), phosphoserine aminotransferase (SerB), and phosphoserine phosphatase (SerC) as well as release of SerA from serine-mediated feedback inhibition by C-terminal deletion of 97 amino acid residues [111, 112] (Fig. 6.6).

During strain development it, however, turned out that these modifications were not sufficient for efficient overproduction. The key was indeed deletion or down-regulation of the serine-consuming reactions catalyzed by serine dehydratase (SdaA) and serine hydroxymethyltransferase (GlyA) [112]. This strain achieved by solely modifying the metabolic reaction network around the serine-node produced up to 86 mM serine with a maximum specific productivity of 1.2 mmol g^{-1} h^{-1} [112].

6.3.1.6 Aromatic Amino Acids

Application range of aromatic amino acids is rather broad ranging from supplement of human and animal nutrition to building blocks for drug manufacturing. Whereas phenylalanine is mainly applied to produce the low-calorie sweetener aspartame, tyrosine is for instance used for the production of the anti-Parkinson's desease drug L-Dopa [18, 82]. Biosynthesis of aromatics occurs via the common intermediate chorismate that is built from erythrose 4-phoshphate (E4P) and phosphoenolpyruvate (PEP) (Fig. 6.7) [42]. The formation of aromatic amino acids is highly demanding with regard to carbon and energy supply as well as reducing power and thus regulated by a highly complex network of metabolic inhibition and repression (Fig. 6.7). Related to this strict control the engineering steps towards the creation of an overproducer for aromatic amino acids is highly challenging and requires several optimization approaches. Main engineering targets thereby comprised the key enzymes 3-deoxy-D-arabino-heptulosonate 7-phosphate synthase (DS) encoded by *aroF*, chorismate mutase (CM) encoded by *csm* and prephenate dehydratase (PD) encoded by *pheA* for phenylalanine, DS and CM for tyrosine [43] and DS and the *trp* operon for tryptophan [48], respectively.

Within the central metabolism, additional engineering strategies focused on a balanced supply of the carbon building blocks PEP and E4P. The limitation of E4P was overcome by amplification of the *tkt* gene encoding transketolase, which functions in directing carbon towards E4P [46, 49]. For improved supply of PEP there have so far no targeted engineering strategies be described for *C. glutamicum* but there were isolates with reduced PEP carboxlase activity described that exhibited improved tryptophan production characteristics [57]. Beyond this, several engineering strategies towards improved production of aromatics in *E. coli* were described, mainly comprising genetic modifications to decrease PEP consumption by competing pathways including inactivation of pyruvate kinase isoenzymes *pykA* and *pykF* [31] or the phosphotransferase system [30, 95] which also appear interesting for further optimization of *C. glutamicum* strains. As serine is needed for the last reaction of tryptophan synthesis increasing the serine availability by amplification of the *serA* gene, the key enzyme in serine biosynthesis, is another requisite for efficient tryptophan production [48]. In addition to balanced supply of the major building blocks, efficient amino acid efflux via the membrane is of outstanding importance especially considering the highly complex feedback regulation of the biosynthetic pathways. Of significant benefit for the production of tryptophan was here a dimished re-uptake of secreted tryptophan from the medium [44, 45]. Ongoing research in the field is promising for further development of optimized cell factories [84, 155].

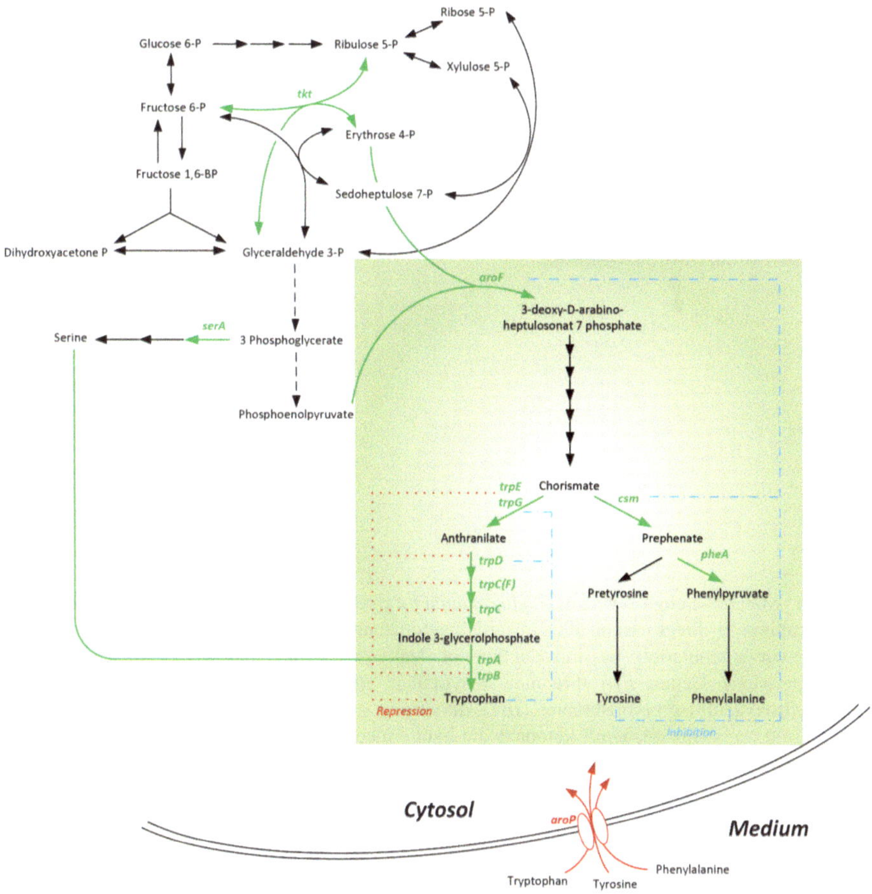

Fig. 6.7 Metabolic engineering of *C. glutamicum* towards overproduction of the aromatic amino acids tryptophan, tyrosine and phenylalanine. The genetic modifications are highlighted in *green* (plasmid-based over expression) and in *red* (genome-based deletion). Pathway regulation by inhibition and repression is indicated. Genes: *aroF* 3-deoxy-D-arabino-heptulosonate 7-phosphate synthase, *aroP* aromatic amino acid uptake system, *csm* chorismate mutase, *pheA* prephenate dehydratase, *serA* 3-phosphoglycerate dehydratase, *tkt* transketolase, *trpAB* tryptophan synthase, *trpC* indole-3-glycerol phosphate synthase, *trpCF* phospho-ribosylanthranilate isomerase, *trpD* anthranilate phosphoribosyltransferase, *trpEG* anthranilate synthase

6.3.2 *Vitamins*

6.3.2.1 Pantothenate

Pantothenate (vitamin B5) is a water-soluble vitamin required in animals and humans to synthesize coenzyme-A. About 4,000 tons of pantothenate is produced annually as a pharmaceutical and a feed additive. In *Escherichia coli* the specific

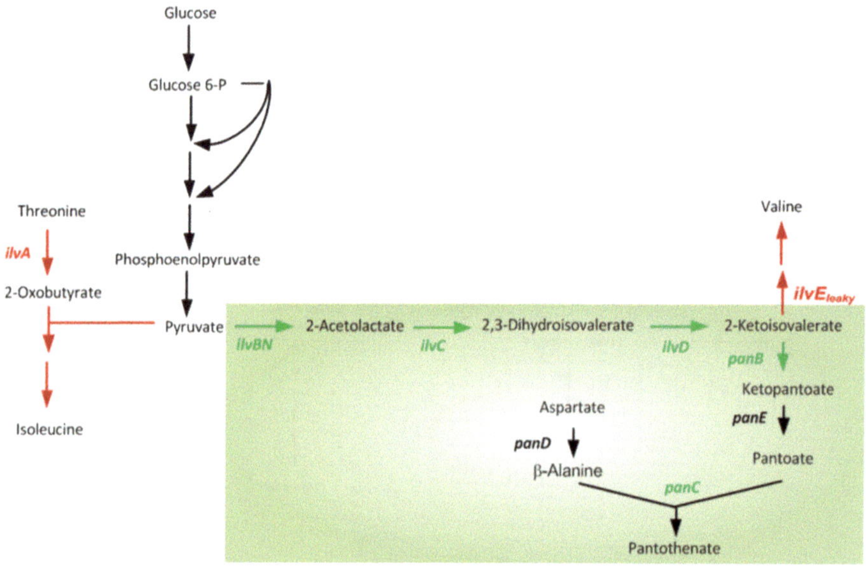

Fig. 6.8 Metabolic engineering of *C. glutamicum* for overproduction of pantothenate. The genetic modifications to direct carbon flux towards pantothenate are highlighted in *red* (chromosomal deletion or attenutation by introduction of leaky mutation) and *green* (plasmid based overexpression). Genes: *ilvA* threonine dehydratase, *ilvBN* acetohydroxy acid synthase, *ilvC* acetohydroxyacid isomeroreductase, *ilvD* dihydroxy acid dehydratase, *ilvE* branched chain amino acid transaminase, *panB* ketopantoate hydroymethyl transferase, *panC* pantothenate synthetase, *panD* aspartate-1-decarboxylase, *panE* ketopantoate reductase

biosynthesis pathway of this vitamin consists of only four steps. The first reaction, catalyzed by ketopantoate-hydroxy-methyl transferase, uses the valine intermediate 2-ketoisovalerate to generate ketopantoate, which is then reduced to pantoic acid. An aspartate-1-decarboxylase activity supplies β-alanine, which is ligated with pantoic acid to yield pantothenate. Based on a first overproducing mutant, a rational strain was recently generated [40]. Shortly, the production strain carried a combination of genetic modifications in the production pathway for optimal redirection of carbon towards pantothenate (Fig. 6.8).

This comprised a deletion of the *ilvA* gene encoding threonine dehydratase to abolish isoleucine synthesis, the promoter down-mutation P-*ilvE*M3 to attenuate *ilvE* gene expression and thereby increase ketoisovalerate availability for product biosynthesis, and two compatible plasmids to overexpress the *ilvBNCD* genes and duplicated copies of the *panBC* operon. The recombinant strain produced 8 mM pantothenate. The occurrence of by-products such as ketoisocaproate, DL-2,3,-dihydroxy-isovalerat, ketopantoate or pantoate points at an imbalanced capacity within the pantothenate biosynthetic pathway suggesting further targets for optimization.

6.3.3 Flavor and Fragrances

6.3.3.1 Pyrazines

Pyrazines, a group of 1,4 dinitrogen substituted benzenes receive increasing interest as flavoring additives. Already at low levels, pyrazines with different ring substituents contribute many aromas and flavors e.g. to potato, popcorn, coffee or pepper. They are widely distributed in plants, animals and microorganisms. Pyrazines are synthesized and degraded by a few bacteria and fungi. The wild type *C. glutamicum* secretes low amounts of di-, tri- and tetramethyl pyrazines into the head space [26]. Isotope labeling studies revealed a close link of pyrazine biosynthesis to the metabolism of branched chain amino acids and acyloins as central building blocks. Subsequently, targeted deletion of ketol-acid reductoisomerase in *C. glutamicum* provided a first mutant with strongly enhanced biosynthesis of pyrazines as compared to the wild type. This included also novel variants with ethyl-, propyl- and butyl-based side chains and opens interesting possibilities to engineer *C. glutamicum* further for aroma production.

6.4 Bio-Based Chemicals and Materials

6.4.1 Diamines

Diamines, exhibiting two amino groups, are used as monomers for polyamides (PA), a class of high performance polymers with an annual production volume of 3,500,000 metric tons. Today, polyamides are still derived from fossil resources and thus suffer from rising oil prices and environmental concerns such as escalating CO_2 production and global warming. For a sustainable bio-production, especially 1,5-diaminopentane (cadaverine) [66, 90] and 1,4-diaminobutane (putrescine) are of high interest [124]. They display promising building blocks for innovative polyamides such as PA5.4 or PA5.10 which reveal excellent material properties [25]. Today, industrial supply of 1,5-diaminopentane is limited due to the difficult chemical synthesis, whereas the petrochemical process for 1,4-diaminobutane involves hazardous chemicals such as hydrogen cyanide or acrylonitrile [73]. Using diamines derived from microbial biosynthesis, polymerization with appropriate bio-blocks such as succinic acid or sebacic acid thus allows the production of completely bio-based polyamides. Beyond existing low-cost standard polymers such as polylactic acid and polyhydroxyalkanoates, this opens novel applications in the automotive industry or in high-value consumer products, meaning that biotechnological processes for these polymers possess enormous ecological and economical potential.

6.4.1.1 Diaminopentane

From a metabolic viewpoint, 1,5-diaminopentane is obtained directly from lysine by decarboxylation, suggesting the lysine hyper-producer *C. glutamicum* as promising production organism. In a first proof of principle, a *C. glutamicum* wild type was engineered by replacing homoserine dehydrogenase with heterologous lysine decarboxylase (*cadA*) from *E. coli* [90]. Whereas this strain still secreted significant amounts of lysine, 1,5-diaminopentane could be obtained in small quantities. Production was recently increased via plasmid-based expression of *cadA* using a strong constitutive promoter and kanamycin as selection pressure [140]. The product yield was, however, rather low. We thus performed systems metabolic engineering of *C. glutamicum* towards competitive industrial production. In contrast to previous work, we focused on a second variant of lysine decarboxylase encoded by constitutively expressed *ldcC* from *E. coli* [153]. The neutral pH optimum of this lysine decarboxylase is obviously more favorable for expression in *C. glutamicum* as compared to CadA which prefers acidic conditions. Moving towards a superior production strain, substantial metabolic engineering of supporting pathways of central metabolism was then carried out through a combination of targets from the newly created diaminopentane biosynthetic pathway and the supply of oxaloacetate as the major building block of the product [66]. Systems wide re-design of the metabolism yielded superior strains with desirable properties (Fig. 6.9). Strain design was thereby strongly guided by the extended knowledge on the key reactions for optimized lysine production and involved release of undesired feedback regulation, improved anaplerotic replenishment and amplified expression of enzymes directly involved in biosynthesis [66]. Multi-omics analysis of the obtained production mutants then suggested rational targets for further improvement. Fluxomics unraveled lysine decarboxylase as bottleneck, since *in vitro* activity and *in vivo* flux through this enzyme were closely correlated.

The combination of the strong constitutive *tuf* promoter and optimized codon usage allowed efficient genome-based *ldcC* expression and resulted in the 1,5-diaminopentane producer *C. glutamicum* DAP-3c, which exhibited a diaminopentane yield of 0.2 mol mol^{-1} on minimal glucose medium. Supplementation with pyridoxal, the cofactor of lysine decarboxylase, even increased the yield to 0.3 mol mol^{-1}. Metabolomics revealed substantial formation of N-acetyl-diaminopentane as undesired by-product. This compound reached levels of more than 25% of that of 1,5-diaminopentane [67] and therefore displayed a highly undesired by-product with respect to carbon yield and product purity. Targeted single deletion of potential candidates annotated as N-acetyltransferases unraveled NCgl1469 as responsible enzyme catalyzing the undesired reaction [67]. The deletion strain, designated as *C. glutamicum* DAP-4, exhibited a complete lack of N-acetyl-diaminopentane accumulation in the medium, accompanied by an increase of the diaminopentane yield by 11%. The mutant strain allowed the production of diaminopentane as the sole product. The deletion did not cause any negative growth effects, since specific growth rate and glucose uptake rate remained unchanged.

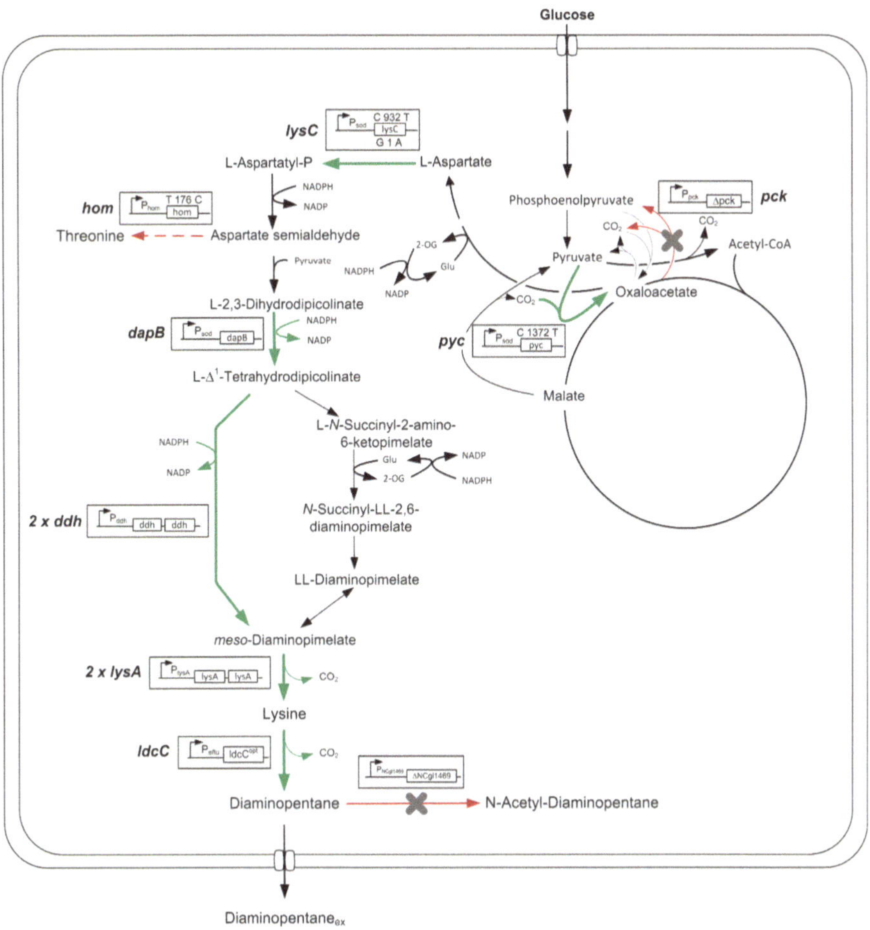

Fig. 6.9 Systems metabolic engineering of *C. glutamicum* for overproduction of 1,5-diaminopentane (cadaverine) [66, 67]. The rational, multi-target strain engineering started from the non-producing wild type *C. glutamicum* ATCC 13032. In addition to heterologous expression of codon optimized *ldcC* from *E. coli*, different endogeneous genes of *C. glutamicum* in product biosynthesis and supporting pathways were modified to increase the flux to 1,5-diaminopentane. All modifications were implemented into the genome providing a stable production strain. *Gray boxes* represent the targeted modification of the respective genes. The symbol "x" indicates gene deletion. *Green arrows* indicate amplification, *red arrows* attenuation or deletion. Genes: *dapB* dihydrodipicolinate reductase, *ddh* diaminopimelate dehydrogenase, *eftu* elongation factor tu, *hom* homoserine dehydrogenase, *ldcC* lysine decarboxylase, *lysA* diaminopimelate decarboxylase, *lysC* aspartokinase, NCgl1469 1,5-diaminopentane transacetylase, *pck* phosphoenolpyruvate carboxykinase, *pyc* pyruvate carboxylase, *sod* superoxide dismutase

In summary, the obtained production strain *C. glutamicum* DAP-4 displays an important step towards an industrially attractive cell factory. The metabolic properties of the mutants created, suggest promising strategies for further improvement, e.g. on the level of product export.

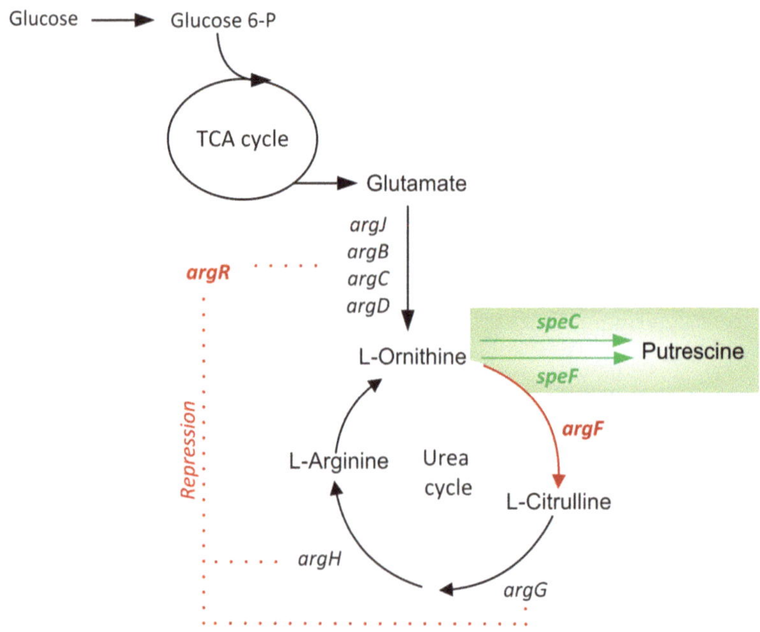

Fig. 6.10 Metabolic engineering of *C. glutamicum* for overproduction of 1,5-diaminobutane (putrescine). The genetic modifications are highlighted in *red* (chromosomal deletion) and *green* (plasmid based overexpression). Genes: *argB* threonine dehydratase, *argC* acetohydroxy acid synthase, *argD* isomero reductase, *argF* dihydroxy acid dehydratase, *argG* branched chain amino acid transaminase, *argH* ketopantoate hydroymethyl transferase, *argR* arginine repressor, *speC* pantothenate synthetase, *speF* aspartate-1-decarboxylase

6.4.1.2 Diaminobutane

The only existing industrial route to 1,4-diaminobutane requires fossil resources and hazardous chemicals such as hydrogen cyanide or acrylonitrile [73]. Bio-based production of this metabolite has been described in different bacteria including *E. coli* or *C. glutamicum* [124]. Due to the fact that biosynthesis of 1,4-diaminobutane involves the glutamate pathway, which is well known and engineered in *C. glutamicum*, this bacterium appears as the promising production organism. Engineering of *C. glutamicum* to produce 1,4-diaminobutane was based on heterologous expression of genes encoding ornithine decarboxylation activity [124]. Among the two alternatives tested, arginine (*speF*) and ornithine decarboxylase (*speC*) from *E. coli*, the latter enzyme appeared rather efficient and provided significantly enhanced yields (Fig. 6.10).

The highest production efficiency was reached by combining overexpression of *speC* with deletion of *argR* encoding the arginine repressor and *argF* encoding the competing enzyme ornithine carbamoyltransferase. In batch culture, the strain produced 6 g L^{-1} 1,4-diaminobutane at a molar product yield of 24%. This pioneering study appears promising towards future production of 1,4-diaminobutane in

C. glutamicum, which is additionally supported by the fact that *C. glutamicum* exhibits a good natural tolerance against 1,4-diaminobutane reflected by undisturbed growth of the organism up to concentrations of about 40 g L^{-1} putrescine [124].

6.4.2 Succinic Acid

Succinic acid represents an important building block for the production of a large spectrum of different compounds that find application as consumer products such as food, agriculture, the health and pharmaceutical products or as well as commodity chemicals [154]. The industrial demand for succinic acid is nowadays covered by the petrochemical production route. With regard to the expected depletion of fossil resources, prices for the raw material are increasing and thus the production costs. Accordingly, the biotechnological production of succinic acid becomes an attractive alternative to the traditional petrochemical route. In *C. glutamicum*, the production and secretion of succinic acid can easily be induced by shifting the cultivation conditions towards oxygen deprivation [52]. Supplementation of the medium with bicarbonate thereby significantly improved the production of succinate by supporting the CO_2-fixation via the anaplerotic enzymes responsible for carbon flux from phosphoenolpyruvate and pyruvate to oxaloacetate [53]. Moreover bicarbonate can be used to modify the proportionally secreted amounts of lactate and succinate [103]. Deletion of the *ldhA* gene, encoding lactate dehydrogenase, however, was required to avoid lactate accumulation during anaerobic succinate production (Fig. 6.11).

A further step towards strain improvement comprised over expression of the anaplerotic enzyme pyruvate carboxylase which resulted in a twofold increased succinate production [52]. Combining the beneficial modifications, i.e. *ldhA* deletion and pyruvate carboxylase over expression, yielded a highly efficient succinic acid producer that accumulated 146 g L^{-1} succinic acid within 46 h [104]. Recent studies identified a possible succinic acid exporter *sucE* in *C. glutamicum* [39] which represents an attractive target for further metabolic engineering towards improved production performance.

6.4.3 Propanediol

1,2-Propanediol (propylene glycol) is a valuable commodity with industrial applications as food additives, pharmaceuticals, cosmetics, and de-icers because of its physical properties and low toxicity. More than half a million tons of 1,2-propanediol are annually produced in the USA from petroleum. Production from renewable resources is desirable due to concerns of global warming and anticipated exhaustion of oil resources. The wild type of *C. glutamicum* produces only traces of 1,2-propanediol [100]. After 130 h of incubation, about 90 μM of the desired product were detected under aerobic conditions. *C. glutamicum* obviously lacks a

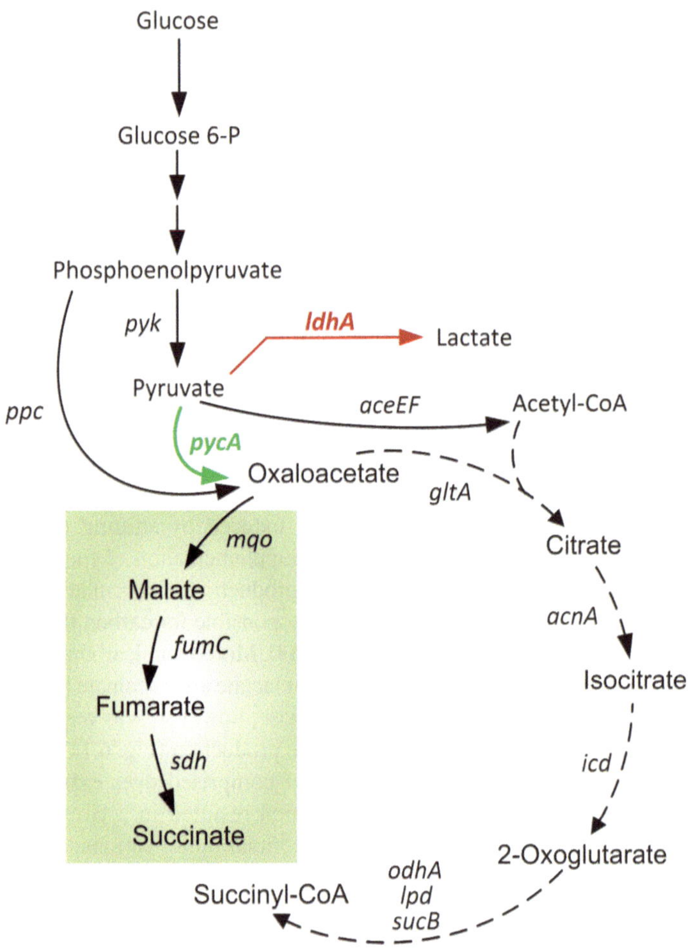

Fig. 6.11 Metabolic engineering of *C. glutamicum* for overproduction of succinate. The genetic modifications are highlighted in *red* (chromosomal deletion) and *green* (plasmid based overexpression). Genes: *aceEF* pyruvate dehydrogenase complex, *acnA* aconitase, *fumC* fumarase, *gltA* citrate synthase, *icd* isocitrate dehydrogenase, *odhA*/*lpd*/*sucB* 2-oxoglutarate dehydrogenase complex, *mqo* malat-quinone-oxidoreductase, *sdh* succinate dehydrogenase, *ppc* phosphoenolpyruvate carboxylase, *pycA* pyruvate carboxylase, *pyk* pyruvate kinase

gene encoding methylglyoxal synthase (Mgs), which catalyzes the first step of 1,2-propanediol synthesis from the glycolytic pathway, but several genes are annotated as putative aldo-keto reductases, which is potentially involved as methylglyoxal reductase in the biosynthetic chain towards 1,2-propanediol (Fig. 6.12).

The expression of the *mgs* gene from *E. coli* gene increased the product yield 100-fold and showed that *C. glutamicum* indeed carries the genes downstream of methylglyoxal synthase. This could be driven further by simultaneous overexpression of *mgs* and cgR_2242, one of the genes annotated as aldo-keto

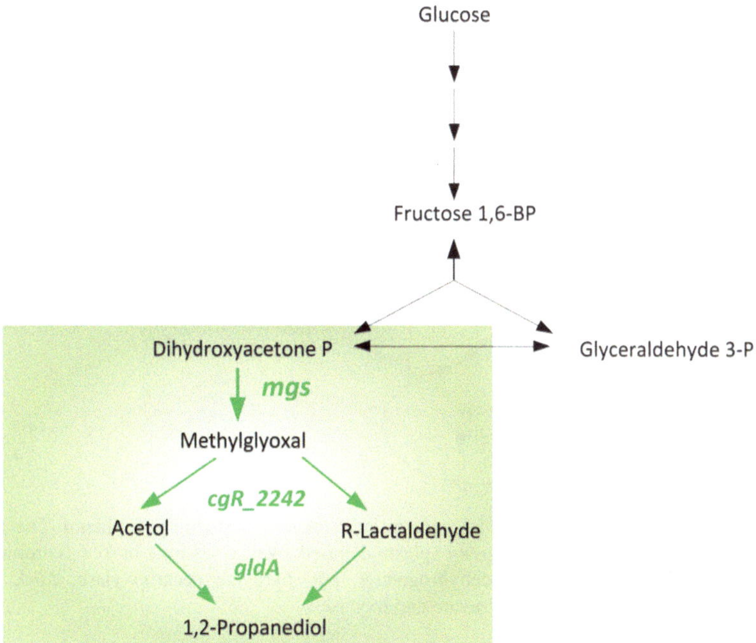

Fig. 6.12 Metabolic engineering of *C. glutamicum* for overproduction of 1,2-propanediol. The genetic modifications are highlighted in *green* (plasmid based overexpression). Genes: CgR_2242 Aldo-keto reductase from *C. glutamicum* R

reductase as well as glycerol dehydrogenase (*gldA*). The obtained mutant showed an enhanced 1,2-propanediol production of 24 mM, a promising starting point for further developments. Acetol was still secreted as major product suggesting further improvement.

6.5 Bio-Fuels

6.5.1 Ethanol

Bioethanol is an important fuel. It is applied worldwide in pure form or as additive in combustion engines and serves as starting material for large scale synthesis of ethyl-tert-butyl-ether, another fuel additive. Except for countries such as Brazil with an extremely advantageous availability of renewable feedstocks, bio-ethanol production is still to date not cost-effective compared to fossil fuels, thus demanding for bioprocesses with increased efficiency. This has initiated metabolic engineering of different bacteria and yeasts for hyper production of bio-based ethanol. Despite tremendous improvements reported e.g. for *S. cerevisiae*, *Z. mobilis* or *E. coli*, remaining problems such as the incapability to completely

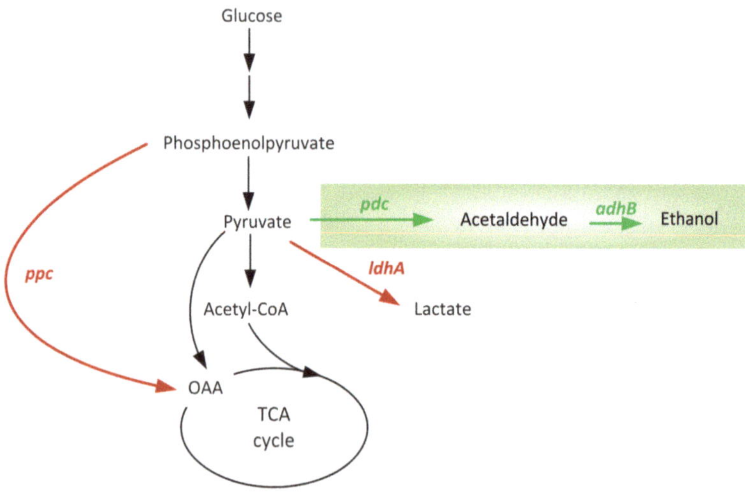

Fig. 6.13 Metabolic engineering of *C. glutamicum* for overproduction of ethanol. The genetic modifications are highlighted in *green* (plasmid based overexpression) or *red* (chromosomal deletion). Genes: *adhB* alcohol dehydrogenase, *pdc* pyruvate decarboxylase, *ldhA* lactate dedydrogenase, *ppc* phosphoenolpyruvate carboxylase

utilize hexoses and pentoses from cheap hemicellulose or cellulose raw materials, reduced solvent tolerance or genetic instability still drive the search for alternative bio-ethanol producers. *C. glutamicum* appeared as promising production strain, since it can be successfully engineered to metabolize a wide variety of relevant carbon sources and shows a natural capability to secrete large amounts of reduced by-products such as lactate under conditions of oxygen limitation [52]. In a recent study, the central metabolism of *C. glutamicum* was engineered to produce ethanol (Fig. 6.13).

The recombinant strain *C. glutamicum* WT-pCRA723 expressed the *Z. mobilis* genes encoding for pyruvate decarboxylase (*pdc*) and alcohol dehydrogenase (*adhB*) under control of the *ldhA* promoter. Under conditions of oxygen limitation it exhibited ethanol production from glucose at a conversion yield 20% with substantial formation of the by-products succinate and lactate. Subsequent deletion of the *ldhA* gene prevented the formation of lactate and increased the yield to 50%. Succinate formation was then eliminated by disruption of the phosphoenolpyruvate carboxylase gene (*ppc*), leading to increased ethanol production, especially when pyruvate was added as supporting substrate. Incubation of the *ldhA* mutant at a cell concentration of about 300 g L^{-1} of wet cells under oxygen-deprivation conditions yielded an ethanol production rate of almost 30 g L^{-1} h^{-1}. This value is substantially higher than that of many other bacteria reported so far. It displays a promising starting point for further optimization. Important properties to be optimized by systems metabolic engineering include increased tolerance of *C. glutamicum* to the solvent or to furfurals or furans present in hemicellulose and cellulose feedstocks [119].

6.5.2 Isobutanol

Among the bio-fuels, increasing attention is being paid to higher alcohols. Those alcohols have four to five carbons and possess a host of qualities making them more suitable as a liquid fuel than ethanol with lower vapor pressure, lower hygroscopicity, and higher energy density [4]. The biosynthetic routes towards higher alcohols share common precursors with amino acid biosynthetic pathways. This suggests *C. glutamicum*, the most important working horse for amino acid production, as interesting host for higher biofuels. Analysis of this host's sensitivity to isobutanol toxicity revealed that *C. glutamicum* shows an increased tolerance to isobutanol relative to *E. coli* [134]. The potential of *C. glutamicum* genes was first explored for synthesis of 2-methyl-1-butanol in *E. coli*. The gene encoding threonine deaminase (*ilvA*) from *C. glutamicum* was successfully used in recombinant *E. coli*. Overexpression of *ilvA* was hereby combined with overexpression of AHAS II (*ilvGM*) from *S. typhimurium* and the native threonine biosynthetic operon (*thrABC*) as well as deletion of competing reactions (Δ*metA*, Δ*tdh*) upstream of threonine production. The resulting strain produced 1.25 g L^{-1} 2-methyl-1-butanol in 24 h, a total alcohol content of 3 g L^{-1}, and yields of up to 0.17 g 2-methyl-1-butanol (g glucose)$^{-1}$.

The first approach, realizing higher alcohol biosynthesis directly in *C. glutamicum* focused on isobutanol production [134]. Valine, which shares the precursor 2-ketoisovalerate with isobutanol, has been previously produced in *C. glutamicum* from glucose by overexpression of the valine biosynthesis pathway (*ilvBNCDE*) and elimination of carbon competing pathways [14, 16]. To convert *C. glutamicum* into an isobutanol producer, a similar strategy was applied involving overexpression of the 2-keto acid synthesis pathway (Fig. 6.14).

This pathway comprises *alsS* (*B. subtilis*) and *ilvCD* (*C. glutamicum*) along with downstream genes for the subsequent decarboxylation (*kivd* from *L. lactis*) and reduction (*adhA* from *C. glutamicum*) of 2-ketoisovalerate to isobutanol. Overexpression of these genes and a native alcohol dehydrogenase, *adhA*, led to the production of 2.6 g L^{-1} isobutanol in 48 h. In addition, other higher chain alcohols such as 3-methyl-1-butanol, 1-propanol, 2-methyl-1-butanol, 1-butanol, and 2-phenylethanol were detected as by-products. Using longer-term batch cultures, isobutanol titers reached 4.0 g L^{-1} after 96 h with the wild type *C. glutamicum* as a host. Upon the inactivation of several genes to direct more carbon through the isobutanol pathway, production was increased to 4.9 g L^{-1} isobutanol in a Δ*pyc* Δ*ldh* background. A recent study describes systems metabolic engineering of *C. glutamicum* for isobutanol production based on a mutant strain which already overproduced the intermediate 2-ketoisovalerate. This mutant carried a deletion of the competing pyruvate dehydrogenase complex and pyruvate:quinone oxidoreductase and transaminase B, in combination with overexpression of the *ilvBNCD* genes of the valine pathway encoding acetohydroxyacid synthase, acetohydroxyacid isomeroreductase, and dihydroxyacid dehydratase [7]. The strain was then tailored towards isobutanol production under oxygen deprivation by inactivation

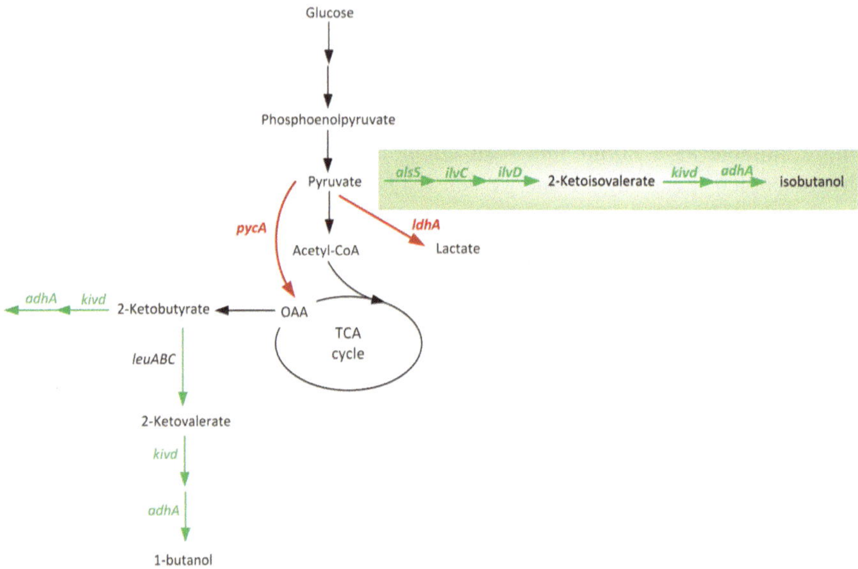

Fig. 6.14 Metabolic engineering of *C. glutamicum* for overproduction of isobutanol. The genetic modifications are highlighted in *green* (plasmid based overexpression) or *red* (chromosomal deletion). Genes: *alsS* acetohydroxy acid synthase, *ilvC* acetohydroxyacid isomeroreductase, *ilvD* dihydroxy acid dehydratase, *kivd* ketoacid decarboxylase, *adhA* acohol dehydrogenase

of L-lactate and malate dehydrogenases, implementation of the ketoacid decarboxylase from *Lactococcus lactis*, alcohol dehydrogenase 2 (ADH2) from *S. cerevisiae*, and expression of the transhydrogenase genes *pntAB* from *E. coli*. The resulting strain produced isobutanol on glucose with a molar yield of 0.6. In fed-batch fermentations with an aerobic growth phase and an oxygen-depleted production phase, *C. glutamicum* Δ*aceE* Δ*pqo* Δ*ilvE* Δ*ldhA* Δ*mdh* (pJC4*ilvBNCD-pntAB*) (pBB1*kivd-adhA*) produced about 13 g L^{-1} isobutanol with an overall molar yield of about 0.48 in the production phase. These results show promise in engineering *C. glutamicum* for higher chain alcohol production using the 2-keto acid pathways. Further systems metabolic engineering work, addressing tolerance issues, eliminating by-products and increasing the conversion yield, are needed towards more efficient production.

6.6 The Next Level – Synthetic Metabolic Engineering

Shortly, all traditional amino-acid producing strains have been created over many years by multiple rounds of random mutagenesis and selection [61]. Due to the unavoidable accumulation of side-mutations during development they typically exhibit undesired growth deficiency, weak stress tolerance or by-product formation, suboptimal phenotypes, which often completely resist subsequent optimization rounds [101, 107]. Despite strong efforts aiming at targeted optimization of different microorganisms for industrial lysine production there is no report on a genetically defined L-Lysine over-producer that can compete with the traditionally generated production strains created in the past 50 years [81]. The recent creation of a wild type based lysine hyper-producer is the first example of a tailor-made strain design by synthetic metabolic engineering applied to *C. glutamicum* [12]. It based on the model-based draft of an optimal blueprint for production, comparing the *in vivo* fluxes from ^{13}C fluxomics in the wild type (start) and the theoretical *in silico* fluxes from computational genome-scale modeling for maximum production (target). This flux design was used to predict a combination of 12 beneficial genetic modifications that optimally support lysine production in *C. glutamicum* (Fig. 6.15). These genetic targets were located in all central metabolic pathways and accounted for the key-characteristics required for optimal lysine-production e.g. (i) amplified expression of enzymes involved in lysine biosynthesis, (ii) improved supply of carbon building blocks, (iii) reduced withdrawal of carbon for competing reactions and (iv) improved co-factor supply. Following this strategy, the metabolism of *C. glutamicum* was systematically modified (Fig. 6.15). The final engineered *C. glutamicum* strain was able to produce lysine with a high yield of 0.55 g per gram of glucose, a titer of 120 g L^{-1} lysine and a productivity of 4.0 g L^{-1} h^{-1} in fed-batch culture. As confirmed by metabolic flux analysis, major carbon fluxes were re-directed in a desired manner towards the optimally pathway usage predicted by *in silico* modeling. The resulting space-time yield of 4.0 g L^{-1} h^{-1} is twice as high than the value of 2.1 g L^{-1} h^{-1} reported for fed-batch fermentation by classical producers [37]. The shortened fermentation time is also beneficial in avoiding reactor contamination which represents a severe threat to production [61]. The excellent performance of *C. glutamicum* Lys-12 also holds for the achieved titer and the carbon yield. Concerning these criteria it lies at the maximum limit of classically derived producers with carbon yields of 40–50% [83] and lysine titers of 80–120 g L^{-1} [1].

Overall, this yielded the first rationally designed lysine producer that can compete with classical production strains and clearly highlights the value of synthetic metabolic engineering [12]. Beyond the iterative analysis and implementation rounds of systems metabolic engineering, synthetic metabolic engineering directly drafts a blueprint of an optimized producer a priori on basis of model predictions. The predicted flux changes are efficiently and globally translated into appropriate genetic engineering strategies to realize optimized phenotypes *in vivo*. Together with the impressive developments recently described for production of valine, threonine, artemisinine or taxol in *E. coli* (see Chap. 5), this provides the next-level of industrial strain engineering on a global and knowledge based scale towards industrially competitive designer bugs.

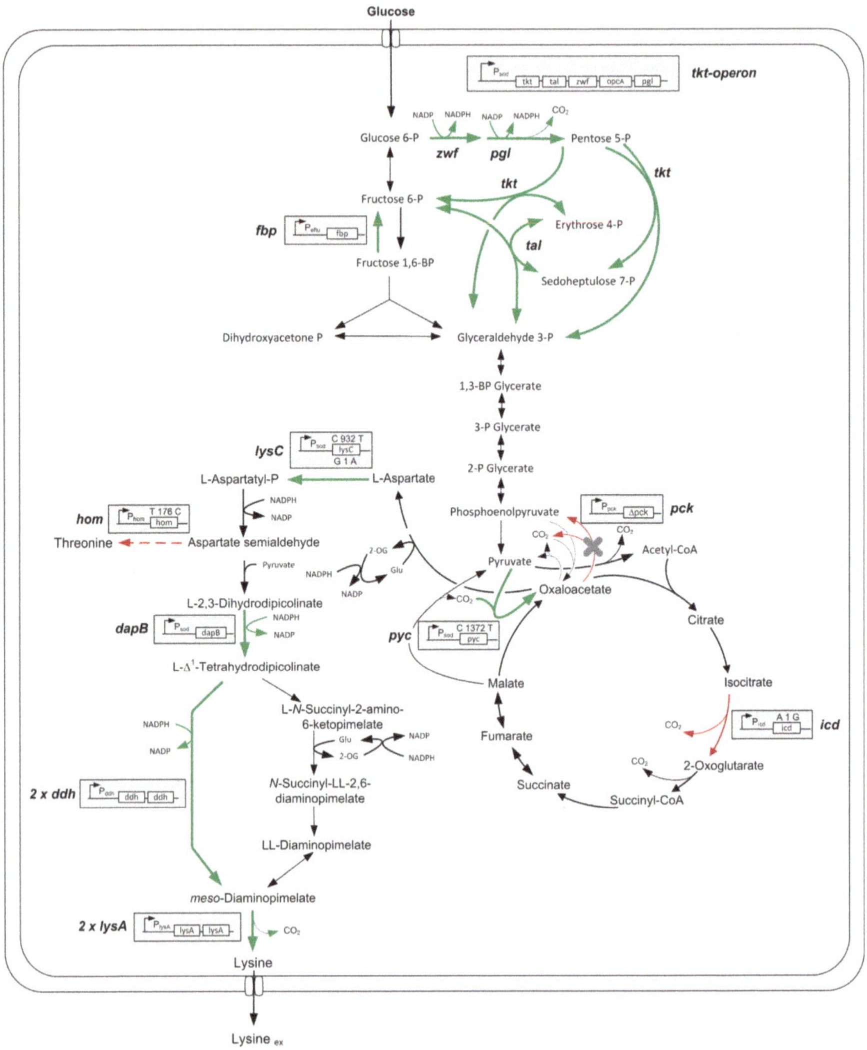

Fig. 6.15 Synthetic metabolic engineering of *C. glutamicum* for overproduction of lysine [12]. The rational, multi-target engineering started from the non-producing wild type *C. glutamicum* ATCC 13032. All modifications were implemented into the genome providing a stable production strain. *Gray boxes* represent the targeted modification of the respective genes. The symbol "x" indicates gene deletion. *Green arrows* indicate amplification, *red arrows* attenuation or deletion. Genes: *dapB* dihydrodipicolinate reductase, *ddh* diaminopimelate dehydrogenase, *eftu* elongation factor tu, *fbp* fructose-1,6-bisphosphatase, *hom* homoserine dehydrogenase, *lysA* diaminopimelate decarboxylase, *lysC* aspartokinase, *pck* phosphoenol-pyruvate carboxykinase, *pgl* phosphogluconolactonoase, *pyc* pyruvate carboxylase, *sod* superoxide dismutase, *tkt* transketolase, *tal* transaldolase, *zwf*/*opcA* glucose 6-phosphate dehydrogenase complex

6.7 Conclusions and Perspectives

The power of systems metabolic engineering strategies together with a strong need for a bio-based and eco-efficient production of chemicals, materials and fuels are major drivers for the development of novel bio-based production processes [51, 66, 124, 134]. The examples presented in this review, underline the tremendous progress to turn well accepted industrial working horses such as *C. glutamicum* into platform cell factories. Hereby, the product portfolio of *C. glutamicum* is continuously extended. Recent additions include e.g. polyhydroxyalkanoates [87] or D-amino acids [137]. As shown for the two examples of lysine production [12] and 1,5-diaminopentane production [66] systems metabolic engineering or synthetic metabolic engineering created industrial strains of *C. glutamicum* with excellent production properties in a well-defined genomic context. Based on a computational genome-scale model and considering biological feasibility, metabolic properties can be investigated through *in silico* and then be used to develop concepts for the improved bio-production of desired compounds. We can expect an even faster and powerful development of industrial production strains with the advent and integration of synthetic biology and more sophisticated models on metabolic and regulatory networks, used as basis to create desired combinations of pathways, enzyme variants, biosynthetic routes and regulatory circuits for the production of natural or non-natural products encoded in artificial genomes, that can then directly be synthesized *in vitro* [114].

References

1. Anastassiadis S (2007) L-lysine fermentation. Recent Pat Biotechnol 1(1):11–24
2. Aristidou A, Penttila M (2000) Metabolic engineering applications to renewable resource utilization. Curr Opin Biotechnol 11(2):187–198
3. Asakura Y, Kimura E, Usuda Y, Kawahara Y, Matsui K, Osumi T, Nakamatsu T (2007) Altered metabolic flux due to deletion of *odhA* causes L-glutamate overproduction in *Corynebacterium glutamicum*. Appl Environ Microbiol 73(4):1308–1319
4. Atsumi S, Hanai T, Liao JC (2008) Non-fermentative pathways for synthesis of branched-chain higher alcohols as biofuels. Nature 451(7174):86–89
5. Barrett E, Stanton C, Zelder O, Fitzgerald G, Ross RP (2004) Heterologous expression of lactose- and galactose-utilizing pathways from lactic acid bacteria in *Corynebacterium glutamicum* for production of lysine in whey. Appl Environ Microbiol 70(5):2861–2866
6. Bartek T, Blombach B, Zönnchen E, Makus P, Lang S, Eikmanns BJ, Oldiges M (2010) Importance of NADPH supply for improved L-valine formation in *Corynebacterium glutamicum*. Biotechnol Prog 26(2):361–371
7. Bartek T, Blombach B, Lang S, Eikmanns BJ, Wiechert W, Oldiges M, Nöh K, Noack S (2011) Comparative ^{13}C metabolic flux analysis of pyruvate dehydrogenase complex-deficient, L-Valine-producing *Corynebacterium glutamicum*. Appl Environ Microbiol 77 (18):6644–6652
8. Becker J, Heinzle E, Klopprogge C, Zelder O, Wittmann C (2005) Amplified expression of fructose 1,6-bisphosphatase in *Corynebacterium glutamicum* increases *in vivo* flux through

the pentose phosphate pathway and lysine production on different carbon sources. Appl Environ Microbiol 71(12):8587–8596
9. Becker J, Klopprogge C, Herold A, Zelder O, Bolten CJ, Wittmann C (2007) Metabolic flux engineering of L-lysine production in *Corynebacterium glutamicum*-over expression and modification of G6P dehydrogenase. J Biotechnol 132(2):99–109
10. Becker J, Klopprogge C, Schröder H, Wittmann C (2009) Metabolic engineering of the tricarboxylic acid cycle for improved lysine production by *Corynebacterium glutamicum*. Appl Environ Microbiol 75(24):7866–7869
11. Becker J, Buschke N, Bücker R, Wittmann C (2010) Systems level engineering of *Corynebacterium glutamicum* – reprogramming translational efficiency for superior production. Eng Life Sci 10(5):430–438
12. Becker J, Zelder O, Häfner S, Schröder H, Wittmann C (2011) From zero to hero-Design-based systems metabolic engineering of *Corynebacterium glutamicum* for l-lysine production. Metab Eng 13(2):159–168
13. Blombach B, Seibold GM (2010) Carbohydrate metabolism in *Corynebacterium glutamicum* and applications for the metabolic engineering of L: -lysine production strains. Appl Microbiol Biotechnol 75:7866
14. Blombach B, Schreiner ME, Holatko J, Bartek T, Oldiges M, Eikmanns BJ (2007) L-valine production with pyruvate dehydrogenase complex-deficient *Corynebacterium glutamicum*. Appl Environ Microbiol 73(7):2079–2084
15. Blombach B, Schreiner ME, Moch M, Oldiges M, Eikmanns BJ (2007) Effect of pyruvate dehydrogenase complex deficiency on L-lysine production with *Corynebacterium glutamicum*. Appl Microbiol Biotechnol 76(3):615–623
16. Blombach B, Schreiner ME, Bartek T, Oldiges M, Eikmanns BJ (2008) *Corynebacterium glutamicum* tailored for high-yield L-valine production. Appl Microbiol Biotechnol 79 (3):471–479
17. Bolten CJ, Schröder H, Dickschat J, Wittmann C (2010) Towards methionine overproduction in *Corynebacterium glutamicum* – methanethiol and dimethyldisulfide as reduced sulfur sources. J Microbiol Biotechnol 20(8):1196–1203
18. Bongaerts J, Krämer M, Müller U, Raeven L, Wubbolts M (2001) Metabolic engineering for microbial production of aromatic amino acids and derived compounds. Metab Eng 3 (4):289–300
19. Borner J, Buchinger S, Schomburg D (2007) A high-throughput method for microbial metabolome analysis using gas chromatography/mass spectrometry. Anal Biochem 367 (2):143–151
20. Brinkrolf K, Brune I, Tauch A (2007) The transcriptional regulatory network of the amino acid producer *Corynebacterium glutamicum*. J Biotechnol 129(2):191–211
21. Brinkrolf K, Schröder J, Pühler A, Tauch A (2010) The transcriptional regulatory repertoire of *Corynebacterium glutamicum*: reconstruction of the network controlling pathways involved in lysine and glutamate production. J Biotechnol 149(3):173–182
22. Brune I, Brinkrolf K, Kalinowski J, Pühler A, Tauch A (2005) The individual and common repertoire of DNA-binding transcriptional regulators of *Corynebacterium glutamicum*, *Corynebacterium efficiens*, *Corynebacterium diphtheriae* and *Corynebacterium jeikeium* deduced from the complete genome sequences. BMC Genomics 6(1):86
23. Buchinger S, Strösser J, Rehm N, Hänssler E, Hans S, Bathe B, Schomburg D, Krämer R, Burkovski A (2009) A combination of metabolome and transcriptome analyses reveals new targets of the *Corynebacterium glutamicum* nitrogen regulator AmtR. J Biotechnol 140 (1–2):68–74
24. Burkovski A (2006) Proteomics of *Corynebacterium glutamicum*: essential industrial bacterium. Methods Biochem Anal 49:137–147
25. Carothers WH (1938) Linear polyamids and their production. US Patent
26. Dickschat J, Wickel S, Bolten CJ, Nawrath T, Schulz S, Wittmann C (2010) Pyrazine biosynthesis in *Corynebacterium glutamicum*. Eur J Org Chem 2010:2687–2695

27. Dominguez H, Lindley ND (1996) Complete sucrose metabolism requires fructose phosphotransferase activity in *Corynebacterium glutamicum* to ensure phosphorylation of liberated fructose. Appl Environ Microbiol 62(10):3878–3880
28. Dominguez H, Rollin C, Guyonvarch A, Guerquin-Kern JL, Cocaign-Bousquet M, Lindley ND (1998) Carbon-flux distribution in the central metabolic pathways of *Corynebacterium glutamicum* during growth on fructose. Eur J Biochem 254(1):96–102
29. Dong X, Quinn PJ, Wang X (2010) Metabolic engineering of *Escherichia coli* and *Corynebacterium glutamicum* for the production of L-threonine. Biotechnol Adv 29(1):11–23
30. Escalante A, Calderon R, Valdivia A, de Anda R, Hernandez G, Ramirez OT, Gosset G, Bolivar F (2010) Metabolic engineering for the production of shikimic acid in an evolved *Escherichia coli* strain lacking the phosphoenolpyruvate: carbohydrate phosphotransferase system. Microb Cell Fact 9:21
31. Gosset G, Yong-Xiao J, Berry A (1996) A direct comparison of approaches for increasing carbon flow to aromatic biosynthesis in *Escherichia coli*. J Ind Microbiol 17(1):47–52
32. Haberhauer G, Schröder H, Pompejus M, Zelder O, Kröger B (2001) *Corynebacterium glutamicum* genes encoding proteins involved in membrane synthesis and membrane transport. Patent WO 01/00805 Patent
33. Hagishita T, Yoshida T, Izumi Y, Mitsunaga T (1996) Efficient L-serine production from methanol and glycine by resting cells of *Methylobacterium* sp. strain MN43. Biosci Biotechnol Biochem 60(10):1604–1607
34. Hasegawa T, Hashimoto K, Kawasaki H, Nakamatsu T (2008) Changes in enzyme activities at the pyruvate node in glutamate-overproducing *Corynebacterium glutamicum*. J Biosci Bioeng 105(1):12–19
35. Hayashi M, Ohnishi J, Mitsuhashi S, Yonetani Y, Hashimoto S, Ikeda M (2006) Transcriptome analysis reveals global expression changes in an industrial L-lysine producer of *Corynebacterium glutamicum*. Biosci Biotechnol Biochem 70(2):546–550
36. Hermann T, Finkemeier M, Pfefferle W, Wersch G, Krämer R, Burkovski A (2000) Two-dimensional electrophoretic analysis of *Corynebacterium glutamicum* membrane fraction and surface proteins. Electrophoresis 21(3):654–659
37. Hirao T, Nakano T, Azuma T, Sugimoto M, Nakanishi T (1989) L-Lysine production in continuous culture of an L-lysine hyperproducing mutant of *Corynebacterium glutamicum*. Appl Microbiol Biotechnol 32:269–273
38. Holatko J, Elisakova V, Prouza M, Sobotka M, Nesvera J, Patek M (2009) Metabolic engineering of the L-valine biosynthesis pathway in *Corynebacterium glutamicum* using promoter activity modulation. J Biotechnol 139(3):203–210
39. Huhn S, Jolkver E, Krämer R, Marin K (2011) Identification of the membrane protein SucE and its role in succinate transport in *Corynebacterium glutamicum*. Appl Microbiol Biotechnol 89:327–335
40. Hüser AT, Chassagnole C, Lindley ND, Merkamm M, Guyonvarch A, Elisakova V, Patek M, Kalinowski J, Brune I, Pühler A, Tauch A (2005) Rational design of a *Corynebacterium glutamicum* pantothenate production strain and its characterization by metabolic flux analysis and genome-wide transcriptional profiling. Appl Environ Microbiol 71(6):3255–3268
41. Ikeda M (2003) Amino acid production processes. Adv Biochem Eng Biotechnol 79:1–35
42. Ikeda M (2006) Towards bacterial strains overproducing L-tryptophan and other aromatics by metabolic engineering. Appl Microbiol Biotechnol 69(6):615–626
43. Ikeda M, Katsumata R (1992) Metabolic engineering to produce tyrosine or phenylalanine in a tryptophan-producing *Corynebacterium glutamicum* strain. Appl Environ Microbiol 58 (3):781–785
44. Ikeda M, Katsumata R (1994) Transport of aromatic amino acids and its influence on overproduction of the amino acids in *Corynebacterium glutamicum*. J Ferment Bioeng 78 (6):420–425
45. Ikeda M, Katsumata R (1995) Tryptophan production by transport mutants of *Corynebacterium glutamicum*. Biosci Biotechnol Biochem 59(8):1600–1602

46. Ikeda M, Katsumata R (1999) Hyperproduction of tryptophan by *Corynebacterium glutamicum* with the modified pentose phosphate pathway. Appl Environ Microbiol 65 (6):2497–2502
47. Ikeda M, Nakagawa S (2003) The *Corynebacterium glutamicum* genome: features and impacts on biotechnological processes. Appl Microbiol Biotechnol 62(2–3):99–109
48. Ikeda M, Nakanishi K, Kino K, Katsumata R (1994) Fermentative production of tryptophan by a stable recombinant strain of *Corynebacterium glutamicum* with a modified serine-biosynthetic pathway. Biosci Biotechnol Biochem 58(4):674–678
49. Ikeda M, Okamoto K, Katsumata R (1999) Cloning of the transketolase gene and the effect of its dosage on aromatic amino acid production in *Corynebacterium glutamicum*. Appl Microbiol Biotechnol 51(2):201–206
50. Ikeda M, Ohnishi J, Hayashi M, Mitsuhashi S (2006) A genome-based approach to create a minimally mutated *Corynebacterium glutamicum* strain for efficient L-lysine production. J Ind Microbiol Biotechnol 33(7):610–615
51. Ikeda M, Mitsuhashi S, Tanaka K, Hayashi M (2009) Reengineering of a *Corynebacterium glutamicum* L-arginine and L-citrulline producer. Appl Environ Microbiol 75(6):1635–1641
52. Inui M, Murakami S, Okino S, Kawaguchi H, Vertes AA, Yukawa H (2004) Metabolic analysis of *Corynebacterium glutamicum* during lactate and succinate productions under oxygen deprivation conditions. J Mol Microbiol Biotechnol 7(4):182–196
53. Inui M, Suda M, Okino S, Nonaka H, Puskas LG, Vertes AA, Yukawa H (2007) Transcriptional profiling of *Corynebacterium glutamicum* metabolism during organic acid production under oxygen deprivation conditions. Microbiology 153(Pt 8):2491–2504
54. Jojima T, Fujii M, Mori E, Inui M, Yukawa H (2010) Engineering of sugar metabolism of *Corynebacterium glutamicum* for production of amino acid L-alanine under oxygen deprivation. Appl Microbiol Biotechnol 87(1):159–165
55. Kalinowski J, Cremer J, Bachmann B, Eggeling L, Sahm H, Pühler A (1991) Genetic and biochemical analysis of the aspartokinase from *Corynebacterium glutamicum*. Mol Microbiol 5(5):1197–1204
56. Kalinowski J, Bathe B, Bartels D, Bischoff N, Bott M, Burkovski A, Dusch N, Eggeling L, Eikmanns BJ, Gaigalat L, Goesmann A, Hartmann M, Huthmacher K, Krämer R, Linke B, McHardy AC, Meyer F, Möckel B, Pfefferle W, Pühler A, Rey DA, Rückert C, Rupp O, Sahm H, Wendisch VF, Wiegrabe I, Tauch A (2003) The complete *Corynebacterium glutamicum* ATCC 13032 genome sequence and its impact on the production of L-aspartate-derived amino acids and vitamins. J Biotechnol 104(1–3):5–25
57. Katsumata R, Kino K (1989) Process for producing amino acids by fermentation. Japan Patent 01,317,395 A
58. Kawaguchi H, Vertes AA, Okino S, Inui M, Yukawa H (2006) Engineering of a xylose metabolic pathway in *Corynebacterium glutamicum*. Appl Environ Microbiol 72 (5):3418–3428
59. Kawaguchi H, Sasaki M, Vertes AA, Inui M, Yukawa H (2008) Engineering of an L-arabinose metabolic pathway in *Corynebacterium glutamicum*. Appl Microbiol Biotechnol 77(5):1053–1062
60. Kawaguchi H, Sasaki M, Vertes AA, Inui M, Yukawa H (2009) Identification and functional analysis of the gene cluster for L-arabinose utilization in *Corynebacterium glutamicum*. Appl Environ Microbiol 75(11):3419–3429
61. Kelle R, Hermann T, Bathe B (2005) L-lysine production. In: Eggeling L, Bott M (eds) Handbook of *Corynebacterium glutamicum*. CRC Press, Boca Raton, pp 465–488
62. Kiefer P, Heinzle E, Wittmann C (2002) Influence of glucose, fructose and sucrose as carbon sources on kinetics and stoichiometry of lysine production by *Corynebacterium glutamicum*. J Ind Microbiol Biotechnol 28(6):338–343
63. Kiefer P, Heinzle E, Zelder O, Wittmann C (2004) Comparative metabolic flux analysis of lysine-producing *Corynebacterium glutamicum* cultured on glucose or fructose. Appl Environ Microbiol 70(1):229–239

64. Kim J, Fukuda H, Hirasawa T, Nagahisa K, Nagai K, Wachi M, Shimizu H (2009) Requirement of de novo synthesis of the OdhI protein in penicillin-induced glutamate production by *Corynebacterium glutamicum*. Appl Microbiol Biotechnol 86:911–920
65. Kim J, Hirasawa T, Sato Y, Nagahisa K, Furusawa C, Shimizu H (2009) Effect of *odhA* overexpression and *odhA* antisense RNA expression on Tween-40-triggered glutamate production by *Corynebacterium glutamicum*. Appl Microbiol Biotechnol 81(6):1097–1106
66. Kind S, Jeong WK, Schröder H, Wittmann C (2010) Systems-wide metabolic pathway engineering in *Corynebacterium glutamicum* for bio-based production of diaminopentane. Metab Eng 12:341–351
67. Kind S, Jeong WK, Schröder H, Zelder O, Wittmann C (2010) Identification and elimination of the competing pathway towards N-acetyl diaminopentane for improved production of diaminopentane by *Corynebacterium glutamicum*. Appl Environ Microbiol 76:5175–5180
68. Kinoshita S, Shigezo U, Shimono M (1957) Studies on the amino acid fermentation: Part I: Production of L-glutamic acid by various microorganisms. J Gen Appl Microbiol 3 (3):193–205
69. Kinoshita S, Nakayama K, Kitada S (1961) Method of producing L-lysine by fermentation. US Patent 2979439
70. Kjeldsen KR, Nielsen J (2009) *In silico* genome-scale reconstruction and validation of the *Corynebacterium glutamicum* metabolic network. Biotechnol Bioeng 102(2):583–597
71. Kohl TA, Tauch A (2009) The GlxR regulon of the amino acid producer *Corynebacterium glutamicum*: detection of the Corynebacterial core regulon and integration into the transcriptional regulatory network model. J Biotechnol 143:239–246
72. Kohlstedt M, Becker J, Wittmann C (2010) Metabolic fluxes and beyond – systems biology understanding and engineering of microbial metabolism. Appl Microbiol Biotechnol 88 (5):1065–1075
73. Korz DJ, Rinas U, Hellmuth K, Sanders EA, Deckwer WD (1995) Simple fed-batch technique for high cell density cultivation of *Escherichia coli*. J Biotechnol 39(1):59–65
74. Kotrba P, Inui M, Yukawa H (2003) A single V317A or V317M substitution in Enzyme II of a newly identified β-glucoside phosphotransferase and utilization system of *Corynebacterium glutamicum* R extends its specificity towards cellobiose. Microbiology 149(Pt 6):1569–1580
75. Krömer JO, Sorgenfrei O, Klopprogge K, Heinzle E, Wittmann C (2004) In-depth profiling of lysine-producing *Corynebacterium glutamicum* by combined analysis of the transcriptome, metabolome, and fluxome. J Bacteriol 186(6):1769–1784
76. Krömer JO, Wittmann C, Schröder H, Heinzle E (2006) Metabolic pathway analysis for rational design of L-methionine production by *Escherichia coli* and *Corynebacterium glutamicum*. Metab Eng 8(353–369)
77. Krömer JO, Bolten CJ, Heinzle E, Schröder H, Wittmann C (2008) Physiological response of *Corynebacterium glutamicum* to oxidative stress induced by deletion of the transcriptional repressor McbR. Microbiology 154(Pt 12):3917–3930
78. Kumar D, Gomes J (2005) Methionine production by fermentation. Biotechnol Adv 23 (1):41–61
79. Kusumoto K, Sakiyama M, Sakamoto J, Noguchi S, Sone N (2000) Menaquinol oxidase activity and primary structure of cytochrome bd from the amino-acid fermenting bacterium *Corynebacterium glutamicum*. Arch Microbiol 173(5–6):390–397
80. Lee H-S (2005) Sulfur metabolism and its regulation. In: Eggeling L, Bott M (eds) Handbook of Corynebacterium glutamicum. CRC Press/Taylor & Francis, Boca Raton, pp 351–371
81. Lee SY, Park JH (2010) Integration of systems biology with bioprocess engineering: L: -threonine production by systems metabolic engineering of *Escherichia coli*. Adv Biochem Eng Biotechnol 88:1065–1075
82. Leuchtenberger W (1996) Amino acids -technical production and use. In: Rehm HJ, Reed G, Pühler A, Stadler P (eds) Biotechnology, vol 6. VCH, Weinheim, pp 465–502
83. Leuchtenberger W, Huthmacher K, Drauz K (2005) Biotechnological production of amino acids and derivatives: current status and prospects. Appl Microbiol Biotechnol 69(1):1–8
84. Li PP, Liu YJ, Liu SJ (2009) Genetic and biochemical identification of the chorismate mutase from *Corynebacterium glutamicum*. Microbiology 155(Pt 10):3382–3391

85. Marx A, Striegel K, de Graaf A, Sahm H, Eggeling L (1997) Response of the central metabolism of *Corynebacterium glutamicum* to different flux burdens. Biotechnol Bioeng 56(2):168–180
86. Marx A, Hans S, Möckel B, Bathe B, de Graaf AA (2003) Metabolic phenotype of phosphoglucose isomerase mutants of *Corynebacterium glutamicum*. J Biotechnol 104 (1–3):185–197
87. Matsumoto K, Yamada M, Leong CR, Jo SJ, Kuzuyama T, Taguchi S (2011) A new pathway for poly(3-hydroxybutyrate) production in *Escherichia coli* and *Corynebacterium glutamicum* by functional expression of a new acetoacetyl-coenzyme a synthase. Biosci Biotechnol Biochem 75(2):364–366
88. McAloon A, Taylor F, Yee W (2000) Determining the cost of producing ethanol from corn starch and lignocellulosic feedstocks. National Renewable Energy Laboratory Report
89. Melzer G, Esfandabadi ME, Franco-Lara E, Wittmann C (2009) Flux design: *in silico* design of cell factories based on correlation of pathway fluxes to desired properties. BMC Syst Biol 3:120
90. Mimitsuka T, Sawai H, Hatsu M, Yamada K (2007) Metabolic engineering of *Corynebacterium glutamicum* for cadaverine fermentation. Biosci Biotechnol Biochem 71(9):2130–2135
91. Mondal S, Das YB, Chatterjee SP (1996) Methionine production by microorganisms. Folia Microbiol (Praha) 41(6):465–472
92. Moon MW, Park SY, Choi SK, Lee JK (2007) The phosphotransferase system of *Corynebacterium glutamicum*: features of sugar transport and carbon regulation. J Mol Microbiol Biotechnol 12(1–2):43–50
93. Morbach S, Kelle R, Winkels S, Sahm H, Eggeling L (1996) Engineering the homoserine dehydrogenase and threonine dehydratase control points to analyse flux towards L-isoleucine in *Corynebacterium glutamicum*. Appl Microbiol Biotechnol 45:612–620
94. Morbach S, Sahm H, Eggeling L (1996) L-isoleucine production with *Corynebacterium glutamicum*: further flux increase and limitation of export. Appl Environ Microbiol 62 (12):4345–4351
95. Munoz AJ, Hernandez-Chavez G, de Anda R, Martinez A, Bolivar F, Gosset G (2011) Metabolic engineering of *Escherichia coli* for improving L: -3,4-dihydroxyphenylalanine (L: -DOPA) synthesis from glucose. J Ind Microbiol Biotechnol 38:1845–1852
96. Murai T, Ueda M, Shibasaki Y, Kamasawa N, Osumi M, Imanaka T, Tanaka A (1999) Development of an arming yeast strain for efficient utilization of starch by co-display of sequential amylolytic enzymes on the cell surface. Appl Microbiol Biotechnol 51(1):65–70
97. Nakayama K, Araki K (1973) Process for producing L-lysine. No US3708395 Patent
98. Nakayama K, Tanaka H, Hagino H, Kinoshita S (1966) Studies on lysine fermentation. V. Concerted feed-back inhibition of aspartokinase and the absence of lysine inhibition on aspartic semialdehyde-pyruvate condensation in *Micoroccus glutamicus*. Agric Biol Chem 30:611–616
99. Nakayama K, Araki K, Kase H (1978) Microbial production of essential amino acid with *Corynebacterium glutamicum* mutants. Adv Exp Med Biol 105:649–661
100. Niimi S, Suzuki N, Inui M, Yukawa H (2011) Metabolic engineering of 1,2-propanediol pathways in *Corynebacterium glutamicum*. Appl Microbiol Biotechnol 90:1721–1729
101. Ohnishi J, Mitsuhashi S, Hayashi M, Ando S, Yokoi H, Ochiai K, Ikeda M (2002) A novel methodology employing *Corynebacterium glutamicum* genome information to generate a new L-lysine-producing mutant. Appl Microbiol Biotechnol 58(2):217–223
102. Ohnishi J, Katahira R, Mitsuhashi S, Kakita S, Ikeda M (2005) A novel *gnd* mutation leading to increased L-lysine production in *Corynebacterium glutamicum*. FEMS Microbiol Lett 242 (2):265–274
103. Okino S, Inui M, Yukawa H (2005) Production of organic acids by *Corynebacterium glutamicum* under oxygen deprivation. Appl Microbiol Biotechnol 68:475–480

104. Okino S, Suda M, Fujikura K, Inui M, Yukawa H (2008) Production of D-lactic acid by *Corynebacterium glutamicum* under oxygen deprivation. Appl Microbiol Biotechnol 78 (3):449–454
105. Parche S, Burkovski A, Sprenger GA, Weil B, Krämer R, Titgemeyer F (2001) *Corynebacterium glutamicum*: a dissection of the PTS. J Mol Microbiol Biotechnol 3(3):423–428
106. Park SD, Lee JY, Sim SY, Kim Y, Lee HS (2007) Characteristics of methionine production by an engineered *Corynebacterium glutamicum* strain. Metab Eng 9(4):327–336
107. Park JH, Lee SY, Kim TY, Kim HU (2008) Application of systems biology for bioprocess development. Trends Biotechnol 26(8):404–412
108. Petersen S, de Graaf AA, Eggeling L, Möllney M, Wiechert W, Sahm H (2000) *In vivo* quantification of parallel and bidirectional fluxes in the anaplerosis of *Corynebacterium glutamicum*. J Biol Chem 275(46):35932–35941
109. Petersen S, Mack C, de Graaf AA, Riedel C, Eikmanns BJ, Sahm H (2001) Metabolic consequences of altered phosphoenolpyruvate carboxykinase activity in *Corynebacterium glutamicum* reveal anaplerotic regulation mechanisms *in vivo*. Metab Eng 3(4):344–361
110. Peters-Wendisch PG, Schiel B, Wendisch VF, Katsoulidis E, Möckel B, Sahm H, Eikmanns BJ (2001) Pyruvate carboxylase is a major bottleneck for glutamate and lysine production by *Corynebacterium glutamicum*. J Mol Microbiol Biotechnol 3(2):295–300
111. Peters-Wendisch P, Netzer R, Eggeling L, Sahm H (2002) 3-Phosphoglycerate dehydrogenase from *Corynebacterium glutamicum*: the C-terminal domain is not essential for activity but is required for inhibition by L-serine. Appl Microbiol Biotechnol 60(4):437–441
112. Peters-Wendisch P, Stolz M, Etterich H, Kennerknecht N, Sahm H, Eggeling L (2005) Metabolic engineering of *Corynebacterium glutamicum* for L-serine production. Appl Environ Microbiol 71(11):7139–7144
113. Pfefferle W, Möckel B, Bathe B, Marx A (2003) Biotechnological manufacture of lysine. Adv Biochem Eng Biotechnol 79:59–112
114. Pleiss J (2006) The promise of synthetic biology. Appl Microbiol Biotechnol 73(4):735–739
115. Radmacher E, Vaitsikova A, Burger U, Krumbach K, Sahm H, Eggeling L (2002) Linking central metabolism with increased pathway flux: L-valine accumulation by *Corynebacterium glutamicum*. Appl Environ Microbiol 68(5):2246–2250
116. Rey DA, Pühler A, Kalinowski J (2003) The putative transcriptional repressor McbR, member of the TetR-family, is involved in the regulation of the metabolic network directing the synthesis of sulfur containing amino acids in *Corynebacterium glutamicum*. J Biotechnol 103(1):51–65
117. Rey DA, Nentwich SS, Koch DJ, Rückert C, Pühler A, Tauch A, Kalinowski J (2005) The McbR repressor modulated by the effector substance S-adenosylhomocysteine controls directly the transcription of a regulon involved in sulphur metabolism of *Corynebacterium glutamicum* ATCC 13032. Mol Microbiol 56(4):871–887
118. Rittmann D, Lindner SN, Wendisch VF (2008) Engineering of a glycerol utilization pathway for amino acid production by *Corynebacterium glutamicum*. Appl Environ Microbiol 74 (20):6216–6222
119. Sakai S, Tsuchida Y, Okino S, Ichihashi O, Kawaguchi H, Watanabe T, Inui M, Yukawa H (2007) Effect of lignocellulose-derived inhibitors on growth of and ethanol production by growth-arrested *Corynebacterium glutamicum* R. Appl Environ Microbiol 73(7):2349–2353
120. Sano K, Ito K, Miwa K, Nakamori S (1987) Amplification of the phosphoenol pyruvate carboxylase gene of *Brevibacterium lactofermentum* to improve amino acid production. Agric Biol Chem 51(2):597–599
121. Sasaki M, Jojima T, Kawaguchi H, Inui M, Yukawa H (2009) Engineering of pentose transport in *Corynebacterium glutamicum* to improve simultaneous utilization of mixed sugars. Appl Microbiol Biotechnol 85:105–115
122. Sato H, Orishimo K, Shirai T, Hirasawa T, Nagahisa K, Shimizu H, Wachi M (2008) Distinct roles of two anaplerotic pathways in glutamate production induced by biotin limitation in *Corynebacterium glutamicum*. J Biosci Bioeng 106(1):51–58

123. Sawada K, Zen-In S, Wada M, Yokota A (2010) Metabolic changes in a pyruvate kinase gene deletion mutant of *Corynebacterium glutamicum* ATCC 13032. Metab Eng 12:401–407
124. Schneider J, Wendisch VF (2010) Putrescine production by engineered *Corynebacterium glutamicum*. Appl Microbiol Biotechnol 88:859–869
125. Schneider J, Niermann K, Wendisch VF (2011) Production of the amino acids l-glutamate, l-lysine, l-ornithine and l-arginine from arabinose by recombinant *Corynebacterium glutamicum*. J Biotechnol 154:191–198
126. Seibold G, Auchter M, Berens S, Kalinowski J, Eikmanns BJ (2006) Utilization of soluble starch by a recombinant *Corynebacterium glutamicum* strain: growth and lysine production. J Biotechnol 124(2):381–391
127. Shiio I, Sano K (1969) Microbial producion of L-lysine. II. production by mutants sensitive to threonine or methionine. J Gen Appl Microbiol 15:267–287
128. Shiio I, Miyajima R, Nakagawa M (1972) Regulation of aromatic amino acid biosynthesis in *Brevibacterium flavum*. I. Regulation of anthranilate synthetase. J Biochem (Tokyo) 72 (6):1447–1455
129. Shiio I, Sugimoto S, Yoshino H, Kawamura K (1990) Isolation and properties of threonine-producing mutants with both, dihydrodipicolinate-synthase defect and feddback-resistant homoserine dehydrogenase from *Brevibacterium flavum*. Agric Biol Chem 54(6):1505–1511
130. Shinfuku Y, Sorpitiporn N, Sono M, Furusawa C, Hirasawa T, Shimizu H (2009) Development and experimental verification of a genome-scale metabolic model for *Corynebacterium glutamicum*. Microb Cell Fact 8:43
131. Shirai T, Nakato A, Izutani N, Nagahisa K, Shioya S, Kimura E, Kawarabayasi Y, Yamagishi A, Gojobori T, Shimizu H (2005) Comparative study of flux redistribution of metabolic pathway in glutamate production by two coryneform bacteria. Metab Eng 7(2):59–69
132. Shirai T, Fujimura K, Furusawa C, Nagahisa K, Shioya S, Shimizu H (2007) Study on roles of anaplerotic pathways in glutamate overproduction of *Corynebacterium glutamicum* by metabolic flux analysis. Microb Cell Fact 6:19
133. Silberbach M, Schäfer M, Hüser AT, Kalinowski J, Pühler A, Krämer R, Burkovski A (2005) Adaptation of *Corynebacterium glutamicum* to ammonium limitation: a global analysis using transcriptome and proteome techniques. Appl Environ Microbiol 71(5):2391–2402
134. Smith KM, Cho KM, Liao JC (2010) Engineering *Corynebacterium glutamicum* for isobutanol production. Appl Microbiol Biotechnol 87(3):1045–1055
135. Sonntag K, Eggeling L, De Graaf AA, Sahm H (1993) Flux partitioning in the split pathway of lysine synthesis in *Corynebacterium glutamicum*. Quantification by ^{13}C- and ^{1}H-NMR spectroscopy. Eur J Biochem 213(3):1325–1331
136. Sonntag K, Schwinde J, de Graaf A, Marx A, Eikmanns B, Wiechert W, Sahm H (1995) ^{13}C NMR studies of the fluxes in the central metabolism of *Corynebacterium glutamicum* during growth and overproduction of amino acids in batch cultures. Appl Microbiol Biotechnol 44:489–495
137. Stäbler N, Oikawa T, Bott M, Eggeling L (2011) *Corynebacterium glutamicum* as a host for the synthesis and export of D-amino acids. J Bacteriol 193:1702–1709
138. Sugimoto S, Nakagawa M, Tsuchida T, Shiio I (1973) Regulation of aromatic amino acid biosynthesis and production of tyrosine and phenylalanine in *Brevibacterium flavum*. Agric Biol Chem 37(10):2327–2336
139. Tateno T, Fukuda H, Kondo A (2007) Production of L-Lysine from starch by *Corynebacterium glutamicum* displaying α-amylase on its cell surface. Appl Microbiol Biotechnol 74 (6):1213–1220
140. Tateno T, Okada Y, Tsuchidate T, Tanaka T, Fukuda H, Kondo A (2009) Direct production of cadaverine from soluble starch using *Corynebacterium glutamicum* coexpressing alpha-amylase and lysine decarboxylase. Appl Microbiol Biotechnol 82(1):115–121
141. Tauch A, Homann I, Mormann S, Ruberg S, Billault A, Bathe B, Brand S, Brockmann-Gretza O, Rückert C, Schischka N, Wrenger C, Hoheisel J, Möckel B, Huthmacher K, Pfefferle W, Pühler A, Kalinowski J (2002) Strategy to sequence the genome of *Corynebacterium*

glutamicum ATCC 13032: use of a cosmid and a bacterial artificial chromosome library. J Biotechnol 95(1):25–38

142. Udaka S (1960) Screening method for microorganisms accumulating metabolites and its use in the isolation of *Micrococcus glutamicus*. J Bacteriol 79:754–755
143. Uteza Y, Rouillot JS, Kobetz A, Marchant D, Pecqueur S, Arnaud E, Prats H, Honiger J, Dufier JL, Abitbol M, Neuner-Jehle M (1999) Intravitreous transplantation of encapsulated fibroblasts secreting the human fibroblast growth factor 2 delays photoreceptor cell degeneration in Royal College of Surgeons rats. Proc Natl Acad Sci USA 96(6):3126–3131
144. Wagner H, Tanner H, Liebetanz E, Mannsfeld SP, Pfeiffer A (1970) Verfahren zur Herstellung von Methionin. German Patent 19 06 405
145. Wendisch VF (2003) Genome-wide expression analysis in *Corynebacterium glutamicum* using DNA microarrays. J Biotechnol 104(1–3):273–285
146. Wendisch VF, de Graaf AA, Sahm H, Eikmanns BJ (2000) Quantitative determination of metabolic fluxes during co-utilization of two carbon sources: comparative analyses with *Corynebacterium glutamicum* during growth on acetate and/or glucose. J Bacteriol 182 (11):3088–3096
147. Willis LB, Lessard PA, Sinskey AJ (2005) Synthesis of L-threonine and branched-chain amino acids. In: Eggeling L, Bott M (eds) Handbook of *Corynebacterium glutamicum*. CRC Press/Taylor & Francis, Boca Raton, pp 511–531
148. Wittmann C (2010) Analysis and engineering of metabolic pathway fluxes in *Corynebacterium glutamicum*. Adv Biochem Eng Biotechnol 120:21–49
149. Wittmann C, Becker J (2007) The L-lysine story: from metabolic pathways to industrial production. In: Wendisch VF (ed) Amino acid biosynthesis – pathways, regulation and metabolic engineering, vol 5, Microbiol Monogr. Springer, Berlin/Heidelberg, pp 39–70
150. Wittmann C, Heinzle E (2001) Application of MALDI-TOF MS to lysine-producing *Corynebacterium glutamicum*: a novel approach for metabolic flux analysis. Eur J Biochem 268 (8):2441–2455
151. Wittmann C, Heinzle E (2002) Genealogy profiling through strain improvement by using metabolic network analysis: metabolic flux genealogy of several generations of lysine-producing *Corynebacteria*. Appl Environ Microbiol 68(12):5843–5859
152. Wittmann C, Kiefer P, Zelder O (2004) Metabolic fluxes in *Corynebacterium glutamicum* during lysine production with sucrose as carbon source. Appl Environ Microbiol 70 (12):7277–7287
153. Yamamoto Y, Miwa Y, Miyoshi K, Furuyama J, Ohmori H (1997) The *Escherichia coli ldcC* gene encodes another lysine decarboxylase, probably a constitutive enzyme. Genes Genet Syst 72(3):167–172
154. Zeikus JG, Jain MK, Elankova P (1999) Biotechnology of succinic acid production and markets for derived industrial products. Appl Microbiol Biotechnol 51:545–552
155. Zhao Z, Ding JY, Li T, Zhou NY, Liu SJ (2011) The ncgl1108 (PheP (Cg)) gene encodes a new L-Phe transporter in *Corynebacterium glutamicum*. Appl Microbiol Biotechnol 90 (6):2005–2013

Chapter 7
Towards a Synthetic Biology of the Stress-Response and the Tolerance Phenotype: Systems Understanding and Engineering of the *Clostridium acetobutylicum* Stress-Response and Tolerance to Toxic Metabolites

Eleftherios T. Papoutsakis and Keith V. Alsaker

Contents

E.T. Papoutsakis (✉)
Molecular Biotechnology Laboratory, Delaware Biotechnology Institute,
The Department of Chemical & Biomolecular Engineering, University of Delaware,
15 Innovation Way, Newark, DE 19711, USA
e-mail: papoutsakis@dbi.udel.edu

K.V. Alsaker
Amyris, Inc., 5885 Hollis Street, Suite 100, Emeryville, CA 94608, USA

C. Wittmann and S.Y. Lee (eds.), *Systems Metabolic Engineering*,

DOI 10.1007/978-94-007-4534-6_7, © Springer Science+Business Media Dordrecht 2012

Abstract We discuss experimental and computational systems-biology approaches to support the development of more toxic-metabolite tolerant strains, with emphasis on solventogenic clostridia. We also discuss ideas that have the potential to move the field towards the development of integrated, predictive models of the metabolic and regulatory networks of the stress response to toxic metabolites. Clostridia are Gram+, obligate anaerobic, endospore forming bacteria of major importance to fermentative production of biofuels and chemicals from renewables. We focus on efforts to understand, model and exploit the stress-response of *Clostridium acetobutylicum* to important toxic metabolites: butanol, butyrate, and acetate. This is a problem of profound and general importance not only in clostridial biotechnologies and systems, but in all microbial systems of interest for bioenergy and chemicals production. Furthermore, the analyses and approaches discussed here are expected to be generalizable and applicable to a much broader set of microorganisms of interest to the production of biofuels and chemicals from renewable sources.

Keywords Synthetic biology • Stress response • Tolerance • Engineering • *Clostridium acetobutylicum* • Toxic metabolites • Systems biology • Metabolic model • Regulatory model • Butanol • Butyrate • Acetate • Bioenergy • Chemicals • Biofuels • Renewable resources • Multigenic phenotypes • Transcriptional stressome • Heat shock proteins • Stress regulon • Membrane modifications • Gene expression • Anaerobic • Cellulosic material • Biorefinery • Evolutionary engineering • DNA libraries • Central metabolism • Nitrogen metabolism

7.1 Introduction

Development of strains with superior tolerance characteristics to specific chemicals and general stressful bioprocess conditions is an important and widely recognized goal not only in the context of production of chemicals and biofuels from carbohydrates, but also for many bioremediation applications [22, 26, 31, 74]. If Green Chemistries are to find wide applications in the production of chemicals and biofuels, Metabolic Engineering strategies must be developed in host strains that have the ability to produce the desirable chemicals at high concentrations. Similarly, bioremediation of toxic chemicals will require metabolically engineered strains, which must be capable of degrading the desirable chemical(s) at high concentrations, i.e., they must be also tolerant to these chemicals. It is likely that the mechanism of tolerance varies widely among classes or even the same class of chemicals such as alcohols or halogenated hydrocarbons. Thus, it is critical to be able to develop strains which are tolerant to the desirable chemical(s). Biofuel production has mainly focused on ethanol, but in the last 3–4 years, butanols (n-, i- or 2-butanol) are attracting considerable interest as biofuels, as they exhibit superior chemical properties in terms of energy content, volatility, and corrosiveness [46]. Other bioprocess-based metabolites that can serve as biofuels or biofuel precursors include higher alcohols (e.g., pentanol, hexanol, propanediol, butanediols), carboxylic acids (butyrate, acetate, propionate), aldehydes, ketones, and hydrocarbons. Carboxylic acids such as butyrate are viewed as excellent feedstocks for chemical conversion to biofuels and other chemicals [62]. Accumulation of many metabolites during fermentation can be toxic to cells, inhibiting cell growth or resulting in death [56]. Thus for applications in biofuel production, understanding the cellular response to various chemicals and general stressful bioprocess conditions is an important goal [56].

Clostridia are Gram+, obligate anaerobic, endospore forming bacteria. Solventogenic and other clostridia are of major importance for developing technologies for biofuel production, and more broadly for biorefinery applications. A major and unique advantage is their ability to utilize a large variety of carbohydrate (hexoses, pentoses, oligosaccharides, xylans, starches) substrates. Among the two sequenced solventogenic clostridia, *C. acetobutylicum* is the only one that contains a full cellulosome, and may thus directly utilize cellulosic material for production of fuels and chemicals.

7.1.1 Chemical Toxicity and Tolerance Are Interconnected Complex, Multigenic Phenotypes

Stress response and tolerance of microorganisms to chemicals is a complex, multigenic trait affected by several process parameters such as pH, temperature, osmotic pressure, and the presence of other molecules [56]. Both the stress response and the tolerant phenotype are the result of several simultaneous cellular programs and mechanisms of action which involve the general stress response, altered energy metabolism, changes in membrane properties, changes in cell wall composition, and the engagement of molecular pumps. In these responses and mechanisms, the mode of action may be independent from each other, and the genes involved in each mechanism or program are many and generally not well understood, let alone modeled at the systems level [10, 37, 43, 44, 55, 58, 66, 69, 70, 90, 94, 97].

Research on the impact of alcohols and ketones on *E. coli* and other Gram-negative (Gram-) organisms has shown that intercalation of solvents within the lipid bilayer increases membrane fluidity [35], and also affects lipid-protein interactions integral to membrane function [38, 39, 69]. More broadly, responses to solvents include active solvent expulsion by molecular pumps, alteration in the composition of membrane-lipid headgroups, and adjustment of the protein content within the cell membrane [37, 68, 69, 94]. Also, solvents and other toxic chemicals induce a stress response in the cells, whereby the expression of select stress proteins (or otherwise known as heat shock proteins, HSPs) is upregulated to alleviate cell damage and specifically denaturation of proteins [89]. *E. coli* metabolism changes in the presence of ethanol and other solvents, resulting in synthesis of lipids rich in unsaturated fatty acids and more rigid membranes to compensate for the fluidizing effect of ethanol [21, 35, 69]. Much less is known about the impact of solvents and toxic chemicals on Gram + organisms such as lactobacilli and clostridia [11].

There has been limited progress in the development of alcohol, or other-solvent tolerant strains. Much effort has been based on the concept that alcohol (and more broadly solvent) tolerance can be controlled by a single gene, or a few genes. It is particularly instructive to note, however, that the tolerant phenotype is the result of a several simultaneous mechanisms of action (including molecular pumps, changes in membrane properties, changes in cell wall composition, altered energy metabolism, changes in cell size and shape) whereby, the mode of action may be independent from each other [10, 37, 43, 44, 55, 58, 66, 68–70, 90, 94, 97]. Indeed, recent research demonstrates that no single gene can uniquely confer higher resistance to a solvent [80]. Tolerance is a multigenic, complex trait, and in order to develop it, new techniques such as global transcription machinery engineering [2] must be developed and employed. Thus, high-throughput genomic strategies for developing complex phenotypes, including tolerance to alcohols and carboxylic acids, are discussed in this chapter.

7.1.2 *The Importance of* C. acetobutylicum *and Its Response and Tolerance to Butanol, Acetate and Butyrate*

C. acetobutylicum is the best model and platform organism for clostridial-based bioprocessing and biorefinery applications [62, 63]. Butanol is desirable as either as a biofuel or a commodity chemical. Among the key factors responsible for the poor process economics is the low butanol titer in the product stream, largely due to the low tolerance of *C. acetobutylicum* and most other organisms to butanol, but also to butyrate and acetate as is now more widely recognized [12]. Butanol toxicity in solventogenic clostridia is attributed to its chaotropic effect on the cell membrane [13, 91], but the actual tolerance mechanisms are much more complex. There is growing evidence that the stress response is related to tolerance [4, 5, 11, 12, 51, 86, 87]. Notably, it has been shown [56] that the upregulation or overexpression of individual heat shock proteins (HSPs) provides tolerance to solvent and possibly acid stress [5, 56, 86]. Optimized expression of these proteins, and possibly a suitable ensemble of HSPs, would provide even better tolerance, given the exquisite regulation and team action of the HSP response. Likewise, the upregulation of efflux pump genes upon solvent stress as assessed by gene-expression studies is consistent with the findings that several of these genes provide tolerance upon overexpression in engineered strains [56]. Thus, a detailed understanding at the genomic level and modeling of the cellular programs and pathways engaged in metabolite stress responses are likely to lead to creative solutions for engineering such programs to achieve superior tolerance [56].

7.2 Genome Engineering Without Models

7.2.1 *Whole-Cell and Evolutionary Engineering*

The microbial diversity present in nature, coupled with fast evolution rates, results in an enormous stock of genetic material that can be utilized to create useful microbial phenotypes. Evolutionary engineering aims to replicate and accelerate natural processes for accelerated and selective strain evolution. Beyond random mutagenesis and selection, Whole-Genome Shuffling (WGS) was developed to generate desirable microbial phenotypes. WGS is the prototypical paradigm of genome engineering whereby not prior genomic or genetic knowledge is needed, and in many ways the first instance where genome engineering has been beneficially employed. WGS has been used to develop desirable microbial phenotypes. In WGS, the DNA of similar but diversified cell populations is shuffled to recombine genes and mutations in order to generate improved bacterial phenotypes that can be screened for any screenable phenotype. Such phenotypes could include tolerance to toxic chemicals and harsh fermentation conditions (e.g., hyperoxia, high or low pH, high osmotic pressure), for increased production of commodity chemicals (such as tylosin) [19, 64, 92, 98], for utilization of substrates or the conversion of specific chemicals to other chemicals. Mutant strains from a parent strain are generated by a chosen mutagenic process, and are then recombined (based on regions of homology) by repeated protoplast fusions. Thus, protoplast fusion techniques must be developed for the desirable strains. Protoplast fusion allows for a high number of recombination events [65] based on high sequence similarity, and thus the process makes possible to combine desirable mutations for screening. Thus, evolution is accelerated and strain improvement is achieved faster. For WGS, the parental strains must have some genetic (mutational) differences but must otherwise be highly homologous as regions of homology are necessary for recombination by protoplast fusion to generate the screenable diverse populations of fusants. Prokaryotic diversity prior to WGS can be achieved by chemostat-mediated adaptation [64], whereby mutations accumulate slowly and the chemostat culture allows for their enrichment in the population, chemical mutagenesis (e.g., using the mutagen nitrosoguanidine (NTG) [64]) or UV radiation [92]. WGS was successfully used with lactobacilli [64, 98], but so far no major reports have emerged for *E. coli*, probably due to the difficulty of creating true protoplasts (a well-known difficulty in Gram- bacteria) for fusing. It should be noted that characterization of strains that underwent WGS can be difficult due to the high number of simple mutations; however, the development of second generation, deep-sequencing technologies (such as the Illumina sequencing by synthesis technology, and the 454 sequencing technology (Roche)) have now made this possible, albeit for now still a relatively expensive proposition. Such mutations can be practically identified by sequencing the entire genome, which is possible but tedious and still expensive.

Another approach is global transcription machinery engineering (gTME), whereby a core transcription factor is mutated via error-prone PCR to generate diversity for screening [1, 2]. gTME was used to enhance ethanol tolerance [2]. A strategy, called multiplex automated genome engineering (MAGE), was reported, whereby synthetic DNAs (~ 90 bp) are integrated in the genome in multiple rounds to increase genetic diversity and screen for desirable traits [93]. More recently, this was extended by what was termed conjugative assembly genome engineering (CAGE) [36]. These approaches are based on generating the genetic diversity that can be obtained from a single genome or similar genomes (e.g., WGS can only used with similar species and only if protoplasts can be constructed and fused), but cannot easily integrate the much larger genetic diversity that exists in unrelated microbes.

7.2.2 Screening Genome-Wide or Metagenomic DNA Libraries for Trait Conferring Genes

Cellular phenotypes have been screened for selected characteristics via the construction of genomic and metagenomic libraries. There are genomic DNA, metagenomic, and cDNA libraries. Libraries can be constructed so that genes are expressed from a constitutive promoter, conditionally (from an inducible promoter), or using the gene's natural promoter(s). Genomic or meatagenomic libraries on vectors (plasmids, cosmids or fosmids) provide a high or tunable copy number of genes that can be screened to identify desirable traits. Since the grand majority of microbes cannot be cultured, metagenomic libraries are employed aiming to capture some of the otherwise unexplorable genetic diversity of specific ecosystems, such as the those that are enriched naturally or by human or external action for specific traits. Such traits could include the ability to tolerate toxic chemicals, carry out specialized biotransformations, or produce desirable chemicals. DNA or cDNA enriched from a desirable macro or micro-ecosystem is cloned on the desirable vector and used to create a metagenomic library in a suitable host.

For genomic libraries, the entire genome is sheared or digested to desirable sizes, cloned into vectors, and transformed into cells for screening [11, 12, 23]. cDNA libraries are constructed by cloning the reverse transcription products of mRNA expression libraries. The hosts are then exposed to some selective or stressful pressure (such as ethanol, butanol or any other toxic chemical, but also to low or high pH, high osmotic pressure, etc.) with the assumption that some gene(s) represented in the library will allow for favorable growth under such stress. The population is eventually taken over with cells carrying vectors with these fragment(s), and sequencing identifies the trait-conferring gene(s). This high-throughput method can be coupled with DNA microarrays to identify and isolate the enriched genes, in a method that has been termed "parallel-gene trait mapping" [11, 23]. Genomic library screens can thus be used to identify genes that improve a desirable phenotype or generate a novel phenotype. Thus, cells can be metabolically engineered to have a desired characteristic, such as higher chemical, oxidative-stress or low-pH tolerance.

As mentioned above, libraries can be constructed with various insert sizes and with or without host-specific promoters [11, 12] that would allow one to express genes at will and the strength desired, plus express genes in both sense and antisense direction [12] aiming to identify genes whose high expression provides the selectable trait or genes whose downregulation (by antisense RNA [12, 88]) provides a benefit or the selectable trait. The benefits of these two approaches were demonstrated by two studies [11, 12] from our laboratory where different types of libraries were employed.

In the first approach, a plasmid-based, sheared-DNA library of *C. acetobutylicum* with insert of an average 3-kb size without any additional promoters was used [11]. This library was screened for tolerance to butanol, but exposing the plasmid-bearing library of cells to butanol. Two screening protocols were used. Protocol I utilized a single round of butanol challenge, while protocol II was based on the serial transfer of stationary-phase cultures into progressively higher butanol concentrations. Protocol II enabled the successful identification of DNA fragments containing several intact genes conferring preferential growth under conditions of butanol stress. Since gene expression in this library is possible only from native promoters, 16 genes that constitute the first cistron of a transcriptional unit were identified. These genes included four transcriptional regulators (CAC0977, CAC1463, CAC1869, and CAC2495). It was found that overexpression of CAC1869 (strain 824(pCAC1869)) led to an 81% improved butanol tolerance relative to the plasmid control strain. 824(pCAC1869) consistently grew to higher cell densities in challenged and unchallenged cultures and exhibited prolonged metabolism. DNA microarray analysis made a high-resolution assessment of the dynamic process of library enrichment (Fig. 7.1). The *C. acetobutylicum* genome (chromosome and pSOL1 megaplasmid) is represented as a circle with the ordinal gene position increasing clockwise direction. pSOL1 genes are between 11 and 12 o'clock. Concentric circles show the relative ranks of gene enrichment in a single transfer starting from the initial inoculum (innermost circle) and moving outwards to the 13th transfer (outermost circle). Enrichment of a gene is assessed by a change from green to red. Several genes (e.g., CAC0742, CAC3289, and CAC3005) quickly increase in signal intensity rank with increasing transfers, while other genes (e.g., CAC1868–CAC1870) are more gradually enriched during the selection process. While there are several genes that were enriched in both biological replicate challenges, some were not, and this, together with the overall gene enrichment patterns suggest that the process involves some stochasticity, but details remain to be explored.

In a second approach, we used a genomic library from sheared *C. acetobutylicum* DNA, whereby inserts can be expressed in both directions from a native thiolase promoter [12]. Serial transfer of library cultures exposed to increasing butyrate concentrations consistently enriched for inserts containing fragments of rRNA genetic loci. The selected library inserts were expressed antisense to the rRNAs,

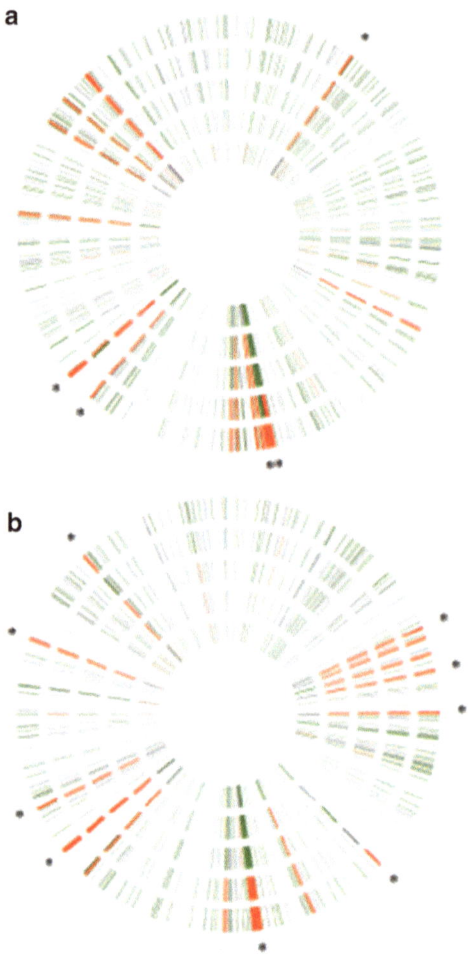

Fig. 7.1 Physical map of the *C. acetobutylicum* genome color coded for the degree of fragment enrichment as determined by DNA microarray analysis for (**a**) biological replicate one and (**b**) biological replicate two of stationary-phase transfer challenges (protocol II; see text and original paper). Plasmid library insert DNA isolated from the initial challenge inoculum was PCR amplified, Cy3 labeled, and hybridized against Cy5-labeled, amplified library insert DNA from the 4th, 7th, 10th, and 13th challenge transfers. After subtraction of background and nonspecific hybridization signals and averaging of replicate spots, gene signal intensities were ranked within a given challenge transfer. Concentric circles represent (starting with the *innermost circle*) the ranks of individual genes from the initial inoculum and 4th, 7th, 10th, and 13th transfers, respectively. Genes were color coded according to their percentile rank in a given transfer, such that the top 5% are *red*, 5–33% are *green*, and 33–100% are *gray*. Genes that did not generate a signal-to-noise ratio greater than three on a given array are *white* (Reproduced by permission from [11])

thus expressing non-coding RNAs (ncRNAs). Different enriched inserts imparted similar butyrate-tolerance characteristics. A minimal tolerance fragment (RDNA7) was identified as the16S-rRNA promoter region. Expressed on plasmid pRD7,

RDNA7 produces the ncRNA ncRNARD7. Expression of ncRNARD7 resulted in superior resistance to butyrate and other carboxylic acids. Our data suggested that by hybridizing against unprocessed rRNA precursors, ncRNARD7 alters rRNA processing, thus resulting in acid tolerance, possibly through a mechanism involving the Ffh protein of the of the signal recognition particle (SRP) system, which, delivers membrane proteins to cell membranes during the translation process. The bidirectionally expressing library of this study generates double diversity by expressing both sense and antisense transcripts off a strong (here the thiolase) promoter. Thus, an important finding of this study is that genomic libraries can be used to identify genes or DNA fragments that can act by an antisense RNA mechanism to achieve a desirable or novel phenotype.

7.2.3 Issues, Problems and Future Directions

WGS is a great technology, but it suffers from two major limitations. It can be applied only in organisms that can be protoplasted, and this means largely for a subset of Gram$^+$ prokaryotes, and perhaps yeasts, although, so far the success in yeast systems has been limited. The second and perhaps more important limitation is that the genetic diversity of WGS is created by combining simple genetic mutations that improves the function of existing proteins, but does not generate novel diversity, since the only genetic space that is engaged in WGS is that of the parent strain.

Metagenomic libraries suffer from a specific and a more general library limitation. The specific one is that unless a host-specific promoter is used with small library fragments, heterologous/ allogeneic genes have a low probability of being expressed due to the inability of the host (typically *E. coli*) to recognize heterologous/allogeneic promoters, but also due to non-optimal codon usage. A recent study and our data (not shown) suggest that indeed only a very small fraction (0.34–2%) of library inserts in a metagenomic library can be expressed in *E. coli* [27]. This would apply by inference in other hosts, in view of the fact that *E. coli* remains the most permissive host known. A generic limitation of all library approaches is that only one library clone can be stably propagated in each individual cell, and thus only single genomic fragments in the library vector can be selected for and identified. While the use of cosmids and fosmids can be used to generate libraries with large genomic fragments (30–40 kb), such libraries would be useful only for screening homologous libraries in a host. Still, interactions between distantly located genes that are present on two different library fragments cannot be identified. Our lab is currently working on overcoming this single-genomic fragment limitation by employing two different strategies so that multiple genetic loci can co-exist and co-express in a host for screening and selection studies. The first is reported in a recent manuscript [57], and is employing

Coexisting/Coexpressing Genomic Libraries (CoGeL) to solve the problem of identifying interactions among distant genetic loci. As a proof of principle, four sets of two genes of the L-lysine biosynthesis pathway distantly located on the *E. coli* chromosome were knocked out. Upon transformation of these auxotrophs with CoGeLs, cells growing without supplementation were found to harbor library inserts containing the knocked-out genes demonstrating the interaction between the two libraries. CoGeLs were also employed to identify genetic loci that work synergistically to create an acid-tolerance phenotype [57]. Finally, our laboratory is developing strategies to allow for the expression of a large fraction of metagenomic libraries by engineering the host's transcriptional machinery using a strategy termed "microbial alloys".

7.3 Use of Transcriptomic Data to Explore the Transcriptional Space of Metabolite Stress and Tolerance

The *C. acetobutylicum* stress response to butanol, acetate and butyrate at the molecular systems level. Our laboratory was the first to develop microarrays for transcriptional analysis of *C. acetobutylicum*. We were also the first to explore the transcriptional program of butanol stress and tolerance [4, 87]. More recently, we published detailed transcriptomic analyses [5] of the response of *C. acetobutylicum* to the three key toxic metabolites: butanol, butyrate and acetate. These studies provide a detailed assessment of the breadth and depth of these stress responses. They assess the impact of these stresses on essential cellular programs, core among which is the stress-response system (HSPs or chaperones and related stress proteins), but also on metabolic pathways, and stress-adapted energy metabolism, but also novel, previously unexplored programs. These are discussed below, but first, I provide a quick review of the HSPs and other stress proteins, and provide context as to of what is known about metabolite toxicity and tolerance from prior studies using non-genomic tools and strategies.

7.3.1 Heat Shock Proteins (HSPs), Other Stress Proteins, and Specialized Stress Regulons

The cellular stress response utilizes HSPs, which are molecular chaperones that assist in folding and refolding of proteins damaged by chemical and other stress [25, 28, 45, 50], as well as several non-HSP stress proteins [56]. Cells also engage specialized sigma factors to transcribe or upregulate their regulons to mitigate stress-induced cell damage [95]. The interplay of HSPs with other stress proteins and proteins engaged in energy generation and core metabolism facilitate protection from and tolerance to chemical and other bioprocessing stresses [56].

Metabolic engineering efforts in our laboratory have shown that an engineered HSP system can be engaged to improve tolerance [86, 87].

Stress systems have been most extensively studied for *E. coli*, the Gram- model organism. HSPs include DnaK (Hsp70), DnaJ (Hsp40), GrpE, GroEL (Hsp60), GroES (Hsp10), the Clp family (Hsp100), and many small stress proteins. In the model Gram + prokaryote *Bacillus subtilis*, HSPs have been classified as: Class I (*dnaK* and *groESL*; HrcA dependent), Class II (regulated by the stress-specific sigma factor σ^B [29]), Class III (*clp*; regulated by CtsR repressor [20]), and Class IV (all other: not induced by HrcA, σ^B, or CtsR [87]). The *dnaK* and *groESL* operon structures in *C. acetobutylicum* are nearly identical to those in *B. subtilis* [8] and have been shown to share the same regulatory mechanisms. The σ^B system has not been identified in clostridia. Several Class I, III and IV genes/proteins have been recently identified in clostridia by the author's group based on gene expression patterns in response to chemical (butanol, butyrate and acetate) stress [5].

7.3.2 Metabolite Toxicity in Clostridia

Metabolites are produced at especially high levels by strict or facultative anaerobes due to the fact that these are primary metabolites necessary for energy generation by the organisms, and thus for survival. In the prototypical example, *C. acetobutylicum* generates ATP through substrate-level phosphorylation and produces the acids acetate and butyrate. These acids cross cytoplasmic membranes in the undissociated (protonated) form and dissociate in the cytoplasmic space [9, 33, 42]. The unprotonated acid cannot cross the membrane and, when it accumulates intracellularly, is toxic to the cells due to membrane uncoupling [9, 30, 33, 41] or anion accumulation [71, 72]. It is now well accepted that butyrate accumulation induces solvent formation [24, 34, 52, 83], possibly as a stress response. Notably, it is well accepted that solventogenic clostridia respond to acid accumulation by shifting from exponential growth and acidogenesis to stationary phase, solventogenesis (production of acetone, butanol, and ethanol), and sporulation. However, butanol is also toxic to the cells by disrupting the transmembrane ΔpH and $\Delta\psi$, lowering intracellular ATP levels, and inhibiting glucose uptake [13, 32, 83]. When solvents are produced, in order to deal with the membrane fluidization effect of solvents, clostridia, like many other organisms, change their metabolic pathways in order to increase saturated fatty-acid chain content [6, 47, 91, 99] apparently in an effort to control membrane fluidity in the presence of solvents [7, 91], a process known as homeoviscous adaptation [79]. Solventogenic-phase and butanol-stressed clostridia also express stress genes, including chaperones [67, 84] [3, 8, 75]. Overexpression of one of these operons, *groESL*, improves butanol tolerance and production, suggesting that solvent tolerance can be improved with enhanced protein stability [86, 87].

7.3.3 *Ontological Analysis of the Transcriptional Stressome in Response to Core Metabolites: Butanol, Butyrate, and Acetate*

Genomic approaches like DNA-microarray analysis provide an excellent opportunity to identify novel genetic programs affected by metabolites and potential associated stresses. We have first used partial genome microarrays to examine butanol stress in four different *C. acetobutylicum* strains at varying stress levels to elucidate what genes were involved in the short- and long-term stress responses to butanol [4, 87]. In a more recent study we used full genome *C. acetobutylicum* microarrays to identify and compare the transcriptional programs associated with the cell response to the key and largely toxic primary metabolites acetate, butyrate, and butanol. We also used such information and computational tools to assess the impact of these metabolites on the metabolic programs of the cells.

In the genome annotation, each *C. acetobutylicum* gene was assigned [59] to a broad functional category (translation, amino acid transport and metabolism, cell motility etc.) based clusters of orthologous groups [82]. A gene may be assigned to multiple clusters, but this does not affect the conclusions from this analysis. For each stress and each time point, the number of genes within a category that was differentially upregulated and downregulated (with a $\geq$95% confidence interval) was counted and divided by the total number of genes in that functional category. The resulting percentages are plotted colorimetrically in Fig. 7.2a. The darker the color, the higher the number of genes in a particular category and time point that are affected by a given stress. The two categories with the highest percentage of differentially upregulated genes for all three stressants are O (posttranslational modifications, protein turnover, and chaperones), and C (energy production and conversion). Enrichments in other functional groups are stressant and time specific. Notably, in the inorganic ion transport and metabolism group (category P), both the acetate (two time points) and butyrate (two time points) stresses had sample points where at least 10% of the genes were differentially upregulated while none of the butanol stress time points did so. The amino acid transport and metabolism group (category E) had relatively high enrichment of both differentially upregulated and downregulated genes. Thus, we expect to find amino acid pathways upregulated and downregulated due to metabolite stress. The nucleotide transport and metabolism group (category F) shows some enrichment of differentially downregulated genes, and we thus expect to find pathways within this functional category to be generally downregulated. We also searched for functional groups whose general gene expression trends (either higher or lower post stress) were significantly different than global trends (Fig. 7.2b). A chi-squared test assessed if the number of genes with higher and lower expression within any given functional category was significantly different than the number of genes that were expressed higher or lower globally. P-values are plotted colorimetrically in Fig. 7.2b. As an examples, 12 time points in functional category J (translation-related genes) had a p-value less than 0.01, indicating that a significant percentage of genes (Fig. 7.2b) had a different trend

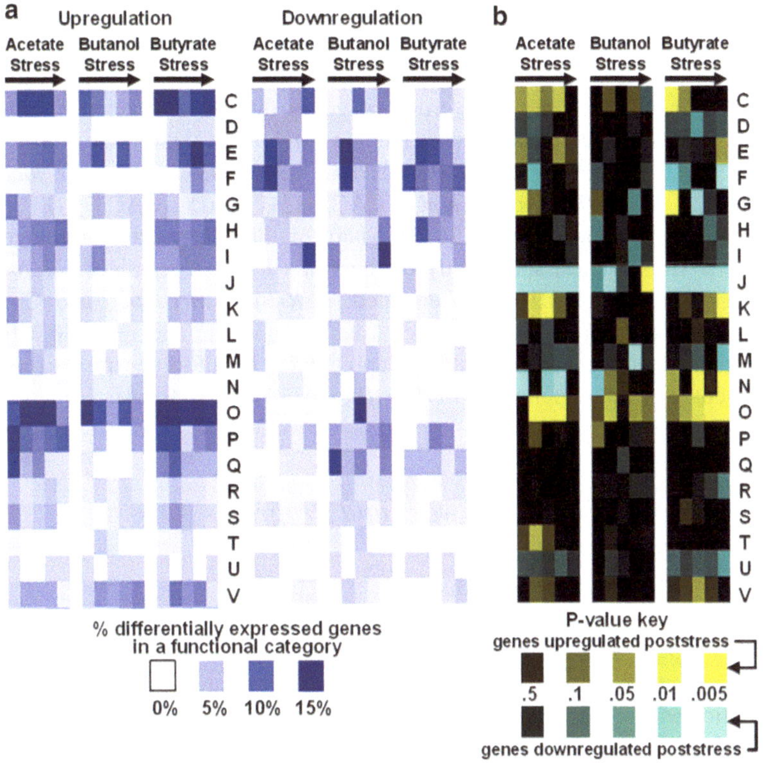

Fig. 7.2 Colorimetric representation of the percentage of genes of functional categories that were either differentially upregulated or downregulated by acetate, butanol, or butyrate stress (**a**). Colorimetric representation of p-values resulting from chi-squared tests that compared the relative numbers of genes in a functional category with higher or lower expression poststress compared to global transcriptome patterns (**b**). Functional categories and time points for which genes were generally expressed higher poststress are plotted in gradients of *black* to *yellow* and lower expression poststress in gradients of *black* to *blue*. RNA samples for microarray hybridization were taken 10, 30, 45, 60, and 120 min poststress. Abbreviations: *C* energy production and conversion, *D* cell cycle control, *E* amino acid transport and metabolism, *F* nucleotide transport and metabolism, *G* carbohydrate transport and metabolism, *H* coenzyme transport and metabolism, *I* lipid transport and metabolism, *J* translation, *K* transcription, *L* replication, recombination and repair, *M* cell wall/membrane biogenesis, *N* cell motility, *O* posttranslational modification, protein turnover, chaperones, *P* inorganic ion transport and metabolism, *Q* secondary metabolite biosynthesis, transport and catabolism, *R* general function prediction only, *S* function unknown, *T* signal transduction, *U* intracellular trafficking and secretion, *V* defense mechanisms. (Reproduced by permission from [5])

(lower expression) when compared to the global transcriptome, which shows the profound impact of these stresses on the protein biosynthesis machinery. This result was not apparent by looking only at the percentages of differentially expressed genes within specific functional categories (Fig. 7.2a). In the motility functional group (category N; Fig. 7.2b), one observes an opposite effect for acetate (downregulation) versus butyrate (upregulation) stress.

Figure 7.3 summarizes some of the most important and relevant genes and pathways that were differentially upregulated and downregulated among the three metabolite stresses, demonstrating a large, and largely unexpected, diversity of stress responses. The only class/group of genes that were affected by all three stresses was the HSP genes. These include *lonA* (CAC0456), *hrcA-grpE-dnaK* (CAC1280-CAC1282), *groESL* (CAC2704, CAC2703), *ctsR-yacH-yacI-clpC* (CAC3192-CAC3189), *hsp90* (*htpG*, CAC3315), *hsp18* (CAC3714), and an HtrA-like serine protease (CAC2433). Many of these genes are orthologous to *B. subtilis* stress genes, specifically of the Class I (HrcA-regulated: *hrcA-grpE-dnaK*, *groESL*), Class III (CtsR-regulated: *ctsR-yacH-yacI-clpC*) and Class IV (*htrA*) genes and correspond to proteins with chaperone and protease activities [77]. Other notable genes differentially upregulated by all three stresses (Fig. 7.3) the solventogenic *sol* operon (*aad-ctfA-ctfB*, CAP0162-CAP0164).

7.3.4 Stress-Specific Pathways and Programs

7.3.4.1 Central Metabolic Pathways

Glycolysis and primary metabolism are differentially affected by the three stresses. As summarized in Fig. 7.4, acetate and butyrate alter the expression of glycolytic genes in a similar fashion which is quite distinct from that of the butanol stress. Acetate stress downregulated expression of the major butyryl-CoA- and butyrate-formation genes, while butyrate stress downregulated expression of acetate-formation genes, thus suggesting that each of these two acid metabolites behave "selfishly". Pyrimidine-biosynthesis genes were downregulated by most stresses, but purine-biosynthesis genes were upregulated by acetate and butyrate [5]. Methionine biosynthesis genes were upregulated by acetate stress, indicating a possibly conserved stress-response mechanism also observed in *E. coli*. Surprisingly, metabolite stress had no apparent effect on the expression of the sporulation-cascade genes.

7.3.4.2 Nitrogen Metabolism

Nitrogen-fixation and utilization genes are part of the acetate-specific stress response. The *nif* locus genes (CAC0253-CAC0260) encode proteins which are involved in nitrogen fixation and its regulation. Some of the genes (CAC0254, CAC0255, and CAC0258) were upregulated following acetate stress. Both *C. acetobutylicum* and *C. beijerinckii* can fix nitrogen with peak activity during acidogenesis [17]. Expression of the nitrogen regulatory gene *nrgB* (CAC0681) is upregulated 3.0-fold 10 min post-acetate stress (93% confidence). The putative operon CAC2448-CAC2450 is differentially upregulated following both acetate and butyrate stresses. CAC2448 encodes a *nirB*-family dehydrogenase, CAC2449 encodes a flavoprotein, and CAC2450 encodes a desulfoferrodoxin. The NirB protein would reduce nitrite to ammonia, but it is

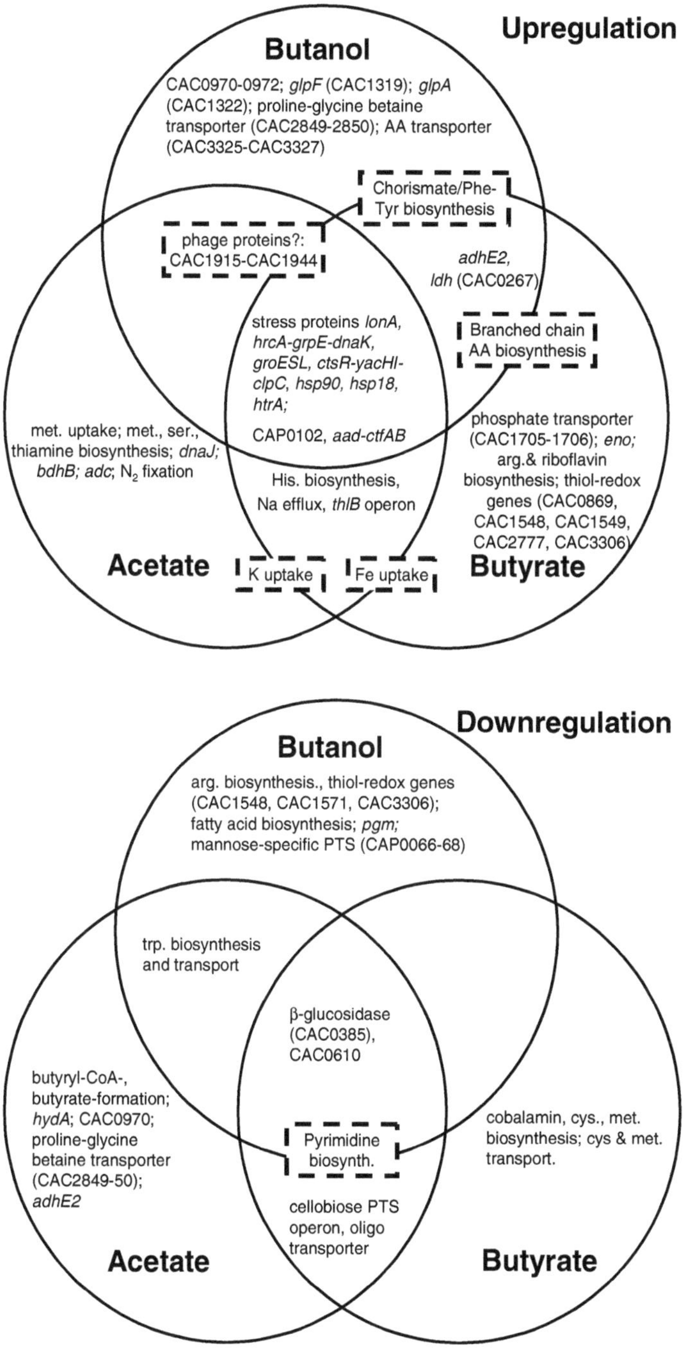

Fig. 7.3 Venn diagrams of selected genes and pathways which were differentially upregulated or downregulated (≥95% confidence for at least two time points per stress experiment) in response to metabolite stress(es). Putative operons and pathways that had genes in adjacent sections of the Venn diagram are marked with dashed boxes. This list is not all inclusive; a complete list of differentially expressed genes is available in the Supplemental Materials. Abbreviations: *arg.* arginine, *cys.* cysteine, *his.* histidine, *met.* methionine, *ser.* serine, *phe.* phenylalanine, *PTS* phosphotransferase system, *trp* tryptophan, *tyr.* tyrosine. (Reproduced by permission from [5])

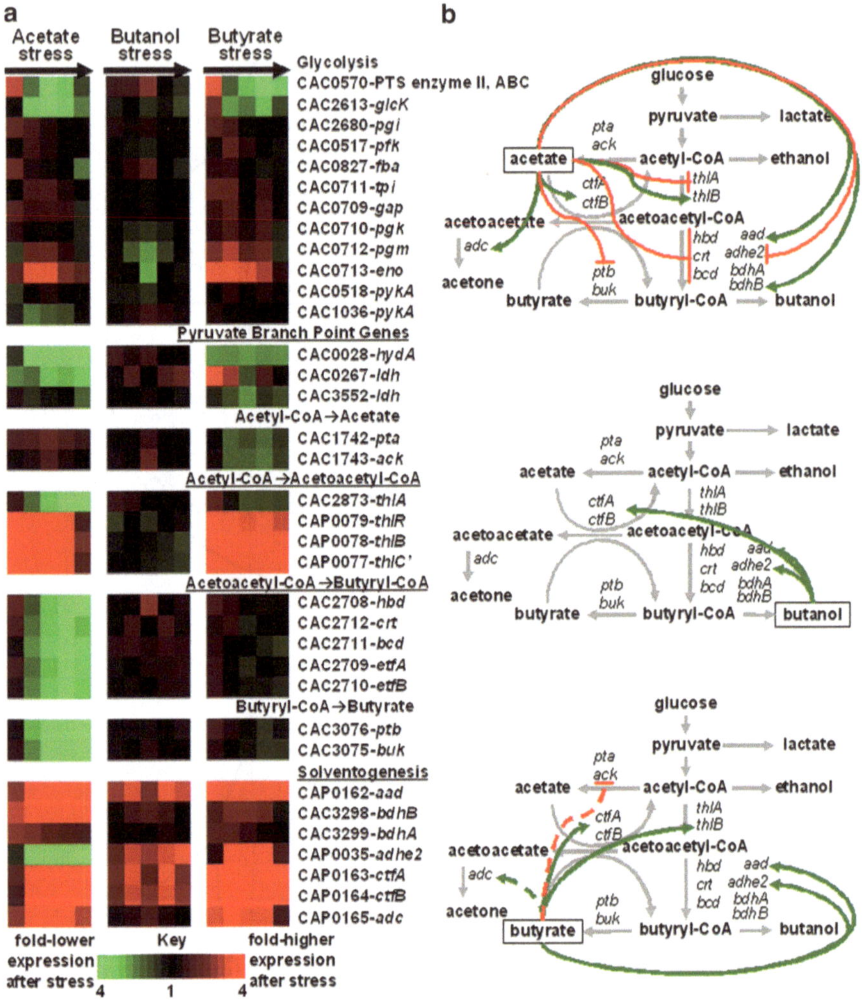

Fig. 7.4 Expression profiles of the *C. acetobutylicum* glycolytic and primary metabolism genes 10, 30, 45, 60, and 120 min after acetate, butyrate, and butanol stress (**a**). Diagrams of the glycolytic and primary metabolism showing which pathways had differential expression (95% confidence interval at least twice for a metabolite stress response; (**b**) *Gray arrows* indicate directions of metabolic pathways. *Green arrows* indicate upregulation of genes; *red lines with bars* indicate downregulation of genes. For simplicity, regulation of glycolysis, lactate formation, hydrogenase, and ethanol formation (via the *aad* and *adhE2* genes) has been omitted. The following up- and downregulation patterns did not pass the 95%-confidence-at-two-time-points criteria but had significant trends and are represented with *dashed lines*: butyrate upregulation of *adc* (>90% confidence, 10–60 min); *pta* (64–67% confidence, 30–60 min); and *ack* (65–72% confidence, 30–60 min). Abbreviation: *PTS* phosphotransferase (Reproduced by permission from [5])

unclear based on sequence homology if this function is conserved. A putative ferredoxin-nitrite reductase (CAC0094) was also upregulated 10 min after both acetate (4.0-fold, 98% confidence) and butyrate (2.1-fold, 87% confidence) stresses. The expression patterns of these genes seem to indicate a need for ammonia following stress. However, expression of ammonia-utilizing genes such as glutamine synthetase (CAC2658) was not necessarily upregulated. Following acetate and butyrate stresses, the genes *gltB* (CAC0764) and *asnB* (CAC2243) whose corresponding proteins putatively convert glutamine to glutamate were expressed slightly higher (data not shown). Also, expression of an ammonium-specific transporter *nrgB* (CAC0682) was downregulated at all time points following acetate and butyrate stresses.

7.3.4.3 Membrane Biosynthesis and Modifications

As discussed, solventogenic and butanol-stressed *C. acetobutylicum* modifies membrane lipid composition to increase saturated fatty acid content. In response to butanol stress, we expected to observe in response to butanol stress the corresponding downregulation of unsaturated fatty-acid biosynthesis genes and upregulation of saturated fatty-acid biosynthesis genes. The characterized microbial genes for producing unsaturated fatty acids include *fabAB* in *E. coli*, *fabM* in *Streptococcus pneumoniae*, and the desaturase *des* gene (whose protein product requires oxygen) in *B. subtilis* [76]. *C. acetobutylicum* does not contain such orthologous unsaturated fatty acid biosynthesis genes [16, 49, 61] and, thus, we were unable to look for downregulation of such genes in our stress-response data. In *C. acetobutylicum*, the major locus for saturated fatty acid biosynthesis are genes CAC3568-3580 [4, 16, 61]. The sequential arrangement of genes in the locus is preserved among *S. pneumoniae*, *S. pyogenes*, *C. acetobutylicum* [61] and *C. tetani* (data not shown) with one exception. In the *Streptococcus* species, the locus begins with the *fabM* gene for unsaturated fatty acid synthesis. In the clostridia, this gene (CAC3580) has been annotated as either an additional copy of *fabK* or as a dioxygenase related to 2-nitropropane dioxygenase [59]. Expression of the entire locus (CAC3569-CAC3580) following butanol stress was differentially downregulated 120 min poststress, and a few of these genes (CAC3570-CAC3572, CAC3574, and CAC3575) were also differentially downregulated 10 min poststress. None of these genes were differentially downregulated following acetate or butyrate stress.

7.3.5 Similarities Between Metabolite-Stress Responses and Batch Fermentation Gene Expression

We have used microarrays to examine the transcriptional changes during the shift from acidogenesis and exponential growth to solventogenesis, stationary phase, and sporulation in batch fermentation [3]. Cultures in fermentors accumulate significant amounts of both acids and solvents (Fig. 7.5a), and we wanted to examine if gene expression patterns during this fermentation were consistent with the acetate, butyrate, and butanol stress responses studied in the study above (Fig. 7.5b). With the time course experiment, genes were assigned to one of 24 clusters (named t0-t23) generated from self-organizing maps (SOMs) [81] from gene-expression data. We counted the number of genes within the SOM-generated clusters that were differentially expressed (≥95% confidence at two time points per stress) either higher or lower by each metabolite stress. The results are plotted colorimetrically in Fig. 7.5b. During the time course fermentation, solventogenesis was initiated at time point E by which time a significant amount of acids (40 mM acetate, 41 mM butyrate) had accumulated. Therefore, the culture probably experienced significant carboxylic acid stress by the onset of solventogenesis. We had observed that many stationary phase-related genes, including solvent formation and sporulation, were significantly upregulated at or near this time point [3]. At least 10% of all genes in clusters t9 and t10 were differentially upregulated by each metabolite stress. In addition, acetate stress upregulated at least 10% of the genes in clusters t8, t22, and t11, and butyrate stress upregulated 19% of the genes in cluster t4. Many of these clusters (t4, t9, t10, and t11) are particularly indicative of higher expression after the onset of sporulation and solventogenesis, by which time a significant amount of acids had accumulated. In cluster t13, 11% of the genes were differentially downregulated by acetate and butyrate stresses. This cluster displays lower expression after the onset of solventogenesis and sporulation. 23% of the genes in cluster t22 were differentially downregulated by butyrate stress, including many genes (CAC0102-CAC0110, CAC0391, CAC0878-CAC0880, CAC0929, CAC0930, CAC2235, and CAC3325-CAC3327) putatively involved in sulfate uptake, cysteine biosynthesis and uptake, and methionine biosynthesis. Since we previously demonstrated that many of these genes are downregulated by butyrate stress but are upregulated by acetate stress, the expression pattern observed in the fermentor (Fig. 7.5) may be a complex combination of these two opposing stresses. In summary, the patterns of gene expression observed in bioreactors and after metabolite stress are consistent.

7.4 Approaches for Modeling General and Specific Stress Responses

Understanding the complex dynamics of stress response and of other cellular processes that are not necessarily organized in metabolic pathways requires integration of methods and models that describe these processes at the levels of RNA and protein expression, protein-protein interactions, and integrated flux regulation. The potential of systems biology approaches is currently limited by difficulties in integrating metabolic measurements across the different functional levels of the cell. The challenge in linking metabolic flux phenotypes with other omics

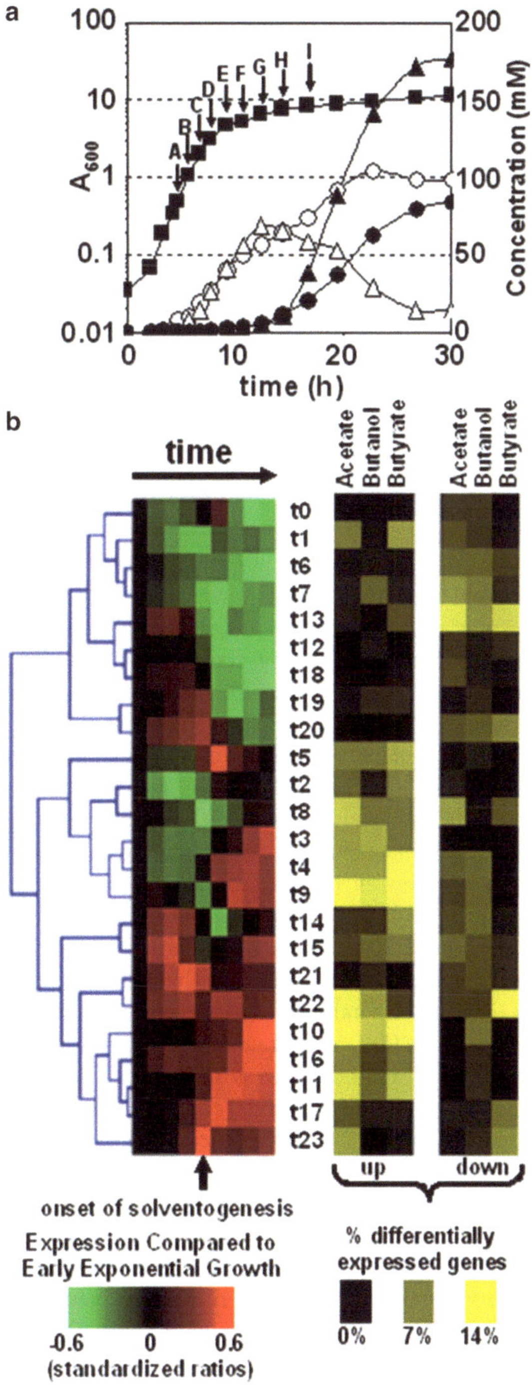

Fig. 7.5 WT fermentor profiles (pH ≥ 5.0) from a previously published study [3] that used microarrays to analyze the time course of gene expression during the shift from acidogenesis and exponential growth to solventogenesis and stationary phase (**a**). Solid squares, A600; *open circles*, acetate; *open triangles*, butyrate; *solid circles*, acetone; *solid triangles*, butanol. Letters correspond to microarray time points. A comparison of gene expression clusters determined from the time course experiment [3] and the percentages of genes within the clusters that were differentially expressed after treatment with acetate, butanol, or butyrate (**b**). The *short arrow* underneath the Eisen plot indicates the onset of sporulation and solventogenesis during the time course experiment. In the Eisen plot, *red* and *green* respectively represent higher and lower expression compared to the first microarray (midexponential) time point

data has been investigated in a recent paper [54] and provides a paradigm for how transcriptional control networks rewire the metabolic flux in a yeast cell by simultaneously monitoring the transcriptome, the metabolome, and the fluxome. If successful here, quantitative genome-scale models that link genomics and physiology (the genotype-phenotype relationship) would be a powerful framework for rational strain design for improved bacterial phenotypes. First, we review a select set of recent studies whereby multiple sets of omics data were used to elucidate and generate a conceptual, at least, model of stress responses to specific chemicals. These studies constitute important paradigms of integrated and largely complete studies that provide exceptional sets of data for mining and modeling.

7.4.1 Integrated Omic Studies of Stress Responses to Chemicals in Microbes

A few studies have been published where omics data combined with computational genomic analyses were shown to provide an enhanced understanding of stress responses at the genome scale. One study examined the oxidative stress response in *B. subtilis* [53] using transcriptomic and proteomic analyses. Two reports examined the response of the metal-reducing *Shewanella oneidensis* to chromate. Using temporal genomic profiling and whole-cell proteomic analyses, the first study [14] revealed that response to chromate shock involves an oxidative stress response, as well as SOS-controlled DNA repair mechanisms. The 2nd study [85], using proteomic analysis, demonstrated differential responses to different levels of sub-lethal chromate challenges. Systems analysis [18] of transcriptomic and proteomic data on stresses imposed on *Desulfovibrio vulgaris Hildenborough*, a sulfate-reducing prokaryote important to metal bioremediation, revealed promoter features and sequences responding to alternate sigma factors and suggested substantially different mechanisms of stress response from those established *in E. coli* and *B. subtilis*. A combined transcriptomic and proteomic analysis of *Methanosarcina acetivorans* [48], an acetate- and methanol-utilizing methane-producing archaeon, revealed a significant stress response to both chemicals, and the engagement of regulatory protein families commonly found in prokaryotes. Using proteomic analysis and computational prediction of promoters, it was possible to determine the RpoS (or σ^S) stress regulon of *Burkholderia pseudomallei* [60]. Finally, combined proteomic, transcriptomic and compound-specific isotope analysis (CSIA), the cellular response of *Polaromonas sp.* strain JS666 to cis-dichloroethene (cDCE) was examined to determine the pathways and programs involved in cDCE utilization [40]. A combined transcriptomic-proteomic study was also published recently on the response of *E. coli* to butanol stress [73], and gene expression data combined with computational analyses were used to map the initial isobutanol response network of *E. coli* [15].

7.4.2 Computational Approaches for Identifying and Modeling Microbial Stress Responses

A framework was recently proposed for analyzing the stress response in yeast cells at the metabolic level [78], showing that the transcriptional activity of metabolic functions in response to stress could be organized and modeled at the genome scale. Metabolic functions were described in terms of elementary modes and computed in a modular fashion using a genome-scale model (GSM), thus enabling the identification of induced or repressed metabolic processes, leading to the identification of stress-specific metabolic backbones. Elementary modes are defined as "minimal sets of reactions that can operate in steady state in a metabolic network" [78]. This modeling approach deals with the impact of stress on metabolic processes without *a priori* knowledge of which stress components are engaged for bringing about the metabolic changes.

Data mining of the sets of genes in key elementary modes can be subsequently used to understand the molecular details that bring about these changes, and thus identify transcription factors (TFs) and their regulons involved in the specific stress responses. Such information can be then used to build the detailed, mechanistic stress response models, such as those discussed next.

Using computational analyses from sequence homology, domain architecture, and genomic context, Wecke et al. [96] developed a detailed model network underlying the cell envelope stress response in *Bacillus licheniformis*, a Gram$^+$ prokaryote. The envelope of bacterial cells is responsible, among other functions, for maintaining its integrity, which is necessary for cell survival. Surveillance of and response to environmental perturbations is carried out by sensing systems and notably by two-component systems (TCS) and extracytoplasmic function (ECF) factors. The integrated system is known as the cell-envelope stress response. In this work, using comparative computational genomic analyses, Wecke et al. [96] identified five TCS and eight ECF factors as potential candidate regulatory systems that could mediate cell envelope stress response in *B. licheniformis*. As an initial screen, they then examined the regulatory network by comparative transcriptomic analysis and regulon identification. This was followed by detailed transcriptional profiling aimed at defining the inducer specificity of each identified stress sensor, thus completing the identification of the cell-envelope stress system in this organism, which proved to be substantially different than that in the model organism *B. subtilis*.

A more structured and mechanistic approach was developed in the identification of the stress response in the Gram$^+$ *Bifidobacterium breve* [100]. These authors used detailed transcriptional analysis, DNA-protein interactions, and GusA reporter fusions, combined with *in silico* analysis to build a molecularly detailed conceptual model. They examined heat, osmotic, and solvent stresses aiming to formulate a general stress response model with substantial mechanistic detail as shown in Fig. 7.6. In this model, HspR controls the SOS response and the ClgR regulon, which in turn regulates and is regulated by HrcA. This model of an interacting regulatory network is similar in nature to the general stress-response model in

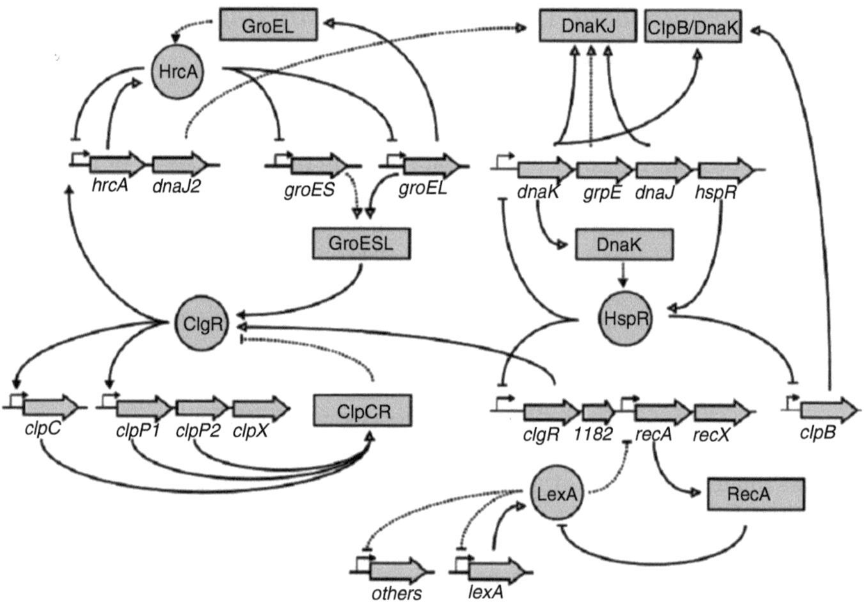

Fig. 7.6 Schematic representation of the stress gene regulatory network of *Bifidobacterium breve* UCC 2003 (From [100] reprinted with permission). *Dotted lines* indicate the predicted interaction, and *closed lines* indicate a proven interaction. A *dash* at the end of a line indicates repression, while a *triangle* at the end of a line indicates activation

Gram$^+$ bacteria discussed in Sect. 7.3.1 but here the connectivity is specific and superior. Mechanistic models like the one of Fig. 7.6 can then be used as a basis to mathematically code a dynamic stress response model, a process which is substantially different in nature from the modeling approach of Schwartz et al. [78] discussed above.

Acknowledgments We thank Sergios Nicolaou for discussions, and Dr. Carles Paredes for helpful discussions and assistance with the chi-squared tests. This work was supported in part by Office of Naval Research (USA) Grant N000141010161, by National Science Foundation (USA) grant CBET-1033926, and by Department of Energy (USA) grant DE-SC0007092. K.V.A. was supported in part by an NIH/NIGMS Biotechnology Training Grant (T32-GM08449).

References

1. Alper H, Stephanopoulos G (2007) Global transcription machinery engineering: a new approach for improving cellular phenotype. Metab Eng 9(3):258–267
2. Alper H, Moxley J, Nevoigt E, Fink GR, Stephanopoulos G (2006) Engineering yeast transcription machinery for improved ethanol tolerance and production. Science 314 (5805):1565–1568

3. Alsaker KV, Papoutsakis ET (2005) Transcriptional program of early sporulation and stationary-phase events in *Clostridium acetobutylicum*. J Bacteriol 187(20):7103–7118
4. Alsaker KV, Spitzer TR, Papoutsakis ET (2004) Transcriptional analysis of *spo0A* overexpression in *Clostridium acetobutylicum* and its effect on the cell's response to butanol stress. J Bacteriol 186(7):1959–1971
5. Alsaker KV, Paredes C, Papoutsakis ET (2010) Metabolite stress and tolerance in the production of biofuels and chemicals: gene-expression-based systems analysis of butanol, butyrate, and acetate stresses in the anaerobe *Clostridium acetobutylicum*. Biotechnol Bioeng 105(6):1131–1147
6. Baer SH, Blaschek HP, Smith TL (1987) Effect of butanol challenge and temperature on lipid composition and membrane fluidity of butanol-tolerant *Clostridium acetobutylicum*. Appl Environ Microbiol 53(12):2854–2861
7. Baer SH, Bryant DL, Blaschek HP (1989) Electron spin resonance analysis of the effect of butanol on the membrane fluidity of intact cells of *Clostridium acetobutylicum*. Appl Environ Microbiol 55(10):2729–2731
8. Bahl H, Müller H, Behrens S, Joseph H, Narberhaus F (1995) Expression of heat shock genes in *Clostridium acetobutylicum*. FEMS Microbiol Rev 17(3):341–348
9. Baronofsky JJ, Schreurs WJ, Kashket ER (1984) Uncoupling by acetic acid limits growth of and acetogenesis by *Clostridium thermoaceticum*. Appl Environ Microbiol 48(6):1134–1139
10. Bernal P, Segura A, Ramos JL (2007) Compensatory role of the cis-trans-isomerase and cardiolipin synthase in the membrane fluidity of *Pseudomonas putida* DOT-T1E. Environ Microbiol 9(7):1658–1664
11. Borden JR, Papoutsakis ET (2007) Dynamics of genomic-library enrichment and identification of solvent tolerance genes for *Clostridium acetobutylicum*. Appl Environ Microbiol 73 (9):3061–3068
12. Borden JR, Jones SW, Indurthi D, Chen Y, Papoutsakis ET (2010) A genomic-library based discovery of a novel, possibly synthetic, acid-tolerance mechanism in *Clostridium acetobutylicum* involving non-coding RNAs and ribosomal RNA processing. Metab Eng 12 (3):268–281
13. Bowles LK, Ellefson WL (1985) Effects of butanol on *Clostridium acetobutylicum*. Appl Environ Microbiol 50(5):1165–1170
14. Brown SD, Thompson MR, Verberkmoes NC, Chourey K, Shah M, Zhou J, Hettich RL, Thompson DK (2006) Molecular dynamics of the *Shewanella oneidensis* response to chromate stress. Mol Cell Proteomics 5(6):1054–1071
15. Brynildsen MP, Liao JC (2009) An integrated network approach identifies the isobutanol response network of *Escherichia coli*. Mol Syst Biol 5:277
16. Campbell JW, Cronan JE Jr (2001) Bacterial fatty acid biosynthesis: targets for antibacterial drug discovery. Annu Rev Microbiol 55:305–332
17. Chen JS, Toth J, Kasap M (2001) Nitrogen-fixation genes and nitrogenase activity in *Clostridium acetobutylicum* and *Clostridium beijerinckii*. J Ind Microbiol Biotechnol 27 (5):281–286
18. Chhabra SR, He Q, Huang KH, Gaucher SP, Alm EJ, He Z, Hadi MZ, Hazen TC, Wall JD, Zhou J, Arkin AP, Singh AK (2006) Global analysis of heat shock response in *Desulfovibrio vulgaris Hildenborough*. J Bacteriol 188(5):1817–1828
19. Dai M, Copley SD (2004) Genome shuffling improves degradation of the anthropogenic pesticide pentachlorophenol by *Sphingobium chlorophenolicum* ATCC 39723. Appl Environ Microbiol 70(4):2391–2397
20. Derre I, Rapoport G, Msadek T (1999) CtsR, a novel regulator of stress and heat shock response, controls *clp* and molecular chaperone gene expression in gram-positive bacteria. Mol Microbiol 31(1):117–131
21. Dombek KM, Ingram LO (1984) Effects of ethanol on the *Escherichia coli* plasma membrane. J Bacteriol 157(1):233–239

22. Fernandes P, Ferreira BS, Cabral JM (2003) Solvent tolerance in bacteria: role of efflux pumps and cross-resistance with antibiotics. Int J Antimicrob Agents 22(3):211–216
23. Gill RT, Wildt S, Yang YT, Ziesman S, Stephanopoulos G (2002) Genome-wide screening for trait conferring genes using DNA microarrays. Proc Natl Acad Sci USA 99 (10):7033–7038
24. Gottwald M, Gottschalk G (1985) The internal pH of *Clostridium acetobutylicum* and its effect on the shift from acid to solvent formation. Arch Microbiol 143:42–46
25. Guisbert E, Yura T, Rhodius VA, Gross CA (2008) Convergence of molecular, modeling, and systems approaches for an understanding of the *Escherichia coli* heat shock response. Microbiol Mol Biol Rev 72(3):545–554
26. Gupta A, Singh R, Khare SK, Gupta MN (2006) A solvent tolerant isolate of *Enterobacter aerogenes*. Bioresour Technol 97(1):99–103
27. Han SS, Lee JY, Kim WH, Shin HJ, Kim GJ (2008) Screening of promoters from metagenomic DNA and their use for the construction of expression vectors. J Microbiol Biotechnol 18(10):1634–1640
28. Han MJ, Yun H, Lee SY (2008) Microbial small heat shock proteins and their use in biotechnology. Biotechnol Adv 26(6):591–609
29. Helmann JD, Wu MF, Gaballa A, Kobel PA, Morshedi MM, Fawcett P, Paddon C (2003) The global transcriptional response of *Bacillus subtilis* to peroxide stress is coordinated by three transcription factors. J Bacteriol 185(1):243–253
30. Herrero AA, Gomez RF, Snedecor B, Tolman CJ, Roberts MF (1985) Growth inhibition of Clostridium thermocellum by carboxylic acids: a mechanism based on uncoupling by weak acids. Appl Microbiol Biotechnol 22(1):53–62
31. Hosokawa K, Park NH, Inaoka T, Itoh Y, Ochi K (2002) Streptomycin-resistant (*rpsL*) or rifampicin-resistant (*rpoB*) mutation in *Pseudomonas putida* KH146-2 confers enhanced tolerance to organic chemicals. Environ Microbiol 4(11):703–712
32. Huang L, Gibbins LN, Forsberg CW (1985) Transmembrane pH gradient and membrane potential in *Clostridium acetobutylicum* during growth under acetogenic and solventogenic conditions. Appl Environ Microbiol 50(4):1043–1047
33. Hüsemann M, Papoutsakis ET (1986) Effect of acetoacetate, butyrate, and uncoupling ionophores on growth and product formation of *Clostridium acetobutylicum*. Biotechnol Lett 8(1):37–42
34. Hüsemann MH, Papoutsakis ET (1988) Solventogenesis in *Clostridium acetobutylicum* fermentations related to carboxylic acid and proton concentrations. Biotechnol Bioeng 32 (7):843–852
35. Ingram LO (1976) Adaptation of membrane lipids to alcohols. J Bacteriol 125(2):670–678
36. Isaacs FJ, Carr PA, Wang HH, Lajoie MJ, Sterling B, Kraal L, Tolonen AC, Gianoulis TA, Goodman DB, Reppas NB, Emig CJ, Bang D, Hwang SJ, Jewett MC, Jacobson JM, Church GM (2011) Precise manipulation of chromosomes in vivo enables genome-wide codon replacement. Science 333(6040):348–353. doi:10.1126/science.1205822
37. Isken S, de Bont JA (1998) Bacteria tolerant to organic solvents. Extremophiles 2(3):229–238
38. Jain MK, Wu NM (1977) Effect of small molecules on dipalmitoyl lecithin liposomal bilayer.3. phase-transition in lipid bilayer. J Membr Biol 34(2–3):157–201
39. Jain MK, Gleeson J, Upreti A, Upreti GC (1978) Intrinsic perturbing ability of alkanols in lipid bilayers. Biochim Biophys Acta 509(1):1–8
40. Jennings LK, Chartrand MM, Lacrampe-Couloume G, Lollar BS, Spain JC, Gossett JM (2009) Proteomic and transcriptomic analyses reveal genes upregulated by cis-dichloroethene in *Polaromonas sp.* strain JS666. Appl Environ Microbiol 75(11):3733–3744
41. Kashket ER (1987) Bioenergetics of lactic acid bacteria: cytoplasmic pH and osmotolerance. FEMS Microbiol Rev 46(3):233–244
42. Kell DB, Peck MW, Rodger G, Morris JG (1981) On the permeability to weak acids and bases of the cytoplasmic membrane of *Clostridium pasteurianum*. Biochem Biophys Res Commun 99(1):81–88

43. Kivistik PA, Putrins M, Puvi K, Ilves H, Kivisaar M, Horak R (2006) The ColRS two-component system regulates membrane functions and protects *Pseudomonas putida* against phenol. J Bacteriol 188(23):8109–8117
44. Kobayashi K, Tsukagoshi N, Aono R (2001) Suppression of hypersensitivity of *Escherichia coli acrB* mutant to organic solvents by integrational activation of the *acrEF* operon with the IS1 or IS2 element. J Bacteriol 183(8):2646–2653
45. Kultz D (2005) Molecular and evolutionary basis of the cellular stress response. Annu Rev Physiol 67:225–257
46. Lee SY, Park JH, Jang SH, Nielsen LK, Kim J, Jung KS (2008) Fermentative butanol production by Clostridia. Biotechnol Bioeng 101(2):209–228
47. Lepage C, Fayolle F, Hermann M, Vandecasteele JP (1987) Changes in membrane lipid composition of *Clostvidium acetobutylicum* during acetone-butanol fermentation: effects of solvents, growth temperature and pH. J Gen Microbiol 133:103–110
48. Li L, Li Q, Rohlin L, Kim U, Salmon K, Rejtar T, Gunsalus RP, Karger BL, Ferry JG (2007) Quantitative proteomic and microarray analysis of the archaeon *Methanosarcina acetivorans* grown with acetate versus methanol. J Proteome Res 6:759–771
49. Lu YJ, Zhang YM, Rock CO (2004) Product diversity and regulation of type II fatty acid synthases. Biochem Cell Biol 82(1):145–155
50. Lund PA (2009) Multiple chaperonins in bacteria–why so many? FEMS Microbiol Rev 33 (4):785–800
51. Mazumder R, Pinkart HC, Alban PS, Phelps TJ, Benoit RE (2000) Low-substrate regulated microaerophilic behavior as a stress response of aquatic and soil bacteria. Curr Microbiol 41 (2):79–83
52. Monot F, Engasser JM, Petitdemange H (1984) Influence of pH and undissociated butyric acid on the production of acetone and butanol in batch cultures of *Clostridium acetobutylicum*. Appl Microbiol Biotechnol 19:422–426
53. Mostertz J, Scharf C, Hecker M, Homuth G (2004) Transcriptome and proteome analysis of *Bacillus subtilis* gene expression in response to superoxide and peroxide stress. Microbiology 150(Pt 2):497–512
54. Moxley JF, Jewett MC, Antoniewicz MR, Villas-Boas SG, Alper H, Wheeler RT, Tong L, Hinnebusch AG, Ideker T, Nielsen J, Stephanopoulos G (2009) Linking high-resolution metabolic flux phenotypes and transcriptional regulation in yeast modulated by the global regulator Gcn4p. Proc Natl Acad Sci USA 106(16):6477–6482
55. Neumann G, Veeranagouda Y, Karegoudar TB, Sahin O, Mausezahl I, Kabelitz N, Kappelmeyer U, Heipieper HJ (2005) Cells of *Pseudomonas putida* and *Enterobacter sp.* adapt to toxic organic compounds by increasing their size. Extremophiles 9(2):163–168
56. Nicolaou SA, Gaida SM, Papoutsakis ET (2010) A comparative view of metabolite and substrate stress and tolerance in microbial bioprocessing: from biofuels and chemicals, to biocatalysis and bioremediation. Metab Eng 12(4):307–331
57. Nicolaou SA, Gaida SM, Papoutsakis ET (2011) Coexisting/Coexpressing Genomic Libraries (CoGeL) identify interactions among distantly located genetic loci for developing complex microbial phenotypes. Nucleic Acids Res 39(22):e152. doi:10.1093/nar/gkr817
58. Nikaido H, Zgurskaya HI (1999) Antibiotic efflux mechanisms. Curr Opin Infect Dis 12 (6):529–536
59. Nölling J, Breton G, Omelchenko MV, Makarova KS, Zeng Q, Gibson R, Lee HM, Dubois J, Qiu D, Hitti J, Wolf YI, Tatusov RL, Sabathe F, Doucette-Stamm L, Soucaille P, Daly MJ, Bennett GN, Koonin EV, Smith DR (2001) Genome sequence and comparative analysis of the solvent-producing bacterium *Clostridium acetobutylicum*. J Bacteriol 183(16):4823–4838
60. Osiriphun Y, Wongtrakoongate P, Sanongkiet S, Suriyaphol P, Thongboonkerd V, Tungpradabkul S (2009) Identification and characterization of RpoS regulon and RpoS-dependent promoters in *Burkholderia pseudomallei*. J Proteome Res 8(6):3118–3131
61. Osterman A, Overbeek R (2003) Missing genes in metabolic pathways: a comparative genomics approach. Curr Opin Chem Biol 7(2):238–251

62. Papoutsakis ET (2008) Engineering solventogenic clostridia. Curr Opin Biotechnol 19 (5):420–429
63. Paredes CJ, Alsaker KV, Papoutsakis ET (2005) A comparative genomic view of clostridial sporulation and physiology. Nat Rev Microbiol 3(12):969–978
64. Patnaik R, Louie S, Gavrilovic V, Perry K, Stemmer WP, Ryan CM, del Cardayre S (2002) Genome shuffling of Lactobacillus for improved acid tolerance. Nat Biotechnol 20 (7):707–712
65. Petri R, Schmidt-Dannert C (2004) Dealing with complexity: evolutionary engineering and genome shuffling. Curr Opin Biotechnol 15(4):298–304
66. Phoenix P, Keane A, Patel A, Bergeron H, Ghoshal S, Lau PC (2003) Characterization of a new solvent-responsive gene locus in *Pseudomonas putida* F1 and its functionalization as a versatile biosensor. Environ Microbiol 5(12):1309–1327
67. Pich A, Naberhaus F, Bahl H (1990) Induction of heat shock proteins during initiation of solvent formation in *Clostridium acetobutylicum*. Appl Microbiol Biotechnol 33(6):697–704
68. Ramos JL, Duque E, Rodriguez-Herva JJ, Godoy P, Haidour A, Reyes F, Fernandez-Barrero A (1997) Mechanisms for solvent tolerance in bacteria. J Biol Chem 272(7):3887–3890
69. Ramos JL, Duque E, Gallegos MT, Godoy P, Ramos-Gonzalez MI, Rojas A, Teran W, Segura A (2002) Mechanisms of solvent tolerance in gram-negative bacteria. Annu Rev Microbiol 56:743–768
70. Ramos A, Macias JR, Gil JA (1997) Cloning, sequencing and expression of the gene encoding elongation factor P in the amino-acid producer *Brevibacterium lactofermentum* (*Corynebacterium glutamicum* ATCC 13869). Gene 198(1–2):217–222
71. Russell JB (1992) Another explanation for the toxicity of fermentation acids at low pH: anion accumulation versus uncoupling. J Appl Bacteriol 73(5):363–370
72. Russell JB, Diez-Gonzalez F (1998) The effects of fermentation acids on bacterial growth. Adv Microb Physiol 39:205–234
73. Rutherford BJ, Dahl RH, Price RE, Szmidt HL, Benke PI, Mukhopadhyay A, Keasling JD (2010) Functional genomic study of exogenous n-butanol stress in *Escherichia coli*. Appl Environ Microbiol 76(6):1935–1945
74. Santos PM, Benndorf D, Sa-Correia I (2004) Insights into *Pseudomonas putida* KT2440 response to phenol-induced stress by quantitative proteomics. Proteomics 4(9):2640–2652
75. Schaffer S, Isci N, Zickner B, Durre P (2002) Changes in protein synthesis and identification of proteins specifically induced during solventogenesis in *Clostridium acetobutylicum*. Electrophoresis 23(1):110–121
76. Schujman GE, de Mendoza D (2005) Transcriptional control of membrane lipid synthesis in bacteria. Curr Opin Microbiol 8(2):149–153
77. Schumann W, Hecker M, Msadek T (2002) Regulation and function of heat-inducible genes in *Bacillus subtilis*. In: Sonenshein AL, Hoch JA, Losick R (eds) *Bacillus subtilis* and its closest relatives: from genes to cells. ASM Press, Washington, DC, pp 359–368
78. Schwartz JM, Gaugain C, Nacher JC, de Daruvar A, Kanehisa M (2007) Observing metabolic functions at the genome scale. Genome Biol 8(6):R123
79. Sinensky M (1974) Homeoviscous adaptation–a homeostatic process that regulates the viscosity of membrane lipids in *Escherichia coli*. Proc Natl Acad Sci USA 71(2):522–525
80. Stephanopoulos G (2007) Challenges in engineering microbes for biofuels production. Science 315(5813):801–804
81. Tamayo P, Slonim D, Mesirov J, Zhu Q, Kitareewan S, Dmitrovsky E, Lander ES, Golub TR (1999) Interpreting patterns of gene expression with self-organizing maps: methods and application to hematopoietic differentiation. Proc Natl Acad Sci USA 96(6):2907–2912
82. Tatusov RL, Galperin MY, Natale DA, Koonin EV (2000) The COG database: a tool for genome-scale analysis of protein functions and evolution. Nucleic Acids Res 28(1):33–36
83. Terracciano JS, Kashket ER (1986) Intracellular conditions required for initiation of solvent production by *Clostridium acetobutylicum*. Appl Environ Microbiol 52(1):86–91

84. Terracciano JS, Rapaport E, Kashket ER (1988) Stress- and growth phase-associated proteins of *Clostridium acetobutylicum*. Appl Environ Microbiol 54(8):1989–1995
85. Thompson MR, VerBerkmoes NC, Chourey K, Shah M, Thompson DK, Hettich RL (2007) Dosage-dependent proteome response of *Shewanella oneidensis* MR-1 to acute chromate challenge. J Proteome Res 6(5):1745–1757
86. Tomas CA, Welker NE, Papoutsakis ET (2003) Overexpression of *groESL* in *Clostridium acetobutylicum* results in increased solvent production and tolerance, prolonged metabolism, and changes in the cell's transcriptional program. Appl Environ Microbiol 69(8):4951–4965
87. Tomas CA, Beamish J, Papoutsakis ET (2004) Transcriptional analysis of butanol stress and tolerance in *Clostridium acetobutylicum*. J Bacteriol 186(7):2006–2018
88. Tummala SB, Welker NE, Papoutsakis ET (2003) Design of antisense RNA constructs for downregulation of the acetone formation pathway of *Clostridium acetobutylicum*. J Bacteriol 185(6):1923–1934
89. VanBogelen RA, Kelley PM, Neidhardt FC (1987) Differential induction of heat shock, SOS, and oxidation stress regulons and accumulation of nucleotides in *Escherichia coli*. J Bacteriol 169(1):26–32
90. Volkers RJ, de Jong AL, Hulst AG, van Baar BL, de Bont JA, Wery J (2006) Chemostat-based proteomic analysis of toluene-affected *Pseudomonas putida* S12. Environ Microbiol 8 (9):1674–1679
91. Vollherbst-Schneck K, Sands JA, Montenecourt BS (1984) Effect of butanol on lipid composition and fluidity of *Clostridium acetobutylicum* ATCC 824. Appl Environ Microbiol 47(1):193–194
92. Wang Y, Li Y, Pei X, Yu L, Feng Y (2007) Genome-shuffling improved acid tolerance and L-lactic acid volumetric productivity in *Lactobacillus rhamnosus*. J Biotechnol 129(3): 510–515
93. Wang HH, Isaacs FJ, Carr PA, Sun ZZ, Xu G, Forest CR, Church GM (2009) Programming cells by multiplex genome engineering and accelerated evolution. Nature 460(7257):894–898
94. Weber FJ, de Bont JA (1996) Adaptation mechanisms of microorganisms to the toxic effects of organic solvents on membranes. Biochim Biophys Acta 1286(3):225–245
95. Weber H, Polen T, Heuveling J, Wendisch VF, Hengge R (2005) Genome-wide analysis of the general stress response network in *Escherichia coli*: sigmaS-dependent genes, promoters, and sigma factor selectivity. J Bacteriol 187(5):1591–1603
96. Wecke T, Veith B, Ehrenreich A, Mascher T (2006) Cell envelope stress response in *Bacillus licheniformis*: integrating comparative genomics, transcriptional profiling, and regulon mining to decipher a complex regulatory network. J Bacteriol 188(21):7500–7511
97. Wei Y, Vollmer AC, LaRossa RA (2001) In vivo titration of mitomycin C action by four *Escherichia coli* genomic regions on multicopy plasmids. J Bacteriol 183(7):2259–2264
98. Zhang YX, Perry K, Vinci VA, Powell K, Stemmer WP, del Cardayre SB (2002) Genome shuffling leads to rapid phenotypic improvement in bacteria. Nature 415(6872):644–646
99. Zhao Y, Hindorff LA, Chuang A, Monroe-Augustus M, Lyristis M, Harrison ML, Rudolph FB, Bennett GN (2003) Expression of a cloned cyclopropane fatty acid synthase gene reduces solvent formation in *Clostridium acetobutylicum* ATCC 824. Appl Environ Microbiol 69 (5):2831–2841
100. Zomer A, Fernandez M, Kearney B, Fitzgerald GF, Ventura M, van Sinderen D (2009) An interactive regulatory network controls stress response in *Bifidobacterium breve* UCC2003. J Bacteriol 191(22):7039–7049

Chapter 8
Model-Based Design of Superior Cell Factory: An Illustrative Example of *Penicillium chrysogenum*

I. Emrah Nikerel, Peter J.T. Verheijen, Walter M. van Gulik, and Joseph J. Heijnen

Contents

Abstract A dynamic model for metabolic reaction network of *Penicillium chrysogenum*, coupling the central metabolism to growth, product formation and storage pathways is presented. In constructing the model, we started from an

---

I.E. Nikerel (✉)
Delft Bioinformatics Lab, Delft University of Technology, Mekelweg 4, 2628 CD Delft, The Netherlands

Department of Biotechnology, Kluyver Centre for Genomics of Industrial Fermentation, Delft University of Technology, Julianalaan 67, 2628 BC Delft, The Netherlands
e-mail: I.E.Nikerel@TUDelft.nl

P.J.T. Verheijen • W.M. van Gulik • J.J. Heijnen
Department of Biotechnology, Kluyver Centre for Genomics of Industrial Fermentation, Delft University of Technology, Julianalaan 67, 2628 BC Delft, The Netherlands

C. Wittmann and S.Y. Lee (eds.), *Systems Metabolic Engineering*,
DOI 10.1007/978-94-007-4534-6_8, © Springer Science+Business Media Dordrecht 2012

existing stoichiometric model, and systematically reduced this initial model to a one compartment model and further eliminated unidentifiabilities due to time scales. Kinetic analysis focuses on a time scale of seconds, thereby neglecting biosynthesis of new enzymes. We used linlog kinetics in representing the kinetic rate equations of each individual reaction. The final parameterization is performed for the final reduced model using previously published short term glucose perturbation data. The constructed model is a self-contained model in the sense that it can also predict the cofactor dynamics. Using the model, we calculated the Metabolic Control Analysis (MCA) parameters and found that the interplay among the growth, product formation and production of storage materials is strongly governed by the energy budget in the cell, which is in agreement with the previous findings. The model predictions and experimental observations agree reasonably well for most of the metabolites.

Keywords Model-based design • *Penicillium chrysogenum* • Dynamic model • Metabolic network • Cell factory • Linlog kinetics • Parameterization • Metabolic Control Analysis (MCA) • Model prediction • Time-scale analysis • Elasticities • Kinetic model • P/O ratio • Post-genomic • Genome-scale • Stimulus response experiments • Metabolic flux • Stoichiometry • Network reconstruction • Compartments • Pseudo equilibrium • Pseudo steady-state

8.1 Introduction

By building a multidisciplinary framework, the emerging field of systems biology aims at achieving a more fundamental knowledge of biological complexity through system-level understanding [13]. Mathematical models of biological systems play a central role in this aim since they allow not only the organization, understanding and interpretation of the data e.g. description of the complex kinetic behavior, but also allow performing exploratory simulations, design and optimization of such systems [12]. Especially in the post-genomic era, where gene perturbations are common practice, these models can be used to predict the effect of such a perturbation in the genotype on the metabolism (phenotype) [24, 37]. Furthermore, once the system is abstracted, it may also serve to further evaluate how the newly produced data fits to our current understanding (both qualitatively and quantitatively) of these systems.

Within this core, the intricate nature of biological systems calls further for the construction of large scale (e.g. genome scale) kinetic models for e.g. improved prediction capability. From a systems biology perspective, these models should not only describe the kinetic behavior of metabolic reaction networks that feature metabolite-enzyme interactions (allosteric feedback or feed forward), inter-compartmental transport, and cofactor coupling, but, they should also ultimately allow combining several pathways (horizontal modeling) and/or "omic" levels (vertical modeling) in the cell. Nevertheless, currently by far most of the available kinetic models are limited to only one pathway or one "omic" level [4, 15, 22].

Being of central interest, the construction of large scale kinetic models is a demanding task due to several challenges in many aspects ranging from the data generation (optimal experimental design), to the selection of enzyme kinetics and parameter identifiability.

One of the initial challenges to address is the availability (for kinetic modeling) of experimental data on a large scale. Although the list of measurable metabolites is growing, the number of measured metabolites is far less than the number of metabolites considered in a genome-scale model. Furthermore, the compartmentation of metabolism in eukaryotic cells poses a problem for the accurate determination of relevant concentrations of intracellular metabolites. This is not sufficiently addressed by current "metabolomics" approaches, which provide total concentration averages over the entire cell. That is a major constraint to current efforts in quantitative physiology and kinetic modeling of metabolic reaction networks, which rely on the accuracy of the information regarding the *in vivo* environment surrounding the enzymes [11]. This issue is less severe for localized reactions like glycolysis reactions, whereas especially for currency metabolites, e.g. ATP, NADH, NADPH, the compartmentation is an important challenge to overcome [35, 36].

In order to obtain data to parameterize or to test the model, perturbations to a well characterized state are to be performed. The design of the applied perturbation depends initially on the scope of the model to be build. Keeping in mind that the cells are composed of several highly interacting "omic" layers, aiming to construct at once a whole-cell model is quite ambitious. The alternative approach is to isolate a selection (preferably one) of these "omic" layers and one of the challenges in obtaining experimental data is to design an experiment in which a given perturbation affects only the metabolome so that the remaining is assumed not to change, bypassing the need to measure e.g. enzyme activities [26, 34]. Stimulus response experiments have been proposed and performed in the past for this purpose [26, 33].

From the modeling point of view, one important task is to decide on the type of kinetics to use. Traditional kinetics suffer from several problems, like high nonlinearity, not being a uniform format, *in vivo/in vitro* dilemma. Furthermore, while for systems with a small number of reactions, the selection of the kinetic format is rather straight forward dictated by the mechanism of the enzyme, for large scale systems, although there are dedicated databases [23], since traditional kinetics do not have a uniform format, the use of these kinetics is not suitable. As an alternative, uniform kinetic formats are more suitable for large scale modeling. A good example of this approach is to use Metabolic Control Analysis (MCA) based approximative kinetics [7], or other formats [16].

So far, many models cover either one omic level or one specific pathway. Rizzi et al. [22] published a model to represent the central metabolism for yeast cells using mechanistically derived enzyme kinetic equations and short-term perturbation data. Using a similar approach, [2] published a mathematical model for *Escherichia coli*. A more focused work has been published by Vaseghi et al. [31], focusing on dynamic modeling on pentose phosphate pathway (PPP). Furthermore, a number of different models have been proposed to study different aspect of the metabolism e.g. for oscillations [9], for ethanol production [6, 25].

It is hence the aim of this work to build up a large scale kinetic model of *P. chrysogenum*, based on the metabolome data from stimulus response experiment, presented in Nasution et al. [18]. Although still focusing on one level (i.e. metabolome), the model aims at encompassing all the major pathways (e.g. Glycolysis, TCA, PP pathway, amino acid and product pathways) in the organism. Furthermore, by building such a model, we aim to evaluate our procedures to set up such a large scale model and our current knowledge on *in vivo* interactions in the cell. Such a dynamic kinetic model is also relevant for industrial conditions where the product pathway and storage pathway dynamics can be studied.

Since its discovery by accident in 1928 *P. chrysogenum* serves as the main source for the industrially and clinically important β-lactam antibiotics, such as penicillin-G, ampicillin, amoxicillin and cephalexin, which are used to treat bacterial infections. This fundamental clinical and applied industrial relevance calls for further research to improve the performance of the antibiotics production. For many years, *P. chrysogenum* has been intensively studied in several laboratories [3, 10, 17, 27]. In our laboratory the organism is studied in many aspects from the determination of the metabolic capabilities in steady state flux regimes [29, 30], to the investigation of short term (<300 s) dynamic response of the system under substrate limitation to a substrate pulse [18], and to the investigation of the cytosolic NADPH metabolism in relation with the penicillin synthesis [14].

The book chapter is organized as follows: after describing our stepwise model reduction approach in the Background section, we describe the characteristics of the case system and this is followed by the constructed model presented in the Results section.

8.2 Background

8.2.1 General Equations

Traditionally, building a kinetic model starts by

$$\frac{d\mathbf{x}}{dt} = \mathbf{S} \cdot \mathbf{v} - \mu \mathbf{x}, \tag{8.1}$$

where $\mathbf{x}$ and $\mathbf{S}$ are the intracellular metabolite vector and the stoichiometric matrix respectively, $\mathbf{v}$ is the rate vector and the second term on the right hand side is the dilution term due to growth, with μ being the growth rate. For stoichiometric matrix **S**, several genome scale stoichiometric models are available (e.g. [5, 21]). To fully define the model, the kinetic function of the rate vector **v** in Eq. 8.1 must be defined. We use linlog kinetics which is a promising approximative kinetic format introduced in [32], with low number of parameters and good approximation quality Eq. 8.2. Furthermore, since the parameters in linlog kinetics appear in linear fashion, estimation of these parameters can be performed via linear regression as explained in [19]

$$\frac{v_i}{J_i^0} = \left[\frac{e_i}{e_i^0}\right] \cdot \left(\mathbf{1} + \varepsilon_{\mathrm{x_j}}^{\mathrm{v_i}} \ln\left(\frac{\mathbf{x}_\mathrm{j}}{\mathbf{x}_\mathrm{j}^0}\right) + \varepsilon_{\mathrm{c_k}}^{\mathrm{v_i}} \ln\left(\frac{\mathbf{c}_\mathrm{k}}{\mathbf{c}_\mathrm{k}^0}\right)\right) \tag{8.2}$$

In Eq. 8.2, v_i, e_i represent the rate and enzyme level of reaction i respectively, x and c are intracellular and extracellular metabolites. All entities in Eq. 8.2 are normalized with respect to a reference (steady) state for rate, enzyme and metabolite levels, the reference state noted with superscript "^{0}", and J_i^0 denotes the flux for reaction i at the reference (steady) state. The parameters (ε^x and ε^c in Eq. 8.2) are introduced as elasticities in MCA, representing the local sensitivity of metabolite (x or c) towards rate v as:

$$\varepsilon_x^v = \left.\frac{\partial v}{\partial x}\right|_{x^0} \frac{x^0}{J^0}$$

8.2.2 *Case Study Considered*

Focusing on *P. chrysogenum*, we considered the stoichiometric network model presented in van Gulik et al. [29] as the model system and for the stoichiometric matrix. The network in van Gulik et al. [29] consists of three compartments (cytosol, mitochondria and peroxisome), 156 intracellular reactions, 155 intracellular metabolites. Though there will be a number of additions and modifications in the reactions (described below), for the sake of consistency, the nomenclature for the reactions is kept the same as in van Gulik et al. [29] wherever possible. The list of reactions and nomenclature for metabolites for the obtained base model is given in the Appendix Tables 8.14–8.16.

In order to parameterize the model, short term (<300 s) perturbation data presented in [18] are used, which is the most comprehensive metabolome data available on *P. chrysogenum*. The data is obtained by perturbing the steady state (presented in Table 8.11) of an aerobic glucose limited chemostat culture with excess glucose. Glycolysis, TCA pathway intermediates and several amino acids as well as the uptake rates of glucose, O_2 and production rate of CO_2 are measured.

8.2.3 Preliminary Considerations for Building the Model

In building the kinetic model, we departed from the network presented in vanGulik et al. [29] and implemented a number of modifications and/or simplifications, considering both the challenges mentioned in the introduction, and dynamic characteristics of the system (e.g. classification of slow and fast pools/reactions). In this section we describe these considerations.

8.2.3.1 Adjustments/Simplifications on the Reaction Network

Delineating Storage from Biomass

In [18], it has been postulated that upon a glucose pulse, the biosynthesis of storage materials is greatly enhanced. To elaborate on this behaviour the biomass definition is changed. As it was defined in [29], the storage materials were part of the biomass formation. In this work, the growth reaction is separated into a synthetic (syn_X, 118.2) and a storage/trehalose part (tre_X, 117.6). Such an explicit definition for storage formation allows us to focus further on determining where the carbon flux (re)directs upon glucose addition during the rapid pulse experiment.

Reducing the Number of Compartments

There is a gap in data availability vis-a-vis the compartmentation. Since the available experimental information for metabolites are whole-cell averages, our current experimental knowledge on metabolites does not allow building kinetic models with distinct compartments, since such information is not achievable in the available data. In order to tackle this problem, we considered the system as one compartment and removed all the transport reactions between the compartments (see Fig. 8.1). Furthermore, if there are two (or more) reactions localized in two (or more) compartments, only one of those reactions is kept in order to avoid duplicate reactions. It should be noted that, all the enzymatic reactions are kept, only compartmentation related transport information is lost.

Lumping the Non-measured Pathway Intermediates

Having only a limited number of measured metabolites, many unmeasured intermediate metabolites have been eliminated by lumping the involved reactions. By this, the model was reduced by eliminating metabolites from the secondary pathways and lumping the pathways in e.g. the amino acid synthesis (r11.i), nucleotide synthesis (r13.i), fatty acid synthesis (r16.i) and the biosynthesis of

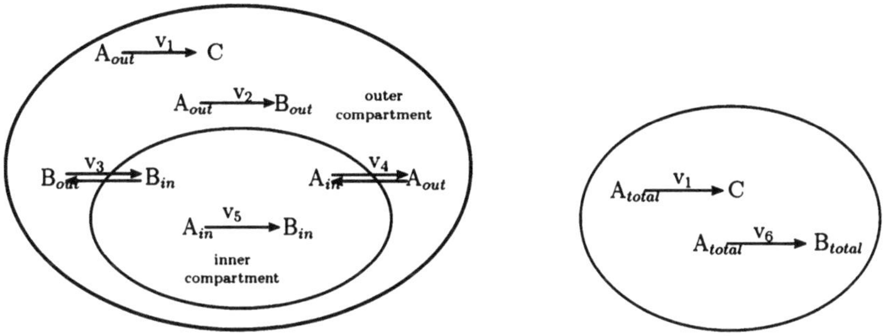

$$\begin{bmatrix} dA_{in}/dt \\ dA_{out}/dt \\ dB_{in}/dt \\ dB_{out}/dt \\ dC/dt \end{bmatrix} = \begin{bmatrix} 0 & 0 & 0 & 1 & -1 \\ -1 & -1 & 0 & 1 & 0 \\ 0 & 0 & 1 & 0 & 1 \\ 0 & 1 & -1 & 0 & 0 \\ 1 & 0 & 0 & 0 & 0 \end{bmatrix} \begin{bmatrix} v_1 \\ v_2 \\ v_3 \\ v_4 \\ v_5 \end{bmatrix} \qquad \begin{bmatrix} dA_{tot}/dt \\ dB_{tot}/dt \\ dC/dt \end{bmatrix} = \begin{bmatrix} -1 & -1 \\ 0 & 1 \\ 1 & 0 \end{bmatrix} \begin{bmatrix} v_1 \\ v_6 \end{bmatrix}$$

Fig. 8.1 An illustrative example on how the compartmentation challenge is tackled. On the *left*, the molecules A and B are present in both *inner* and *outer* compartments and the reaction from A to B occurs in both compartments (v_2 and v_5). These molecules are transported between the compartments via transport reactions v_3 and v_4. On the *right*, the *inner* compartment is removed together with the transport reactions. For the molecules A and B, only total concentrations are considered. Note that v_1 is unchanged, v_3 and v_4 vanished and v_2 and v_5 are replaced by v_6

glycogen and polysaccharides. Similarly, although the penicillin biosynthesis is of main interest, since the corresponding intermediates have not been measured, the penicillin pathway has been lumped.

8.2.3.2 Time Scale Analysis

In a cell, many processes (metabolite conversion, enzyme synthesis, gene expression, etc.) occur simultaneously, following a time hierarchy. The resulting co-existence of multiple time-scales among these processes has a number of implications for which it is important to classify the time scales in cellular processes.

In a recent study, the turn-over time in metabolite pools has been proposed for such classification [20]. The metabolite pools are ranked according to their turn-over times (τ's) calculated as the ratio of their steady state levels to the incoming (or outgoing) total flux ($\tau = \frac{X^0}{J^0_{IN}}$). This ranking leads to the classification such as the one presented in Fig. 8.2, which further leads to model reduction [20]. For genome scale modeling, providing the systematic tools for classification and reduction of a, e.g. genome-scale stoichiometric matrix into smaller sub-systems based on time scale separation is of high value. In what follows, we describe our approach of *a priori* reducing the stoichiometric matrix and the rate equations (**S** and **v** in Eq. 8.1) from an initially available larger scale model, considering the time-scale classification based on experimental data of the perturbed steady state.

Fast pools → **Slow pools**

Pseudo equilibrium	Pseudo steady state	Dynamic	Frozen
$\tau \approx 0$	$\tau \ll t_{exp}$	$\tau \backsim t_{exp}$	$\tau \approx \infty$

Fig. 8.2 Classification of cellular processes according to time scale analysis. t_{exp} denotes the experimental time window and the τ denotes the turn-over time

$$\xrightarrow{v_1} x_1 \underset{v_{2,b}}{\overset{v_{2,f}}{\rightleftharpoons}} x_2 \xrightarrow{v_3} \qquad\qquad \xrightarrow{v_1} p_1 \xrightarrow{v_3}$$

$$\frac{dx_1}{dt} = v_1 - v_{2,f} + v_{2,b} \qquad\qquad \frac{dp_1}{dt} = v_1 - v_3$$

$$p_1 = x_1 + x_2$$

$$\frac{dx_2}{dt} = v_{2,f} - v_{2,f} - v_3 \qquad\qquad K_{eq} = \frac{x_2}{x_1}$$

Fig. 8.3 Example system, illustrating how to treat the reactions that operate at pseudo-equilibrium. On the *left panel*, the reaction v_2 operates at pseudo-equilibrium, so that $v_{2f}, v_{2b} < < v_1, v_3$. By defining p_1, we eliminate the pseudo-equilibrium reaction (and its kinetic expression) by eliminating v_2 from the mass balance, and obtain a lumped pseudo-equilibrium pool. Having p_1, the individual x_1 and x_2 metabolites can be calculated using the equilibrium relation

Considering Reactions at Pseudo Equilibrium

Assuming that a selected reaction operating at pseudo equilibrium conditions at steady state does not move away significantly from pseudo equilibrium during dynamic conditions allows replacing the kinetic function of that reaction by its equilibrium constant. This leads to a reduction and modification of the stoichiometric matrix. This reduction procedure consists of lumping the metabolite pools that are connected by the reactions at pseudo equilibrium, into one larger pool. The procedure is exemplified in Fig. 8.3 for a simple example system.

While for small systems, determining the pool composition to eliminate the reactions at pseudo equilibrium is obvious, for large systems the pool composition is given by Eq. 8.3,

$$\mathbf{p} = \mathbf{xP} \cdot \mathbf{x_{peq}} \quad \text{with} \quad \mathbf{xP} = \left[\text{null}\left(\mathbf{S}_{\text{peq}}^{\text{T}}\right)\right]^{\text{T}}, \tag{8.3}$$

where **xP** is the matrix defining the composition of the near equilibrium pools (**p**) as a linear combination of the pseudo equilibrium metabolites $\mathbf{x_{peq}}$, and $\mathbf{S}_{peq}$ is the sub-matrix created by taking the columns of the stoichiometric matrix **S**, that contains the reactions at pseudo equilibrium. The mass balances for the resulting pseudo equilibrium pools (**p**) is obtained by pre-multiplying **S** with **xP**,

$$\frac{d\mathbf{p}}{dt} = \mathbf{S}_{\text{lumped}} \cdot \mathbf{v}_{\text{dyn}} \quad \text{with} \quad \mathbf{S}_{\text{lumped}} = \mathbf{xP} \cdot \mathbf{S}. \tag{8.4}$$

In Eq. 8.4, $\mathbf{v}_{\text{dyn}}$ does not contain any reactions that operate at pseudo equilibrium state.

Lastly, the calculation of individual metabolite concentrations present in the equilibrium pools is important. This can be formulated by first taking the equilibrium relations into account,

$$K_{eq} = \frac{\prod_{i=1}^{i=pr} (x_i)^{\gamma_i}}{\prod_{j=1}^{j=rc} (x_j)^{\gamma_j}} \tag{8.5}$$

where the term at the numerator is defined as the multiplication of metabolite levels for metabolites as product of the reaction at equilibrium, taken to the power of their stoichiometric coefficient (γ_i) and the denominator contains the multiplication of metabolites as substrate for the reaction at equilibrium, again taken to the power of their stoichiometric coefficients (γ_j). The indices i and j runs from 1 to pr and rc, number of products and reactants respectively. Eliminating the K_{eq} by considering the equilibrium relation at reference state, taking the natural logarithm of the resulting ratio and rearrangement results in Eq. 8.6. The pool composition equation (Eq. 8.3) and equilibrium relations (Eq. 8.6) completely define the individual pools belonging to the lumped equilibrium pools

$$0 = \mathbf{S}^{\text{T}}_{\mathbf{peq}} \ln\left(\frac{x_{peq}}{x^0_{peq}}\right). \tag{8.6}$$

It should be realized that considering the equilibrium relations has another important implication, a number of the unmeasured metabolites can be calculated using these equilibrium relations by making use of Eq. 8.6 and obtaining linear-logarithmic algebraic relations as

$$\ln\left(\frac{\text{x}_{\text{peq}}}{\text{x}^0_{\text{peq}}}\right)^{\text{unmeas.}} = \mathbf{B}\ln\left(\frac{\text{x}_{\text{peq}}}{\text{x}^0_{\text{peq}}}\right)^{\text{meas.}}. \tag{8.7}$$

where, **B** is a matrix obtained by rearranging $\mathbf{S}^{\text{T}}_{\mathbf{peq}}$.

Considering Pools at Pseudo Steady State

After considering pseudo equilibrium reactions, fast pseudo steady state metabolite pools are considered. These pools are characterized by short turn-over time (much shorter than the experimental time). For these fast pools, after a very short transient, the net production rate equals the net consumption rate hence the left hand side of Eq. 8.1 can be neglected. Using linlog kinetics for $\mathbf{v}$ in Eq. 8.1, yields further linear-logarithmic algebraic relations between fast and slow pools,

$$\ln\left(\frac{\mathrm{x}_{\mathrm{fast}}}{\mathrm{x}^0_{\mathrm{fast}}}\right) = \mathbf{A} \cdot \ln\left(\frac{\mathrm{x}_{\mathrm{slow}}}{\mathrm{x}^0_{\mathrm{slow}}}\right). \quad (8.8)$$

where, x are the fast and slow metabolites, and the matrix $\mathbf{A}$ is obtained from right hand side of Eq. 8.1 by rearranging the elasticity sub-matrices for fast and slow metabolites [20]. The implications of establishing algebraic relations between the fast and slow pools in terms of model reduction of both stoichiometric and elasticity matrix, and parameter identifiability has already been discussed in our previous communication [20]. It should be noted that in reducing models, first the pseudo equilibrium reactions have to be simplified, and this should be followed by further reducing the model by removing the pseudo steady state pools.

Considering the pseudo steady state pools and Eq. 8.8, leads also to lumped reactions. Since the parameters are linear in linlog kinetics, explicit rate equations for the lumped reactions can also be obtained [20]

$$\frac{v_i}{J_i^0} = \left[\frac{e_i}{e_i^0}\right]\left(1 + \delta^{v_i}_{x_{slow}} \ln\left(\frac{\mathrm{x}_{\mathrm{slow}}}{\mathrm{x}^0_{\mathrm{slow}}}\right)\right) \quad \text{where } \delta = \delta(\varepsilon, \alpha). \quad (8.9)$$

Where the $\delta^{v_i}_{x_{slow}}$ is the reduced elasticity matrix describing the normalized sensitivity coefficients of the slow metabolites $\mathrm{x}_{\mathrm{slow}}$ on the rate v_i.

Considering Conserved Moieties

Lastly, dependent metabolites due to conserved moieties should be removed from the stoichiometric matrix. It should be noted that this step is a mathematically exact transformation, unlike the previous time scale based model reduction steps which introduced an approximation. These conserved moieties can be detected by analyzing the dependencies in the rows of the stoichiometric matrix [8].

8.2.4 *Final Parameterization of the System*

8.2.4.1 Postulation of Reaction Kinetics

After having adjusted the stoichiometric matrix, lumped the unmeasured pathway intermediates and reduced the reaction network using time scale considerations, we

Table 8.1 List of black-box equations

C-bal:	$6\,q_S + 8\,q_{PAA} + q_{CO_2} + \mu_{act} + 16\,q_{Pen} + q_{\text{storage}} = 0$
Deg of red bal.:	$24\,q_S - 4\,q_{O_2} + 4.21\,\mu_{\text{act}} + 74\,q_{Pen} + 36\,q_{PAA} + 4\,q_{\text{storage}} = 0$
ATP bal.:	$P/O \cdot (2\,q_{O_2}) + \Sigma q_i^{ATP} + K_x \cdot \mu_{act} + K_P \cdot q_{Pen} + m_{ATP} = 0$

postulated a kinetic expression for each of the remaining irreversible reactions. Using approximative linlog kinetics for the rate expressions allows representing the enzyme-metabolite kinetic interactions by an elasticity matrix. Furthermore, the use of linlog kinetics results in explicit algebraic relations (as Eq. 8.6 and Eq. 8.8) that are linear in the logarithm of the metabolites, which is important in reducing the models and obtaining expressions for the lumped kinetics. In order to fully define each rate with linlog kinetics, steady state fluxes ($\mathbf{J}^0$), metabolite concentrations ($\mathbf{X}^0$) at the reference state and elasticity parameters (ε's) are to be defined. The reference parameters ($\mathbf{J}^0$, $\mathbf{x}^0$) are available from the steady state before the perturbation; therefore only elasticities need to be parameterized.

Information on the presence or absence of mass action or allosteric effects on enzyme kinetics was obtained from literature surveys, biochemical knowledge and database search and this information is implemented as non-zero entries of the elasticity matrix. The remaining ε-values need to be parameterized using the metabolite time traces from the pulse experiment.

8.2.4.2 Considering Black-Box Equivalency: Adjusting P/O

Having reduced the model considerably, the next step is to assess the black-box equivalency, i.e. by making use of the carbon, degrees of reduction and ATP balances presented in the Introduction and the measured rates q_S, q_{O2} and q_{CO2} we calculate the growth (μ_{act}), product formation (q_{Pen}) and storage formation ($q_{storage}$) rates both for initial and reduced model. The original model from [29] and the decompartmentalized model should give the same calculated uptake/secretion rates (black-box equivalency). It is important to note that the P/O parameter must be adapted because of the decompartmentalization (see Sect. 8.3).

In Table 8.1, μ_{act} describes the actual growth rate, which is the sum of the wash-out rate (D) and the synthesis rate of the lysed biomass (q_{ExPept}), i.e. $\mu_{act} = D + q_{ExPept}$. The synthesis of storage material (noted as $q_{storage}$) is defined as the sum of the trehalose synthesis rate (q_{tre}) and the synthesis rate of polysaccharides (q_{psachh}), i.e. $q_{\text{storage}} = q_{\text{tre}} + q_{\text{psacch}}$. The other parameters for ATP balance (K_x, K_P, m_{ATP}) are taken from [28].

8.2.4.3 A Priori Parameter Identifiability Analysis

Having reduced the model based on time scale analysis, we can perform *a priori* parameter identifiability analysis. Since the procedure for such an analysis has already been discussed in [20] for pseudo steady state pools, here we describe the case for pseudo equilibrium reactions.

Since a number of metabolites are algebraically linked via equilibrium relations, not all the parameters are separately identifiable. Furthermore, for the new equilibrium pools, since they replace the individual metabolite pools, only the elasticities with respect to those lumped pools can be identified. Taking the previous example case, the aim is to express the elasticity of a rate v with respect to the lumped equilibrium pool p in terms of the individual metabolite concentrations

$$p = p(x_1, x_2)$$
$$\varepsilon_p^v = \left.\frac{\partial v}{\partial p}\right|_{p^0} \cdot \frac{p^0}{J^0}$$

using the chain rule

$$\frac{\partial v}{\partial p} = \frac{\partial v}{\partial x_1} \cdot \frac{\partial x_1}{\partial p} + \frac{\partial v}{\partial x_2} \cdot \frac{\partial x_2}{\partial p}$$
$$\varepsilon_p^v = \varepsilon_{x_1}^v \frac{p^0/x_1^0}{\partial p/\partial x_1} + \varepsilon_{x_2} \frac{p^0/x_2^0}{\partial p/\partial x_2}. \tag{8.10}$$

The coefficients $\partial p/\partial x_i$ are the known coefficients from the linear pool relations, defined as **xP** in Eq. 8.3. It is clear from Eq. 8.10 that only ε_p^v can be determined, but not $\varepsilon_{x_1}^v$ or $\varepsilon_{x_2}^v$.

Generalizing the notation for equilibrium pools containing more metabolite pools, the elasticity for the lumped pool can be expressed in terms of the elasticities for the individual metabolites as:

$$\varepsilon_{p_j}^{v_k} = \sum_i \varepsilon_{x_i}^{v_k} \frac{p_j^0/x_i^0}{\partial p_j/\partial x_i} \qquad \begin{array}{l} i = 1, \ldots, \#\ \text{metabolites} \\ j = 1, \ldots, \#\ \text{pools} \\ k = 1, \ldots, \#\ \text{reactions} \end{array}.$$

Lastly, there is a practical identifiability problem, concerning the rates of product formation and growth. Since a change in the biomass or penicillin levels caused by the glucose perturbation is much smaller than the noise in the measurements of these two, there is no way to quantify the change in the rates of product formation or growth. In this model, we assumed the growth rate to be a function of the ATP (with an elasticity of 1) and the penicillin production rate is assumed to be function of the precursors, again with elasticities of 1.

8.2.4.4 Estimating the Elasticities

The final step in building the model is to estimate the ε-parameters of the model, based on experimental data via rigorous non-linear fitting procedure, where the squared error between the experimental data and model simulation is minimized for m metabolites at p data points,

$$\min \sum_i \sum_j \frac{\left(x_{ij}^{sim} - x_{ij}^{exp}\right)^2}{\sigma_i^{2,exp}} \qquad x_{sim} = \int_{t_i}^{t_f} (dx/dt)\, dt \qquad \begin{array}{l} i = 1, \ldots, p \\ j = 1, \ldots, m \end{array} \tag{8.11}$$

In order to solve this non-linear parameter estimation problem, a decent initial estimate is required. An advantage of using linlog kinetics is that the parameters appear in linear fashion (Eq. 8.2), allowing to use linear algebra in the initial parameter estimation using the concentration time traces, which can be further fed to the non-linear parameter estimation problem as the initial estimate. Such parameter estimation scheme is presented in detail in our previous communication [19]. Starting from Eq. 8.11, inserting linlog kinetics (Eq. 8.2) for $\mathbf{v}$, setting $\mathbf{e}/\mathbf{e}^0 = \mathbf{I}$ yields

$$\begin{aligned} \frac{d\mathbf{x}(\mathrm{t})}{dt} &= \mathbf{S}\mathbf{J}^0\left(\mathbf{i} + \mathbf{E}_{\mathrm{x}}^0 \ln\left(\frac{\mathbf{x}(\mathrm{t})}{\mathbf{x}^0}\right) + \mathbf{E}_{\mathrm{c}}^0 \ln\left(\frac{\mathbf{c}(\mathrm{t})}{\mathbf{c}^0}\right)\right) \\ \frac{d\mathbf{c}(\mathrm{t})}{dt} &= \mathbf{S}\mathbf{J}^0\left(\mathbf{i} + \mathbf{E}_{\mathrm{x}}^0 \ln\left(\frac{\mathbf{x}(\mathrm{t})}{\mathbf{x}^0}\right) + \mathbf{E}_{\mathrm{c}}^0 \ln\left(\frac{\mathbf{c}(\mathrm{t})}{\mathbf{c}^0}\right)\right) \cdot c_{\mathrm{X}} + D \cdot (\mathbf{c}_{\mathbf{feed}} - \mathbf{c}(\mathrm{t})) \end{aligned}$$

and integration with respect to time and rearranging yields

$$\begin{aligned} \Delta\mathbf{x}\big|_{t_i}^{t_{i+1}} &= \mathbf{S}\mathbf{J}^0\mathbf{E}_{\mathrm{x,c}}^0 \int_{\mathbf{t_i}}^{\mathbf{t_{i+1}}} \ln(\mathbf{x}/\mathbf{x}^0), \ln(\mathbf{c}/\mathbf{c}^0)\mathbf{dt} \\ \Delta\mathbf{c}\big|_{\mathbf{t_i}}^{\mathbf{t_{i+1}}} &- \left(D \int_{\mathbf{t_i}}^{\mathbf{t_{i+1}}} (\mathbf{c}_{\mathbf{feed}} - \mathbf{c})\mathbf{dt} + \mathbf{S_c}\,\mathbf{J}^0\,\mathbf{i}c_{\mathbf{X}}\Delta\mathbf{t}\right) = \ldots \\ &\mathbf{S_c}\,\mathbf{J}^0\,\mathbf{E}_{\mathrm{x,c}}^0 \int_{\mathbf{t_i}}^{\mathbf{t_{i+1}}} ln(\mathbf{x}/\mathbf{x}^0), \ln(\mathbf{c}/\mathbf{c}^0) \cdot c_{\mathbf{X}}. \end{aligned} \tag{8.12}$$

For the integration, linear interpolation between successive measurements is used. Equation 8.12 is easily recognized as a linear system of equations of the general form $\mathbf{a} = \mathbf{Y} \cdot \mathbf{b}$. Here, $\mathbf{a}$ is a vector containing $\Delta\mathbf{x}$, $\Delta\mathbf{c}$, Δt, D, $\mathbf{c}_{feed}$ and $\mathbf{S}$, $\mathbf{J}^0$; $\mathbf{Y}$ is the design matrix containing $\int \ln(\mathbf{x}/\mathbf{x}^0, \mathbf{c}/\mathbf{c}^0)dt$ and $\mathbf{b}$ is the parameter vector that contains the non-zero elements of $\mathbf{E}_x^0$ and $\mathbf{E}_c^0$.

8.3 Results

We constructed a kinetic model of our model organism *P. chrysogenum* starting from the stoichiometric network published in van Gulik et al. [29] and using previously published short-term perturbation data in Nasution et al. [18] for parameter estimation. The initial network contains 167 reactions and 169 intracellular metabolites. In this section we applied the steps presented above and present the resulting model. For the sake of bookkeeping, each step in model reduction/modification is given a name and the demographics of the resulting model (size of the stoichiometric matrix, number of independent metabolites etc.) is shown in Table 8.9.

8.3.1 Adjustment/Simplification on the Network

8.3.1.1 Adjusting Biomass Definition

To calculate how the carbon flux redirects into storage upon glucose perturbation, the biomass formation is separated into a synthetic (syn_X, 118.2) and a storage trehalose part (tre_X, 117.6) (Step 1). The steady state metabolic flux of the trehalose formation (tre) is then equal to:

$$q_{tre} = \frac{v_{tre} \cdot \mu_{old}}{1 - v_{tre}}$$

in which v_{tre} is the stoichiometric coefficient of tre in the original biomass synthesis reaction (118.1), μ_{old} is the biomass growth rate based on the old definition and q_{tre} is the biomass specific storage material part of the biomass synthesis. The stoichiometric coefficient v_{tre} is equal to 0.00107 on the basis of 1 Cmol X. Since the biomass definition has been changed and solely consists of the synthetic part now, all biomass specific net conversion rates need to be adjusted according to:

$$q_{i,new} = \frac{q_{i,old}}{1 - v_{tre,X}}$$

in which $q_{i,old}$ is the specific biomass net conversion rate of i based on the old biomass definition and $q_{i,new}$ for the new biomass definition. By redefining the biomass composition, the number of reactions increased to 168 and the number of extracellular metabolites is increased to 20, increasing the degrees of freedom of the system to 20 (Table 8.9).

8.3.1.2 Removing the Peroxisome Compartment

Having only the cell-average concentrations and lacking information on concentration ratios between the compartments for *P. chrysogenum*, we first removed the relatively small peroxisome compartment, which includes the biosynthesis of penicillin, and several related by-products. By removing the peroxisome, the number of reactions decreased to 157 and there are 155 metabolites located in two compartments. Removing the peroxisome had no effect on the degrees of freedom of the network (Table 8.9).

Table 8.2 Metabolites that are present in the cytosol and mitochondria

Pathway	Metabolite(s)
Glycolysis	pyr
TCA metabolites	citr, iCitr, mal, OAA, fum
Amino Acid Synthesis	glu, ile, val, aKG, thr
Cofactors	ATP, ADP,
	NADH, NAD,
	NADPH, NADP
	AcCoA, HCoA
Miscellaneous	Pi, H, H_2O, CO_2, O_2

8.3.1.3 Removing the Mitochondria Compartment

As the last step in preliminary adjustment/reduction of the network, we removed the mitochondria compartment and all its intercompartmental transport reactions. Similar to the peroxisome, the enzymatic reactions in the mitochondria are kept in the model. After removing the mitochondria, 139 reactions and 132 metabolites remained in the model. As expected, removing the compartments reduced the number of conserved moieties from 13 in the initial model to 6 in the one compartment model.

8.3.1.4 Removing the By-Products Related to Penicillin Production

After removing the compartments we removed the reactions leading to by-products with respect to the penicillin biosynthesis, e.g. 6APA, OPC, 8HPA, since these have not been measured and are known to be minor carbon sinks. Apart from decrease in the number of reactions and metabolites (129 and 128 respectively), removing these by-products resulted in the decrease of the degrees of freedom of the system to 6 (Table 8.9).

8.3.1.5 Eliminating Reactions in Parallel Pathways

As a result of removing the compartments, a number of metabolites appear in both the cytosol and the mitochondria. A list of such metabolites is given in Table 8.2.

Removing the mitochondria compartment results in parallel reactions/pathways in the stoichiometric model and these should be eliminated. Inspecting the resulting network, it is detected that this metabolic network contains five parallel routes:

1. r4.5 (mitochondria) with r4.6 (cytosol)
2. r4.11 (mitochondria) with r5.3 (cytosol)
3. r4.1 (mitochondria) + r5.2 (cytosol) with ATP hydrolysis (cytosol)
4. glycolysis + TCA cycle + PPP (cytosol) + l6.1 + l6.2 + ATP hydrolysis (cytosol)
5. l6.2 and l6.3

r4.5 (mitochondria) with r4.6 (cytosol)
These two routes are parallel because the reactions are identical apart from the location and hence one of them (r4.5) is removed.

r4.11 (mitochondria) with r5.3 (cytosol)
The reaction r5.3, in which *OAA* is converted into *mal*, proceeds in the opposite direction as r4.11. Studying the metabolic fluxes shows the metabolic flux of r5.3 is 25% of that of r4.11 [29]. Furthermore the cytosolic malate flux is used to drive the transport step r10.14, which transports mitochondrial citrate across the mitochondrial membrane while moving malate the opposite way. Since this transport step has been removed, there is no need for r5.3 anymore and hence it is removed.

r4.1 (mitochondria) + r5.2 (cytosol) with ATP hydrolysis (cytosol)
In reaction r5.2, citr is used to produce AcCoA and has a metabolic flux that is 19% of that with r4.1. Hence r5.2 will be removed from the stoichiometric model to keep the TCA cycle intact.

glycolysis + TCA cycle + PPP (cytosol) + l6.1 + l6.2 + ATP hydrolysis (cytosol)
The fourth parallel route consists of many reactions. A closer look on these reactions revealed that there are alternative pathways for the biosynthesis of NADPH which are located in the oxidative branch of the PPP and the isocitrate dehydrogenase reaction. The main biosynthesis of NADPH is catalyzed by the enzymes in the oxidative PPP, i.e. 75% of the total NADPH flux, and hence the NADPH linked isocitrate dehydrogenase (r4.6) has been removed.

l6.2 and l6.3
This parallel route is caused by the oxidative phosphorylation of NADH:mit and NADH:cyt which can be eliminated by lumping them based on their steady state flux distribution (see below Sect. 8.3.1.7).

8.3.1.6 Lumping Pathways Containing Unmeasured Pools

Having only a limited amount of measured metabolites, many unmeasured intermediate compounds, e.g. RNA, chor, bIM, ery, etc. have been eliminated by lumping the involved reactions. The lumped reactions and the pools that are removed by implementing such a reduction is summarized in Table 8.3. By this, the number of intracellular metabolites has been reduced to 74, and the network holds 75 reactions located in one compartment.

8.3.1.7 Adjusting ATP-Stoichiometry Parameters

The reactions describing the oxidative phosphorylation in van Gulik et al. [29] have to be updated/adjusted. This is because (1) the mitochondria compartment has been removed during reduction, and (2), in van Gulik et al. [28] the experimentally

Table 8.3 The lumping scheme considering the unmeasured metabolites

Name	Biosynth. of	Lumped reactions	Eliminated metabolites
11.1	g6p	r1.1, r9.3	glc^{in}
11.2	3PG	r1.5, r1.6	13PG
11.3	PEP	r1.7, r1.8	2PG
111.1	val	r11.23, r11.24	aKI
111.2	leu	r11.23, r11.25, r10.19, r11.26	aKI, bIM
111.3	trp	r11.27, r11.30, r11.31	chor, PRPP
111.4	tyr	r11.27, r11.29	chor
111.5	phe	r11.27, r11.28	chor
111.6	arg	r11.4, r11.5, r11.6, r10.7	carbP, ctl
111.7	cys	r11.11, r11.12, r11.15, r7.2	Ac, PAPS, H2S
111.8	homcys	r11.11-r11.14, r7.2, r11.18	Ac, PAPS, H2S, AcHomser
111.9	thr	r11.18, 11.19	homser
111.10	his	r11.31, r11.32	PRPP
113.2	GMP	r11.31, r8.3, r13.1, r13.3	PRPP, FTHF, IMP
113.3	AMP	r11.31, r13.1, r8.3, r13.2	PRPP, FTHF, IMP
113.4	UTP	r11.31, r13.4, 13.5	PRPP, UMP
117.1	t6P	r13.9, r17.5, r17.6	UDP, UDPglc
117.3	g1P		Introducing g1P
117.4	psacch	r17.1	g6P replaced by g1P in this reaction
117.7	tre_{ext}		Introducing tre transport step
118.2	biomass	r11.33, r12.1, r12.2, r13.7, r14.1, r8.3, r8.4, r13.8, r16.1-r16.16, r17.2-r17.4, r17.8, mu0.05.21	AApool, Aaprotsyn, PROT, CMP, RNA, FTHF, SAM, A, ino, gcl3P, steaCoA, olCoA, linCoA, phospht, CDPDAcgcl, CMP, pHeta, Phchol, Phino, Phser, TRIA, meva, lano, ergo, ESE, chit, m1P, man, ery
119.1	penG	r19.1-r19.4	ACV, iPN, PAACoA

determined P/O parameter was 1.84 and glycolytic and mitochondrial NADH was distinguished. In the one compartment model, there is only one type of NADH hence P/O needs to be adapted to obtain the same theoretical yields for the reduced model as the original model.

We first considered the stoichiometric network presented in van Gulik et al. [29], which consists of 155 metabolites and 156 reactions located in three compartments (cytosol, mitochondria and peroxisome). A number of changes have been performed in order to adjust the theoretically defined P/O ratio (being 2.5) for the three compartment model to an experimentally determined and compartment averaged value for a single compartment model. While the detailed procedure is explained in the Appendix, the important step is to average the rate of oxidation of cytosolic and mitochondrial NADH based on their steady state flux distribution (70.5–29.5% respectively) in the three compartment model. This procedure yields the P/O ratio in the one compartment model as 1.623 mol ATP/atom O. It should be noted that, the remaining ATP-balance parameters i.e. K_x, K_p, and m_{ATP} are kept unchanged while considering black-box balances (see below section 8.3.5.1).

Table 8.4 Pseudo equilibrium relations

Glycolysis		
1	r1.2	g6P ⇄ f6P
2	r1.4	f16P ⇄ 2 GAP
3	l1.2	GAP + NAD + Pi + ADP ⇄ 3PG + ATP + H + NADH
Pentose phosphate pathway		
4	r2.3	Ribu5P ⇄ Rib5P
5	r2.4	Ribu5P ⇄ Xylu5P
6	r2.5	Rib5P + Xylu5P ⇄ GAP + sed7P
7	r2.6	GAP + sed7P ⇄ E4P + f6P
8	r2.7	E4P + Xylu5P ⇄ f6P + GAP
TCA cycle		
9	r4.2	AcCoA + H_2O + OAA ⇄ citr + H + HCoA
10	r4.3	citr ⇄ iCitr
11	r4.9	FAD + succ ⇄ FADH2 + fum
12	r4.10	fum + H_2O ⇄ mal
13	r4.11	mal + NAD ⇄ H + NADH + OAA
Transport of 1-C compounds		
14	r8.2	H + METHF + NADH ⇄ MYTHF + NAD
Amino acid synthase		
15	r11.1	αKG + H + NADPH + NH_4 ⇄ glu + H_2O + NADP
16	r11.10	ser + THF ⇄ gly + H_2O + METHF
Synthesis of glycogen and polysaccharides		
17	l17.3	g6P ⇄ g1P

8.3.2 Time-Scale Analysis

Considering Pseudo Equilibrium Reactions

As a first step for the time scale analysis, reactions at pseudo equilibrium are to be considered. To classify a given reaction as pseudo equilibrium or not, we either used the information from the biochemistry (e.g. transamination or isomerization reactions) or in case where the transient data is available, parity plots (Π substrate vs Π product) is used to verify whether a given reaction operates at pseudo equilibrium or not.

Using this information, the reactions listed in Table 8.4 are considered to operate in pseudo equilibrium. In order to eliminate these reactions, we defined the pseudo equilibrium pools in Table 8.5 by using Eq. 8.3.

In Table 8.5, the first two pools represent the metabolites in the upper part of the glycolysis and PPP, p_3 to p_6 represent the TCA metabolites, and p_8 covers the transamination of aKG to glu. Lastly, using the equilibrium relations and the equilibrium constants (in Table 8.16), the levels of a number of unmeasured metabolites can be calculated from available measured metabolite data. Using the equilibrium relations for the reactions in Table 8.4, we calculated the metabolite levels of, GAP, NAD, NADH, NADP, NADPH, Xylu5P, Rib5P, Ribu5P, E4P, OAA, CoA, AcCoA, succCoA, FAD, $FADH_2$, THF, METHF, MYTHF. The profiles of fold changes are presented in Fig. 8.4 and Table 8.11 contains the reference steady state concentrations.

Table 8.5 The list of pseudo equilibrium pools defined to eliminate reactions that operate in pseudo equilibrium

$p_1 = g6P + f6P + 8\,f16P + 4\,3PG + 4\,GAP + 2\,Rib5P + 2\,Ribu5P + 2\,Xylu5P + 3\,E4P + g1P$
$p_2 = g6P + f6P + 2\,f16P + 3PG + GAP + Rib5P + Ribu5P + Xylu5P + E4P + sed7P + g1P$
$p_3 = OAA + iCitr + citr + 3PG + METHF + THF + NAD$
$p_4 = OAA + iCitr + citr + fum + mal + succ$
$p_5 = OAA + iCitr + citr + fum + mal + FAD$
$p_6 = iCitr + citr + AcCoA$
$p_7 = AcCoA + HCoA$
$p_8 = aKG + glu$
$p_9 = gly + ser$
$p_{10} = ATP\text{-}\ 3PG$
$p_{11} = aKG + NADP$
$p_{12} = THF + gly$

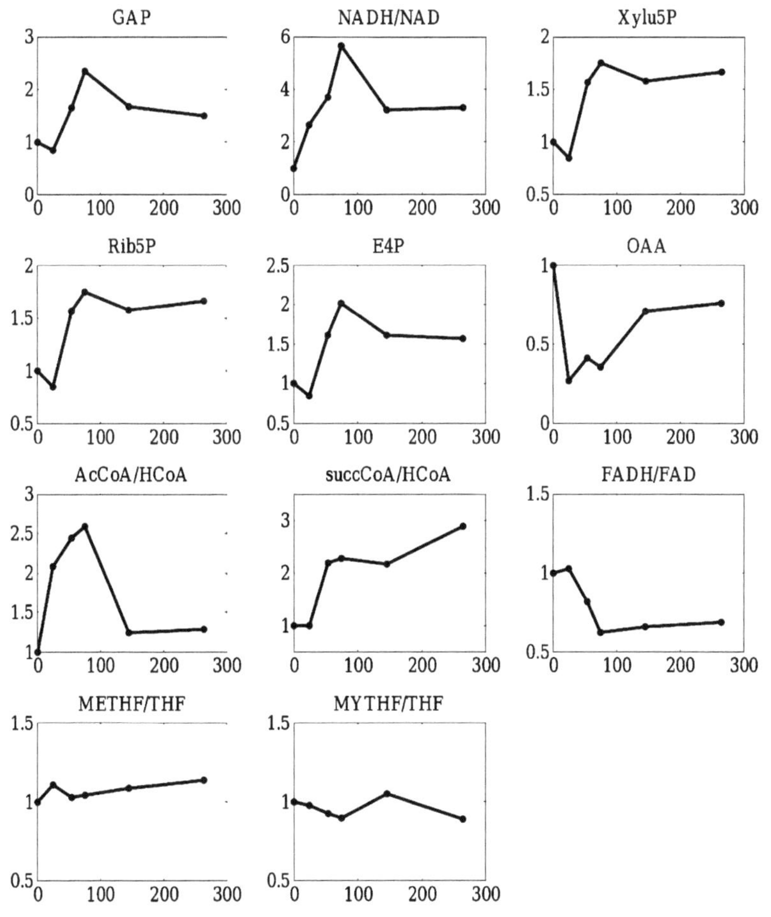

Fig. 8.4 Calculated fold changes of non-measured metabolites using the equilibrium relations as function of time (s). The steady state values are presented in Table 8.11

Considering Pseudo Steady State Metabolites

The next step after processing pseudo equilibrium reactions is to consider the pseudo steady state pools. Following Nikerel et al. [20], we use the turn over time for classification of the pools. Table 8.6 presents the turn over time for each remaining metabolite pool. From this table, considering the first sample was taken at 5th second, we deduce that the metabolite pools pyr, PEP, 6PG, p_{11} and p_3 have lower turn-over time hence these are fast pools compared to the remaining. Focusing on the composition of p_{10}, since it contains ATP, which is used or produced by several reactions in the network, the total influx to this pool is very high. Similarly for p_3, this pool contains NAD, therefore the total influx to this pool is high.

Having classified the fast metabolites, equations according to Eq. 8.8 are to be constructed. Using pep as example an explicit algebraic relation in terms of all the effectors can be expressed as:

$$\ln\left(\frac{\text{pep}}{\text{pep}^0}\right) = \begin{bmatrix} \ln\left(\frac{p_1}{p_1^0}\right) \\ \ln\left(\frac{p_2}{p_2^0}\right) \\ \ln\left(\frac{p_3}{p_3^0}\right) \\ \ln\left(\frac{p_{10}}{p_{10}^0}\right) \\ \ln\left(\frac{\text{phe}}{\text{phe}^0}\right) \\ \ln\left(\frac{\text{trp}}{\text{trp}^0}\right) \\ \ln\left(\frac{\text{tyr}}{\text{tyr}^0}\right) \end{bmatrix}^T \cdot \begin{bmatrix} \alpha_1 \\ \alpha_2 \\ \alpha_3 \\ \alpha_4 \\ \alpha_5 \\ \alpha_6 \\ \alpha_7 \end{bmatrix} \tag{8.13}$$

and the α's are related to ε-parameters as:

$$\alpha_1 = -\frac{-0.320\varepsilon_{52} - 0.760\varepsilon_{55} + 95.9\varepsilon_7 - 185.\varepsilon_9 - 1.41\varepsilon_{58}}{\Gamma}$$

$$\alpha_2 = -\frac{-0.107\varepsilon_{52} - 0.253\varepsilon_{55} + 24.0\varepsilon_7 - 46.3\varepsilon_9 - 0.470\varepsilon_{58}}{\Gamma}$$

$$\alpha_3 = -\frac{24.0\varepsilon_7}{\Gamma}$$

$$\alpha_4 = -\frac{-24.0\varepsilon_7 - 23.1\varepsilon_{11}}{\Gamma}$$

$$\alpha_5 = \frac{0.470\varepsilon_{59}}{\Gamma}$$

Table 8.6 The turn-over time τ (s) for each metabolite (ordered in decreasing order). The steady state levels of the pools are given in μmol/gDW and the intracellular fluxes are given in μmol/gDW/s

	X^0	J^0_{IN}	τ		X^0	J^0_{IN}	τ
pyr	0.22	0.23	0.9	AMP	0.27	$9.20 \cdot 10^{-4}$	300
pep	0.24	0.24	1.0	pro	0.95	$3.13 \cdot 10^{-3}$	300
6PG	0.26	0.09	2.7	p_{12}	2.09	0.01	330
$\mathbf{p_{10}}$	6.79	2.06	3.3	psacch	76.41	0.21	370
$\mathbf{p_3}$	5.39	0.86	6.3	p_8	55.06	0.14	380
p_{11}	3.93	0.31	12	orn	1.53	$3.25 \cdot 10^{-3}$	470
p_1	16.52	1.18	14	lys	1.24	$2.52 \cdot 10^{-3}$	490
p_6	2.92	0.17	17	p_9	7.82	0.02	510
p_7	2.30	0.12	20	his	0.72	$1.31 \cdot 10^{-3}$	550
p_2	9.20	0.4	23	asn	1.49	$2.15 \cdot 10^{-3}$	690
cys	0.10	$3.64 \cdot 10^{-3}$	27	arg	2.30	$3.25 \cdot 10^{-3}$	710
GMP	0.02	$6.85 \cdot 10^{-4}$	32	aAd	1.81	$2.52 \cdot 10^{-3}$	720
p_4	7.16	0.15	48	asp	16.3	$1.75 \cdot 10^{-2}$	930
p_5	9.09	0.15	60	penG	3.23	$3.24 \cdot 10^{-3}$	1000
met	0.14	$2.19 \cdot 10^{-3}$	63	tre	0.53	$5.28 \cdot 10^{-4}$	1000
phe	0.19	$2.35 \cdot 10^{-3}$	81	thr	5.95	$5.66 \cdot 10^{-3}$	1100
ile	0.33	$2.45 \cdot 10^{-3}$	140	t6P	0.57	$5.28 \cdot 10^{-4}$	1100
homcys	0.37	$2.19 \cdot 10^{-3}$	170	gln	28.70	$1.79 \cdot 10^{-2}$	1600
leu	0.73	$4.00 \cdot 10^{-3}$	180	CTP	0.96	$4.43 \cdot 10^{-4}$	2200
trp	0.11	$5.33 \cdot 10^{-4}$	200	PAA	7.81	$3.24 \cdot 10^{-3}$	2400
tyr	0.27	$1.27 \cdot 10^{-3}$	210	UTP	1.36	$4.43 \cdot 10^{-4}$	3100
val	2.08	$7.84 \cdot 10^{-3}$	270	ala	21.80	$6.69 \cdot 10^{-3}$	3300

$$\alpha_6 = \frac{0.107\varepsilon_{53}}{\Gamma}$$

$$\alpha_7 = \frac{0.253\varepsilon_{56}}{\Gamma}$$

with

$$\Gamma = -0.107\varepsilon_{51} - 0.253\varepsilon_{54} - 0.470\varepsilon_{57} + 24.0\varepsilon_8 - 23.1\varepsilon_{10}.$$

Since the time traces for the pep and slow pools are available, the entries of the **A** matrix (≡α) can be estimated via linear regression. Furthermore, since Eq. 8.8 is linear in logarithmic space, standard linear algebra tools can be used to inspect the identifiability of the α-parameters. This is important, especially for p_{10} and p_3. These pools, containing ATP and NAD respectively, take part in several reactions, so that on the left hand side of Eq. 8.8, there would be many more terms (30 for p_{10} (ATP) and 21 for p_3 (NAD)). This causes an overfit in linear regression, not only because of the multitude of the effectors, but also the time profiles of some of the effectors are already calculated using equilibrium relations.

Table 8.7 The calculated Π-matrix for the pep pool

η	$\pi(\alpha_1)$	$\pi(\alpha_2)$	$\pi(\alpha_3)$	$\pi(\alpha_4)$	$\pi(\alpha_5)$	$\pi(\alpha_6)$	$\pi(\alpha_7)$
1	0	0	0	0	0	0	0
3.9	0	0	0	0	0	0	0
7.3	0	0	0	0	0	0	0
13	0	0	0	0	0	0	0
32	0	0	0	0	0	0	0
89	0	0.44	0.15	0.17	0	0	0.26
161	0.88	0.55	0.78	0.81	0.82	0.83	0.73

In order to determine which of the α-parameters are identifiable given the experimental data, we followed the data driven approach to the parameter identifiability procedure described in Nikerel et al. [20], and calculated Π-matrices for the **A** matrix. Focusing again on pep, the calculated Π-matrix is given in Table 8.7.

From this table, we can immediately see that not all the α-parameters are identifiable $(\eta_6, \eta_7 \gg \eta_1)$ and that the parameters α_1, α_3, α_4, α_5, and α_6 are highly collinear since their corresponding value is close to 1. This is already expected, since two of the effectors p_3 and p_{10} are already classified as fast metabolites.

We proceeded by first combining all the fast pools into one equation therefore leaving only slow metabolites on the right hand side of the equation 8.8. This allows to estimate the entries of the **A** matrix in one round. Furthermore, considering the Π-matrix for the resulting **A** matrix, we iteratively removed unidentifiable α-parameters until only the identifiable parameters are left.

The final form of the **A** matrix is given in Eq. 8.14 and the entries are presented in Table 8.8. Using the estimated values for the **A** matrix, the predicted profiles for the fast metabolites are given in Fig. 8.5.

$$\begin{bmatrix} \ln\left(\frac{pep}{pep^0}\right) \\ \ln\left(\frac{pyr}{pyr^0}\right) \\ \ln\left(\frac{6PG}{6PG^0}\right) \\ \ln\left(\frac{p_{10}}{p_{10}^0}\right) \\ \ln\left(\frac{p_3}{p_3^0}\right) \end{bmatrix} = \begin{bmatrix} \ln\left(\frac{p_1}{p_1^0}\right) \\ \ln\left(\frac{p_2}{p_2^0}\right) \\ \ln\left(\frac{p_{11}}{p_{11}^0}\right) \\ \ln\left(\frac{lys}{lys^0}\right) \\ \ln\left(\frac{phe}{phe^0}\right) \\ \ln\left(\frac{trp}{trp^0}\right) \\ \ln\left(\frac{val}{val^0}\right) \\ \ln\left(\frac{t6P}{t6P^0}\right) \end{bmatrix}^T \cdot \begin{bmatrix} \alpha_1 & \alpha_8 & \alpha_{13} & \alpha_{18} & \alpha_{20} \\ \cdot & \cdot & \alpha_{14} & \cdot & \cdot \\ \cdot & \cdot & \alpha_{15} & \cdot & \alpha_{21} \\ \cdot & \alpha_9 & \alpha_{16} & \cdot & \alpha_{22} \\ \alpha_5 & \cdot & \cdot & \cdot & \cdot \\ \cdot & \alpha_{10} & \cdot & \cdot & \cdot \\ \cdot & \alpha_{11} & \cdot & \cdot & \cdot \\ \cdot & \alpha_{12} & \alpha_{17} & \alpha_{19} & \cdot \end{bmatrix} \tag{8.14}$$

Table 8.8 The entries of the A Matrix linking the fast metabolites to the slow ones via algebraic relations (Eq. 8.8)

	pep		pyr		6PG		p_{10}		p_3
α_1	−0.2995	α_8	−0.1148	α_{13}	0.9	α_{18}	−0.1703	α_{20}	−0.1794
α_5	0.3039	α_9	−0.529	α_{14}	−1.5855	α_{19}	−0.1655	α_{21}	0.18
		α_{10}	−1.9614	α_{15}	0.72			α_{22}	0.58
		α_{11}	−1.349	α_{16}	−2.0684				
		α_{12}	0.24	α_{17}	0.41				

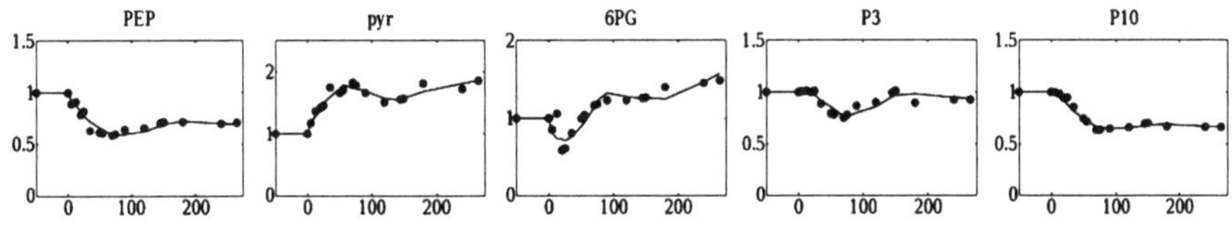

Fig. 8.5 Comparison of the predicted time traces (*continuous lines*) of the fast metabolites using Eq. 8.8 and the estimated parameters with the experimental data (points). The *x-axis* represents time in seconds and the *y-axis* represents the corresponding metabolite relative to its reference state

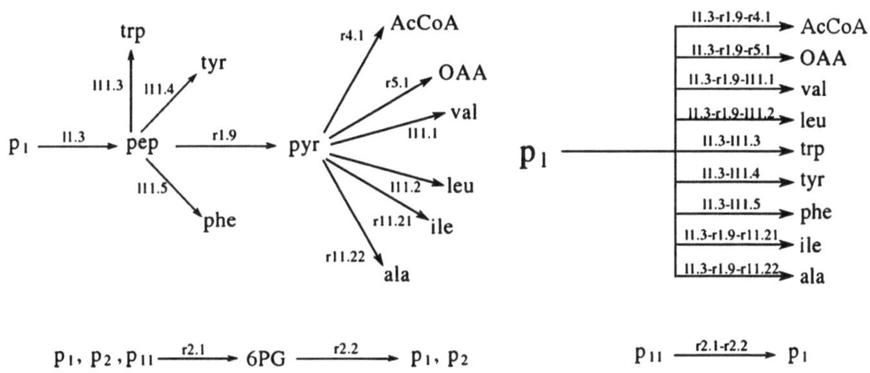

Fig. 8.6 Lumping the rates around the fast metabolites. *Upper panel* is the simplified scheme of the rates around the fast metabolites pep and pyr. The *lower panel* represents the simplified scheme where the reactions around pep and pyr are lumped

Lastly, the reactions around the fast metabolites should be lumped and the rate expressions for these newly defined lumped reactions are to be derived. For the PEP, pyr and 6PG pools the initial and lumped rates are depicted in Fig. 8.6.

Considering the Conserved Moieties

This model contains six conserved moieties namely ATP + ADP, NAD + NADH, NADP + NADPH, FAD + $FADH_2$, AcCoA + HCoA + succCoA, THF + METHF + MYTHF. For the unmeasured metabolites, we assumed the total moiety sum to be 1 mM. For the ATP + ADP, the sum is also taken as the experimentally observed values, presented in Fig. 8.7.

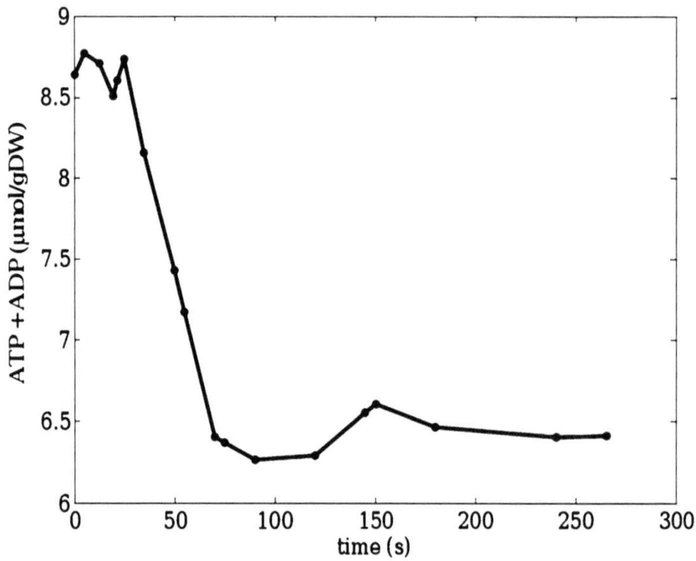

Fig. 8.7 The experimentally observed ATP + ADP sum, taken from [18]

Table 8.9 Demographics of the stoichiometric network after each reduction

		#Metabolites				
	#rxn	IC	EX	d.o.f	CM	#comp.
Original model	167	169	19	11	13	3
Biomass redefined	168	169	20	12	13	3
Remove peroxisome	157	155	20	12	10	2
Remove mitochondria	139	132	20	12	6	1
Remove the byproducts	129	128	14	6	6	1
Lump reactions with unmeasured metabolites	75	74	14	6	6	1
Remove PEQ reactions	51	57	7	6	6	1
Remove PSS metabolites (Final Model)	48	52	7	6	6	1

The final stoichiometric model still contains 6° of freedom and the resulting metabolic reaction network contains 48 rates, 7 extracellular and 52 intracellular metabolites. The characteristics of the current metabolic network can be found in Table 8.9.

8.3.3 Selecting Non-zero Elasticities

Before the final parameterization, it is needed to determine the non-zero elasticities for the remaining dynamic reactions. To achieve this, we used available databases on enzyme kinetics [23] and visited the biochemistry books. For each reaction, the selected effectors (both from mass action, or allosteric) are described in Table 8.10, together with the corresponding nomenclature.

Table 8.10 Summary of the non-zero elasticities

#ri.i	Representative transformation	Substrates	Effectors	$\varepsilon \neq 0$
l1.1	glc → g6P	glc(ε_3)+, ATP(ε_2)+, g6P(ε_1)−		3
r1.3	f6P → f16P	f6P(ε_4)+, ATP(ε_6)+, f16P(ε_5)−		3
l1.3	3PG → PEP	3PG(ε_7)+, PEP(ε_8)−		2
r1.9	PEP → pyr	PEP(ε_{10})+, ATP(ε_{11})−	f16P(ε_9)+	3
r2.1	g6P → 6PG	g6P(ε_{12})+, NADP(ε_{14})+	ATP(ε_{13})−	3
r2.2	6PG → Ribu5P	6PG(ε_{15})+, NADP(ε_{17})+	ATP(ε_{16})−	3
r4.1	pyr → AcCoA	NAD(ε_{20})−, pyr(ε_{18})+, AcCoA(ε_{21})−,	ATP(ε_{19})−	4
r4.4	iCitr → aKG	iCitr(ε_{22})+, NAD(ε_{23})+		2
r4.7	aKG → succCoA	NAD(ε_{26})+, αKG(ε_{24})+, HCoA(ε_{27})−	OAA(ε_{25})−	4
r4.8	succCoA → succ	succ(ε_{28})+, AcCoA(ε_{29})+, HCoA(ε_{30})−		3
r5.1	pyr → OAA	pyr(ε_{31})+	AcCoA(ε_{33})+, asp(ε_{32})−	3
l6.7	$H_2O_2FADH_2H_2O$	FAD(ε_{32})−, ATP(ε_{34})−		2
l6.9	$H_2O_2NADHH_2O$	NAD(ε_{37})−, ATP(ε_{36})−		2
r8.1	CO_2 → gly	glu(ε_{38})+, THF(ε_{40})+, METHF(ε_{39})−		3
l11.1	pyr → val	pyr(ε_{41})+, val(ε_{43})−	ile(ε_{42})−	3
r11.2	glu → gln	glu(ε_{45})+, gln(ε_{44})−		2
l11.2	pyr → leu	AcCoA(ε_{50})+, pyr(ε_{46})+, leu(ε_{48})−	val(ε_{49})−, ile(ε_{47})−	5
r11.3	glu → pro	ATP(ε_{53})+, glu(ε_{51})+, pro(ε_{52})−		3
l11.3	PEP → trp	E4P(ε_{55})+, PEP(ε_{54})+, trp(ε_{56})−		3
l11.4	PEP → tyr	E4P(ε_{58})+, PEP(ε_{57})+, tyr(ε_{59})−		3
l11.5	PEP → phe	E4P(ε_{61})+, PEP(ε_{60})+, phe(ε_{62})−		3
r11.7	glu → aAd	AcCoA(ε_{66})+, αKG(ε_{63})+, aAd(ε_{65})−	lys(ε_{64})−	4
l11.7	ser → cys	ser(ε_{68})+, ATP(ε_{69})+, cys(ε_{67})−		3
r11.8	glu → lys	aAd(ε_{71})+, lys(ε_{70})−		2
l11.8	asp → homcys	asp(ε_{72})+, ATP(ε_{75})+, homcys(ε_{76})−	met(ε_{73})−, thr(ε_{74})−	5
r11.9	3PG → ser	3PG(ε_{77})+, NAD(ε_{81})+, glu(ε_{79})+, αKG(ε_{78})−, ser(ε_{80})−		5
l11.9	asp → thr	asp(ε_{82})+, thr(ε_{84})−	cys(ε_{83})−	3

(continued)

Table 8.10 (continued)

#ri.i	Representative transformation	Substrates	Effectors	$\varepsilon \neq 0$
l11.10	gln → his	ATP(ε_{87})+, Rib5P(ε_{85})+, his(ε_{86})−		3
l11.11	glu → orn	glu(ε_{88})+, orn(ε_{89})−,		2
l11.12	orn → arg	orn(ε_{90})+, ATP(ε_{91})+		2
r11.16	glu → asp	OAA(ε_{92})+, asp(ε_{93})−		2
r11.17	asp → asn	asp(ε_{95})+, asn(ε_{94})−		2
r11.20	met → homcys	met(ε_{96})−, THF(ε_{97})−, homcys(ε_{98})+		3
r11.21	pyr → ile	pyr(ε_{99})+, ile(ε_{100})−	val(ε_{101})−	3
r11.22	pyr → ala	pyr(ε_{102})+, ala(ε_{103})−		2
r11.34	q_{ExPept}	ATP(ε_{104})		1
l13.2	→ GMP	ATP(ε_{108})+, Rib5P(ε_{105})+, asp(ε_{106})+, GMP(ε_{109})−	AMP(ε_{107})−	5
l13.3	→ AMP	Rib5P(ε_{110})+, asp(ε_{111})+, AMP(ε_{112})−		3
l13.4	→ UTP	Rib5P(ε_{113})+, asp(ε_{114})+, UTP(ε_{115})−		3
r13.6	→ CTP	UTP(ε_{117})+, CTP(ε_{116})−		2
r15.1	m_{ATP}	ATP(ε_{118})+		1
l17.1	g6P → t6P	g6P(ε_{119})+, t6P(ε_{120})−	AMP(ε_{121})−	3
l17.4	g1P → psacch	g1P(ε_{123})+, AMP(ε_{125})+, psacch(ε_{124})−	f16P(ε_{122})+	4
r17.7	t6P → tre	t6P(ε_{126})+		1
l17.6	q_{tre}	tre(ε_{127})+		1
l18.2	μ	ATP(ε_{128})+		1
l19.1	→ PenG	cys(ε_{129})+, val(ε_{130})+, aAd(ε_{131})+, ATP(ε_{134})+, PenG(ε_{133})−, PAA(ε_{132})+		6
r9.14	$q_{psacchh}$	psacch(ε_{135})		1
r9.8	q_{PenG}	PenG(ε_{136})		1
r9.11	q_{PAA}	PAA_{ext} (ε_{137})		1
r9.1	H_{ext} →	ATP(ε_{138})		1
			Total # ε	138

Table 8.11 Characteristics of the reference steady state. Steady state metabolite concentrations (X^0) are given in μmol/gDW, and intracellular fluxes (J^0) are given in μmol/gDW/s

$D = 0.05\ h^{-1}$, $C_x = 6$ gDW/L, $Glucose_{feed} = 15$ g/L							
q-rates (mmol/Cmol/h) $q_{O2} = 47.5$, $q_{CO2} = 50.3$, $\mu = 50.14$, $q_S = 19.5$							
X^0 (μmol/gDW)							
g6P	4.66	mal	3.37	ser	5.73	HCoA	2.29
f6P	0.72	succ	0.24	trp	0.11	METHF	2.48×10^{-3}
f16P	0.9	ala	21.8	tyr	0.27	THF	8.43×10^{-5}
GAP	0.07	arg	2.3	tre	40	CTP	0.96
3PG	0.58	asn	1.49	thr	5.95	UTP	1.36
PEP	0.24	asp	16.3	val	2.08	GMP	0.02
Pyr	0.22	cys	0.1	aAd	1.81	homcys	0.37
6PG	0.26	gln	28.7	PAA	1.95		
Ribu5P	0.07	glu	53	penG	12.9		
Rib5P	0.09	gly	2.09	g1P	0.71		
Xylu5P	0.14	his	0.72	t6P	0.57		
sed7P	0.35	ile	0.33	psacch	305.63		
E4P	0.01	leu	0.73	AMP	0.27		
aKG	2.06	lys	1.24	ATP	7.38		
OAA	0.38×10^{-3}	met	0.14	NADP	1.87		
iCitr	0.11	orn	1.53	NAD	1.9		
citr	2.8	phe	0.19	FAD	2.17		
fum	0.65	pro	0.95	AcCoA	0.01		
Extracellular metabolites (C^0, mmol/L)							
glc_f	0.103	$ExPept_f$	18.39	$psacch_f$	33.22	tre_f	0.23
$biomass_f$	213.47	$penG_f$	1.4	PAA_f	3.39		
Steady state fluxes (J^0, μmol/gDW/s)							
11.1	0.1911	r11.2	0.0179	111.10	0.0013	r13.6	4.43×10^{-4}
r1.3	0.1138	11.3-r1.9-111.2	0.0040	111.11	0.0032	r15.1	0.9609
r2.1-r2.2	0.0944	r11.3	0.0031	111.12	0.0032	117.1	5.28×10^{-4}
11.3-r1.9-r4.1	0.1714	11.3-111.3	5.33×10^{-4}	r11.16	0.0175	117.4	0.2076
r4.4	0.1329	11.3-111.4	0.0013	r11.17	0.0021	r17.7	5.28×10^{-4}
r4.7	0.1170	11.3-111.5	0.0023	r11.20	0.0022	117.6	5.28×10^{-4}
r4.8	0.1170	r11.7	0.0025	11.3-r1.9-r11.21	0.0025	118.2	0.4928
11.3-r1.9-r5.1	0.0277	111.7	0.0036	11.3-r1.9-r11.22	0.0067	119.1	0.0032
16.7	0.1170	r11.8	0.0025	r11.34	0.0094	R9.14	0.0767
16.9	0.8208	111.8	6.95×10^{-4}	113.2	6.85×10^{-4}	R9.8	0.0032
r8.1	0.0013	r11.9	0.0142	113.3	9.20×10^{-4}	R9.11	0.0032
11.3-r1.9-111.1	0.0078	111.9	0.0057	113.4	4.43×10^{-4}	R9.1	0.3868

8.3.4 *Characterization of Reference Steady State*

Having considered the above explained simplifications for the network and reduced it to its final form, here the characteristics of the reference steady state are described and presented in Table 8.11. Intracellular fluxes are calculated using Metabolic Flux Analysis and considering the stoichiometric network and the ATP balance. The levels of intracellular metabolites are either measured, or calculated using the equilibrium relations.

For intracellular metabolite levels of PenG and PAA, since there was no information available, we assumed that the intracellular and extracellular concentrations are the same. In calculating the intracellular metabolite levels, the specific cell volume is assumed to be 2.3 mL/gDW (Table 8.11).

8.3.5 *Final Parameterization*

8.3.5.1 Black-Box Equivalency

After the modifications and simplifications on the stoichiometric network, redefining the growth equations, the next step is to investigate black-box equivalency. Using the three balances (carbon, degrees of reduction and ATP balance, presented in Table 8.1) relations among nine rates namely, q_{O2}, q_{CO2}, q_S, μ, q_{ExPept}, q_{tre}, q_{psacch}, q_{Pen} and q_{PAA} are established. Here the q_{ExPept} represents the lysed biomass, therefore can be included in actual growth rate as $\mu_{act} = D + q_{ExPept}$, and it was further assumed $q_{ExPept} = 0.0864\ \mu$ based on steady state flux distribution. Further, the two rates of synthesis for storage materials can be added and $q_{storage}$ is defined as $q_{storage} = q_{psacch} + q_{tre}$. Lastly, q_{PAA} is equal to the q_{PenG} since PAA is only used in this pathway using these definitions. The carbon and degrees of reduction balances become then:

$$6q_S + q_{CO_2} + \mu_{act} + 8q_{Pen} + q_{\text{storage}} = 0$$

$$24q_S - 4q_{O_2} + 4.37\mu_{\text{act}} + 38q_{Pen} + 4q_{storage} = 0$$

Lastly, the ATP balance can be written with the updated P/O ratio as

$$1.623 \cdot (2q_{O_2}) + \Sigma q_i^{ATP} + 0.43 \cdot \mu_{\text{act}} + 93 \cdot q_{\text{Pen}} + 0.0039 = 0.$$

The second term in the ATP balance is the net ATP consumption by stoichiometrically known processes which can be calculated from the (reduced) stoichiometric matrix.

Having the measurements for q_{O2}, q_{CO2} and q_S we obtained the remaining rates (Table 8.12). From this table, it is seen that the black-box equivalency is indeed achieved.

8.3.5.2 A Priori Parameter Identifiability Analysis

Following the method presented in the background for lumped equilibrium pools and the method described in [20] for pseudo steady state metabolites, *a priori* parameter identifiability analysis has been carried out.

Table 8.12 The measured and calculated q-rates (in μmol/gDW/s) for the full and reduced model

	Full model	Reduced model
q_S	0.191	0.191
q_{CO2}	0.508	0.507
q_{O2}	0.475	0.476
μ_{act}	0.490	0.486
q_{Pen}	3.24×10^{-3}	3.52×10^{-3}
q_{tre}	0.51×10^{-3}	1.11×10^{-3}

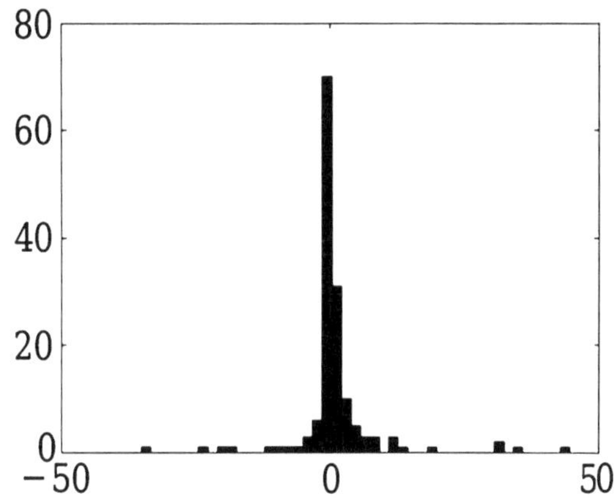

Fig. 8.8 The distribution of the estimated elasticities

8.3.5.3 Non-linear Estimation of the Elasticities

The final parameterization of the model has been performed by a non-linear minimization algorithm (Nelder-Mead simplex method), where the squared error (weighted by standard deviation) between the experimental measurement and model prediction is minimized. We generated an initial estimate for the nonlinear parameter estimation, following Nikerel et al. [19]. The distribution of the estimated parameters is presented in Fig. 8.8 and the complete list of estimated parameters is given in the Appendix.

The obtained model predicts the uptake and secretion rates, which are presented in Fig. 8.9. We can see from this figure that, given the decrease in ATP level, the growth rate and penicillin production decrease, and the synthesis of t6P increases.

Furthermore, the obtained model simulates the effect of the glucose perturbation. The comparison of the simulation for the final model with the experimental data is presented in Figs. 8.10, 8.11, 8.12, 8.13, 8.14, 8.15, 8.16, 8.17, 8.18, and 8.19, which show metabolite concentrations and reaction rates in fold changes relative to steady state values (Table 8.11). Overall, the model adequately mimics the dynamics of the metabolite levels. Following the glucose pulse, the metabolites in the upper part of the glycolysis are typically increasing and the

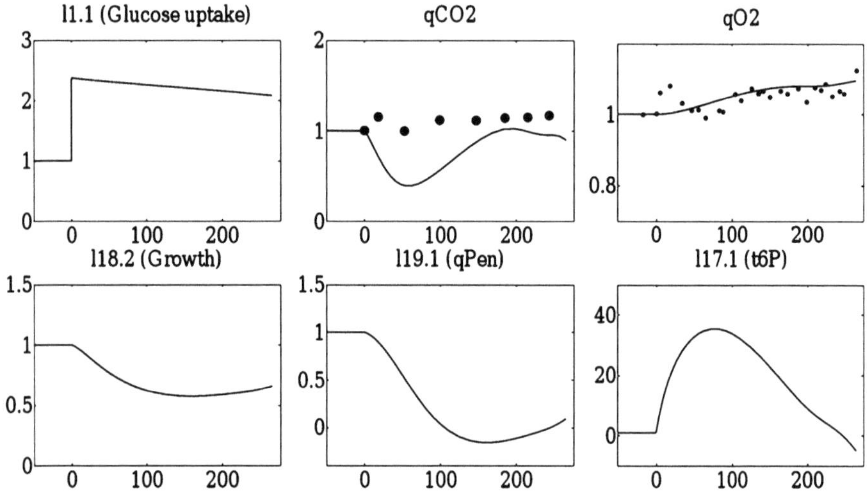

Fig. 8.9 The model prediction for the extracellular rates. The *x-axis* represents time in seconds and the *y-axis* represents the corresponding extracellular rate relative to its reference state. For q_{CO2} and q_{O2}, off-gas measurements are represented as points

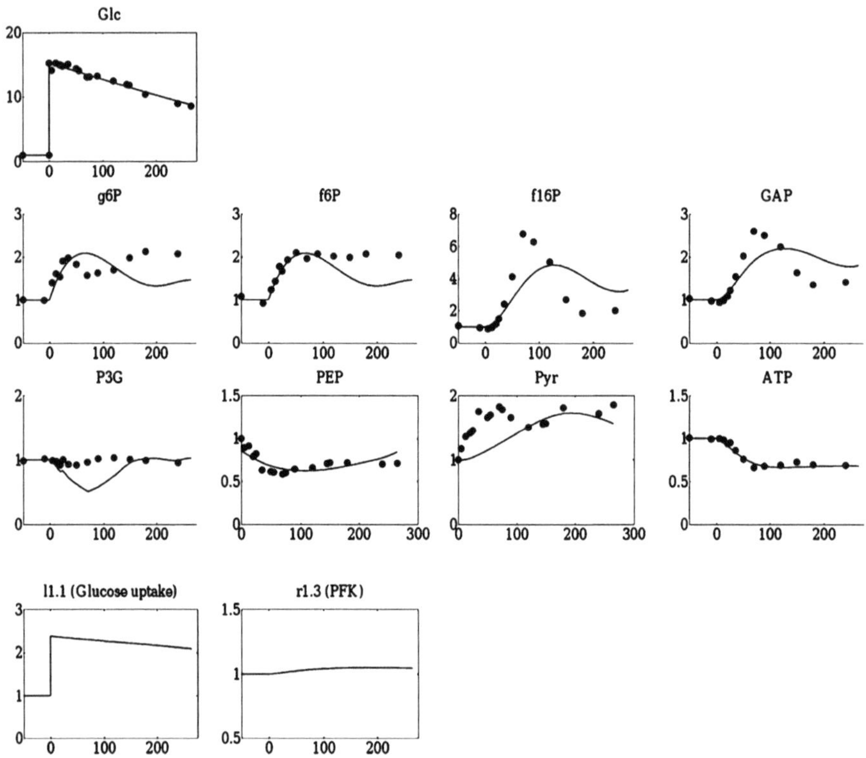

Fig. 8.10 Available data (*points*) and model prediction (*continuous lines*) for glycolysis metabolites and rates. The *x-axis* represents time in seconds and the *y-axis* represents the corresponding metabolite or rate relative to its reference state (no units)

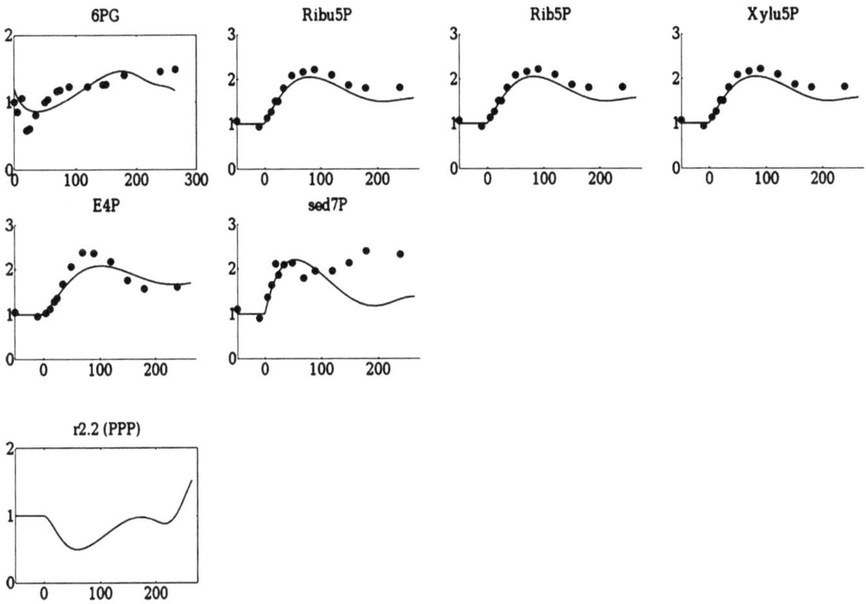

Fig. 8.11 Available data (*points*) and model prediction (*continuous lines*) for PPP metabolites and rates. The *x-axis* represents time in seconds and the *y-axis* represents the corresponding metabolite or rate relative to its reference state (no units)

ones in the lower part are decreasing. In PPP, while the metabolite levels are increasing following the changes in g6P, gap and f6P, the rate towards PPP decreases following the dynamics of NADPH and ATP. Among the TCA intermediates, OAA is decreasing since the C4-pool (malate, fumarate, succinate) is more reduced. Further, AcCoA/HCoA ratio is increased expectedly, since while the glycolytic flux is increased, TCA burden is almost constant (only 10% increase in the q_{O2} level). Lastly, the NADH/NAD ratio is expectedly increasing due to the increased reduced state of the TCA.

8.3.6 Calculation of the MCA Parameters

Using the estimated parameters, we calculated the control coefficients using Eq. 8.15. The significant flux control coefficients for penicillin production and storage synthesis are presented in Table 8.13. The calculated control coefficients for penicillin production ($\mathbf{C^J}_{\mathbf{Pen}}$) points to the fact that increasing the capacity of the oxidative phosphorylation would greatly increase the penicillin production and also improving any ATP consuming reaction would decrease the penicillin production, since this heavily requires ATP.

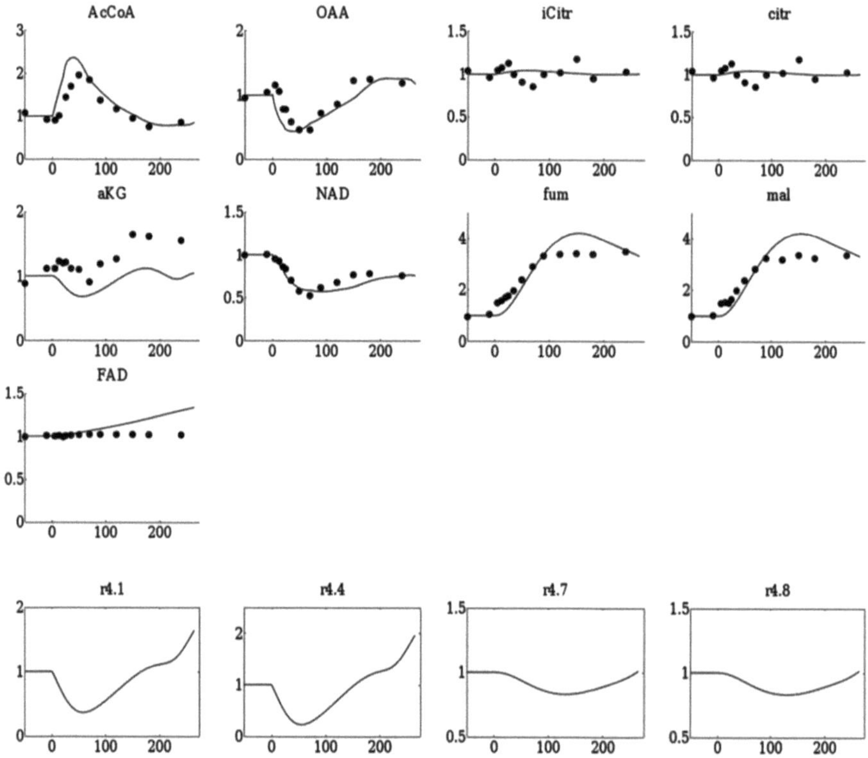

Fig. 8.12 Available data (*points*) and model prediction (*continuous lines*) for TCA metabolites and rates. The *x-axis* represents time in seconds and the *y-axis* represents the corresponding metabolite or rate relative to its reference state (no units)

As for the synthesis of storage materials, the calculated flux control coefficients point to the fact that, should the enzymes for oxidative phosphorylation engineered for improvement, the synthesis rate of storage materials would decrease, which is in agreement with the expectation since the excess carbon upon perturbation redirects to storage given the energy shortage. In agreement with this, increasing the flux to ATP consuming reactions would enhance storage production (Table 8.13).

$$\mathbf{C}^x = -\left(\mathbf{S}\mathbf{J}^0\mathbf{E}^x\right)^{-1}\mathbf{S}\mathbf{J}^0 \quad \mathbf{C}^J = \mathbf{I} + \mathbf{E}^x\mathbf{C}^x \tag{8.15}$$

8.4 Discussion and Conclusions

We constructed a dynamic model for our model organism *P. chrysogenum* describing the dynamic behavior of intracellular metabolites upon glucose perturbation. In this section, we discuss the main points regarding the construction of the dynamic model.

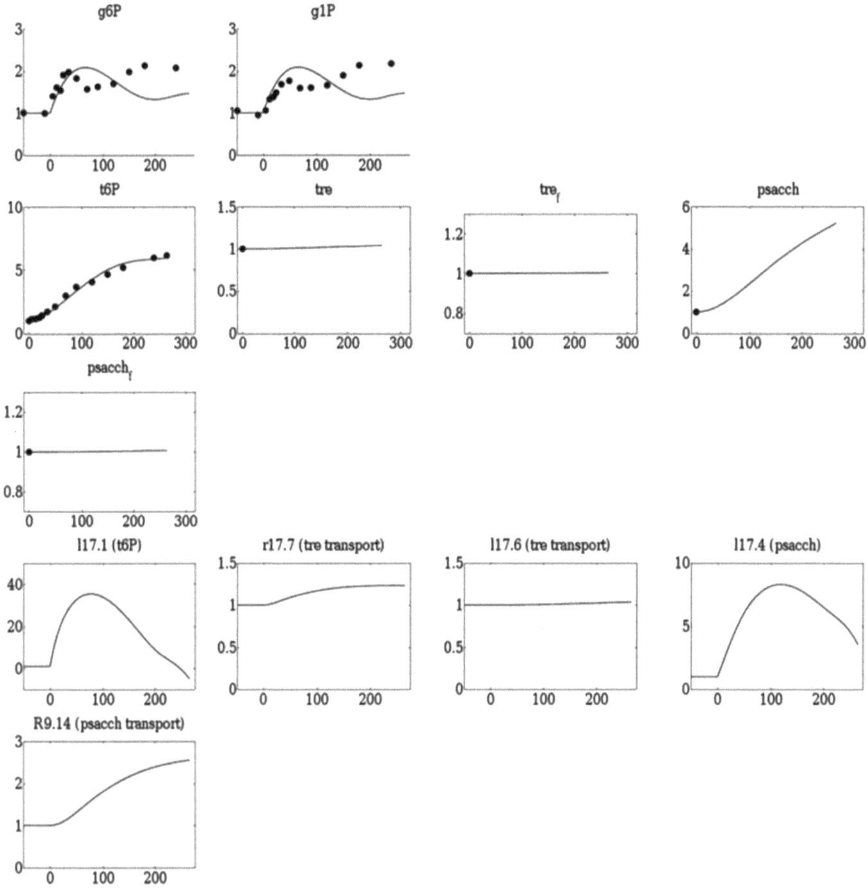

Fig. 8.13 Available data (*points*) and model prediction (*continuous lines*) for metabolites and rates on Storage pathway. The x-axis represents time in seconds and the y-axis represents the corresponding metabolite or rate relative to its reference state (no units)

Construction of large (genome) scale models for metabolic reaction networks is an intricate task, not only because of the size of the system, but also because of the interactions involved, lack of measurements, time hierarchy and identifiability issues. We approached this task by *a priori* reducing the model, i.e. we simplified the model as much as possible, without losing the actual dynamic information. This study can be seen as a demonstration of how our proposed framework can be used to build genome scale dynamic models.

One of the main reductions made is to remove the compartments. By doing so, the compartmentation information on e.g. rates is lost and the cell becomes a "bag of reactions". By employing the black-box equivalency before the parameter estimation, we conclude that the original and the reduced model are equivalent, in predicting the extracellular rates.

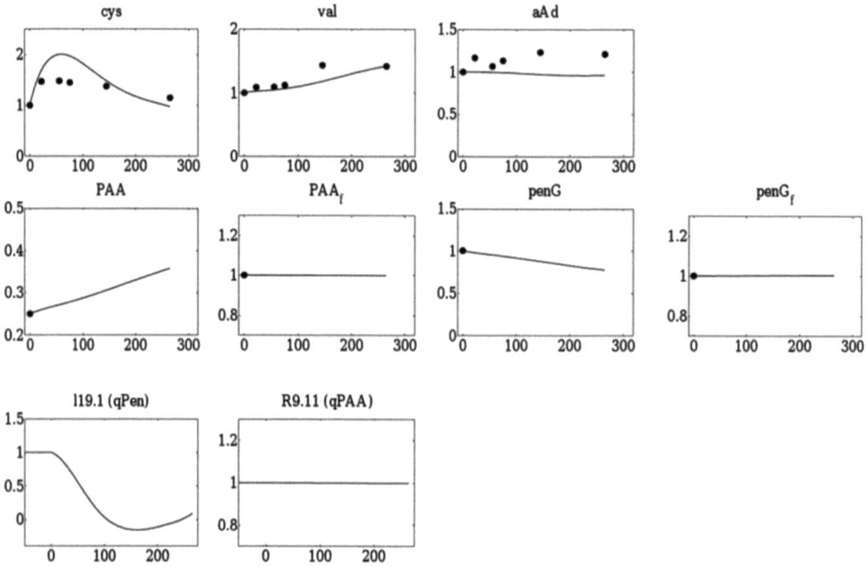

Fig. 8.14 Available data (*points*) and model prediction (*continuous lines*) for metabolites and rates on penicillin pathway. The *x-axis* represents time in seconds and the *y-axis* represents the corresponding metabolite or rate relative to its reference state (no units)

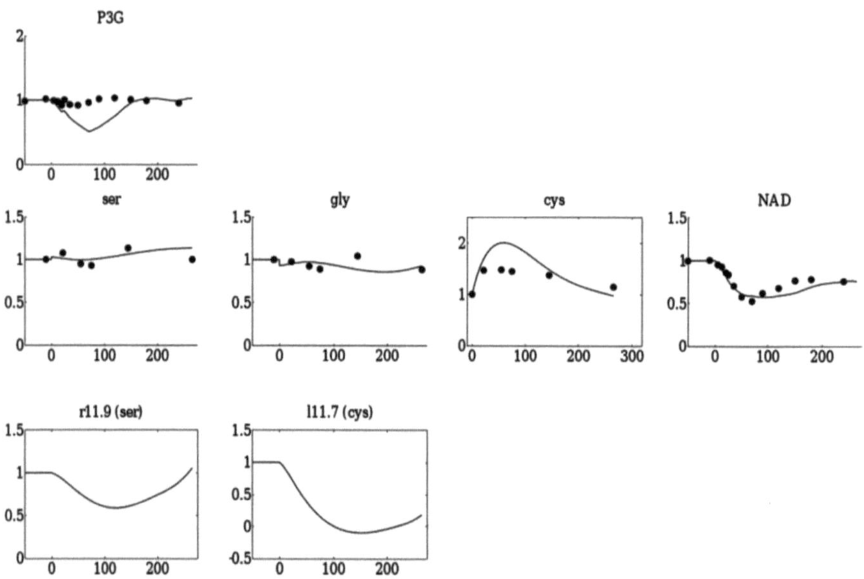

Fig. 8.15 Available data (*points*) and model prediction (*continuous lines*) for metabolites and rates for Amino acids from 2PG + 3PG. The *x-axis* represents time in seconds and the *y-axis* represents the corresponding metabolite or rate relative to its reference state (no units)

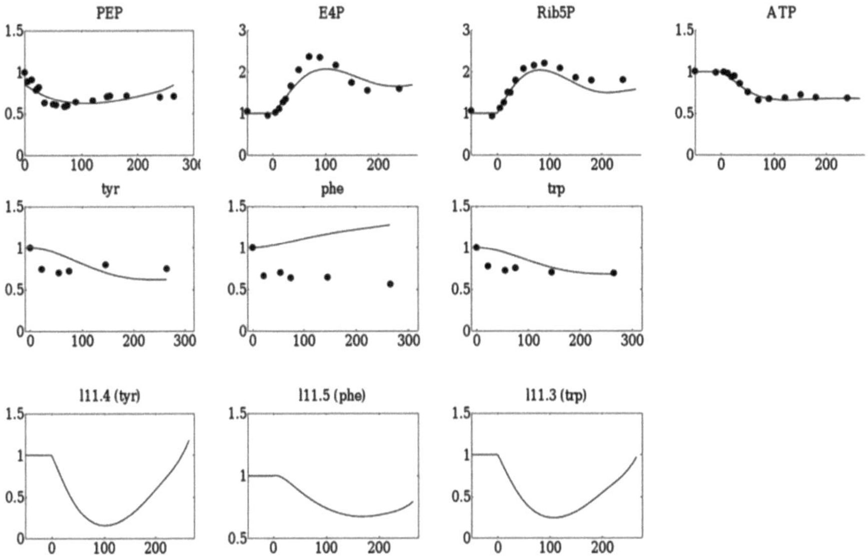

Fig. 8.16 Available data (*points*) and model prediction (*continuous lines*) for metabolites and rates for Amino acids from PEP. The *x-axis* represents time in seconds and the *y-axis* represents the corresponding metabolite or rate relative to its reference state (no units)

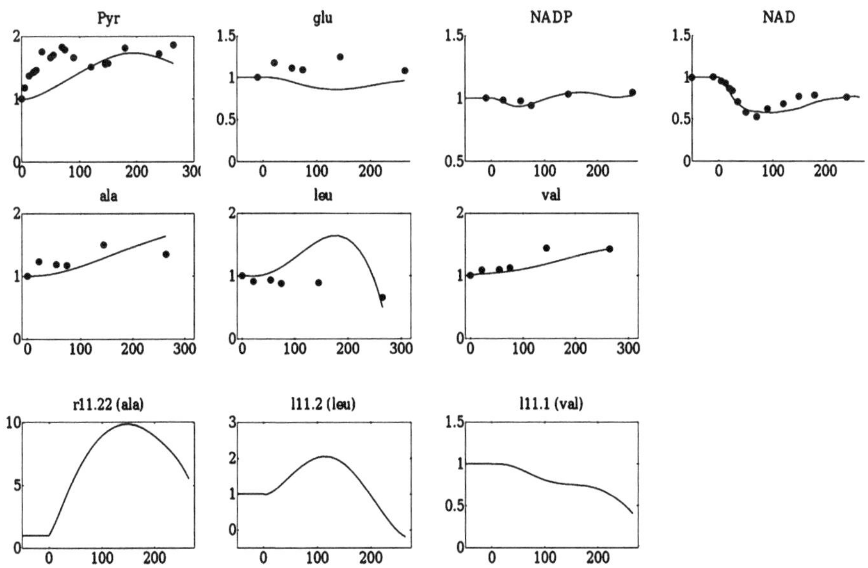

Fig. 8.17 Available data (*points*) and model prediction (*continuous lines*) for metabolites and rates for Amino acids from pyr. The *x-axis* represents time in seconds and the *y-axis* represents the corresponding metabolite or rate relative to its reference state (no units)

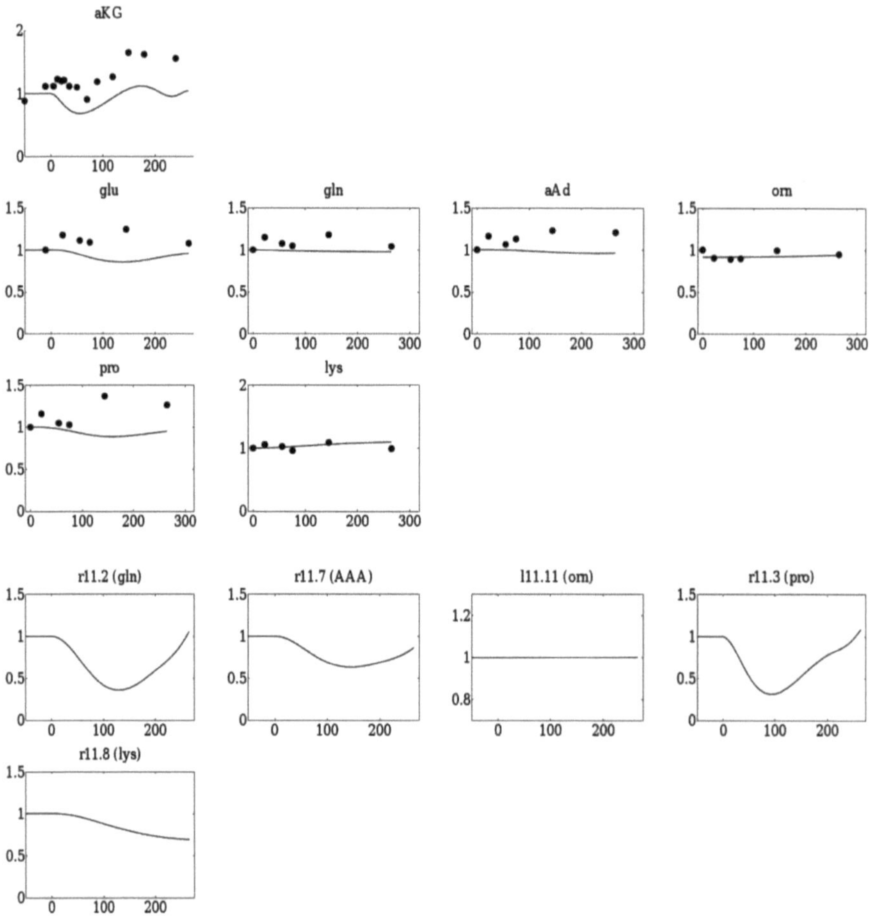

Fig. 8.18 Available data (*points*) and model prediction (*continuous lines*) for metabolites and rates for Amino acids from aKG. The *x-axis* represents time in seconds and the *y-axis* represents the corresponding metabolite or rate relative to its reference state (no units)

Another major model reduction is performed by considering time-scale analysis, pseudo equilibrium reactions and pseudo steady state pools. This step is crucial both conceptually in order to have a better understanding on dynamics of the system, and practically, such an analysis clearly separates the fast and slow metabolites, avoiding stiffness in forward simulations and reduces parameter identifiability problem.

The time scale analysis yielded a number of lumped equilibrium pools or reactions. One important issue to consider is that, a certain order in time scale analysis should be followed, i.e. pseudo equilibrium pools should be processed first, simply because, since the actual forward and backward reaction rates in a pseudo equilibrium rate is unknown, the turn-over time might be underestimated if these reactions are not processed first.

At this point it is worth to note that, fast metabolites are not necessarily the metabolite pools that have a low concentration in the cell e.g. pep, but it can as well be a metabolite with relatively higher concentration yet has a higher influx,

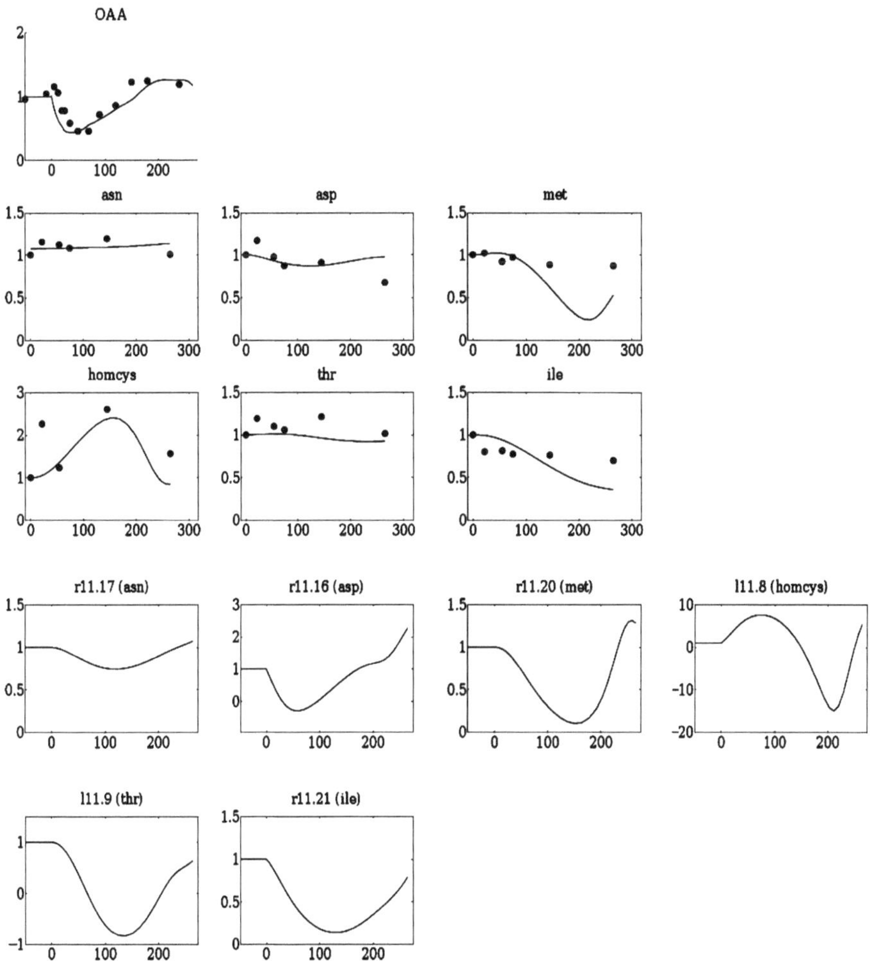

Fig. 8.19 Available data (*points*) and model prediction (*continuous lines*) for metabolites and rates for Amino acids from OAA. The *x-axis* represents time in seconds and the *y-axis* represents the corresponding metabolite or rate relative to its reference state (no units)

Table 8.13 Calculated MCA parameters. For the sake of clarity, only significant control coefficients are shown

	C_{μ}^{J}		C_{tre}^{J}		C_{pen}^{J}
r1.3 (PFK)	1.73	r1.3 (PFK)	−0.97	qPAA	1.0
l6.7 (Oxy. Phos)	−0.89	l6.7 (Oxy. Phos)	0.5		
l18.2 (mu)	0.32	l18.2 (mu)	0.52		
		r17.7 (qtre)	1.0		

e.g. P10. In that sense, the fastness or slowness of metabolites is a system property rather than the property of the individual pools.

Because ATP and NAD appear as fast metabolites, the resulting final model is a complete model in the sense that it does not require any additional input such as externally provided cofactor concentrations and can predict those.

Another feature of the time scale analysis is that when linlog kinetics is used *a priori* parameter identifiability analysis is possible, making the parameter estimation problem more transparent and easier to handle. The large scale model building scheme proposed in this work is advantageous in the sense that the parameter identifiability analysis can be performed *a priori*, i.e. before the final parameterization, as opposed to the classical approach of having an estimate for the large model and assessing such identifiability analysis *a posteriori*. The time scale analysis and linlog kinetics yield **xP** and **A** matrices for pseudo equilibrium pools and pseudo steady state pools respectively. These matrices provide also the linking between the identifiable and original parameters. Any further experimental design should also focus on the unidentifiable parameters and therefore consider the entries of those matrices.

Seeing the lumped pool definitions and the algebraic relations between the fast and slow metabolites, it is clear that, a subpart of the pools (e.g. P1, P2, t6P) contribute more to the dynamics then others, simply because they either contain a large number of metabolites in the case of P1 and P2 that contains all the metabolites of the upper part of glycolysis and PPP, or these pools determine the levels of ATP and NAD that effects several reactions. This fact would leads to a number of manipulations in parameter estimation and formulation of the objective function. The contribution of those pools to the total sum of square can be weighed by an arbitrary number. In the current parameter estimation problem, each lumped pool is weighted by the number of metabolites that is contained in the specific pool, e.g. total sum of square for P1 is weighted by 10, whereas the one for P8 is weighted by 2. This calls for further investigation on how error on the lumped pools propagates into the individual metabolites or vice versa. This should be considered in future studies.

In this work, we redefined the biomass and considered the trehalose formation as a separate reaction, which was formerly part of the biomass in van Gulik et al. [29]. This allowed focusing separately on growth and formation of storage material. Considering the storage as a separate product renders e.g. the MCA analysis results more transparent for further comment.

An important point to discuss is the performance of the model in simulating the effect of glucose perturbation. The model predicts well the glycolysis intermediates and TCA cycle metabolites. Taken the time scale simplifications along, the model predicts the levels of ATP and NAD as well. However, there are also a number of metabolite profiles that the model fails to mimic. In most cases, the trend is captured but not the absolute change (e.g. aKG), in few other cases, the trend is also in opposite direction (e.g. phe). This fact points out that, in cases where at least the trend cannot be represented, additional effectors should be added as non-zero elasticities to the elasticity matrix, since the available interactions cannot explain the change in the corresponding data.

The currently obtained model can be considered as an initial model that can be further refined by either adding new interactions or implementing new data or both. It should be noted that this model would be the basis for further model based experimental design. The newly designed experiments should be designed in such a way to resolve the identifiability problems. Furthermore, given that the parameters in this model are elasticities defined within the framework of the MCA, the flux and concentration control coefficients are calculated. Here it is important to see that since ATP is considered to be a fast metabolite, any results that would be obtained from MCA, should be analyzed with care. This is simply because of the fact that, the time scale analysis shows that ATP dynamics are dependent on a number of slow metabolites hence the effect would be implicit.

The model and the inspection of the control coefficients points to the fact that, the "energy budget" in the cell determines which of the three products (growth, penicillin production, storage) would the carbon go. This is reflected in the obtained MCA parameters. In order to grasp the meaning of the control parameters presented in Table 8.13, it should be marked that ATP (P10) is linked to t6P and P1 via pseudo steady state relations, in such a way that any decrease in these two pools would increase the ATP (P10) level ($\alpha_{18} = -0.170$, $\alpha_{19} = -0.165$). $C_{PFK}{}^{\mu} = 1.73$ points to the fact that, in case the activity of this enzyme is increased, both P1 and t6P level would decrease, resulting in a higher P10 level, hence increased growth rate, provided that the growth is assumed to be only effected by ATP. Similarly, in order to increase the storage formation rate, the PFK enzyme level should be decreased ($C_{PFK}{}^{qtre} = -0.97$), which on one side directly determines the t6P level and indirectly effects the ATP level (via ATP-t6P relation). These findings are in agreement with the previous postulations in [18], where the authors claimed that under substrate excess conditions, the carbon redirects to storage.

In the current model, it is not possible to measure changes in the rates of growth and q_{ExPept} and their elasticities cannot be observed. Therefore these rates are assumed to be a function of ATP only and the elasticity of ATP in these rates are set manually to 1. Furthermore, most of the amino acids are produced by one reaction and is consumed by the growth reaction. It should be noted that, in a number of cases, the model predicts that the level of the amino acid increases while the production rate decreases. Considering the fact that growth and lysed biomass formation is a function of ATP only, this increase in the amino acid level is simply because the production rate decreases less than growth rate.

In building the model, we encountered several metabolites that were not measured. Naturally, as more measurements are available, a number of assumptions can then be checked and relaxed afterwards if necessary. This is important both for unmeasured metabolites and also metabolites that can be found in multiple compartments. With the advances in this area e.g. [1], the removal of compartments can be relaxed hence the obtained model can even be more realistic.

Overall, this work can be considered as a successful first attempt for construction of large scale dynamic model using *P. chrysogenum* as test case. The model provides some more insights on the dynamics of the system, points to genetic targets for increasing a specific flux and allows linking the dynamic metabolite concentrations to the dynamic change of the fluxes. A short list of what should be done in the future covers the testing stage of the model under alternative perturbations, incorporation of new measured metabolites and checking back the assumptions and relaxing if necessary.

Appendix

Nomenclature Used for Metabolites in the Model

Below is the nomenclature list for metabolites used in the model, following the names in van Gulik et al. [29]

13PG	1,3-Bisphosphoglycerate	glc	Glucose
2PG	2-Phosphoglycerate	gln	Glutamine
3PG	3-Phosphoglycerate	glu	Glutamate
6APA	6-Aminopenicillinic acid	gly	Glycine
6Pgluct	6-Phosphogluconate	GMP	Guanosine monophosphate
8HPA	8-Hydroxypenillic acid	H	Proton
aAd	a-Aminoadipate	H2O	Water
AApool	Pool of free amino acids	H2S	Hydrogen sulfide
AAprotsyn	Amino acids for protein synthesis	HCoA	Coenzyme A
Ac	Acetate	his	Histidine
AcCoA	Acetyl coenzyme A	homcys	Homocysteine
AcHomser	o-Acetylhomoserine	homser	Homoserine
ACV	d-(a-Aminoadipyl)-cysteinylvaline	iCitr	Isocitrate
ADP	Adenosine-5-diphosphate	ile	Isoleucine
aKG	a-Ketoglutarate	IMP	Inosine monophosphate
aKI	a-Ketoisovalerate	ino	Inositol
ala	Alanine	iPN	Isopenicilline
AMP	Adenosine-5-monophosphate	lano	Lanosterol
arg	Arginine	leu	Leucine
asn	Asparagine	linCoA	Linoleoyl coenzyme A
asp	Aspartate	lys	Lysine
ATP	Adenosine-5-triphosphate	m1P	Mannitol-1-phosphate
bIM	b-Isopropylmalate	mal	Malate
biomass	Average biomass composition of glucose-limited cultures	man	Mannitol
		met	Methionine
carbP	Carbamoylphosphate	METHF	Methylene tetrahydrofolate
CDPDAcgcl	Cytidine diphosphate-diacylglyerol	meva	Mevalonate
chit	Chitine	MYTHF	Methyltetrahydrofolate
chor	Chorismate	NAD	Nicotinamide adenine dinucleotide (oxidized)
citr	Citrate		
CMP	Cytidine monophosphate	NADH	Nicotinamide adenine dinucleotide (reduced)
CO2	Carbon dioxide		
ctl	Citruline	NADP	Nicotinamide adenine dinucleotide phosphate
CTP	Cytidine triphosphate		
cys	Cysteine	NADPH	Nicotinamide adenine dinucleotide phosphate
E4P	Erythrose-4-phosphate		

(continued)

(continued)

ergo	Ergosterol	NH4	Ammonia
ery	Erythritol	O2	Oxygen
ESE	Ergosterolester	OAA	Oxaloacetate
ExPept	Excreted peptides	OPC	6-Oxopiperidine-2-carboxylic acid
f16P	Fructose-1,6-bisphosphate		
f6P	Fructose-6-phosphate	PAA	Phenylacetic acid
FAD	Flavine adenine dinucleotide (oxidized)	PAACoA	Phenylacetyl coenzyme A
		PAPS	3-Phosphoadenosine-5-phosphosulfate
FADH2	Flavine adenine dinucleotide (reduced)		
FTHF	Formyltetrahydrofolate	penG	Penicillin-G
fum	Fumarate	PEP	Phosphoenolpyruvate
g6P	Glucose-6-phosphate	PHchol	Phosphatidylcholine
GAP	3-Phosphoglyceraldehyde	phe	Phenylalanine
gcl3P	3-Phosphoglycerol	PHeta	Phosphatidylethanolamine
pro	Proline	PHino	Phosphatidylinositol
PROT	Protein	phospht	Phosphatidate
PRPP	a-5-Phosphoribosylpyrophosphate	PHser	Phosphatidylserine
psacch	Polysaccharides	Pi	Orthophosphate
pyr	Pyruvate		
Rib5P	Ribose-5-phosphate		
Ribu5P	Ribulose-5-phosphate		
RNA	Ribose nucleic acid		
SAM	S-Adenosylmethionine		
sed7P	Sedoheptulose-7-phosphate		
ser	Serine		
succ	Succinate		
succCoA	Succinyl coenzyme A		
t6P	Trehalose-6-phosphate		
THF	Tetrahydrofolate		
thr	Threonine		
tre	Trehalose		
trp	Tryptophane		
tyr	Tyrosine		
UDP	Uridine-5-diphosphate		
UDPglc	Uridine-5-diphosphoglucose		
UMP	Uridine monophosphate		
UTP	Uridine triphosphate		
val	Valine		
Xylu5P	Xylulose-5-phosphate		

Data for the Kinetic Model

In the following, three tables are specified the finally used reactions, the input data, the assumed equilibrium constants and data on metabolite concentrations.

Table 8.14 List of reactions in the reduced model

Glycolysis	
111.1	$ATP_{cyt} + H_{ext} + glc_{ext} \rightarrow ADP_{cyt} + 2\ H_{cyt} + g6P_{cyt}$
r1.3	ATP + f6P → ADP + f16P + H
r1.9	ADP + H + PEP → ATP + pyr
PPP	
r2.1	g6P + H_2O + NADP → 6Pgluct + 2 H + NADPH
r2.2	6Pgluct + NADP → CO_2 + NADPH + Ribu5P
TCA	
r4.1	HCoA + NAD + pyr → AcCoA + CO2 + NADH
r4.4	iCitr + NAD → aKG + CO_2 + NADH
r4.7	aKG + HCoA + NAD → CO_2 + NADH + succCoA
r4.8	ADP + Pi + succCoA ↔ ATP + HCoA + succ
Anaplerotic pathways	
r5.1	ATP + CO_2 + H_2O + pyr → ADP + 2 H + OAA + Pi
Oxidative Phosphorylation	
16.7	1.104 ADP + 1.104 H + FADH2 + 0.5 O_2 + 1.104 Pi → 1.104 ATP + 2.104 H_2O + FAD
16.9	1.6225 ADP + 2.6225 H + NADH + 0.5 O_2 + 1.6225 Pi → 1.6225 ATP + 2.6225 H_2O + NAD
Transfer of 1 C compounds	
r8.2	METHF + NADH + H → MYTHF + NAD
Transport across the cell membrane	
R9.1	$ATP_{cyt} + H_2O_{cyt} \leftrightarrow ADP_{cyt} + H_{ext} + Pi_{cyt}$
R9.11	$PAA_{ext} \rightarrow PAA_{cyt}$
R9.13	$ExPept_{cyt} \rightarrow ExPept_{ext}$
R9.14	$psacch_{cyt} \rightarrow psacch_{ext}$
R9.2	$2\ H_{ext} + Pi_{ext} \rightarrow 2\ H_{cyt} + Pi_{cyt}$
R9.8	$penG_{cyt} \rightarrow penG_{ext}$
Amino acid synthesis	
111.1	2 H + NADPH + glu + 2 pyr → CO_2 + H_2O + NADP + aKG + val
111.2	AcCoA + H + NADPH + NAD + glu + 2 pyr → 2 CO_2 + HCoA + NADH + NADP + aKG + leu
111.3	3 ATP + E4P + NADPH + 2 PEP + Rib5P + gln + ser → 3 ADP + CO_2 + GAP + H_2O + 3 H + NADP + 6 Pi + glu + pyr + trp
111.4	ATP + E4P + NADPH + NAD + 2 PEP + glu → ADP + CO_2 + NADH + NADP + 4 Pi + aKG + tyr
111.5	ATP + E4P + H + NADPH + 2 PEP + glu → ADP + CO_2 + H_2O + NADP + 4 Pi + aKG + phe
111.7	4 ATP + 2H + 4 NADPH + SO4 + ser → 4 ADP + H_2O + 4 NADP + 4 Pi + cys
111.8	5 ATP + 4H + 6 NADPH + SO4 + asp → 5 ADP + H_2O + 6 NADP + 5 Pi + homcys
111.9	2 ATP + H_2O + H + 2 NADPH + asp → 2 ADP + 2 NADP + 2 Pi + thr
111.10	5 ATP + CO_2 + 3 H_2O + NADPH + 2 NAD + NH4 + Rib5P + gln → 5 ADP + 10H + 2 NADH + NADP + 6 Pi + aKG + his

(continued)

Table 8.14 (continued)

l11.11	ATP + 2 glu + NADPH + H $\rightarrow$ ADP + aKG + orn + NADP + Pi
l11.12	4 ATP + CO_2 + gln + 3 H_2O + orn + asp $\rightarrow$ 4 ADP + glu + 7H + 4 Pi + arg + fum
r11.3	ATP + glu + H + 2 NADPH $\rightarrow$ ADP + H_2O + 2 NADP + Pi + pro
r11.7	AcCoA + glu + H_2O + NAD $\rightarrow$ aAd + CO_2 + HCoA + NADH
r11.8	aAd + 2 ATP + glu + H_2O + NAD + 2 NADPH $\rightarrow$ 2 ADP + aKG + H + lys + NADH + 2 NADP + 2 Pi
r11.9	3PG + glu + H_2O + NAD $\rightarrow$ aKG + H + NADH + Pi + ser
r11.10	ser + THF $\rightarrow$ gly + H_2O + METHF
r11.16	glu + OAA $\leftrightarrow$ aKG + asp
r11.17	asp + 2 ATP + H_2O + NH4 $\rightarrow$ 2 ADP + asn + H + 2 Pi
r11.20	homcys + MYTHF $\leftrightarrow$ met + THF
r11.21	glu + 2H + NADPH + pyr + thr $\rightarrow$ aKG + CO_2 + H_2O + ile + NADP + NH4
r11.22	glu + pyr $\rightarrow$ aKG + ala
r11.34	0.134 ala + 0.07 arg + 0.0386 asn + 0.0386 asp + 0.02 cys + 0.0509 gln + 0.0509 glu + 0.132 gly + 0.0119 his + 0.0307 ile + 0.0559 leu + 0.0348 lys + 0.0103 met + 0.0266 phe + 0.0537 pro + 0.044 ser + 0.0467 thr + 0.01 trp + 0.0203 tyr + 0.12val $\rightarrow$ 4.53 ExPept + H_2O
Nucleotide Biosynthesis	
l13.2	10 ATP + CO_2 + 9 H_2O + 2 METHF + 3 NAD + Rib5P + asp + 3 gln + gly $\rightarrow$ 10 ADP + GMP + 19H + 3 NADH + 10 Pi + 2 THF + fum + 3 glu
l13.3	9 ATP + CO_2 + 6 H_2O + 2 METHF + 2 NAD + Rib5P + 2 asp + 2 gln + gly $\rightarrow$ 9 ADP + AMP + 16H + 2 NADH + 9 Pi + 2 THF + 2 fum + 2 glu
l13.4	6 ATP + 2 H_2O + NAD + Rib5P + asp + gln $\rightarrow$ 6 ADP + 6H + NADH + 4 Pi + UTP + glu
r13.6	ATP + gln + H_2O + UTP $\leftrightarrow$ ADP + CTP + glu + 2H + Pi
Glycogen and polysaccharides synthesis	
l17.1	ATP + H_2O + 2 g6P $\rightarrow$ ADP + H + 2 Pi + t6P
l17.4	0.167 ATP + 0.167 g1P + 0.167 H_2O $\rightarrow$ 0.167 ADP + 0.167H + 0.333 Pi + psacch
l17.6	$tre_{cyt} \rightarrow tre_{ext}$
r17.7	H_2O + t6P $\rightarrow$ Pi + tre
PenG biosynthesis	
l19.1	8 ATP + 4 H_2O + O_2 + cys + val + PAA $\rightarrow$ 8 ADP + 8H + 8 Pi + penG
ATP Hydrolysis	
r15.1	ATP + H_2O $\rightarrow$ ADP + H + Pi
Biomass formation	
l18.2	0.01077 $f6P_{cyt}$ + 0.43797284 ATP_{cyt} + 0.365703983 H_2O_{cyt} + 0.265589 $psacch_{cyt}$ + 0.09173 $NADPH_{cyt}$ + 0.064991 $AcCoA_{cyt}$ + 0.015249336 gln_{cyt} + 0.011598612 $NADH_{cyt}$ + 0.011012043 ala_{cyt} + 0.009260646 gly_{cyt} + 0.007813 $O_{2,cyt}$ + 0.007058727 val_{cyt} + 0.007049897 leu_{cyt} + 0.006940612 ser_{cyt} + 0.005614911 thr_{cyt} + 0.005328166 pro_{cyt} + 0.005257028 arg_{cyt} + 0.005003177 asp_{cyt} + 0.004457085 lys_{cyt} + 0.004395004 ile_{cyt} + 0.004260689 phe_{cyt} + 0.004244145 met_{cyt} + 0.0036256830 asn_{cyt} + 0.0024294300 his_{cyt} + 0.002269 GAP_{cyt} + 0.002184598 tyr_{cyt} + 0.00186666 AMP_{cyt} + 0.001390635 GMP_{cyt} + 0.00124 $E4P_{cyt}$ + 0.001187388 $Rib5P_{cyt}$ + 0.001006 THF_{cyt} + 0.000898564 CTP_{cyt} + 0.000890068 trp_{cyt} + 0.000445003 cys_{cyt} + 0.000403 $g6P_{cyt}$ $\rightarrow$ 0.001006 $MYTHF_{cyt}$ + 0.00303 $homcys_{cyt}$ + 0.003031405 glu_{cyt} + 0.004632 $CO_{2\,cyt}$ + 0.011598612 NAD_{cyt} + 0.064991 $HCoA_{cyt}$ + 0.09173 $NADP_{cyt}$ + 0.341110356 H_{cyt} + 0.43797284 ADP_{cyt} + 0.452434968 Pi_{cyt} + $biomass_{ext}$

Table 8.15 Assumed equilibrium constants

K_{eq} rxn	Value	K_{eq} rxn	Value
r1.4	2.54510^{-6} M	r4.2	2.2410^{6}
r1.5	0.487	r4.3	0.041
r1.6	2.2710^{3}	r4.8	0.949
r1.7	0.0920	r4.9	0.85
r2.3	0.833	r4.11	2.4110^{-5} M
r2.4	1.82	r8.2	4.510^{10}
r2.5	0.48	r11.1	9.8910^{-7}
r2.6	0.37	r11.10	10.72
r2.7	0.0337		

Table 8.16 Assumed metabolite concentrations

Metabolite	Assumed value
Pi:cyt	10 mM
NH4:cyt	0.1 mM
SO4:cyt	1 mM
H:cyt	10^{-7} M
arg:cyt	1 mM

P/O Ratio Calculations

The reactions describing the oxidative phosphorylation in [29] have to be updated/adjusted not only because the mitochondria compartment has been removed, but also, in [28], the experimentally determined P/O parameter was 1.84. The P/O parameter is changed in such a way to have the same theoretical yields for the reduced model as the original model. Using the Herbert-Pirt substrate utilization equation as a quality assessment to these reduction steps, we compared the Herbert-Pirt equation for the three compartment and the one compartment model.

We first considered the stoichiometric network presented in [29], which consists of 155 metabolites and 156 reactions located in three compartments (cytosol, mitochondria and peroxisome). The Herbert-Pirt substrate utilization relation can be written as:

$$q_S = 5.17 q_{\text{Pen}} + 0.288\mu + 1.37\,10^{-3}$$

The P/O value is calculated from reaction r6.1 and r6.4. The first adjustment is to multiply the theoretical stoichiometric coefficients of mitochondrial and cytosolic protons, with the ratio of the estimated P/O ratio to the maximum P/O ratio

$$8.360H_m + NADH_m + 0.5O_2 \xrightarrow{r6.1} 7.360H_c + H_2O_c + NAD_m$$
$$5.416H_m + NADH_c + 0.5O_2 \xrightarrow{r6.2} 4.416H_c + H_2O_c + NAD_c$$
$$4.416H_m + FADH_{2,m} + 0.5O_2 \xrightarrow{r6.3} 4.416H_c + H_2O_c + FAD_m$$

Subsequently the formation of ATP which is driven by the inward flux of cytosolic protons into the mitochondria (r6.4) is lumped with the water transport and the ADP/ATP shuttle (r10.3 and r10.1 respectively) in order to eliminate all mitochondrial reactants with exception of H_m.

$$\begin{array}{lcl} ADP_m + 4H_c + Pi_m & \xrightarrow{r6.4} & ATP_m + 3H_m + H_2O_m \\ ADP_c + ATP_m & \xrightarrow{r10.1} & ADP_m + ATP_c \\ H_2O_m & \xrightarrow{r10.3} & H_2O_c \\ \hline ADP_c + 4H_c + Pi_m & \xrightarrow{l6.4} & ATP_c + 3H_mH_2O_c \end{array}$$

The next step is to eliminate the H in reactions r6.1-3 using l6.4. This yields the following lumped reactions:

$$\begin{array}{lcl} 1.840ADP_c + 2.840H_c + NADH_m + 0.5O_{2,c} + 1.840Pi_c & \xrightarrow{l6.5} & 1.840ATP_c + 2.84H_2O_c + NAD_m \\ 1.104ADP_c + 2.104H_c + NADH_c + 0.5O_{2,c} + 1.104Pi_c & \xrightarrow{l6.6} & 1.104ATP_c + 2.104H_2O_c + NAD_c \\ 1.104ADP_c + 1.104H_c + FADH_{2,m} + 0.5O_{2,c} + 1.104Pi_c & \xrightarrow{l6.7} & 1.104ATP_c + 2.104H_2O_c + FAD_m \end{array}$$

Since the model does not contain any inner compartments, reactions l6.5 and l6.6 are parallel to each other, and are therefore lumped by averaging based on their steady state flux distribution (70.5–29.5% respectively). This yields the following lumped reaction:

$$1.623ADP + 2.623H + NADH + 0.5O_2 + 1.623P_i \xrightarrow{l6.8} 1.623ATP + 2.623H_2O + NAD$$

This shows that the P/O ratio in the simplified model became 1.623 mol ATP/mol O. It should be noted that, the remaining ATP-balance parameters i.e. K_x, K_{PenG}, m_{ATP} are kept unchanged in considering black-box balances

After modifying the model, we calculated the substrate Herbert-Pirt equation for the one compartment model.

$$q_S = 6.32q_{\text{Pen}} + 0.291\mu + 2.22q_{\text{tre}} + 1.74\,10^{-3}$$

We conclude that the obtained Herbert Pirt relation for the one compartment model is in agreement with the one for the three compartment model given the standard deviation in the ATP stoichiometry parameters.

Additional Details on the Identifiable Elasticities

List of identifiable elasticities as a function of the original parameters

l1.1	εP1 = ε1	r5.1	εP6 = ε33
	εP2 = ε1		εP7 = ε33
	εP10 = ε2		εpyr = ε31
	εglc_f = ε3		εasp = ε32

(continued)

(continued)

r1.3	εP1 = ε4 + 1/8 ε5	l6.7	εP5 = ε35
	εP2 = ε4 + 1/2 ε5		εP10 = ε34
	εP10 = ε6	l6.9	εP3 = ε37
l1.3	εP1 = 1/4 ε7		εP10 = ε36
	εP2 = ε7	r8.1	εP3 = ε39 + ε40
	εP3 = ε7		εP8 = ε38
	εP10 = −ε7		εP12 = ε40
	εPEP = ε8	l11.1	εpyr = ε41
r1.9	εP1 =1/8 ε9		εile = ε42
	εP2 =1/2 ε9		εval = ε43
	εP10 = ε11	r11.2	εP8 = ε45
	εPEP = ε10		εgln = ε44
r2.1	εP1 = ε12	l11.2	εP6 = ε50
	εP2 = ε12		εP7 = ε50
	εP10 = ε13		εpyr = ε46
	εP11 = ε14		εile = ε47
r2.2	εP10 = ε16		εleu = ε48
	εP11 = ε17		εval = ε49
	ε6Pgluct = ε15	r11.3	εP8 = ε51
r4.1	εP3 = ε20		εP10 = ε53
	εP6 = ε21		εpro = ε52
	εP7 = ε21	l11.3	εP1 =1/3 ε55
	εP10 = ε19		εP2 = ε55
	εpyr = ε18		εPEP = ε54
r4.4	εP3 = ε22 + ε23		εtrp = ε56
	εP4 = ε22	l11.4	εP1 =1/3 ε58
	εP5 = ε22		εP2 = ε58
	εP6 = ε22		εPEP = ε57
r4.7	εP3 = ε25 + ε26		εtyr = ε59
	εP4 = ε25	l11.5	εP1 =1/3 ε61
	εP5 = ε25		εP2 = ε61
	εP7 = ε27		εPEP = ε60
	εP8 = ε24		εphe = ε62
	εP11 = ε24	r11.7	εP6 = ε66
r4.8	εP6 = ε29		εP7 = ε66
	εP7 = ε29 + ε30		εP8 = ε63
	εP10 = ε28		εP11 = ε63
			εlys = ε64
			εaAd = ε65
l11.7	εP9 = ε68	l13.2	εP1 =1/2 ε105
	εP10 = ε69		εP2 = ε105
	εcys = ε67		εP10 = ε108
r11.8	εlys = ε70		εasp = ε106
	εaAd = ε71		εAMP = ε107
l11.8	εP10 = ε75		εGMP = ε109
	εasp = ε72	l13.3	εP1 =1/2 ε110
	εmet = ε73		εP2 = ε110
	εthr = ε74		εasp = ε111
	εhomcys = ε76		εAMP = ε112

(continued)

(continued)

r11.9	εP1 = 1/4 ε77	l13.4	εP1 = 1/2 ε113
	εP2 = ε77		εP2 = ε113
	εP3 = ε77 + ε81		εasp = ε114
	εP8 = ε78 + ε79		εUTP = ε115
	εP9 = ε80	r13.6	εCTP = ε116
	εP10 = −ε77		εUTP = ε117
	εP11 = ε78	r15.1	εP10 = ε118
l11.9	εasp = ε82	l17.1	εP1 = ε119
	εcys = ε83		εP2 = ε119
	εthr = ε84		εt6P = ε120
l11.10	εP1 = 1/2 ε85		εAMP = ε121
	εP2 = ε85	l17.4	εP1 = ε123 + 1/8 ε122
	εP10 = ε87		εP2 = ε123 + 1/2 ε122
	εhis = ε86		εpsacch = ε124
l11.11	εP8 = ε88		εAMP = ε125
	εorn = ε89	r17.7	εt6P = ε126
l11.12	εP10 = ε91	l17.6	εtre = ε127
	εorn = ε90	l18.2	εP10 = ε128
r11.16	εP3 = ε92	l19.1	εP10 = ε134
	εP4 = ε92		εcys = ε129
	εP5 = ε92		εval = ε130
	εasp = ε93		εaAd = ε131
r11.17	εasn = ε94		εPAA = ε132
	εasp = ε95		εpenG = ε133
r11.20	εP3 = ε97	R9.14	εpsacch = ε135
	εP12 = ε97	R9.8	εpenG = ε136
	εmet = ε96	R9.11	εPAA_f = ε137
	εhomcys = ε98	R9.1	εP10 = ε138
r11.21	εpyr = ε99		
	εile = ε100		
	εval = ε101		
r11.22	εpyr = ε102		
	εala = ε103		
r11.34	εP10 = ε104		

Complete List of Estimated Elasticities

#ri.i	Representative transformation	Identifiable elasticities
l1.1	glc → g6P	P1(−0.01), P2(−0.01), P10(0.01), glc$_f$(0.51)
r1.3	f6P → f16P	P1(−0.01), P2(−0.01), P10(−0.15)
r2.2	6PG → Ribu5P	P1(−1.00), 6PG(0.01), P11(2.11)
r4.1	pyr → AcCoA	P1(0.01), P2(0.02), P6(−17.60), P3(−0.01), pyr(0.02)
r4.4	iCitr → aKG	P3(5), P4(0.07), P5(1.16), P6(0.01)
r4.7	aKG → succCoA	P7(11.61), P11(0.01), P3(0.01)

(continued)

(continued)

#ri.i	Representative transformation	Identifiable elasticities
r4.8	succCoA → succ	P8(1.28), P9(0.01), P11(0.01), P3(−0.01)
r5.1	pyr → OAA	P4(−0.01), P5(−0.01), pyr(0.01)
l6.7	$H_2 + O_2 \rightarrow H_2O$ (FAD)	P5(−0.01), P10(−0.01)
l6.9	$H_2 + O_2 \rightarrow H_2O$ (NADH)	P3(0.25), P10(−0.25)
r8.1	CO_2 → gly	P8(0.01), P12(−1.67), P3(4.59)
l11.1	pyr → val	pyr(0.30), val(−11.99), ile(−3.65)
r11.2	glu → gln	P8(4.87), gln(−0.07)
l11.2	pyr → leu	P6(1.21), P7(0.01), ile(6.30), leu(3.24), val(−1.96), pyr(4.09)
r11.3	glu → pro	P8(8.66), pro(−8.95), P10(0.686)
l11.3	PEP → trp	PEP(2.87), trp(−0.01)
l11.4	PEP → tyr	P1(0.40), P2(0.12), tyr(−0.07), PEP(5.17)
l11.5	PEP → phe	P1(0.59), P2(0.26), phe(−1.89), PEP(3.26)
r11.7	glu → aAd	P6(0.01), P7(0.01), P8(2.05), lys(−1.86), aAd(−0.01)
l11.7	ser → cys	P9(1.11), cys(−0.01), P10(2.64)
r11.8	glu → lys	aAd(0.01), lys(−3.63)
l11.8	asp → homcys	asp(43.03), P10(−35.1), homcys(−0.91), met(18.7), thr(7.08)
r11.9	3PG → ser	P2(0.01), P8(2.33), P3(0.68)
l11.9	asp → thr	P11(−0.96), asp(12.57), cys(0.50), thr(−0.01)
l11.10	gln → his	P1(1.81), P2(0.26), P10(6.79)
l11.11	glu → orn	P8(0.01), orn(−0.01),
l11.12	orn → arg	orn(0.01), P10(0.01), gln(0.61), arg(−0.76)
r11.16	glu → asp	P4(0.83), P5(0.83), asp(−0.01), P3(12.77)
r11.17	asp → asn	asp(2.29), asn(−0.01)
r11.20	met → homcys	met(−0.01), homcys(−1.03)
r11.21	pyr → ile	P1(−0.76), P2(0.16), ile, val(−0.76), pyr(−0.84)
r11.22	pyr → ala	P1(5.32), P2(0.01), pyr(8.67), ala(−2.52)
r11.34	qExPept	P10(1)
l13.2	→ GMP	P1(2.43), P2(3.71), P4(−0.01), P5(−2.62), P10(−0.01), asp(30.69), GMP(−0.38)
l13.3	→ AMP	P1(0.20), P2(11.41), P4(−0.34), P5(−0.01), asp(53.4), AMP (−20.31)
l13.4	→ UTP	P1(2.54), P2(−5.37), asp(0.461), UTP(−0.74)
r13.6	→ CTP	UTP(1.10), CTP(−3.60)
r15.1	m_{ATP}	P10(1.00)
l17.1	g6P → t6P	P1(31.85), P2(35.38), t6P(−23.03)
l17.4	g1P → psacch	P1(7.05), P2(0.132), psacch(−0.01), AMP(0.53)
r17.7	t6P → tre	t6P(0.134)
l17.6	q_{tre}	tre(0.91)
l18.2	m	P10(1.00)
l19.1	→ PenG	cys(1.00), val(1.00), aAd(1.00), P10(1.00), PenG(−0.50), PAA (1.00)
R9.14	q_{psacch}	psacch(1.00)
R9.8	q_{PenG}	PenG(1.00)
R9.11	q_{PAA}	PAA_{ext}(1.00)
R9.1	H_{ext} →	P10(1.00)

References

1. Canelas A, van Gulik WM, Heijnen JJ (2008) Determination of the cytosolic free NAD/NADH ratio in *Saccharomyces cerevisiae* under steady-state and highly dynamic conditions. Biotechnol Bioeng 100(4):734–743
2. Chassagnole C, Noisommit-Rizzi N, Schmid JW, Mauch K, Reuss M (2002) Dynamic modeling of the central carbon metabolism of *Escherichia coli*. Biotechnol Bioeng 79 (1):53–73
3. Christensen B, Nielsen J (2000) Metabolic network analysis of *Penicillium chrysogenum* using ^{13} C-labeled glucose. Biotechnol Bioeng 68(6):652–659
4. Demin O, Goryanin I (2008) Kinetic modelling in systems biology, Chapman & hall/CRC mathematical & computational biology. Taylor & Francis Ltd, Hoboken
5. Duarte NC, Herrgard MJ, Palsson BO (2004) Reconstruction and validation of S*accharomyces cerevisiaei* ND750, a fully compartmentalized genome-scale metabolic model. Genome Res 14(7):1298–1309
6. Galazzo JL, Bailey J (1990) Fermentation pathway kinetics and metabolic flux control in suspended and immobilized *Saccharomyces cerevisiae*. Enzyme Microb Technol 12 (3):162–172
7. Heijnen JJ (2005) Approximative kinetic formats used in metabolic network modeling. Biotechnol Bioeng 91(5):534–545
8. Heinrich R, Schuster S (1996) The regulation of cellular systems. Springer
9. Hynne F, Danø S, Sørensen P (2001) Full-scale model of glycolysis in *Saccharomyces cerevisiae*. Biophys Chem 94(1–2):121–163
10. Jørgensen H, Nielsen J, Villadsen J, Mollgaard H (2004) Metabolic flux distributions in *Penicillium chrysogenum* during fed-batch cultivations. Biotechnol Bioeng 42(2):117–131
11. Kell DB (2004) Metabolomics and systems biology: making sense of the soup. Curr Opin Microbiol 7(3):296–307
12. Kitano H (2002) Computational systems biology. Nature 420(6912):206–210
13. Kitano H (2002) Systems biology: a brief overview. Science 295(5560):1662–1664
14. Kleijn RJ, Liu F, van Winden WA, van Gulik WM, Ras C, Heijnen JJ (2007) Cytosolic NADPH metabolism in penicillin-G producing and non-producing chemostat cultures of *Penicillium chrysogenum*. Metab Eng 9(1):112–123
15. Kofahl B, Klipp E (2004) Modelling the dynamics of the yeast pheromone pathway. Yeast 21 (10):831–850
16. Liebermeister W, Klipp E (2006) Bringing metabolic networks to life: convenience rate law and thermodynamic constraints. Theor Biol Med Model 3:41
17. Müller WH, van der Krift TP, Krouwer AJ, Wüsten HA, van der Voort LH, Smaal EB, Verkleij AJ (1991) Localization of the pathway of the penicillin biosynthesis in *Penicillium chrysogenum*. EMBO J 10(2):489–495
18. Nasution U, van Gulik WM, Proell A, van Winden WA, Heijnen JJ (2006) Generating short-term kinetic responses of primary metabolism of *Penicillium chrysogenum* through glucose perturbation in the bioscope mini reactor. Metab Eng 8(5):395–405
19. Nikerel İE, van Winden WA, van Gulik WM, Heijnen JJ (2006) A method for estimation of elasticities in metabolic networks using steady state and dynamic metabolomics data and linlog kinetics. BMC Bioinformatics 7(540)
20. Nikerel İE, van Winden WA, Verheijen PJT, Heijnen JJ (2009) Model reduction and a *priori* kinetic parameter identifiability analysis using metabolome time series for metabolic reaction networks with linlog kinetics. Metab Eng 11:20–30
21. Palsson BO, Schellenberger J, Park JO, Conrad TM (2010) BiGG: a Biochemical Genetic and Genomic knowledgebase of large scale metabolic reconstructions. BMC Bioinformatics 11:213
22. Rizzi M, Baltes M, Theobald U, Reuss M (1997) *In vivo* analysis of metabolic dynamics in *Saccharomyces cerevisiae*: II. Mathematical model. Biotechnol Bioeng 55(4):592–608

23. Schomburg I, Chang A, Schomburg D (2002) BRENDA, enzyme data and metabolic information. Nucleic Acids Res 30:47–49
24. Serrano L, Di Ventura B, Lemerle C, Michalodimitrakis K (2006) From *in vivo* to *in silico* biology and back. Nature 443(7111):527–533
25. Teusink B, Passarge J, Reijenga CA, Esgalhado E, van der Weijden CC, Schepper M, Walsh MC, Bakker BM, van Dam K, Westerhoff HV, Snoep JL (2000) Can yeast glycolysis be understood in terms of *in vitro* kinetics of the constituent enzymes? Testing biochemistry. Eur J Biochem 267(17):5313–5329
26. Theobald U, Mailinger W, Reuss M, Rizzi M (1993) *In vivo* analysis of glucose-induced fast changes in yeast adenine nucleotide pool applying a rapid sampling technique. Anal Biochem 214(1):31–37
27. Thykaer J, Christensen B, Nielsen J (2002) Metabolic network analysis of an adipoyl-7-ADCA-producing strain of *Penicillium chrysogenum*: elucidation of adipate degradation. Metab Eng 4(2):151–158
28. van Gulik WM, Antoniewicz MR, de Laat WT, Vinke JL, Heijnen JJ (2001) Energetics of growth and penicillin production in a high-producing strain of *Penicillium chrysogenum*. Biotechnol Bioeng 72(2):185–193
29. van Gulik WM, de Laat WTAM, Vinke JL, Heijnen JJ (2000) Application of metabolic flux analysis for the identification ofmetabolic bottlenecks in the biosynthesis of penicillin-G. Biotechnol Bioeng 68(6):602–618
30. van Winden WA, van Gulik WM, Schipper D, Verheijen PJT, Krabben P, Vinke JL, Heijnen JJ (2003) Metabolic flux and metabolic network analysis of *Penicillium chrysogenum* using 2D [13 C, 1 H] COSY NMR measurements and cumulative bondomer simulation. Biotechnol Bioeng 83(1):75–92
31. Vaseghi S, Baumeister A, Rizzi M, Reuss M (1999) *In vivo* dynamics of the pentose phosphate pathway in *Saccharomyces cerevisiae*. Metab Eng 1:128–140
32. Visser D, Heijnen JJ (2003) Dynamic simulation and metabolic re-design of a branched pathway using linlog kinetics. Metab Eng 5(3):164–176
33. Visser D, Schmid JW, Mauch K, Reuss M, Heijnen JJ (2004) Optimal re-design of primary metabolism in *Escherichia coli* using linlog kinetics. Metab Eng 6(4):378–390
34. Visser D, van Zuylen G, van Dam J, Oudshoorn A, Eman M, Ras C, van Gulik WM, Frank J, van Dedem G, Heijnen JJ (2002) Rapid sampling for analysis of \it in vivo kinetics using the BioScope: A system for continuous-pulse experiments. Biotechnol Bioeng 79:674–681
35. Weckwerth W (2003) Metabolomics in systems biology. Annu Rev Plant Biol 54:669–689
36. Weckwerth W (2010) Metabolomics: an integral technique in systems biology. Bioanalysis 2 (4):829–836
37. Westerhoff HV, Bruggeman FJ (2007) The nature of systems biology. Trends Microbiol 15 (1):45–50

Chapter 9
Bridging Omics Technologies with Synthetic Biology in Yeast Industrial Biotechnology

António Roldão, Il-Kwon Kim, and Jens Nielsen

Contents

Abstract Metabolic engineering, defined as the practice of manipulating cells' genetic and regulatory processes for improving cellular performance, has been recently integrating systems and synthetic biology with other technologies, e.g. molecular biology, physiology, biochemistry, analysis science, bioinformatics and biochemical engineering. The aim is to surmount cells' performance-related limitations, e.g. metabolic limitation of cellular processes, network rigidity and global regulation, in a comprehensive, rational and high-throughput manner. Using this multi-level/integrated approach, the (re)design and (re)construction of microbial systems for the development of novel products with significant impact on current global problems, e.g. depletion of energy resources and global warming, is becoming a reality.

In the last two decades, technological platforms for systems biology, e.g. genomics, transcriptomics, proteomics, metabolomics and fluxomics, and synthetic biology approaches, e.g. synthetic biological parts, devices and systems, have been

A. Roldão • Il.-K. Kim • J. Nielsen (✉)
Department of Chemical and Biological Engineering, Chalmers University of Technology, Kemivägen 10, 412 96 Gothenburg, Sweden
e-mail: nielsenj@chalmers.se

C. Wittmann and S.Y. Lee (eds.), *Systems Metabolic Engineering*,
DOI 10.1007/978-94-007-4534-6_9, © Springer Science+Business Media Dordrecht 2012

implemented and efficiently used as tools for metabolic engineering of high-value bioproducts. One successful story on how metabolic engineering has been used to enhance the synthesis of microbial-based molecules is the production of biofuels. In this review, synthetic and systems biology technologies for yeast metabolic engineering will be described in detail. In addition, two case studies related with biofuel production (ethanol and 1-butanol) in yeast *S. cerevisiae* will be presented.

Keywords Systems biology • Genomics • Metabolomics • Fluxomics • Proteomics • Transcriptomics • Illumina sequencing • Microarrays • RNAseq • Data mining • Genome-scale metabolic models • Synthetic biology • Targeted gene expression • Multi-gene assembly • Synthetic pathway • Promoter libraries • Riboregulators • Riboswitches • Ribozymes • Protein scaffolds • Promoter shuffling • Promoter binding proteins • Antisense RNA • Metabolic engineering • Yeast • *Saccharomyces cerevisiae* • Industrial biotechnology • High-value bioproducts • Renewable resources • Biofuels • Ethanol • Butanol

9.1 Introduction

A living cell can be viewed as a multi-cluster dynamical system in which the different intracellular compartments and compounds are fully inter- and intra-connected for maximizing cell performance. Spurred on by the increase in the "omics" data and the advances in high-throughput experiments and bioinformatics, systems biology has emerged as a powerful tool to assist bioprocess engineers in solving such complex network of interactions. This allows not only reducing the timeline for moving candidate bioproducts into the market by fast generation of flexible cell factories but also the exploitation of a wide range of strategies for process optimization via metabolic engineering, thereby contributing significantly to the oriented design/construction of novel biological products and production systems from many high-producing microorganisms. Indeed, during the last decades many successful applications of metabolic engineering employing systems biology for microbial process development/optimization have been reported. This includes the production of amino acids, e.g. L-valine [258], biopharmaceuticals, e.g. lovastatin [22], antioxidants, e.g. lycopene [5], fine chemicals, e.g. succinic acid [191], and biofuels, e.g. butanol [50, 66]. Other examples are reviewed in [159, 295].

Microorganisms such as yeast have been long used successfully for the production of food and alcoholic beverages. More recently, the potential of yeast has been under exploitation not only for the generation of high-value products with impact on energy savings, e.g. ethanol, and medicine, e.g. vaccines, for example, but also for the implementation of genetic, molecular and cell biology studies for boosting the knowledge on fundamental cellular processes such as the cell cycle and metabolism [242]. Yeasts *per se* are also of interest for biotechnological industries where they can be used for water remediation [268] or as a probiotic [215].

Major yeast cell factories used for the production of industrial biotechnology products are *S. cerevisiae*, *Pichia pastoris* and *Hansenula polymorpha*. Yeast *S. cerevisiae* are commonly used for the production of recombinant proteins as vaccines, e.g. human papillomavirus vaccine (Gardasil®, Merck & Co., USA), commodity chemicals, e.g. L-Lactic acid [381], active pharmaceutical ingredients, e.g. β-amyrin [210], fine chemicals, e.g. patchoulol [6], and biofuels, e.g. butanol [318], for example (other products generated by engineered *S. cerevisiae* are extensively reviewed in [172]). Similar to *S. cerevisiae*, *Pichia pastoris* is able to produce heterologous proteins as vaccines, e.g. norovirus [47] and hepatitis B virus vaccine [365]. Other prominent products expressed in *P. pastoris* are industrial enzymes [1], receptors [18] and bacterial toxins [203]. *Hansenula polymorpha* has been used for the production of vaccines, e.g. hepatitis B virus vaccine (Hepavax-Gene®, Berna Biotech Korea Corp), peptides, e.g. glutathione [338], commodity chemicals, e.g. 1,3-propanediol [145], polyunsaturated fatty acids, e.g. γ-linolenic acid [170] and ethanol [144], for example. Less representative microbial cell factories include *Candida galli* [21], *Arxula adeninivorans* [49], *Pichia methanolica* [380] and *Kluyveromyces marxianus* [302].

The ability to retrofit efficiently yeast to generate natural or synthetic products requires a high-level understanding of the intracellular network of interactions between compartments and compounds, and its dependency on the extracellular macro-environment. In order to achieve such comprehensive knowledge of the cell, extensive libraries of genes, metabolites, profiling of RNA transcripts, enzymes and protein structures have been created at the same pace as more genomes were sequenced and gene products characterized. With the aid of mathematical frameworks, e.g. flux balance analysis and ^{13}C-based metabolic flux analysis, and computational algorithms/toolboxes, e.g. elementary metabolite unit framework and BioMet toolbox, it is possible to integrate the data deposited in these libraries and extract key information/trends to allow a comprehensive knowledge of how intracellular pools, extracellular environment and components of a functional biological system (e.g. genes, enzymes, metabolites and proteins) are regulated. The ultimate goal is to use this "decodified" information and implement knowledge-oriented genetic engineering strategies for retrofitting of microorganisms, resulting in optimized product yields, titers and productivities.

With a greater understanding of the properties and functions of principal cellular elements, one has been trying to create new synthetic biological constructs, e.g. biological circuits, sensors and actuators, and synthetic biological pathways with predictable and reliable properties based on engineering principles (see Fig. 9.1 for details). Although its numerous definitions, synthetic biology can be generally accepted as an emerging field aiming at designing synthetic biological (non-natural) building blocks and materials, parts, devices and systems. It focuses on the re-design of existing biological systems through engineering principles and designing tools, using for that the knowledge extracted from systems biology tools [305]. For an easier and more effective application of synthetic biology, strong efforts have been put on building/implementing the standardization of biological parts, devices and systems, e.g. BioBricks [107] and BioFab [96], and promoting synthetic biology competitions like iGEM [165].

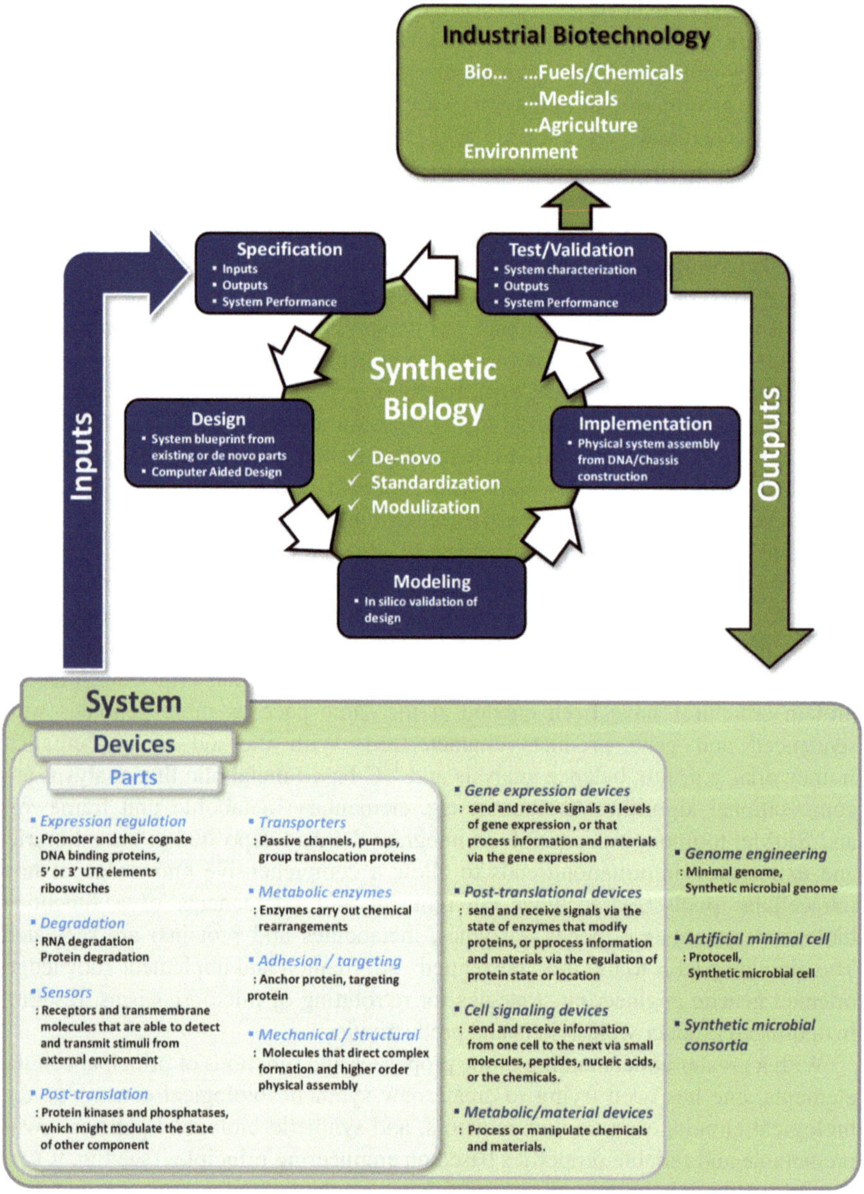

Fig. 9.1 The concept of synthetic biology: a comprehensive description. Synthetic biology aims at putting the engineering principles on biological research. The pipeline for engineering design, consisting of specification, design, modelling, implementation and testing/validation, is introduced into the design and construction of biological systems. Synthetic biology starts from specification of the exciting natural-synthetic biological hierarchy [325]; then, assembles the parts into devices and systems through a step-by-step process according to the pipeline for engineering design [176]. The standardization of biological parts helps to improve the controllability and predictability of designed modules [375]

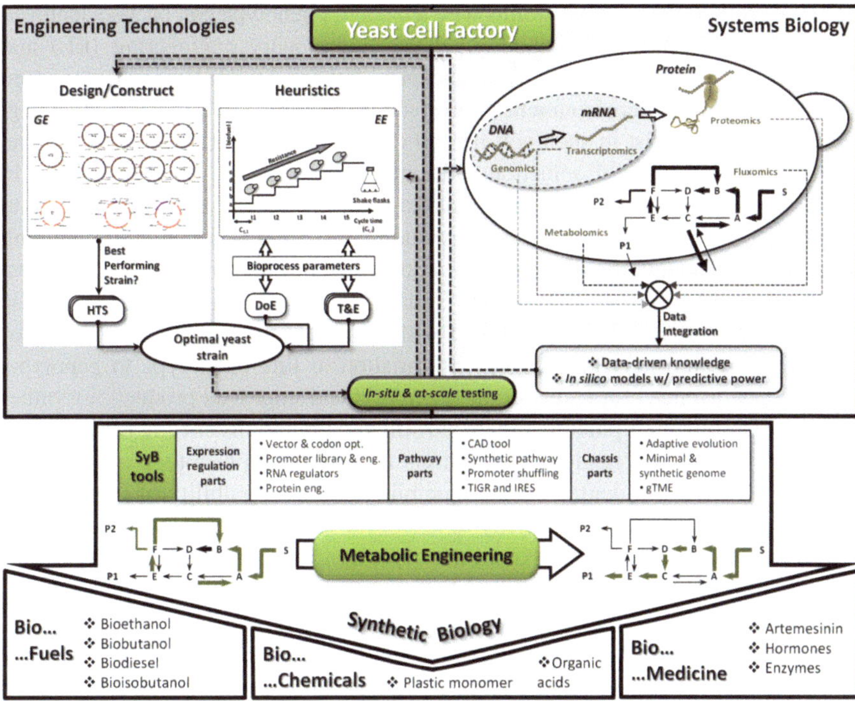

Fig. 9.2 Bridging engineering technologies with systems and synthetic biology towards the development of yeast cell factories for industrial biotechnology applications. Abbreviations: *GE* genetic engineering, *EE* evolutionary engineering, *HTS* high throughput screening, *DoE* design of experiments, *T&E* trial-and-error, *SyB* synthetic biology

The combination of synthetic biology with systems biology for metabolic engineering has been recently under evaluation because of its competence to create novel synthetic systems, empowering engineered microorganisms to gain new abilities which are not naturally inherent to them. It allows fine-tuning of the metabolic and gene regulatory circuits toward achieving a desired metabolic phenotype [231]. In other words and as described in Fig. 9.2, current metabolic engineering has been trying to create/modify microorganisms with potential to generate high-value industrial biotechnology products through synthetic biology (by designing and constructing synthetic parts, devices and systems) and systems biology (analysis for specification and debugging) [33]. Examples of the application of systems and synthetic biology for metabolic engineering include the antimalaria drug artemisinin [290], diagnostic tools for diseases such as HIV and hepatitis viruses [41, 88, 89], biofuels [25, 26], synthetic biosensing systems [214] and, in the long run, to create artificial cells [117]. The creation of artificial cells controlled by chemically synthesized genomes, although in its embryonic stage, may turn out to be the technology of the future in view of the enormous potential of these cells in many research areas, e.g. biomedicine and biomaterials.

The development of yeast strains with a desired phenotype can be also achieved using adaptive evolution methods such as evolutionary engineering (EE) and random mutagenesis (Fig. 9.2). The concept is that some organisms when exposed to a different macro-environment evolve/adapt their intrinsic characteristics, e.g. metabolism, to subsist in such non-natural conditions. This selective pressure can be triggered by external selection inducers, e.g. chemicals or radiation, as in random mutagenesis or by "natural"-based inducers as in EE. Many successful examples of microorganisms' phenotype tuning are reported in literature, ranging from "natural" adaptation of *S. cerevisiae* to high concentrations of toxic products, e.g. acetic acid [362] to "non-natural" adaptation of *S. cerevisiae* to a specific carbon source [199, 358]. If combined with systems and synthetic biology tools, adaptive evolution may provide key information to link phenotype to genotype, opening new avenues for faster strain generation and improved product biosynthesis [369]. Despite their potential, there are still key limitations encumbering the full application of EE and random mutagenesis for microbial phenotype optimization [90]. This include the degree of reproducibility and irreversibility of the outcome of the adaptive process, the capacity to assess the mechanistic changes of the host cell during the adaptive process and the capacity to predict the tolerance of host cells to product, by-product and metabolic intermediates accumulation (some of them toxic) during the adaptive process, for example.

In summary, synthetic and systems biology are two emerging biotechnology disciplines with tremendous potential for driving the development of a wide range of high-value products with application in areas such as life science, materials science, chemistry, or nanotechnology. Undoubtedly, these two multidisciplinary fields will have an important role in the future of industrial biotechnology and subsequently on the economy of the early twenty first century. Indeed, according to one of the recent reports on "Global industry market of Synthetic Biology" published by BCC research, the global market for synthetic biology and systems biology by analogy generated in 2008 around 170 million euros and for the year 2013 the estimate may cross the 1.7 billion euros [43]. The chemicals and energy segment had in 2008 the largest market share (around 35%), a value that is expected to increase up to 67% in 2013. The predictions for the biotechnology and pharmaceuticals segment are not so optimistic; in fact, the market share of this segment is expected to decrease from the 34% in 2008 to approximately 25% in 2013. Despite these encouraging numbers, there are much to be know and issues to handle in systems and synthetic biology for metabolic engineering. This includes:

- *Systems biology – cell-wide fluxomics tools such as flux balance analysis (FBA) and 13C-metabolic flux analysis (MFA)*: dynamic intracellular behaviour is not considered, constrains' number is much smaller than reactions' number in consequent of insufficient analyticals for large-scale analysis of metabolic networks, disregard transport limitations between metabolite pools, unpredictable effect of the addition of other nutrient sources on metabolite labelling and fluxes [105];

- *Synthetic biology – global perspective*: un-definability, unpredictability, complexity, incompatibility and variability of the system as a whole [187] due to insufficient fundamental biological knowledge. Technical, e.g. cloning and assembly of synthetic DNA sequences, biophysical and systematic, e.g. predicting how the new metabolic pathways and the cellular micro- and macro-environment interact, quantitative and computational [331] as well as ethical, social and political issues still impair the full implementation of synthetic biology in industrial biotechnology [135].
- *Metabolic engineering*: cost-effectiveness of novel carbon sources, ideal genes and metabolic routes for product formation, the appropriate microorganism, technology for design and construct of new metabolic pathways, optimization strategy (solution space – iterative and combinatorial process) [169].

In this review, synthetic and systems biology technologies for yeast metabolic engineering will be described in detail. In addition, current and potential yeast *S. cerevisiae* products generated under the mark of synthetic and systems biology and with market in industrial biotechnology will be presented.

9.2 Platforms for Bioprocess Development in Yeast

Robust and accurate genetic and metabolic engineering assisting technologies are critical for the generation of sustainable yeast cell factories capable of producing natural or synthetic products with considerable interest to industrial biotechnology. From the micro, e.g. genes, enzymes and proteins, to the macro-system (cell as a whole), each phase of this complex process of interactions requires precise analysis, e.g. computational-based methods, and correct interpretation.

9.2.1 "Omics"-Based Technologies

The ultimate aim of "omics" technologies, e.g. genomics, transcriptomics, proteomics, metabolomics and fluxomics, is to provide reliable and global quantitative analysis of the system, e.g. genes, metabolic pathways and protein synthesis mechanisms, so that product formation (yields, productivities and titers) can be maximized.

9.2.1.1 From Genes to Proteins

Genomics

Genomics aims at studying the genome of organisms, with emphasis on the determination and comparison of entire DNA sequences, intra-genomic phenomena,

Table 9.1 A list of technologies for yeast genome analysis

Technology	Aim	Reference
Real-time PCR or multiplex PCR	Single or multiple gene analysis	[51, 195]
Polony sequencing or SOLiD sequencing	High-throughput sequencing	[221, 343]
Multiplex ELISA-based OLA MALDI-MS and MALDI-TOF-MS	SNM analysis	[87, 175, 333]
Flow cytometry	Estimate ploidy levels	[337]
Southern Blot	Specific DNA sequence identification	[141s]
Gel electrophoresis and pulse-field gel electrophoresis	DNA separation	[19, 306]
Hierarchical shotgun sequencing and whole-genome shotgun sequencing	Genome sequencing	[4, 121]
Sequencing by hybridization	Mapping nucleotides on DNA strands	[208]
Electron microscopy and nanopore sequencing	Identifying, sequencing and/or detecting DNA molecules	[120, 167]
In silico RFLP	Identify variations in homologous DNA sequences	[281]
Quantification of single-stranded DNA via slot-blot hybridization	Mapping yeast origins of replication	[106]
In situ hybridization	Comparative genomics	[204]
Fusion genes microarray	Detection of fusion transcripts	[313]
Tiling DNA microarrays	Detection of the expression of transcripts or splice forms	[166]
Avadis, FuncExpression, High-Throughput GoMiner and GREAT	Analysis of DNA sequencing data	http://www.geneontology.org/GO.tools.microarray.shtml

Abbreviations: *ELISA* enzyme-linked immunosorbent assay, *MALDI* matrix assisted laser desorption ionization, *TOF* time-of-flight, *MS* mass spectrometry, *PCR* polymerase chain reaction, *RFLP* restriction fragment length polymorphism, *SNM* single nucleotide mutations, *OLA* oligonucleotide ligation assay

e.g. heterosis, and function/interaction of genes within the entire genome's network. Numerous techniques and tools for analysis of gene expression data sets are available in the literature and are described below (other examples are provided in Table 9.1).

A major breakthrough in molecular biology research was the development of polymerase chain reaction (PCR) for analysis of single/specific genes. It allows high-throughput cycle sequencing as most of the PCR machines support well plates configuration, e.g. DNA Engine Tetrad 2 from Bio-Rad Laboratories (CA, USA), generating millions of copies of specific DNA sequences in short time-frames and, thus, bypassing the need of bacteria for DNA amplification. The downside of PCR methods is their low genome coverage and high investment cost (equipment and primers). Genome sequencing methods such as the chain termination method developed by Frederick Sanger [301] and the dye-terminator sequencing developed by Leroy Hood and co-workers [314] along with automated high-throughput DNA sequence analyzers, e.g. MegaBACE™ from GE Healthcare Biosciences (NJ, USA), set the stage for fast and practical DNA sequencing. The "Sanger method" relies on the use of dideoxynucleotide triphosphates (ddNTPs) as DNA chain terminators for the detection in automated sequencing machines. The dye-terminator sequencing method is a technical variation of the "Sanger method" in which the ddNTPs are labelled with a different fluorescent dye (each fluorescing at a different wavelength) thus allowing the sequencing in a single reaction rather than four reactions as in the "Sanger method". The demand for cheaper and high-throughput DNA sequencing has led to the development of novel technologies such as massively parallel signature sequencing (MPSS) [56], pyrosequencing [291] or Illumina's (Solexa) sequencing (http://www.illumina.com). The MPSS is a bead-based method which combines non-gel-based signature sequencing with *in vitro* cloning of DNA templates on micro-bead platforms, and allows the assessment of functional gene-interactions. Pyrosequencing or "sequencing by synthesis" relies on real-time detection of pyrophosphate release on nucleotide incorporation for accurate and consistent analysis of DNA sequences. Illumina sequencing is a reversible terminator-based platform that enables the detection of single, fluorescent labelled nucleotides as they are incorporated into growing DNA strands. It is ideal for genome resequencing, identification of mutations (single nucleotide polymorphisms, indels and copy number variants) and *de novo* genome sequencing. In a recent study, Kvitek and co-workers used Illumina's sequencing by synthesis technology to identify the mutations arising during yeast evolution experiments under glucose-limited conditions [186]. By combining this information with competitive fitness assays, they were able to attribute a specific yeast cell phenotype to single/specific mutations as well as to assess the influence of intergenic interactions on the adaptation process. Massive/parallel DNA sequencing can be also performed using microarrays [81]. The principle of DNA microarrays is simple and encloses four main steps: (1) immobilization of the DNA to sequence on a small, solid support (microarray or chip) containing several thousands of different genes immobilized at fixed locations; (2) hybridization of the DNA with a very large set of short, labelled probes; (3) microarray/chip scanning for assessment of hybridization patterns; and (4) data analysis and computation of target DNA sequence. Other applications of DNA microarrays include mapping SNMs, genotype or re-sequencing of mutant genomes, and comparative genomics [251]. Its cost, expediency, quantitative accuracy and ease of data generation all make DNA microarrays an alternative to high-throughput DNA sequencing methods. Hybrid technologies such as bisulfite sequencing [254] and ChIP-Seq [194] which combine massive/parallel DNA sequencing methods with bisulfite-treated genomic DNA and chromatin immunoprecipitation (ChIP) respectively, are extremely useful to identify interactions between DNA and proteins with minimal disturbance to yeast physiology. One example of the utility of ChIP-Seq is the work performed by Lefrancois and co-workers. Using Illumina's (Solexa) sequencing platform, they were able to map three yeast DNA binding proteins (Ste12, Cse4 and RNA PolII) across the yeast genome in a single experiment. Other hybrids include ChIP-chip, a modified ChIP procedure with DNA microarray analysis that allows the mapping of DNA-protein interactions [285] and of gene-transcriptional regulators association [193] across the entire yeast genome.

In yeast comparative genomics, the aim is to understand how function and evolution processes impact on genomes by linking genome structure and function to a specific yeast strain/species phenotype. Comparative genomics starts with genome-wide sequencing, which can be performed using high-throughput DNA sequencing technologies such as those mentioned above. Then, by applying computational methods such as alignment (pairwise and multiple comparisons), phylogenetic reconstruction or coalescent theory (reviewed in [137]), key features from DNA sequences can be extracted. This includes the identification of SNMs and their genome-wide distribution, modelling and visualization of evolutionary events at a genomic scale (gene duplication and recombination), and detection of selective sweeps (a region of low genetic diversity centred on a selected gene). Two studies demonstrate the usefulness of comparative genomics for yeast metabolic engineering. In the first study, Illumina's (Solexa) sequencing followed by comparative genome analysis was used to identify metabolism-specific SNMs between two *S. cerevisiae* strains (CEN.PK113-7D and S288C) that could be involved in pathway enrichment, e.g. galactose uptake pathway and ergosterol biosynthetic pathway, and thus explain the phenotypic differences observed between the two strains [252]. By linking genotype to phenotype, this work provided key information for engineering yeast strains with improved carbon flux through the mevalonate pathway and therefore generate triterpene high-producer yeast strains [211]. More recently, whole-genome sequencing followed by comparative analysis was complemented with transcriptome and metabolome analysis for comprehensive characterization of yeast strains evolved on galactose as sole carbon source [146]. This system-level approach was able to analyze several different mutants and identify consensus mutations, in this case in the *RAS2* gene, as well as the single mutation, $RAS2^{Tyr112}$, most contributing for the improved galactose uptake in mutated strains.

Transcriptomics

Transcriptomics, or the study of the transcriptome, is a comprehensive way of cataloguing all RNA molecules, e.g. mRNA, tRNA and other non-coding RNA, produced by the genome of a microorganism under specific environmental conditions. RNA profiling can provide valuable information about which genes and/or sequences are being expressed, their function and regulation. It also shed some light on important gene interactions across the genome, potential gene targets for metabolic engineering and functional differences between yeast strains. Several methods/techniques for accurate and robust transcriptome analysis exist and are described below (other examples are provided in Table 9.2).

Table 9.2 Examples of technologies for transcriptome, metabolome and fluxome analysis

Technology	Aim	References
Complementary DNA sequencing	Characterize cellular mRNA profiles	[2]
SAGE, MS-PET, DeepSAGE, 5'RATE and MPSS	Large-scale transcriptome analysis	[56, 123, 240, 243, 344]
Cap analysis of gene expression	High-throughput identification of sequence tags and transcriptional starting points	[310]
Gene identification signature	Transcriptome characterization	[241]
Open sourced software packages such as CLASSIFY, FunCluster, GARBAN, GOdist, dChip, ErmineJ, GoMiner, Matrix2png, TM4 and MAGIC Tool	Statistical/graphical analysis of transcriptome data	http://www.geneontology.org/GO.tools.microarray.shtml
Commercial software packages such as GeneSpring, Array-Pro Analyzer and ArrayStar	Statistical/graphical analysis of transcriptome data	http://www.geneontology.org/GO.tools.microarray.shtml
CE-LIF, OrbiTrap and isotope dilution-MS	Metabolite profiling/metabolomics	[75, 147, 363]
ESI-MS and ESI-Fourier transform MS	Metabolic fingerprinting/metabonomics	[143, 219]
Optimization frameworks such as OptKnock, OptReg, OptStrain, OMNI, OptGene, regulatory on/off minimization, thermodynamics-based MFA	Fluxome data analysis	[20, 52, 60, 139, 266, 267, 311]
Open sourced software packages such as COBRA toolbox and OpenFLUX	Fluxome data analysis	[37, 272]

Abbreviations: *5'-RATE* robust analysis of 5'-transcript ends, *CE-LIF* capillary electrophoresis with laser-induced fluorescence, *ESI* electrospray ionization, *MFA* metabolic flux analysis, *MPSS* massive parallel signature sequencing, *MS* mass spectrometry, *MS-PET* multiplex sequencing of paired-end ditags, *SAGE* serial analysis of gene expression

In the beginning of the transcriptome era, RNA molecules were quantified by calorimetric methods using either the Schneider method or upgraded versions of it [42, 138, 166]. With advances in genetics and molecular biology, these techniques were replaced by highly sensitive gene-by-gene methods such as Northern blot [12] and quantitative real time PCR (qRT-PCR). Although being a landmark in transcriptomics and, still, prime methods in many laboratories worldwide, both technologies are low-throughput. To overcome this limitation and allow a nearly-complete transcriptome analysis, sequencing-based tag-counting methods were developed. The number of possible target transcripts *per* run significantly increased, thus enabling a reduction in overall sequencing cost. A possible downside of these methods is the size and complexity of the data generated, factors that may impair easy and fast transcriptome analysis. Examples of these hybrid technologies include alternative splicing arrays like cross-linking immunoprecipitation (CLIP) [361] and genome tiling arrays [131]. By combining CLIP with high-throughput Illumina sequencing, Wolf and co-workers were able to identify transcriptional regulatory networks with impact on filamentous growth, in this case the binding of Khd1 to repetitive sequences of *FLO11* mRNA. In the other study, ChIP-chip technology was used by Hanlon and co-workers to identify the genome-wide distributions of transcriptional regulators such as Tup1 and Ssn6, thus providing ground for further research on the localization of yeast repressor complexes. Large-scale transcriptome analysis can be also performed by gene expression arrays such as Affymetrix or Agilent [62]. A recently developed technology is RNA-Seq which uses high-throughput sequencing techniques such as Illumina or pyrosequencing for transcriptome profiling. Its major advantage over other methods is the capacity to screen entire transcriptomes in a faster and more accurate, quantitative and high-throughput way (for more details on RNA-Seq see the following review [348]). One example of RNA-Seq application in yeast is the work performed by van Dijk and co-workers on the genome-wide characterization of non-coding RNAs [343]. Using RNA-Seq, they were able to identify novel sensitive unstable transcripts with strong impact on the regulation of gene expression in *S. cerevisiae*.

With such long list of techniques/methods available for transcriptome analysis, it is important to assess the reproducibility and sensitiveness of these technologies so that rational-based decisions about which to choose could be made. In a recent study, our group has compared several platforms for transcriptomics and found that they correlate well with each other [62].

One of main challenges in transcriptomics is data analysis and the subsequent degree of statistical confidence of the results. For transcriptome data mining, several types of statistical analysis/methods are available in literature, including: (1) statistical significant tests [236] – identify genes with different expression patterns in *k*-samples; (2) functional themes based on statistical or gene set enrichment analysis [320] – function of differentially expressed genes in *k*-samples; (3) clustering, classification and dimension reduction tools [299] – revealing global trends; and (4) k-familywise error rate [109] and false-discovery rate [319] – increase statistical power. BioMet Toolbox [74], academic/open-sourced, and Affymetrix Microarray Suite v5.0 (Affymetrix Inc., CA, USA), commercial, are two examples of software packages used for fast and accurate statistical analysis of transcriptome-wide data, and visualization of key features extracted from such analysis (other examples are listed in Table 9.2). The selection of the right method for transcriptome quantification and analysis still remains not trivial, and much "ground" is still to be explored for a global-comprehensive analysis of the cell at the transcriptomic level.

Proteomics

The third level in the global-understanding of cell's biological processes consists in the study of proteins, their intra- and extra-cellular function and state modifications, network of interactions and expression regulation patterns. Although transcriptomics provides valuable information about the expression of individual genes, in-depth characterization of gene products is essential as commonly the transcript and protein levels are weakly correlated [212]. The reason underneath such discrepancy relies on specific regulatory mechanisms related to translational control such as post-translational modifications and translation efficiency [247, 248]. In these two studies, our group demonstrated that protein-mRNA correlation is highly dependent on the gene function and gene codon composition, i.e. genes having similar cellular functions or codon frequencies induce similar transcript-protein levels. Other studies from our group reinforce the importance of a global/integrated view of the transcriptome and proteome for comprehensive understanding of cellular functions and regulation. In the first study, by being able to map the network of interactions around the key protein kinase Snf1 in yeast (a key regulator of carbon metabolism), we not only demonstrated the usefulness of combining quantitative proteomics with transcriptome analysis but also provided a valuable tool for identifying metabolic engineering targets [339]. In the follow-up study, phospho-proteomics in combination with transcriptome analysis allowed Zhang and co-workers to map the interaction of Snf1 with a second key nutrient sensing pathway regulator, TORC1, and subsequently to identify the mechanisms by which these two kinases regulate/control nutrient sensing and metabolic pathways under carbon or nitrogen limited conditions [378]. Proteome analysis alone is equally valuable for metabolic engineering. For example, it can be used to identify limiting nodes in metabolic pathways, e.g. low enzyme activity, thus allowing the re-route of pathways for maximum product formation [284] as well as to categorize proteins and protein complexes from whole genomes, associating them to specific known or unknown biological functions [190].

Classical proteome analysis methods include sodium dodecyl sulphate-polyacrilamide gel electrophoresis (SDS-PAGE) and two-dimensional gel electrophoresis (2DE) [347]. The separation principle by SDS-PAGE is protein size; upon coated with negatively charged SDS molecules and applied to an electric field, proteins migrate according to their molecular weight. In 2DE methods, proteins are separated first according to their charge by isoelectric focusing (first dimension) and second according to their size by SDS-PAGE (second dimension). Despite their conceptual simplicity and potential, these technologies enclose several limitations such as the low reproducibility and visualization/identification/ separation of proteins with low abundance or very high and low molecular weights [122]. Foremost, they are semi-quantitative and not sensitive enough due to their protein detection principle, e.g. Coomasie blue and silver staining [224]. Classical 2DE gels can be combined with protein labelling strategies for simultaneously analysis of up to three protein samples. This technique, called differential in-gel electrophoresis (DIGE), is very useful in comparative proteome analysis since it reduces experimental variation at the same time that increases accuracy as spot detection/matching becomes easier due to the use of fluorescent dyes [188]. There are software (mostly commercial ones) for detecting, matching, quantifying and analysing statistically the spots on a 2D gel; these include amongst others the Dimension from IMGEN Technologies (VA, USA), Progenesis SameSpots from Nonlinear Dynamics Limited (Newcastle, UK) and ProXPRESS 2D from PerkinElmer (MA, USA).

Low-throughput (LTP) and semi-quantitative techniques for proteome analysis include Western blot (combination of SDS-PAGE technology with immunostaining using antibodies), enzyme-linked immunosorbent assay (antibody-based technique), BCA protein assay (UV absorption spectroscopy), flow cytometry (fusion of cellular proteins with fluorescent dyes) or transmission electron microscopy [345]. Static or dynamic light scattering and analytical size-exclusion chromatography [173], SDS-capillary gel electrophoresis [218] and 2D-fluorometry [329] are examples of more sensitive/quantitative methods for protein characterization but still LTP.

Recent methods based on mass spectrometry (MS) techniques have allowed high-throughput proteome analysis. This include electrospray ionization-MS (ESI-MS) [124], matrix-assisted laser-desorption/ ionization-MS (MALDI-MS) [173], MALDI - time-of-flight (MALDI-TOF) [44], liquid chromatography-MS (LC-MS) [152] and gas chromatography-MS (GC-MS) [69] for example. By using stable isotope labelling of peptides, e.g. SILAC, ICAT, iTRAQ and IMAC [108, 129, 249, 292], label-free/computational-based methods, e.g. PAI and T3PQ [126, 157] or a combination of both approaches [213], a more detailed and quantitative analysis of proteins from highly complex samples is achieved [77]. By combining MS with technologies such as zwitterionic chromatography-hydrophilic interaction liquid chromatography (ZIC-HILIC) [17], tandem mass spectrometry [290], nanoflow high-performance liquid chromatography (HPLC) [109] and field-flow fractionation [286], it is possible to analyse glycosylated proteins, obtain amino acid sequence information and characterize most phosphoproteins from a whole-cell lysate in a single experiment.

Alternatives to MS-based methods include nuclear magnetic resonance [217], highly precise but extremely costly technique due to isotopes labelling and low sensitivity (μmole), and the shotgun proteomics which combines a 2D chromatography based separation system (HPLC) with multidimensional protein identification technology (MudPIT) and tandem mass spectrometry [118].

9.2.1.2 Metabolomics and Fluxomics

Metabolomics

Metabolomics or the quantitative study of the profile of cellular metabolites under specific culture conditions constitutes the fourth key element in a systems biology approach. With the information of metabolome analysis, a detailed map of the underlying molecular network can be designed and further used to evaluate network's response under genetic/environmental perturbations (functional genomics and toxicity assessment), to identify key regulatory/limiting nodes (pathway discovery) and to characterize cellular functions such as gene, transcript and protein levels (metabolic characterization). In addition, through the quantification of all intra- and extra-cellular metabolites, e.g. carbohydrates, fatty acids and amino acids, the activity of several different enzymes can be measured which can therefore assist the design of rational and efficient metabolic engineering strategies for process optimization.

The following examples demonstrate the potential of metabolomics for yeast research. In the first study, Kümmel and co-workers aimed at understanding the mechanisms responsible for the different phenotypes exhibited by two hexokinase 2 mutant yeast strains (CEN.PK and S288C) grown under glucose repressing conditions [184]. By combining metabolomics with proteome and physiological analysis, they were able to identify the metabolic pathways contributing most for the observed differences (pathways involved in glucose signalling and regulation) and thus confirm the key role of genetic background in cell's phenotype. In a second study, Corte and co-workers evaluated the toxicity of four chemicals (ethanol, sodium hypochlorite, sodium chloride and sulfur dioxide) in *S. cerevisiae* and, subsequently, in their metabolism [72]. Using Fourier transform infrared spectroscopy, they were able to estimate the concentrations at which those chemicals became lethal to yeast cells and demonstrate that the toxicity/mortality induced by them does not correlate with the corresponding metabolic fingerprint. Nonetheless, further studies are still required to shed some light into the underlying mechanisms conferring resistance of yeast cells towards the chemicals tested. Metabolomics can also be used to characterize cellular functions. Using metabolome and 13C-flux analysis, Christen and Sauer showed for several yeast species that perturbations in cell's extracellular environment are not necessarily reflected at the intracellular level, i.e. cell metabolism [68]. This corroborates well with authors second conclusion that intracellular metabolic profiling of yeasts is species dependent. In another example, Klimacek and co-workers used quantitative metabolomics (LC/MS-MS) in combination with thermodynamic analysis to find correlations between the poor performance of three yeast strains in utilizing xylose as main carbon source (two mutants and one wild-type) and the activity of specific enzymes along the PPP and/or glycolysis [177]. Net fluxes identified PPP- and glycolysis-related enzymes, e.g. xylose reductase and phosphofructokinase, as key players in the regulation of xylose utilization, excluding from this lot the coenzyme and ATP pools.

A list of analytical technologies for metabolomics is provided below (other examples are provided in Table 9.2). For metabolite separation, chromatographic methods are commonly used, e.g. GC, LC, CE and HPLC. For metabolite detection, several techniques can be used and the selection of the correct one depends on parameters such as the number of metabolites to analyse, the time for their analysis and the demanded accuracy. For target analysis (few metabolites), spectrophotometric, chromatographic or spectroscopy based methods, e.g. NMR, provide good selectivity and sensitivity. For metabolite profiling (some selected metabolites), methods combining two or more techniques are preferred. Examples of these include GC-MS, CE-MS, LC-MS and LC-NMR, all biphasic techniques enclosing an initial separation step prior to detection for easier and more accurate metabolite analysis. Although being quantitative and selective, some of these techniques may enclose resolution and sensitiveness problems as well as limitations at the level of the molecules they are able to analyse, e.g. MS cannot be used to analyze metabolites that cannot ionize. For very high number of metabolites (metabolomics and metabolic fingerprinting), MS combined with chromatographic methods and/or different ionization methods, e.g. LC-ion trap MS and GC-TOF, NMR, Fourier transform infrared spectroscopy, IR and Raman spectroscopy [91] are examples of techniques that provide automated, rapid, high-throughput global metabolite analysis. For metabolome data analysis, targeted profiling or chemometrics approaches are commonly used. Targeted profiling is a two phase process where, first, the spectra of individual metabolites are chemically modelled and stored in a database, and then compared to the unknown metabolites mixture spectrum for identification and quantification of metabolites of interest. In chemometric approaches, data-analytical disciplines such as multivariate statistics are commonly used to identify and quantify all metabolites simultaneously and from complex mixtures. In addition, pattern recognition methods such as cluster analysis, principal-component analysis or partial least-squares allow the categorization of identified metabolites into different groups according to their influence on cell's physiology, thus providing key information for the comprehensive understanding of cell's metabolic regulation.

A central topic in metabolomics and, by analogy, in other "omics"-technologies is sample preparation and the potential variability therein resulting. Important rules for capturing the accurate picture of a microbial system at the time of sample collection include: (1) the collection time as well as the time-frame between sample collection and freezing must be minimized at all times; (2) solution must be homogenised; and (3) methods for sample quenching, extraction and preservation must be optimized. While the first two points are mainly user-related, the third depends on sample type and metabolites to analyse, and most importantly on available quenching/extraction methods. The importance of selecting the best extraction method has been recently evaluated in our group [346]. Using GC-MS based comparative metabolomics, we tested six different extraction methods and discovered that metabolites recovery efficiency varies quite significantly among the methods evaluated.

Fluxomics

Although metabolomics provides a comprehensive knowledge of microorganism's intra- and extra-cellular metabolites, it fails in understanding the control and functional regulation of metabolic networks [282]. The measurement of metabolite turnover rates, i.e. flux analysis or fluxomics is important to fill this gap, providing complementary information for the global and complete characterization of metabolic networks and subsequently the phenotype of the microorganism. With such technology, one can predict the impact of genetic modifications on the global physiological behaviour of the microorganism and therefore design rationally novel microbial systems with improved metabolic capacity [371]. In two examples from our group, we were able to demonstrate the usefulness of cell-wide fluxome analysis for yeast metabolic engineering by identifying novel target genes with strong positive impact on the production of succinate and sesquiterpenes [20, 250].

In the first study, a series of deletions and over-expressions of specific genes identified by fluxome analysis as key players in the succinic acid pathway enabled us to enhance succinate production in *S. cerevisiae*. In the second study, the effect of deleting NADPH-dependent glutamate dehydrogenase encoded by GDH1 on the flux distributions of *S. cerevisiae*'s genome-scale metabolic model was assessed. The deletion of GDH1 increased cubebol production, a natural sesquiterpene alcohol, in almost 85%.

Reliable and accurate quantification of intracellular fluxes cannot be accessed directly from the measurements of inputs and outputs on the basis of an assumed network stoichiometry [71]. It requires the introduction of labelled precursors into the network followed by the quantification of label redistribution into other metabolites. Two labelling-based approaches can be used to assist the estimation of multiple fluxes through metabolic networks: static labelling and dynamic labelling [106, 283]. Static labelling is commonly used in steady-state flux models whereas dynamic labelling is used in kinetic models. Steady-state flux models include flux balance analysis (FBA) and 13C-metabolic flux analysis (13C-MFA), amongst others. While the first targets the identification of the best metabolic pathways for process performance, the second aims at understanding the conceptual operation of a metabolic network. By integrating FBA with metabolic pathway analysis, potential pathways for cellular metabolism and product synthesis optimization can be quantitatively assessed [304]. At large-scale (genome-scale metabolic models), the analysis of high-throughput fluxome data is performed using toolboxes such as the BioMet toolbox [74] and the RAVEN Toolbox (Reconstruction, Analysis and Visualization of mEtabolic Networks) (manuscript under review).

Cell-wide quantitative kinetic models such as dynamic FBA (dFBA) and 13C-MFA are commonly used to evaluate the impact of extracellular perturbations on the function and regulation of cell's metabolic network [15, 114]. Two other alternatives are metabolic control analysis (MCA) and the cybernetics approach. MCA is able to predict the effect of specific enzymes on a target metabolic pathway [142] whereas the cybernetics approach can evaluate the impact of multi-enzyme systems (kinetics and synthesis) on a simplified pathway network [234]. To solve dynamic flux models at bearable computational times without losing the labelling information, static optimization approach for dFBA [211] and the elementary metabolite unit framework for the 13C-dMFA [370] can be used. Other methods/algorithms for fluxome data analysis are extensively reviewed in [197] and [257].

9.2.2 Synthetic Biology in Yeast

Yeasts are eukaryotic micro-organisms which have occupied an important position in Human development. For example, *S. cerevisiae* has been used in baking and in fermenting alcoholic beverages for thousands of years, *S. pombe* as well as *S. cerevisiae* are important model organism in modern cell biology research, *C. albicans* cause human infection, and *S. cerevisiae*, *P. pastoris* and *H. polymorpha* have recently been used for the production of bioethanol, organic acids and enzymes [303]. Most efforts in synthetic biology have been on the bacterium *E. coli*, but owing to its importance, there are increasing interest in synthetic biology of yeast in particular in the context of metabolic engineering [183]. In the next sections, synthetic biology methods and its application on metabolic engineering of yeast will be described.

9.2.2.1 Synthetic Biology Tools for Controlling Target Gene Expression

The successfulness of metabolic engineering depends on the controllability of gene expression related with the metabolic pathways of a specific product. Various biological expression regulation parts such as promoters and their cognate binding proteins, open reading frames (ORF), 5′- and 3′-non-translation regions, and riboregulators are needed in both prokaryotes and eukaryotes to drive/regulate the expression of a target protein, balancing single or multiple gene expression with optimal metabolic fluxes for maximum bioprocess performance [151]. A number of different approaches can be used to develop these expression regulation parts and devices, including: (i) native and synthetic promoter library; (ii) modification of promoter's recognition binding protein; (iii) alteration of 5′- or 3′-non-translation region including ribosomal binding site; and (iv) development of synthetic riboregulators.

Native and Synthetic Promoter Libraries

A simple approach for regulating gene expression is to develop an optimal promoter. To such desideratum, native yeast promoter libraries have been created by selecting various promoters from several sets of data, e.g. published articles, transcriptomics and proteomics, and analysing the activity of reporters like cat assay, gal assay, luciferase assay and fluorescence assay [40, 262, 334]. An alternative approach is to construct synthetic promoter libraries using mutation and selection techniques based on the native promoter. For example, constructing the synthetic promoter library by site-directed or random mutation using error-prone PCR of the promoter consensus region in prokaryotes [8, 161] or eukaryotes [133, 162, 238, 239], building the synthetic promoter library having incorporated various synthetic transcription factor binding sites in *E. coli* [73] and in *S. cerevisiae* [228], and establishing the synthetic promoter library designed and constructed based on the mathematical analysis of the known promoters in prokaryotes [268]. Although numerous promoters have been developed via these approaches, they do not allow for rational *de novo* promoter design. To overcome this disadvantage, Ellis *et al.* developed a new approach to generate predictable synthetic promoter library in *S. cerevisiae* through a series of processes including construction of synthetic regulatory promoter library, generating feed-forward loop networks with different predicted input-output characteristics based on the library, and producing a predictable timer acting synthetic gene network. Also, they successfully applied this synthetic gene network timer to control the timing of yeast sedimentation [92].

Promoter Binding Proteins

A second promoter parts-based approach consists in creating a synthetic transcriptional factor (TF) library through random mutation of the natural TFs. For that, two kinds of successful technologies can be implemented: (i) global transcription machinery engineering (gTME); and (ii) zinc-finger library engineering (ZFLE).

Global transcription machinery engineering aims at reprogramming gene transcription for eliciting cellular phenotypes, significantly impacting on the whole transcriptional control process and improving the transcription and expression of target genes [9, 10]. The gTME was described for engineering TFIID, a transcription factor that recognizes and binds to the promoter and consists of TATA-binding protein and 14 other associated factors, in *S. cerevisiae* as a way to reprogram the global transcription level [9].

Synthetic zinc finger proteins (ZFPs) that are fused to appropriate transcriptional effector domains (activators or repressors) have been shown to function as artificial transcription factors for the regulation of certain genes in various organisms [39, 308].

The Cys2His2 zinc finger protein is one of the most common DNA-binding motifs in eukaryote, being an ideal scaffold for designing specific DNA sequence recognition proteins because of its DNA recognition ability via simple mode. Combining a zinc finger motifs with an effector domains, this synthetic module can act as chimeric DNA modifying enzymes and transcription factors engineering technologies [252]. For example, ZFLE library has been successfully used to screen for components that induce diverse phenotypes of interest such as stress resistance and cellular differentiation in *S. cerevisiae* and mammalian cells respectively [259], over-expression of recombinant proteins in *S. cerevisiae* and mammalian cells [48, 261], resistance to the anti-cancer agent Taxol [192] and shifting *S. cerevisiae*'s galactose dependent *GAL4* promoter to an independent type [256].

5′- and 3′-Untranslated Region (UTR)

Transcriptional regulation has been controlled by a DNA binding protein containing transcription regulator proteins and RNA polymerase complex, RNA regulators and its binding DNA sequence, i.e. *cis*-acting elements such as promoter, enhancer, silencer and locus-control element [223]. In eukaryotes, upon mRNA editing, the mature mRNA has a tripartite structure made up of a 5′-UTR, a open reading frame and a 3′-UTR. It is known that UTRs are responsible for important functions like mRNA transport to cytosol, translational regulation [341], subcellular localization [158] and mRNA stability [34].

Regulation by UTRs is facilitated by binding of specific RNA-binding proteins to nucleotide motifs located in 5′-UTRs and 3′-UTRs or interactions between sequence elements located in the UTRs and specific complementary regulatory non-coding RNAs containing antisense RNA, micro RNA (miRNA), short interfering RNAs (siRNAs), riboregulators, riboswitches, and ribozyme. Undeniably, 5′- and 3′-UTRs as well as riboregulators play a key role as synthetic expression regulating parts and devices.

The integration of synthetic tetracycline binding aptamer into 5′-UTRs leads to ORF's down-regulation in *S. cerevisiae* [182]. In addition, it is known that hairpin formed by integration of synthetic cassette can protect mRNA from endonucleolytic cleavage upon transcription [63]. Lastly, depending on the hairpin structure, hairpin-containing mRNA may exhibit half-lives 3–10 times higher than those mRNAs without no hairpin, thereby resulting in increased mRNA and protein levels [64]. Using this knowledge, Smolke *et al.* demonstrated that, depending on inducer concentration, a 4-fold increase in reporter expression can be achieved by integrating RNA cleavage sites and synthetic hairpin cassettes with 5′-UTRs, intergenic region between two reporter ORFs, and 3′-UTRs [315, 316].

Recently, Smolke group reported a novel class of RNA-based control modules, based on the mechanism concerning RNT1 encoding RNase III protein Rnt1p and 3′-UTR including Rnt1p recognition and cleavage sequence, in *S. cerevisiae* ([28, 29]). To construct the RNA based synthetic control module, Rnt1p responding synthetic control modules library was generated through fluorescent based *in vivo* screening and RNA structural engineering methods based on the prediction of multiple secondary structure conformations with similar free energies and interactions with flanking sequences. From the library, a set of hairpins covering a large gene regulatory range from 8% to 85% were selected. Finally, these selected synthetic control modules were applied to the ergosterol biosynthesis pathway to confirm the systematic titration of flux. This approach could be provided as a useful gene expression regulation tool for yeast metabolic engineering because of its easy control and predictable tuning properties. Further, this group extended this synthetic module to a new class of RNA sensing-actuation devices by combining it with ligand respond riboswitch [30]. Using this new device, it was proved that the Rnt1p switch platform could be used for rapid optimization and tailoring of complex genetic networks in yeast.

Riboregulators

Many types of riboregulators being able to interact with other RNA molecules, DNA, proteins and small molecules and including antisense RNA, miRNA, siRNAs, riboregulators, riboswitches and ribozyme are present in both prokaryotic and eukaryotic cells [155]. These riboregulators play an important role in transcription regulation [61], mRNA processing [104], rRNA processing [270], translation regulation [76], and RNA degradation [82]. Due to their versatile characteristics, riboregulators have been used as synthetic parts and devices for conditional gene expression systems [35, 297].

Antisense RNA is a single-stranded RNA and regulates translation of a protein encoded by complementary mRNA. The protein translation regulation by antisense can be performed by physically blocking the translation machinery via hybridization [352].

In yeast, it has been revealed via transcriptome study under a variety of growth conditions that several regulatory non-coding RNAs exist in yeasts and are regulated by a certain condition. For example, antisense RNA such as *spo*6 antisense transcript in *S. pombe* and *RME*2 in *S. cerevisiae* down-regulate meiotic protein *spo*6^{+} and meiotic regulator *IME*4 respectively during vegetative growth by pairing with its sense mRNA located coding region and blocking translation elongation rather than initiation [65].

Using antisense RNA regulation mechanisms synthetic gene regulation modules have been successfully developed. For example, allosteric riboregulators (fully tunable and modular antiswitches) were designed by combining antisense RNA and theophylline binding riboswitch [36]. In the absence of theophylline, the antisense domain is covered up in an 'antisense stem'. On the other hand, in the presence of theophylline, antisense domain can be exposed by conformational change via ligand binding at the aptamer domain and bind with a target mRNA to affect translation. This research gave an opportunity how to design a variety of antisense based synthetic gene expression regulation module. i.e. plenty of synthetic riboregulators can be generated by combining antisense RNA complement to a target gene and a variety of riboswitch sense its ligand and used for metabolic engineering.

Riboswitches are a type of RNA regulators capable of controlling a target gene in a ligand-dependent manner and without the need for proteins [207]. Riboswitches have an aptamer domain which recognizes the ligand. Also, they can be used for an expression platform which couples ligand binding to changes in gene expression originated from translation initiation, transcriptional termination or cleavage of mRNA [208, 355–357]. A variety of methods have been used to generate synthetic riboswitches that bind to non-endogenous small ligands and regulate a counterpart gene expression [208]. One method consists in the incorporation of aptamer sequences into 5′-UTRs to reduce the translation of downstream coding regions in eukaryotes, thereby increasing the strength of RNA secondary structure via increased binding potential of the ligand [125, 134, 325, 353]. Other methods include the enhancement of ligand-dependent transcription in yeast using synthetic aptamer-based transcriptional control system [61], *in vitro* control of ligand-dependent RNA splicing using synthetic aptamer [171] and regulation of eukaryotic translation in trans using antiswitches generated by combining riboswitches and antisense RNA [36, 154–156]. Owing to their capability to respond to non-endogenous small molecules, riboswitches can be powerful synthetic gene regulation module controlled by ligand concentration to reprogramming cellular behavior [336].

Ribozymes are catalytic RNA which catalyse peptidyl transfer in ribosome and RNA splicing process consisting of phosphodiester cleavage and ligation [376].

The representative example is GlcN6P metabolite-responsive *glmS* ribozyme, which regulates bacterial *glmS* gene encoding L-glutamine:D-fructose-6-phosphate aminotransferase through binding of glucosamine-6-phosphate (GlcN6P) concentration dependent activated *glmS* ribozyme to 5′-UTR of *glmS* transcript and down-regulation of *GlmS* protein via degrading its mRNA by the ribozyme [359]. Based on this mechanism, the ribozyme can act as a metabolite-responsive synthetic regulator for metabolic engineering.

For example, synthetic gene regulation module using ribozyme was constructed in *S. cerevisiae* [23]. The authors constructed synthetic gene regulation module via incorporating cis-acting ribozyme into 3′-UTR of reporter *CAT* gene and found that reporter gene expression was dramatically down-regulated up to 90%. Additionally, insertion of antisense sequence into 3′-UTR affected expression of the CAT gene, reducing it.

Yen et al. developed an analogous RNA-based gene regulation system for mammalian cells via integration of self-cleaving RNA motifs, which its ribozyme activity could be inhibited by analogous RNA, into 5′-UTR of a gene or vector [368]. In detail, a highly efficient ribozyme sequence was obtained, showing constitutive activity in mammalian cells, following screening and optimizing natural and engineered ribozymes. As next step, analogues RNA including 2-O-methyl derived RNA, phosphorothiolate-2-O-methyl derived RNA and morpholino oligonucleotide were screened to use as a regulator ligand. Then, the ribozyme and analogous RNA based controller was tested in animal cell lines and live animals. Results show that the optimized ribozyme, integrated in upstream of the translational start of reporter gene, cleaves reporter mRNA before translation can occur and stimulated the mRNA degradation and loss of gene expression. Adding toyocamycin (ribozyme inhibitor) leaded to terminate mRNA cleavage and to up-regulation of reporter expression up to 250-fold. This example shows that synthetic regulation module consisting of ribozyme and ligand binding parts can be used as predictable and controllable expression regulation synthetic biology tool for metabolic engineering.

Small interfering RNAs (siRNA) are RNA regulators consisting of 21- to 25-nucleotide double-stranded RNAs found in plants, animals, fungi and ciliates, and are responsible for posttranscriptional silencing process called as RNA interference (RNAi) [225]. RNAi was initiated via siRNA formation by cleavage of long length double strand RNA by RNA endonuclease Dicer. Then, siRNA become a part of RNA-induced silencing complex (RISC) following formation of RISC loading complex (RLC). RISC binding siRNA cleaved the target mRNA hybridized with siRNA [113]. Owing to the characteristics of siRNA as a repressor to a specific target gene, siRNA is nowadays used for molecular regulation tools as a genetic knock-down regulator to develop anticancer agent [112].

Recently, it was reported that, although model budding yeast *S. cerevisiae* does not have siRNA, some other budding yeast like *Saccharomyces castellii* and *Candida albicans* owned RNAi process including siRNA, RNA endonuclease Dicer homologue and Argonaute homologue, and introducing these RNAi components to *S. cerevisiae* stimulated to knockdown endogeneous retrotransposon [83].

This means that siRNA based synthetic regulator can be applied to budding yeast. Actually, some research concerning siRNA based synthetic circuit has been applied to human kidney cell [287]. In this study, general Boolean logic for making decision dependent on endogenous molecular inputs encoded by the existence of siRNA as a mediator was constructed and proved that arbitrary Boolean decision-making was successfully achieved through sensory up- and down-regulation of input signal by siRNA mediator.

Protein Engineering

The improvement of a specific biochemical reaction (metabolic engineering) can be achieved by modifying the quality of a target protein through protein engineering, i.e. engineer and optimize enzyme kinetics on substrate and product, non-natural substrates specificity, stability, and enantioselectivity [206]. In other word, engineered proteins might be used as a synthetic parts and devices for expression regulation via quality control and synthetic pathway.

Traditionally, protein engineering relies on two different approaches [326]. The first approach is the rational design of proteins by site-directed mutagenesis, using the knowledge about structure and kinetic property of the target enzyme and computational design [318]. Examples in yeast include the improvement of lactate production in *S. cerevisiae* via site-directed mutagenesis of *L. plantarum LDH* gene [53] and development of glucoamylase in *Aspergillus awamori* via molecular dynamics simulations and site-directed mutagenesis for improving thermostability [203]. This approach can be feasible by using inexpensive and easy accessible site-directed mutagenesis techniques. However, its major weakness is the need for detailed protein structural information sometimes unavailable or available but difficult to predict the effects of a mutation [38]. The second approach is the directed evolution of proteins through random mutagenesis, using error-PCR or DNA shuffling [294]. For example, direct evolution could generate chimeric methionine adenosyltransferase encoded by *MAT* gene having enhanced activity by using DNA shuffling of *MAT* genes from *Escherichia coli*, *Saccharomyces cerevisiae*, and *Streptomyces spectabilis* for improved SAM production in recombinant *Pichia pastoris* [148] and lipase B proteins having enhanced capability to hydrolysing diethyl 3-(3′,4′-dichlorophenyl) glutarate (*DDG*) and generated through DNA shuffling of three homologous lipases from *Candida antarctica*, *Hyphozyma*, and *Crytococcus tsukubaensis* [321]. This approach can be performed without prior protein structural knowledge and prediction of what effect a given mutation will have and can apply on non-natural reactive synthetic enzyme. But, to get a desired mutant, cost and time consumable high-throughput approach, which is inapplicable to all proteins, should be required [13]. To overcome the drawbacks of these two approaches and to employ advantages, integration method with these two approaches would be used for protein engineering [13].

9.2.2.2 Synthetic Biology Tools for Assembling and Optimizing Multi-genes

In metabolic engineering, strain improvement can be achieved by controlling the quantity of the target enzyme (overexpressing rate-limiting genes and attenuating or deleting genes with negative impact on cell metabolism), and/or the quality of the target enzyme, eliminating by-products with significant influence on product formation via protein engineering. Although the successful rate on up- or down-regulating a set of target genes belonging to a specific multigene-pathway is high, failed metabolic engineering episodes are also common mostly because of unbalanced metabolic fluxes among multiple genes. In other words, it is important to fine-tune the expression level of multiple genes. Some useful strategies for fine-tuning the expression level of multiple genes are available, including multiple genes promoter shuffling, tunable intergenic regions (TIGRs) library, intergenic ribosomal entry sequence (IRES) of yeast and protein scaffolds.

Multiple Genes Promoter Shuffling (MGPS)

The ultimate goal of metabolic engineering is to improve strain performance by optimizing the metabolite flux distribution and thereby product formation [205]. In metabolic pathways consisting of multi-enzymes, it should be needed to optimize the expression of multiple genes at once in order to achieve optimal metabolite flux to a desired product.

To optimize the expression levels of multiple genes simultaneously in *S. cerevisiae*, a MGPS strategy has been proposed [205]. It consists of three steps: (1) promoters with different strengths or profiles are selected based on the transcriptome data from global expression analyses and validated using reporter genes under various conditions; (2) selected promoters are fused with a specific number of target genes (6 in this manuscript); and (3) each fused promoter-gene is concatenate with different combination and tested. Through MGPS, these authors found the optimal combination for ethanol production to be *GND*2-*TAL*1-*HXK*2-*TKL*1-*HXK*2-*PYK*1. In spite of demand of variety promoters, this approach could give less-intensive labor, greater flexibility and fast optimization of the metabolic pathway consisting of several genes.

Tunable Intergenic Region

An alternative approach for optimal expression of multiple genes in prokaryote is the TIGRs [265]. The sequences containing control elements, e.g. mRNA secondary structures, RNase cleavage sites and ribosome binding site sequestering sequences, is inserted into the intergenic regions between ORFs of bacterial operons, thus directing the processing and segmental stability of a transcript containing multiple coding regions [317, 318]. Depending on the nature of the intergenic region between the coding regions, this type of directed mRNA processing induces differential production of the proteins encoded in the operon. Appling TIGRs on two reporter genes and one operon encoding the heterologous mevalonate biosynthetic pathway, it was shown that TIGRs can vary over 100-fold ranges of relative repression and 7-fold increase in mevalonate production, respectively. Although this technology could be applied just in prokaryotes because of limitation of applying just to operon based multiple gene expression, this approach could have potential to optimize multiple gene expression in yeast through combination with internal ribosomal entry sequence (IRES) approach.

Internal Ribosomal Entry Sequence

The translation of most yeast mRNAs is cap-dependent and encloses the following steps: (1) recruitment of the small ribosomal subunits by the 5′-terminal cap structure (7mGpppN) to initiate its translation in association with eIF4A, 4G, and 4E complex; (2) recognition of the first start codon following scanning the 5′-UTR by 40S subunit; and (3) binding 60S ribosomal subunit with 40S subunit at the starting site followed by starting translation [140].

Several alternative cap-independent modes of translation have recently been described [296]. Among them, IRES dependent translation was found in the 5′-UTR of several cellular mRNA. In a recent study, it was found that integration IRES, existed in the cricket paralysis virus, in front of *URA*3 open reading frame located in the second position of dicistronic mRNAs efficiently guided the translation of a second *URA3* cistron in *S. cerevisiae*, thereby conferring uracil independent growth [332]. Interestingly, while in wild type yeasts IRES did not function properly, IRES functioned properly under the constitutive expression of the eIF2 kinase *GCN2* or under presence of two initiator tRNAmet genes disrupted. Both of these conditions have been shown to lower the amounts of ternary eIF2-GTP initiator tRNA*met* complexes. Furthermore, in the presence of edeine, interfering agent for recognition of start codon by ribosome, tRNA*met*-independent initiation was able to occur in translation-competent extracts prepared from *S. cerevisiae*. These findings allow the use of IRES as a synthetic expression control devices for multiple gene expression in yeast. Recently, it has been proved that the IRES have potential for synthetic expression of regulation parts, enabling the expression of multiple genes simultaneously in yeast [307, 366].

Protein Scaffolds and Fusion Proteins

Recently, an attractive approach for increasing metabolic flux of a desired product while simultaneously decreasing metabolic burden has been proposed – protein scaffold approach [85]. Using protein-protein interaction domains and ligands from metazoan cells (mouse SH3 and PDZ domains and rat GBD) [84, 86, 232], the protein scaffolds containing the first enzyme, acetoacetyl-CoA thiolase (AtoB) of *E. coli*, the codon-optimized second and third enzymes, hydroxy-methylglutaryl-CoA synthase (HMGS) and hydroxymethylglutaryl-CoA reductase (HMGR) from *S. cerevisiae* was constructed. The natural modularity of these domains enabled the stoichiometry optimization of three mevalonate biosynthetic enzymes, thereby increasing 77-fold the product titer with low enzyme expression and reduced metabolic burden.

Although this approach could have a difficulty applying enzymes consisted of multi subunit or having important domain or function at the N-termini or C-termini, the protein scaffold approach provides several advantages: (1) prevents the loss of intermediates to diffusion or competing pathways; (2) keeps unstable intermediates from solvent and reduces the transit times of intermediates; (3) effective concentrations of metabolic intermediates; (4) circumvents unfavorable equilibrium and kinetics imposed by bulk-phase metabolite concentrations; (5) balances relative fluxes of individual enzymes for optimization of production levels by varying the number of interaction domain reaps; and (6) prevents intermediates accumulation to toxic levels owing to optimized flux [85].

Fusion proteins between synthetic pathway enzymes can also prevent the loss of metabolic intermediates by diffusion, degradation or conversion through competitive pathways [6, 59]. In Albertsen's work, several fusion protein variants generated by coupling yeast's farnesyl diphosphate synthase (FPPS) with patchoulol synthase (PTS) of plant origin (*Pogostemon cablin*) were combined with traditional metabolic engineering to further increase the production of patchoulol. Results showed that the fusion proteins in *S. cerevisiae* increased the production of patchoulol (main sesquiterpene produced by PTS) up to 2-fold. This approach showed that the spatial reorganization of vital metabolic enzymes could function as a useful synthetic biology tool for deceasing metabolic burden or increasing metabolic flux to a desired product without increasing protein expression [6].

9.2.2.3 Synthetic Pathway

In yeast metabolic engineering, complementation with a heterogeneous gene is a route to design more efficient metabolic pathways for high-value bioproducts [54, 149, 237, 271]. Although synthetic pathway by introducing heterogeneous genes could not show good effect because of different enzyme reaction kinetics [11], protein expression failure caused by differences of expression system such as codon usage bias [128], protein modification system [78], protein degradation system [340], protein maturation system [113] and protein localization system [280], side effect of new intermediate generated by new pathway [11], or new intermediate or new product degradation system [174], synthetic pathways developed via synthetic biology have been recently applied more and more for efficient metabolic engineering strategies [232]. One of the most representative examples of the application of this technology for process development is the microbial production of the antimalarial drug artemisinin [168]. Artemisinin, a sesquiterpene lactone endoperoxide from *Artemisia annua* L, is an antimalarial agent effective against multidrug-resistant *Plasmodium* spp. [94]. The semi-synthesis of artemisinin or any derivative from microbial-derived artemisinic acid (immediate precursor) may represent a cost-effective, environmentally friendly and reliable approach [293]. High yields of artemisinic acid in *E. coli* have been successfully achieved through synthetic biology [169]. The proposed strategy is based on two synthetic pathways: (1) the farnesyl pyrophosphate (FPP) synthetic pathway enclosing two synthetic operons, the MevT operon (acetyl-CoA to mevalonate) encoding three enzymes (acetoacetyl-CoA synthase from *E. coli* and the hydroxymethylglutaryl CoA synthase and reductase from *S. cerevisiae*) and the MBIS operon (mevalonate to FPP) encoding five enzymes (mevalonate kinase, phosphomevalonate kinase and mevalonate diphosphate decarboxylase from *S. cerevisiae*, and the isopentenyl diphosphate isomerase and FPP synthase from *E. coli*); and (2) artemisinic acid synthetic pathway containing amorphadiene synthase, a novel cytochrome P450 monooxygenase (CYP71AV1) from *A. annua* L (performs a three-step oxidation of amorpha-4,11-diene to artemisinic acid), an expression regulation part and devices containing codon optimization of amorphadiene synthase from *A. annua* L, a library based engineering of the intergenic regions of the polycistronic message of MevT operon [266] and finally synthetic protein scaffolds of MevT operon [85]. This amorphadiene-producing *E. coli* was further engineered to produce artemisinic acid, which can be then converted directly to artemisinin in two chemical steps [294]. Recently, engineered *S. cerevisiae* were reported to be able to produce artemisinic acid titres up to 100 mg/l using an engineered mevalonate pathway, amorphadiene synthase, and CYP71AV1 from *A. annua* L that performs a three-step oxidation of amorpha-4,11-diene to artemisinic acid [288, 289].

9.2.2.4 Synthetic Biology Tools for Engineering Artificial Genomes in Yeast

Although well characterized prokaryotic or eukaryotic microorganisms such as *E. coli* and *S. cerevisiae* have been the gold standards for metabolic engineering, an ideal host microorganism is supposed to have a simplified but robust genome, i.e. minimal genome or synthetic genome to be predictable and controllable [111, 375].

A minimal genome can be defined as the smallest set genes which have the ability to maintain normal cellular functions under ideal conditions of nutrients supply and stress-free environments [180]. To make a minimal genome, two different strategies have been suggested: (1) top-down strategy which aims to delete the non-essential segments from existing natural genomes through site-specific homologues recombination mediated genome deletion or transposon integration; and (2) the bottom-up strategy which construct a minimal genome by *de novo* synthesis and assembly of essential genes [374].

The process of creating a novel organism based on a minimal genome involves three steps: (1) identification of the set of essential genes necessary for basic metabolism and replication; (2) generating the minimal genome; and (3) provide or create the necessary non-genetic components for successful gene expression [68]. To define the set of essential genes, large-scale gene inactivation experiments

[119, 163, 179, 359], comparative genomics [181, 229] and network-based prediction [377] can be used. Making use of the information coming from these technologies, several genome reduction projects have been successfully performed in *S. cerevisiae* [227], *S. pombe* [117], *E. coli* [268], *B. subtilis* [16, 227, 354] and *C. glutamicum* [323], for example.

Recently, researchers in the J. Craig Venter Institute created the first synthetic cell (JCVI-syn1.0) using a multi-step process enclosing: (1) computer aided design of a synthetic chromosome based on the data of essential gene sets; (2) chemical synthesis of the partial DNA fragments; (3) assemble of fragments into *S. cerevisiae* via homologous recombination for the synthesis of a complete genome [115]; and (4) transplantation of the complete genome into a *M. capricolum* recipient cell to create a new *M. mycoides* cells that are controlled only by the synthetic chromosome [116]. Based on artificial minimal genome and synthetic microorganism, the more ample biological information could be accumulated and the more efficient and predictable microorganism factory for a valuable product could be generated.

9.3 Yeasts in Industrial Biotechnology

The increasing industrialization and motorization of the world has led to a steep rise for the demand of petroleum-based fuels [3]. Today, fossil fuels take up 80% of the primary energy consumed in the world, of which 58% alone is consumed by the transportation sector [95]. Soaring energy costs caused by the exhaustion of fossil fuels and increased awareness of global warming as result of greenhouse gas emissions, e.g. CO_2, methane and nitrous oxide, have motivated the development of alternative, renewable, sustainable, efficient and cost-effective energy sources with lesser emissions. Among many alternatives, biomass-derived fuels, e.g. ethanol, methanol, biodiesel, hydrogen and methane, are the most economic/environment friendly energy source [46]. They are eco-friendly because microorganisms have the capacity to recycle the CO_2 generated during biofuel consumption for biomass synthesis, thereby leading to near-zero greenhouse gas emissions [79]. In addition, they are economically viable because the costs associated with the overall production workflow (from raw material pretreatment to fermentation and product recovery technologies) can be significantly reduced when compared to the petrochemical industry based platform [80].

9.3.1 *Case Study I – Bioethanol*

Bioethanol is the current biofuel standard [14]. The fermentation of available carbohydrates into bioethanol is described by:

$$C_nH_{2n}O_n \rightarrow \frac{n}{3}C_2H_5OH + \frac{n}{3}CO_2 + Heat$$

Bioethanol has an energy content lower than gasoline (21.7 *vs.* 32 MJ/l, respectively) but higher octane level (109 compared to 95 of gasoline) [97], can be blended with gasoline allowing its use in current spark-ignition engines and induces lower greenhouse gas emissions due to its improved combustion capacity in normal vehicles [245].

Traditional raw materials for bioethanol production are derived from sucrose-containing feedstock, e.g. sugarcane, and starch materials, e.g. wheat and corn. An alternative carbohydrate source derives from plant biomass – lignocellulosic material. This raw material is enriched in a mixture of cellulose (crystalline glucose polymer – 30–60% of feedstock dry matter), hemicellulose (amorphous polymers of five- and six-carbon sugars – 20–40% of feedstock dry matter) and lignin (non-fermentable poly aromatic compounds – 15–25% of feedstock dry matter) [32]. The conversion of lignocellulosic material into bioethanol requires a pretreatment process which, first, separates cellulose and hemicellulose from lignin, and then breaks-down cellulose-constituent molecules into simple sugars (saccharification) that could be easily up taken by microorganisms, e.g. yeast and bacteria (see Fig. 9.3). However, during this process, several inhibitor compounds such as furfural, acetic acid and phenols are generated, significantly affecting the performance of the fermentation process. In order to constrain the impact of these inhibitors on the fermentative production of ethanol from lignocellulosic material, different systems-level approaches have been used. For example, Petersson and co-workers used microarrays (Affymetrix chips) and comparative expression analysis in order to identify key enzymes with capacity to reduce 5-hydroxymethyl furfural (HMF), a furan derivative and inhibitor of lignocellolosic-based fermentations, to 5-hydroxymethylfurfuryl alcohol, a non-inhibitor compound [264]. By comparing two *S. cerevisiae* strains with natural but different HMF reduction capacity, they were able to identify the reductase gene responsible for HMF reduction, the NADPH-dependent alcohol dehydrogenase ADH6p. When overexpressing ADH6p in yeast, engineered strains showed increased conversion of HMF under both aerobic and anaerobic conditions. In a second study, Hasunuma and co-workers demonstrated the usefulness of metabolome analysis for metabolic pathway optimization. Using gas chromatography and mass spectrometry, they evaluated the impact of acetic acid, an inhibitor of bio-ethanol present in lignocellulosic hydrolysates, on *S. cerevisiae* fermentations having xylose as carbon source [136]. Metabolome analysis revealed that the fluxes through the pentose phosphate pathway (PPP) were negatively affected by the addition of acetic acid, leading to the accumulation of intermediary metabolites such as ribulose-5-phosphate and ribose-5-phosphate. Fluxes could be re-established by overexpressing transaldolase, a PPP-related enzyme. In the follow-up study, these authors combined metabolic pathway engineering (enhancement of transaldolase and formate dehydrogenase activity) with crossbreeding strategies to generate an ethanol high-producer strain with improved tolerance towards inhibitory compounds such as acetate and formate [300]. In another study, Guadalupe Medina and co-workers combined genetic engineering and medium optimization to generate an engineered *S. cerevisiae* strain capable of partially converting acetate to ethanol with residual glycerol formation under

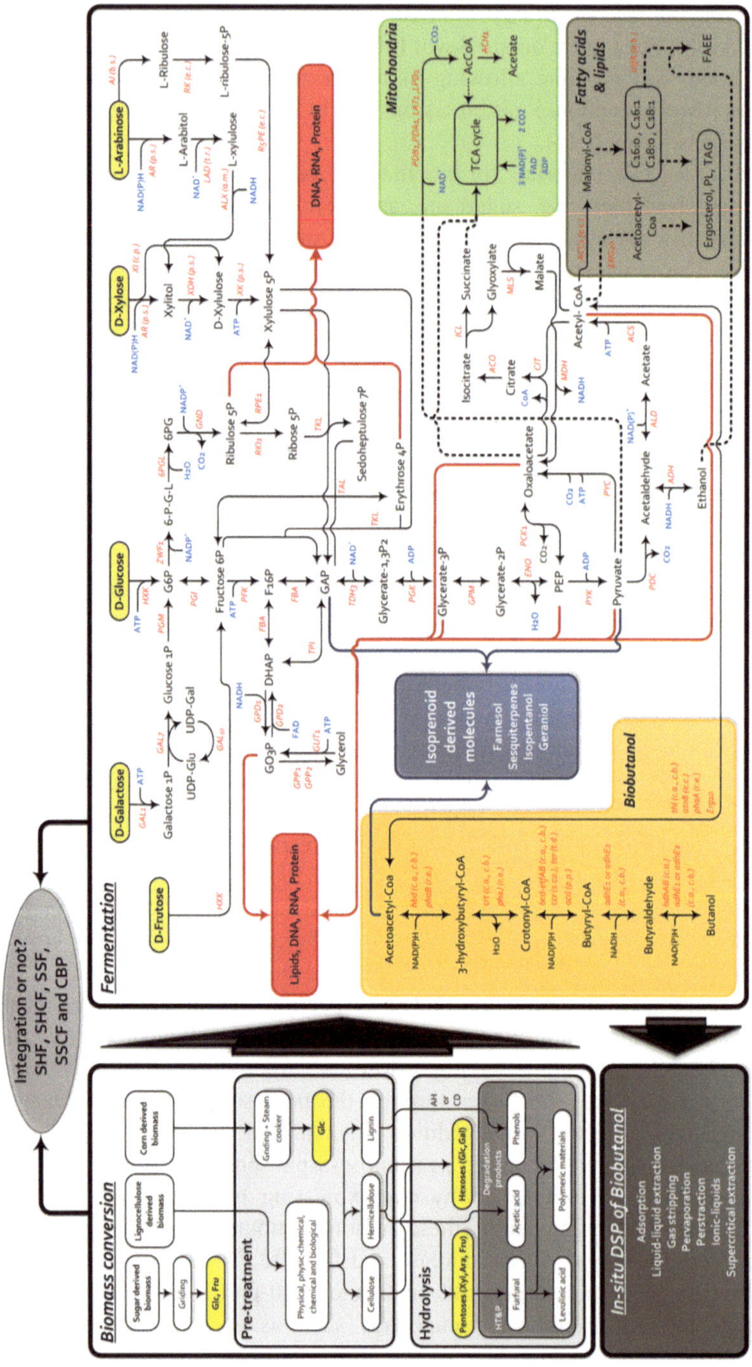

Fermentation
D-Frutose
D-Galactose
D-Glucose
D-Xylose
L-Arabinose
Lipids, DNA, RNA, Protein
DNA, RNA, Protein
Mitochondria
TCA cycle
Fatty acids & lipids
Isoprenoid derived molecules
Farnesol
Sesquiterpenes
Isopentanol
Geraniol
Biobutanol
Acetoacetyl-Coa
3-hydroxybutyryl-CoA
Crotonyl-CoA
Butyryl-CoA
Butyraldehyde
Butanol
Integration or not? SHF, SHCF, SSF, SSCF and CBP
Biomass conversion
Pre-treatment
Hydrolysis
In-situ DSP of Biobutanol
Adsorption
Liquid-liquid extraction
Gas stripping
Pervaporation
Perstraction
Ionic-liquids
Supercritical extraction

Fig. 9.3 A schematic representation of the entire production process of bio-butanol in yeast *Saccharomyces cerevisiae*: from upstream processing to yeast metabolic engineering and downstream technologies. The yellow box indicates hexose/6-carbon (glucose and galactose) and pentose/5-carbon (frutose, arabinose and xylose) sugars. The blue box and lines indicate isoprenoid-derived molecules and metabolites involved in their production, respectively. The red lines indicate the metabolites involved in lipid, DNA, RNA and protein synthesis. The grey box describes fatty acid ethyl esters (FAEE) pathway. The orange box describes the bio-butanol pathway. The dashed lines indicate multiple steps and the letters in italics denote enzymes. The origin of the enzymes is reported next to their abbreviations and inside brackets; enzymes without specified origin are of *S. cerevisiae*. Abbreviations: *6PG* 6-phosphogluconate, *6-P-G-L* 6-phosphogluconolactone, *6PGL* 6-phosphogluconolactonase [222], *a.b. A. baylyi*; a.m., *Ambrosiozyma monospora, ACC1* acetyl-CoA carboxylase [379], *acd* acyl-CoA dehydrogenase [244], *ACH1* acetyl-CoA hydrolase [298], *ACO* aconitate hydratase 1 [298], *ACS* acetyl-CoA synthetase [298], *ADH* alcohol dehydrogenase [298], *adhE1 and adhE2* bifunctional aldehyde/alcohol dehydrogenase [244], *ADP* adenosine diphosphate, *AH* acid hydrolysis, *AI* L-arabinose isomerase [160], *ALD* acetaldehyde dehydrogenase[298], *ALX* L-xylose reductase [45], *AR* aldose reductase [45], *Ara* arabinose, *atfA* wax ester synthase/acyl-CoA:diacylglycerol acyltransferase [371], *atoB* acetyl-CoA acetyltransferase [309], *ATP* adenosine triphosphate, *b.s. Bacillus subtilis, bcd-etfAB* butyryl-CoA dehydrogenase – electron transferring flavoprotein alpha and beta-subunit [309], *bdhAB* butanol dehydrogenase [244], *BPG* 1,3-bisphosphoglycerate, *c.a. Clostridium acetobutylicum, c.b. Clostridium beijerinckii, c.p. Clostridium phytofermentans, C16:0* palmitic acid, *C16:1* palmitoleic acid, *C18:0* stearic acid, *C18:1* oleic acid, *CBP* consolidated bioprocessing, *ccr* butyryl-CoA reductase [50], *CD* carbohydrate degradation, *CIT* citrate synthase [298], CO_2 carbon dioxide, *CoA* coenzyme A, *crt* crotonase [309, 317], *DHAP* dihydroxyacetone phosphate, *e.c. Escherichia coli, ENO* enolase [298], *ERG10* acetyl-CoA acetyltransferase [379], *F16P* fructose 1,6-bisphosphate, *FAD* flavin adenine dinucleotide (oxidized), *FAEE* fatty acids ethyl esters, *FBA* fructose bisphosphate aldolase [298], *Fru* fructose, *G6P* glucose 6-phosphate, *Gal* galactose, *GAL1* galactokinase [57], *GAL7* galactose 1-phosphate uridyltransferase[57], *GAL10* UDP-galactose 4-epimerase [57], *GAP* glyceraldehyde 3-phosphate, *Glc* glucose, *GND* 6-phosphogluconate dehydrogenase [160], *GO3P* glycerol 3-phosphate, *GPD1 (cytoplasm) and GPD2 (mitochondria)* glycerol-3-phosphate dehydrogenase [298], *GPM* phosphoglycerate mutase [298], *GPP1 and GPP2* glycerol 3-phosphatases [246], *GUT1* glycerol kinase [263], *hbd* 3-hydroxybutyryl-CoA dehydrogenase [110, 317], *HT&P* high temperature and pressure, *HXK* hexokinase [298], *ICL* isocitrate lyase [298], *LAD* L-arabitol dehydrogenase [45], *LAT1* pyruvate dehydrogenase E2 component [298], *LPD1* dihydrolipoamide dehydrogenase [298], *MDH* malate dehydrogenase [298], *MLS* malate synthase [298], NAD^+ nicotinamide adenine dinucleotide (oxidized), *NADH* nicotinamide adenine dinucleotide (reduced), *NADP+* nicotinamide adenine dinucleotide phosphate (oxidized), *NADPH* nicotinamide adenine dinucleotide phosphate (reduced), *p.p. Pseudomonas putida*, *p.s. Pichia stipitis*, *PCK1* phosphoenolpyruvate carboxykinase [298], *PDA1* pyruvate dehydrogenase E1 component subunit alpha [298], *PDB1* pyruvate dehydrogenase E1 component subunit beta [298], *PDC* pyruvate decarboxylase [298], *PEP* phosphoenolpyruvate, *PFK* phosphofructokinase [298], *PGI* phosphoglucose isomerase [160], *PGK* phosphoglycerate kinase [298], *PGM* phosphoglucomutase [57], *PL* phospholipid, *phaA* acetoacetyl-CoA thiolase/synthase [50], *phaB* 3-hydroxybutyryl-CoA dehydrogenase [50], *phaJ* crotonyl-CoA hydratase [50], *PYC* pyruvate carboxylase [298], *PYK* pyruvate kinase [298], *r.e. Ralstonia eutrophus*, *R5PE* L-ribulose-5-phosphate 4-epimerase [160], *RK* L-ribulokinase [160], *RKI1* ribose 5-phosphate ketol-isomerase [160], *RPE1* D-ribulose-5-phosphate 3-epimerase [160], *s.co. Streptomyces collinus*, *SHF* separate hydrolysis and fermentation, *SHCF* separate hydrolysis and co-fermentation, *SSF* simultaneous saccharification and fermentation, *SSCF* simultaneous saccharification and co-fermentation, *t.d. Treponema denticola*, *t.r. Trichoderma reesei*, *TAG* triacylglycerol, *TAL* transaldolase [160], *ter* trans-enoyl-CoA reductase [50], *TDH3* glyceraldehyde 3-phosphate dehydrogenase [298], *thl* thiolase [243, 317], *TKL* transketolase [160], *TPI* triosephosphate isomerise [298], *UDP-Glu* uridine 5-diphosphate glucose, *UDP-Gal* uridine 5-diphosphate galactose, *XDH* xylitol dehydrogenase [160], *XK* D-xylulokinase [160], *XI* xylose isomerase [55], *Xyl* xylose, *ZWF1* glucose 6-phosphate dehydrogenase [45]

anaerobic conditions [127]. An alternative to the aforementioned metabolic/genetic engineering strategies is random mutagenesis. For example, Zarif and co-authors used UV radiation and nitrous acid to generate mutant *S. cerevisiae* strains with improved tolerance towards furfural and, subsequently, enhanced ethanol production [372]. Under selective pressure – presence of 0.2% (v/v) furfural - one mutant strain showed to be capable of producing almost 40% more ethanol than the parental strain. Other examples are extensively reviewed in [7, 253].

Yeast *S. cerevisiae* is one of the most frequently used microorganisms for ethanol production. *S. cerevisiae* can efficiently convert hexose sugars, e.g. glucose, into ethanol at yields up to 0.45–0.48 g/g of raw material and volumetric and specific productivities within 1–3 $g_{bioethanol}$/l.h and 2 $g_{bioethanol}/g_{biomass}$.h, respectively, but not pentose sugars, e.g. xylose (reviewed in [130]). To overcome this limitation, *S. cerevisiae* can be genetically engineered to enclose the enzymes involved in the xylose pathway (xylose reductase, xylitol dehydrogenase, xylulokinase and xylose isomerase) from different species of yeasts, e.g. *Pichia stipitis* [46], and bacteria, e.g. thermofilic anaerobic bacteria *Clostridium phytofermentans* [55] – see Fig. 9.3. The introduction of these enzymes in *S. cerecisiae* does not guarantee *per se* an efficient utilization of xylose for ethanol production. The quality and/or quantity of target enzyme(s) must be controlled so that metabolic pathway performance and subsequently product formation are maximized. In metabolic engineering this can be achieved using regulating expression tools such as promoter binding proteins and protein engineering, and gene expression fine-tuning approaches, for example. There are several studies in literature reporting the use of regulating expression tools for strain engineering. For example, Watanabe and co-workers were able to generate a novel $NADP^+$ dependent xylitol dehydrogenase (XDH) mutant from *P. stipitis* via multiple site-directed mutagenesis, which upon introduction of an extra zinc atom presented higher thermostability and catalytic efficiency when compared to wild type XDH [350]. Then, by co-expressing this mutant enzyme with the wild type xylose reductase (XR) from *P. Stipitis* in recombinant *S. cerevisiae*, xylitol excretion (undesirable by-product) was reduced in 86% and ethanol production increased in 41% when compared to the reference strain expressing the wild-type XDH and XR [351]. In another study from the same authors, a second mutant enzyme involved in the xylose utilization pathway was generated via site-directed mutagenesis, the NADH dependent xylose reductase (XR) [349]. Engineered *S. cerevisiae* strains co-expressing the XR mutant with the wild type XDH increased ethanol production in 20% and reduced xylitol excretion in 52% when compared to strains expressing wild type XR and XDH. In two other examples, global transcription machinery engineering (gTME) was used to engineer *S. cerevisiae* strains with improved xylose fermentation capacity and tolerance towards lignocellulosic hydrolysates [200, 202]. Using error-prone PCR, Liu and co-workers generated a library of mutated global transcription factor Spt15p which, upon transformation into *S. cerevisiae*, was screened for dominant mutations conferring increased xylose conversion to ethanol and tolerance to corn cob hydrolysate. Engineered strains were able to efficiently utilize xylose as carbon source, either alone or mixed with glucose, and produce ethanol up to 12 g/L. The desired phenotype results from the combined effect of three separate mutations in the SPT15 gene containing amino acid mutation Phe177Ser, Tyr195His, and Lys218Arg. Multi-gene optimization approaches such as multiple genes promoter shuffling (MGPS) and evolutionary engineering commonly induce a strong, positive impact on the global metabolic network and subsequently on product formation. In a recent study, MPGS was used by Lu and co-workers to fine-tune the expression of multiple genes simultaneously in *S. cerevisiae* for ethanol production in xylose-based fermentations [206]. By shuffling the promoters for GND2 and HXK2 with the genes for transaldolase (TAL1), transketolase (TKL1), and pyruvate kinase (PYK1), they were able to identify the optimal combination of promoters-genes for ethanol production (GND2-TAL1-HXK2-TKL1-HXK2-PYK1). In two other studies, evolutionary engineering was used to increase the consumption rate of a mixture of 5- and 6-carbon sugars under anaerobic fermentation. Engineered strains showed increased specific consumption rates of glucose, xylose and arabinose, thus inducing a significant reduction in the time required for complete fermentation of these sugars [185, 360].

Other studies demonstrate the usefulness of cell-wide genome, transcriptome and fluxome analysis in yeast metabolic engineering for ethanol production. For example, Teixiera and co-workers used a chemogenomics approach (genome-wide analysis followed by clustering analysis) to identify novel candidate genes for increased tolerance of yeast *S. cerevisiae* towards ethanol [330]. Ma and Liu used a recently developed pathway-based qRT-PCR array assay (quantitative comparative transcriptomics) to reveal candidate genes and key regulators for increased tolerance of yeast cells towards ethanol [209]. In another study, Bro and co-workers used the *S. cerevisiae*'s genome-scale metabolic model to identify a key enzyme, the $NADP^+$-dependent glyceraldehyde-3-phosphate dehydrogenase, with capacity to enhance ethanol production and reduce glycerol formation under a mixture of pentose and hexose sugars as main carbon source and anaerobic conditions [58].

9.3.2 Case Study II – Biobutanol

Despite its very large production, bioethanol does not represent the optimal fuel for several reasons [318]: (1) low energy density; (2) hygroscopic; (3) corrosive and therefore cannot be transported in pipelines; (4) costly to distil; and (5) cannot be blended into gasoline in high amounts. Among the alternatives, 1-butanol is particularly attractive because it can be blended directly into gasoline and other hydrocarbon products [260]. When compared to bioethanol, 1-butanol is more hydrophobic, has lower octane rating (96 *vs.* 109 of bioethanol) but higher energy density (29.2 *vs.* 21.7 MJ/l of bioethanol), is safer because of its higher flash point and lower vapour pressure, generates fewer volatile organic compound emissions and is far less corrosive making its distribution through pipelines and filling stations feasible.

The process of fermenting hexose or pentose sugars to biobutanol can be represented by:

Reaction 1: $C_nH_{2n}O_n \rightarrow \frac{n}{6}C_4H_8O_2 + \frac{n}{3}CO_2 + \frac{n}{3}H_2$
Reaction 2: $\frac{n}{6}C_4H_8O_2 + \frac{n}{3} \rightarrow H_2\frac{n}{6}C_4H_{10}O + \frac{n}{6}H_2O$

The maximum theoretical yield of glucose or xylose fermentation to 1-butanol is 0.41 $g/g_{Glucose\ or\ Xylose}$.

In 2008 Steen and co-workers provided the first demonstration of butanol production in *S. cerevisiae*. Through genetic and metabolic engineering techniques, up to 2.5 mg/l of butanol were successfully produced by substituting a variety of isozymes for different reactions in the metabolic pathway of *S. cerevisiae* [318] (see Fig. 9.3). The performance of 1-butanol pathway in yeast *S. cerevisiae* can be theoretically enhanced by using different combinations of the pathway genes from different microorganisms on different vectors and different promoters. In addition, *in silico* pathway optimization may prove to be useful for the metabolic engineering of *S. cerevisiae* to produce 1-butanol [216, 364]. For example, Wu and co-workers used an *in silico* approach to design, evaluate and screen novel enzymatic pathways for 1-butanol biosynthesis. By combining the Biochemical Network Integrated Computational Explorer (BNICE) framework with structured-based screening methods (pathway length, thermodynamic feasibility, metabolic flux analysis and molecular docking), they were able to generate 9 candidate pathways for 1-butanol synthesis. In a second study, Matsuda and co-workers used an *in silico* approach based on flux balance analysis to compare the potential of *S. cerevisae* with *E. coli* for 1-butanol production and thus identify key nodes in the global metabolic network of *S. cerevisiae* limiting the production of 1-butanol at high titers [216]. Metabolic simulations revealed that the central metabolism of *S. cerevisiae* responds quite differently from that of *E. coli* when producing 1-butanol, having lower flexibility (cell compartmentalization and lack of shortcut reactions) and thus impairing the required flux distributions for efficient 1-butanol biosynthesis. Based on this comparative analysis, gene supplementation is pointed out by the authors as a potential strategy for metabolic engineering of *S. cerevisiae*. Although increases in product titers and experimental validation of *in silico* metabolic routes for 1-butanol biosynthesis will certainly be necessary in the near future, these studies represent an important proof of principle.

An important issue to consider is the use of other carbon sources than glucose for fermentation purposes. From an economical point of view, it would be preferable to use a combination of pentose and hexose sugars, e.g. glucose-xylose and xylose-galactose, rather than mono sugar-based substrates [102]. Nonetheless and similar to bioethanol, microorganisms tend to use hexose sugars as their main and solely carbon source. Selected cases where biobutanol production is based on mono-sugars include: (1) glucose fermentation by *E. coli* [110], *P. putida* and *B. subtilis* [244]; and (2) galactose fermentation by *S. cerevisiae* [318] (see Table 9.3 for more details). Although further research in this area is still required, lessons from bioethanol production in yeast may assist the development of *S. cerevisiae* strains with improved capacity to convert both 5- and 6-carbon sugars into 1-butanol.

Of paramount importance for molecular biologists and bioprocess engineers is how to increase the tolerance of microorganisms such as yeasts to high concentrations of butanol. Butanol is known to affect cellular membrane fluidity by changing its composition in phospholipids and fatty acids. This triggers a series of anomalous physicochemical events, e.g. disruption of solute transport,

Table 9.3 Fermentation processes for butanol production: detailed up- and down-stream bioprocess description

		Process							
Microorganism	Mutations	Sugar source	Sugar/Conc. (g/l)	Fermentation mode	Yield (g/g)	Productivity (g/l.h)	Maximum titer (g/l)	*In situ* product recovery	References
C.acetobutylicum P262	–	–	–	–	–	–	13.26	–	[342]
	Δlyt-1	–	–	–	–	–	14.2		
C.acetobutylicum P262	–	Whey	L/29	Batch	0.28	0.16	8.1	–	[93]
		permeate	L/58.3		0.19	0.21	11	GS	
C.acetobutylicum P262	–	Aspen wood	–/69–78	Batch	0.17–0.2	0.16–0.21	11.7–15.1	–	[255]
		Pine sawmill residue	–/53		0.2–0.21	0.16–0.16	10.52–11.16		
		Corn stover	–/81		0.185	0.208	15		
C.acetobutylicum ATCC 824	–	Synthetic	Glc/69.8	Batch	0.138	0.222	9.6	–	[220]
	adc, ctfAB	medium	Glc/79.1		0.164	0.299	13.9		
	adc, ctfA		Glc/80		–	–	11.1		
	Δadc, ctfAB		Glc/80		–	–	10.7		
C.acetobutylicum P262	–	Whey	L/23.7	Continuous	0.225	2.2	5.4	–	[276]
		permeate	L/53.8		0.117	1.2	2.8	Liq-Liq	
C.acetobutylicum ATCC 824	ΔsolR	Corn mash	Glc/80	Batch	0.247	0.247	17.8	–	[233]
C. beijerinckii NRRL B592	–	Synthetic medium	Glc/50	Continuous	0.182	0.16	9.1	–	[230]
C.acetobutylicum ATCC 824	–	Synthetic	Glc/80	Batch	0.146	0.225	11.7	–	[132]
	aad, Δbuk	medium			0.209	0.310	16.75		
	Δbuk				0.236	0.350	18.9		
C.acetobutylicum P262	–	DSPS	Starch/39.7	Continuous	0.139	0.712	5.52	–	[31]
				Batch	0.106	0.543	4.21		
C.beijerinckii BA101	–	Corn starch	Glc/60	Batch	0.317	0.58	19	–	[273]
			Glc/60	Continuous	–	1.74[a]	8.7[a]	–	
			Glc/150	Batch	–	–	51.5[a]	Perv	
			Glc/500	Fed-batch	–	0.98[a]	165.1[a]		

(continued)

Table 9.3 (continued)

Microorganism	Mutations	Process: Sugar source	Sugar/Conc. (g/l)	Fermentation mode	Yield (g/g)	Productivity (g/l.h)	Maximum titer (g/l)	*In situ* product recovery	References
C.beijerinckii BA101	–	Corn starch	Glc/44.6	Batch	0.267	0.199	11.9	GS	[99]
C.acetobutylicum ATCC 824	–	Synthetic medium	Glc/–	Fed-batch	–	0.229	11	–	[335]
	ΔgroESL				–	0.118	13		
	groESL				–	0.155	17		
C. saccharoperbutylacetonicum N1-4	–	Synthetic medium	Glc + ButA/50 + 5	Fed-batch	0.346	–	17	–	[328]
C.beijerinckii BA101	–	Synthetic medium	Glc/45.4	Batch	0.262	0.198	11.9	–	[100]
			Glc/500	Fed-batch	0.303	0.755	151.7	GS	
C. acetobutylicum ATCC 55025	–	Corn	Glc+ButA/54+3.6	Continuous	0.42	4.6	24.192	–	[150]
C.beijerinckii BA101	–	Corn starch	Starch/37.2	Batch	0.384	0.199	14.3	–	[101]
			Glc[b]/60	Continuous	0.200	0.166	11.98		
C.saccharohutylicum DSM 138	–	Sago starch	Starch/30	Continuous	0.170	0.255	5.09	–	[196]
C.acetobutylicum P260	–	CFAX sugars	Glc/60	Batch	0.21	0.18	12.8	GS	[275]
			Xyl/60		0.11	0.09	6.67		
			Gal/60		0.1	0.08	6.1		
			Ara/60		0.15	0.126	9.1		
			Glc + Xyl + Gal + Ara/ 15 *per* sugar		0.286	0.17	12		
		CFAX	CFAX/60		0	0	0		
		OSX	OSX/60		0	0	0		
C. saccharoperbutylacetonicum N1-4	–	Synthetic medium	Glc + ButA/20 + 10	Batch	0.393	0.313	7.5	–	[327]
C. beijerinckii BA101	–	–	Glc/60	Continuous	0.108	0.194	6.46	–	[103]
		D Corn	–/45		0.131	0.1	5.89		
		Sac D Corn	–/60		0.106	0.19	6.33		

C. beijerinckii P260	–	WS	Glc/59.4	Batch	0.22	0.18	12.8	–	[274]
		Pt WS	Xyl/17.8		0.24	0.12	6.05	–	
			Glc + Xyl/19 + 17		0.19	0.09	8.09	–	
			Xyl/16.8		0.29	0.16	7.4	–	
			Glc + Xyl/36 + 13		0.21	0.11	10.75	–	
			Xyl/50.4		0.25	0.25	12.6	GS	
C. beijerinckii BA101	–	–	Glc/47	Batch	0.28	0.22	13.2	–	[274]
		CFH[c]	–/23.6		0.271	0.073	6.4		
		CFH[d]	–/25		0.26	0.09	6.5		
C. beijerinckii P260	–	WS	Glc + Xyl + Gal + Ara/ 430 in total	Fed-batch	0.268	0.216	115.2	–	[274]
C. acetobutylicum M5	aad, ΔctfAB, adc	Synthetic medium	Glc/80	Batch	0.154	0.278	11.1	–	[312]
E.coli BW25113	thl, hbd, crt, bcd, etf, adhE2, ato, ΔadhE, ΔldhA, ΔfrdBC, Δfnr, ΔpflB, Δpta	TB medium	Gly/20	Batch	0.028	0.023	0.55	–	[24]
E.coli JM109	thiL, hbd, crt, bcd, etfB, etfA, adhE1	Synthetic medium	Glc/40	Batch	0.008	0.005	0.32	–	[153]
	thiL, hbd, crt, bcd, etfB, etfA, adhE				0.03	0.02	1.2		
S. cerevisiae BY4742	ERG10, hbd, crt, ccr, adhE2	Synthetic medium	Gal/20	Batch	1.25×10^{-4}	3.5×10^{-5}	0.0025	–	[318]
C.beijerinckii NCIMB 8052	–	Synthetic medium	Glc + ButA/30 + 2	CRIC	0.44f	0.4	13.4	–	[189]

(continued)

Table 9.3 (continued)

		Process							
Microorganism	Mutations	Sugar source	Sugar/Conc. (g/l)	Fermentation mode	Yield (g/g)	Productivity (g/l.h)	Maximum titer (g/l)	*In situ* product recovery	References
C.acetobutylicum EA 2018	ptb	Synthetic medium	Glc/60	Batch	0.24	0.3	14.4	–	[164]
P. putida S12	crt, bcd, etfAB, hbd, thiL, adhE1	TB medium	Gly/5	Batch	0.0244	0.0017	0.122	–	[245]
B. subtilis BK1.0					0.0048	0.0003	0.024		
E. coli BL21Star	atoB, gapA, hbd, crt, bcd, etfAB, adhE2				0.116	0.012	0.580		
C. beijerinckii P260	–	–	Glc/58.3	Batch	0.217	0.246	12.64	–	[279]
		Pt CSH[e]	Glc + Xyl + Gal + Ara/ 33 + 23 + 4 + 1		0.24	0.171	14.5		
		Wd CSH	Glc + Xyl + Gal + Ara/ 50 + 7 + 2 + 1		0.173	0.11	10.4		
		CSH + SGH (1:1)	Glc + Xyl + Gal + Ara/ 39 + 15 + 2 + 4		0.296	0.146	12.3		
C. beijerinckii P260	–	BSH	Glc + Xyl + Gal + Ara/ 8 + 6 + 2 + 6	Batch	0.193	0.063	4.254	–	[278]
		Wd BSH (1:2)	Glc + Xyl + Gal + Ara/ 35 + 6 + 0,5 + 3		0.235	0.25	10.452		
		BSH + WSH (1:1)	Glc + Xyl + Gal + Ara/ 26 + 13 + 2 + 5		0.296	0.33	13.62		
		Pt BSH[e]	Glc + Xyl + Gal + Ara/ 40 + 16 + 1 + 6		0.285	0.266	18.1		

C. acetobutylicum DSM 792	–	Synthetic medium	L/30	Continuous packed bed reactor	0.252	4.43	4.59	–	[235]
			L/15		0.138	0.65	1.2		
L. buchneri B-30929	thl	Synthetic medium	Glc/–	Batch	–	–	0.066	–	[198]
L. lactis LM0230					–	–	0.028		
E. coli DH1	phaA.hbd, crt, araC, ter. adhE2, laclq, aceE.aceF.lpd, laclq	Synthetic medium	Glc/25	Batch	0.186	0.0583	4.65	–	[50]

Abbreviations: *aceE.aceF.lpd* Genes encoding the pyruvate dehydrogenase complex, *adc* Acetoacetate decarboxylase gene, *ato* Acetyl-CoA acetyltransferase, *aad* Aldehyde/alcohol dehydrogenase, *adhE* Alcohol dehydrogenase, *adhE1* Alcohol/aldehyde dehydrogenase E1 gene, *adhE2* Alcohol/aldehyde dehydrogenase E2 gene, *Ara* Arabinose, *bcd* Butyryl-CoA dehydrogenase, *BSH* Barley straw hydrolysate, *buk* Butyrate kinase gene, *ButA* Butyrate, *ccr* Butyryl-CoA dehydrogenase, *CFAX* Corn fiber arabinoxylan, *CFH* Corn fiber hydrolysate, *CSH* Corn stover hydrolysate, CRIC Continuous reactor with immobilized cells, *crt* Crotonase, *ctfAB* Genes for two subunits of CoA-transferase, *DSPS* Debered-sweet-potato-slurry, *D Corn* Degermed corn, *ERG10* Native thialose, *etf* Electron transferring flavoprotein, *fnr* Fumurate and nitrate reductase, *frdBC* Fumarate reductase, *Gal* Galactose, *gapA* Glyceraldehyde 3-phosphate dehydrogenase, *GS* Gas stripping, *Glc* Glucose, *Gly* Glycerol, *hbd* Hydroxybutyryl-CoA dehydrogenase, *L* Lactose, *laclq* Mutant form of the Lac repressor gene, *ldhA* Lactate dehydrogenase A, *Liq-Liq* Liquid-liquid extraction, *lyt-1* Pleiotropic autolysis-deficient mutant, *OSX* Oat spelt xylan, *solR* Open reading frame 1, *pflB* Pyruvate formate lyase, *phaA* β-ketothiolase, *pta* Phosphate acetyltransferase, *ptb* Phosphotransbutyrylase gene, *Perv* Pervaporation, *Pt BSH* Pretreated barley straw hydrolysate, *Pt CSH* Pretreated corn stover hydrolysate, *Pt WS* Pretreated wheat straw, *Sac D Corn* Saccharified degermed corn, *SGH* Switchgrass hydrolysate, *ter Trans*-enoyl-CoA reductase, *thiL* Thiamine monophosphate kinase, *thl* Acetoacetyl-CoA thiolase, *Xyl* xylose, *Wd BSH* Water diluted barley straw hydrolysate, *Wd CSH* Water diluted corn stover hydrolysate, *WS* Wheat straw, *WSH* Wheat straw hydrolysate

[a] Total solvent produced (acetone + butanol + ethanol)

[b] Added to retrograded starch

[c] Sulfuric acid pretreatment

[d] Enzyme pretreatment

[e] Lime pretreatment

[f] Yield on glucose

transmembrane pH gradient, intracellular ATP level, glucose uptake and post-translational modifications of membrane proteins, with considerable negative impact on host cell metabolism and survival [98]. The potential of different hosts for butanol production was recently investigated [178]. A total of 24 different microorganisms were screened and it was found that a barrier to growth exists around 1–2% (v/v) butanol (diploid *S. cerevisiae* strains) with few microorganisms tolerating 2–3% (v/v) butanol (some strains of *Lactobacillus*). Promising strategies for increasing yeast *S. cerevisiae* tolerance to biobutanol include genetic engineering and adaptive evolution, e.g. random mutagenesis and evolutionary engineering, for example (see Fig. 9.2). In a recent study, Kimura and co-workers were able to increase the tolerance of genetically engineered *S. cerevisiae* strains towards n-butanol by manipulating the fatty acid content in the cells [367]. By replacing the usual desaturase genes with a rat elongase 2 gene (*rELO2*) that elongate C16 fatty acids to C18, the content of oleic acid (C18:1n-9) in the cell drastically increases and so thus the tolerance of engineered strains towards alcohols such as ethanol, n-propanol, 2-propanol and n-butanol. Fine-tuning of the expression levels of both elongase (*rELO2*) and desaturase (*OLE1*) genes is pointed out by the authors as a promising strategy to control the fatty acid composition of the cell and, thereby, increase further the tolerance of yeast *S. cerevisiae* towards n-butanol. Although adaptive evolution has been already and successfully applied for improving the tolerance of microorganisms such as *Clostridium* and *E. coli* towards 1-butanol [27, 70, 324], there is no study in literature reporting the use of evolutionary engineering and/or random mutagenesis in *S. cerevisiae*.

9.4 Concluding Remarks and Future Trends

The integration of synthetic biology approaches with systems biology technologies is of paramount importance to close the gap between high-value products development at lab-scale and industrial biotechnology. This biphasic but interactive process may significantly reduce the timeline for moving candidate bio-products into flexible production platforms, thereby reducing overall production cost. Moreover, novel micro-environments for bioprocess optimization, e.g. metabolic engineering, can be exploited, paving the way for the oriented design/construction of non-natural biological products and production systems from many high-producing microorganisms such as yeast.

During the last two decades, synthetic biology has significantly contributed for the development of novel and efficient metabolic engineering strategies, supplying researchers with the necessary molecular tools, e.g. 5′- or 3′-UTRs, synthetic riboregulators, tunable intergenic regions library and intergenic ribosomal entry sequence, for overcoming traditionally metabolic engineering challenges such as metabolic limitation of cellular processes, network rigidity and global regulation. Concomitantly, "omics" technologies, e.g. genomics, transcriptomics, proteomics, metabolomics and fluxomics, have been providing rich-information about the global behaviour of the system, boosting the comprehensive quantitative analysis of the cell, e.g. genes, metabolic pathways and protein synthesis mechanisms, and thus driving the optimization of product yields, productivities and titers. Taken together, these biology-based emerging disciplines have tremendous potential for helping/solving current society issues such as those environmentally (depletion of energy resources and global warming – e.g. biofuels), chemically, e.g. commodity and fine chemicals, or human, e.g. biopharmaceuticals, related. Besides these, synthetic and systems biology may also play an important role in the biomaterials, bioinformatics and nanotechnology areas.

One of the gold standard microorganisms for synthetic and systems biology is *S. cerevisiae*. With the expected advances in genetic and molecular biology techniques as well as in computer-aided screening of novel yeast metabolic targets, one may assist in the forthcoming years to an exponential increase in industrial yeast-based products. In fact, many yeast-derived products nowadays generated at lab-scale, e.g. biobutanol, biodiesel, bioethanol, malic acid, α-amylase, sesquiterpenes, antibiotics and human insulin, have potential to undergo production at industrial scale.

Undoubtedly, systems and synthetic biology fields will play an important role in the future of industrial biotechnology and subsequently on the economy of the early twenty first century. Nonetheless, much ground is still to explore in order to create a "universal" and efficient yeast cell platform for the development of products of interest to the biotechnological industry.

Acknowledgements The authors acknowledge the financial support received from the EU Framework VII project SYSINBIO (www.sysbio.se/sysinbio), the European Research Council, Knut and Alice Wallenberg Foundation and the Chalmers foundation.

References

1. Abad S, Nahalka J, Winkler M et al (2011) High-level expression of Rhodotorula gracilis D-amino acid oxidase in Pichia pastoris. Biotechnol Lett 33(3):557–563
2. Adams M, Kelley J, Gocayne J et al (1991) Complementary DNA sequencing: expressed sequence tags and human genome project. Science 252(5013):1651–1656
3. Agrawal AK (2007) Biofuels (alcohols and biodiesel) applications as fuels for internal combustion engines. Prog Energy Combust Sci 33:233–271
4. Akao T, Yashiro I, Hosoyama A et al (2011) Whole-genome sequencing of sake yeast saccharomyces cerevisiae Kyokai no. 7. DNA Res 18:423–434
5. Albermann C (2011) High versus low level expression of the lycopene biosynthesis genes from Pantoea ananatis in Escherichia coli. Biotechnol Lett 33(2):313–319
6. Albertsen L, Chen Y, Bach LS et al (2011) Diversion of flux toward sesquiterpene production in Saccharomyces cerevisiae by fusion of host and heterologous enzymes. Appl Environ Microbiol 77(3):1033–1040
7. Almeida JRM, Modig T, Petersson A et al (2007) Increased tolerance and conversion of inhibitors in lignocellulosic hydrolysates by Saccharomyces cerevisiae. J Chem Technol Biotechnol 82(4):340–349
8. Alper H, Fischer C, Nevoigt E et al (2005) Tuning genetic control through promoter engineering. Proc Natl Acad Sci U S A 102(36):12678–12683

9. Alper H, Moxley J, Nevoigt E et al (2006) Engineering yeast transcription machinery for improved ethanol tolerance and production. Science 314(5805):1565–1568
10. Alper H, Stephanopoulos G (2007) Global transcription machinery engineering: a new approach for improving cellular phenotype. Metab Eng 9(3):258–267
11. Alper H, Stephanopoulos G (2009) Engineering for biofuels: exploiting innate microbial capacity or importing biosynthetic potential? Nat Rev Microbiol 7(10):715–723
12. Alwine JC, Kemp DJ, Stark GR (1977) Method for detection of specific RNAs in agarose gels by transfer to diazobenzyloxymethyl-paper and hybridization with DNA probes. Proc Natl Acad Sci U S A 74(12):5350–5354
13. Andreas B (2011) Improving on nature's enzymes. Chemistry & Industry Magazine, vol 5. The Society of Chemical Industry (SCI), Wiley
14. Antoni D, Zverlov VV, Schwarz WH (2007) Biofuels from microbes. Appl Microbiol Biotechnol 77(1):23–35
15. Antoniewicz MR, Kraynie DF, Laffend LA et al (2007) Metabolic flux analysis in a nonstationary system: fed-batch fermentation of a high yielding strain of E. coli producing 1,3-propanediol. Metab Eng 9(3):277–292
16. Ara K, Ozaki K, Nakamura K et al (2007) Bacillus minimum genome factory: effective utilization of microbial genome information. Biotechnol Appl Biochem 46(Pt 3):169–178
17. Aricescu AR, Lu W, Jones EY (2006) A time- and cost-efficient system for high-level protein production in mammalian cells. Acta Crystallogr D Biol Crystallogr 62(Pt 10):1243–1250
18. Asada H, Uemura T, Yurugi-Kobayashi T et al (2011) Evaluation of the Pichia pastoris expression system for the production of GPCRs for structural analysis. Microb Cell Fact 10:24
19. Asadollahi MA, Maury J, Moller K et al (2008) Production of plant sesquiterpenes in Saccharomyces cerevisiae: effect of ERG9 repression on sesquiterpene biosynthesis. Biotechnol Bioeng 99(3):666–677
20. Asadollahi MA, Maury J, Patil KR et al (2009) Enhancing sesquiterpene production in Saccharomyces cerevisiae through in silico driven metabolic engineering. Metab Eng 11 (6):328–334
21. Ashengroph M, Nahvi I, Zarkesh-Esfahani H et al (2011) Candida galli strain PGO6: a novel isolated yeast strain capable of transformation of isoeugenol into vanillin and vanillic acid. Curr Microbiol 62(3):990–998
22. Askenazi M, Driggers EM, Holtzman DA et al (2003) Integrating transcriptional and metabolite profiles to direct the engineering of lovastatin-producing fungal strains. Nat Biotechnol 21 (2):150–156
23. Atkins D, Gerlach WL (1994) Artificial ribozyme and antisense gene expression in Saccharomyces cerevisiae. Antisense Res Dev 4(2):109–117
24. Atsumi S, Cann AF, Connor MR et al (2008) Metabolic engineering of Escherichia coli for 1-butanol production. Metab Eng 10(6):305–311
25. Atsumi S, Higashide W, Liao JC (2009) Direct photosynthetic recycling of carbon dioxide to isobutyraldehyde. Nat Biotechnol 27(12):1177–1180
26. Atsumi S, Wu TY, Eckl EM et al (2010) Engineering the isobutanol biosynthetic pathway in Escherichia coli by comparison of three aldehyde reductase/alcohol dehydrogenase genes. Appl Microbiol Biotechnol 85(3):651–657
27. Atsumi S, Wu TY, Machado IM et al (2010) Evolution, genomic analysis, and reconstruction of isobutanol tolerance in Escherichia coli. Mol Syst Biol 6:449
28. Babiskin AH, Smolke CD (2011) A synthetic library of RNA control modules for predictable tuning of gene expression in yeast. Mol Syst Biol 7:471
29. Babiskin AH, Smolke CD (2011) Engineering ligand-responsive RNA controllers in yeast through the assembly of RNase III tuning modules. Nucleic Acids Res 39(12):5299–5311
30. Babiskin AH, Smolke CD (2011) Synthetic RNA modules for fine-tuning gene expression levels in yeast by modulating RNase III activity. Nucleic Acids Res 39(19):8651–8664
31. Badr HR, Toledo R, Hamdy MK (2001) Continuous acetone-ethanol-butanol fermentation by immobilized cells of Clostridium acetobutylicum. Biomass Bioenergy 20(2):119–132

32. Balat M (2011) Production of bioethanol from lignocellulosic materials via the biochemical pathway: a review. Energy Convers Manag 52(2):858–875
33. Barrett CL, Kim TY, Kim HU et al (2006) Systems biology as a foundation for genome-scale synthetic biology. Curr Opin Biotechnol 17(5):488–492
34. Bashirullah A, Cooperstock RL, Lipshitz HD (2001) Spatial and temporal control of RNA stability. Proc Natl Acad Sci U S A 98(13):7025–7028
35. Bauer G, Suess B (2006) Engineered riboswitches as novel tools in molecular biology. J Biotechnol 124(1):4–11
36. Bayer TS, Smolke CD (2005) Programmable ligand-controlled riboregulators of eukaryotic gene expression. Nat Biotechnol 23(3):337–343
37. Becker SA, Feist AM, Mo ML et al (2007) Quantitative prediction of cellular metabolism with constraint-based models: the COBRA Toolbox. Nat Protoc 2(3):727–738
38. Bedford MR, Partridge GG (2010) Enzymes in farm animal nutrition, 2nd edn. In: Bedford M and Partridge G (eds) CABI, Wallingford, UK.
39. Beerli RR, Dreier B, Barbas CF 3rd (2000) Positive and negative regulation of endogenous genes by designed transcription factors. Proc Natl Acad Sci U S A 97(4):1495–1500
40. Bell PJ, Davies IW, Attfield PV (1999) Facilitating functional analysis of the Saccharomyces cerevisiae genome using an EGFP-based promoter library and flow cytometry. Yeast 15 (16):1747–1759
41. Benner SA (2004) Chemistry. Redesigning genetics. Science 306(5696):625–626
42. Benthin S, Nielsen J, Villadsen J (1991) A simple and reliable method for the determination of cellular RNA content. Biotechnol Tech 5(1):39–42
43. Bergin J (2009) Synthetic biology: emerging global markets. BCC Research LLC, MA, USA.
44. Berrow NS, Bussow K, Coutard B et al (2006) Recombinant protein expression and solubility screening in Escherichia coli: a comparative study. Acta Crystallogr D Biol Crystallogr 62(Pt 10):1218–1226
45. Bettiga M, Bengtsson O, Hahn-Hagerdal B et al (2009) Arabinose and xylose fermentation by recombinant Saccharomyces cerevisiae expressing a fungal pentose utilization pathway. Microb Cell Fact 8:40
46. Bhatti HN, Hanif MA, Qasim M et al (2008) Biodiesel production from waste tallow. Fuel 87 (2961–2966)
47. Bisht H, Chugh DA, Raje M et al (2002) Recombinant dengue virus type 2 envelope/hepatitis B surface antigen hybrid protein expressed in Pichia pastoris can function as a bivalent immunogen. J Biotechnol 99(2):97–110
48. Blancafort P, Magnenat L, Barbas CF 3rd (2003) Scanning the human genome with combinatorial transcription factor libraries. Nat Biotechnol 21(3):269–274
49. Boer E, Breuer FS, Weniger M et al (2011) Large-scale production of tannase using the yeast Arxula adeninivorans. Appl Microbiol Biotechnol 92(1):105–114
50. Bond-Watts BB, Bellerose RJ, Chang MCY (2011) Enzyme mechanism as a kinetic control element for designing synthetic biofuel pathways. Nat Chem Biol 7(4):222–227
51. Bonetta L (2005) Prime time for real-time PCR. Nat Methods 2(4):305–312
52. Bordel S, Nielsen J (2010) Identification of flux control in metabolic networks using non-equilibrium thermodynamics. Metab Eng 12(4):369–377
53. Branduardi P, Sauer M, De Gioia L et al (2006) Lactate production yield from engineered yeasts is dependent from the host background, the lactate dehydrogenase source and the lactate export. Microb Cell Fact 5:4
54. Branduardi P, Smeraldi C, Porro D (2008) Metabolically engineered yeasts: 'potential' industrial applications. J Mol Microbiol Biotechnol 15(1):31–40
55. Brat D, Boles E, Wiedemann B (2009) Functional expression of a bacterial xylose isomerase in Saccharomyces cerevisiae. Appl Environ Microbiol 75(8):2304–2311
56. Brenner S, Johnson M, Bridgham J et al (2000) Gene expression analysis by massively parallel signature sequencing (MPSS) on microbead arrays. Nat Biotechnol 18(6):630–634
57. Bro C, Knudsen S, Regenberg B et al (2005) Improvement of galactose uptake in Saccharomyces cerevisiae through overexpression of phosphoglucomutase: example of transcript analysis as a tool in inverse metabolic engineering. Appl Environ Microbiol 71 (11):6465–6472

58. Bro C, Regenberg B, Forster J et al (2006) In silico aided metabolic engineering of Saccharomyces cerevisiae for improved bioethanol production. Metab Eng 8(2):102–111
59. Bulow L (1990) Preparation of artificial bifunctional enzymes by gene fusion. Biochem Soc Symp 57:123–133
60. Burgard AP, Pharkya P, Maranas CD (2003) Optknock: a bilevel programming framework for identifying gene knockout strategies for microbial strain optimization. Biotechnol Bioeng 84(6):647–657
61. Buskirk AR, Landrigan A, Liu DR (2004) Engineering a ligand-dependent RNA transcriptional activator. Chem Biol 11(8):1157–1163
62. Canelas AB, Harrison N, Fazio A et al (2010) Integrated multilaboratory systems biology reveals differences in protein metabolism between two reference yeast strains. Nat Commun 1:145
63. Carrier TA, Keasling JD (1997) Engineering mRNA stability in E. coli by the addition of synthetic hairpins using a 5′ cassette system. Biotechnol Bioeng 55(3):577–580
64. Carrier TA, Keasling JD (1999) Library of synthetic 5′ secondary structures to manipulate mRNA stability in Escherichia coli. Biotechnol Prog 15(1):58–64
65. Chen HM, Neiman AM (2011) A conserved regulatory role for antisense RNA in meiotic gene expression in yeast. Curr Opin Microbiol 14(6):655–659
66. Chen X, Nielsen KF, Borodina I et al (2011) Increased isobutanol production in Saccharomyces cerevisiae by overexpression of genes in valine metabolism. Biotechnol Biofuels 4:21
67. Cho MK, Magnus D, Caplan AL et al (1999) Policy forum: genetics. Ethical considerations in synthesizing a minimal genome. Science 286(5447):2087, 2089–2090
68. Christen S, Sauer U (2011) Intracellular characterization of aerobic glucose metabolism in seven yeast species by 13 C flux analysis and metabolomics. FEMS Yeast Res 11(3):263–272
69. Christensen B, Nielsen J (1999) Isotopomer analysis using GC-MS. Metab Eng 1(4):282–290
70. Connor MR, Cann AF, Liao JC (2010) 3-Methyl-1-butanol production in Escherichia coli: random mutagenesis and two-phase fermentation. Appl Microbiol Biotechnol 86(4): 1155–1164
71. Converti A, Perego P (2002) Use of carbon and energy balances in the study of the anaerobic metabolism of Enterobacter aerogenes at variable starting glucose concentrations. Appl Microbiol Biotechnol 59(2–3):303–309
72. Corte L, Rellini P, Roscini L et al (2010) Development of a novel, FTIR (Fourier transform infrared spectroscopy) based, yeast bioassay for toxicity testing and stress response study. Anal Chim Acta 659(1–2):258–265
73. Cox RS 3rd, Surette MG, Elowitz MB (2007) Programming gene expression with combinatorial promoters. Mol Syst Biol 3:145
74. Cvijovic M, Olivares-Hernandez R, Agren R et al (2010) BioMet Toolbox: genome-wide analysis of metabolism. Nucleic Acids Res 38(Web Server):W144–W149
75. Dada O, Ramsay L, Dickerson J et al (2010) Capillary array isoelectric focusing with laser-induced fluorescence detection: milli-pH unit resolution and yoctomole mass detection limits in a 32-channel system. Anal Bioanal Chem 397(8):3305–3310
76. De Backer MD, Raponi M, Arndt GM (2002) RNA-mediated gene silencing in non-pathogenic and pathogenic fungi. Curr Opin Microbiol 5(3):323–329
77. de Godoy LM, Olsen JV, Cox J et al (2008) Comprehensive mass-spectrometry-based proteome quantification of haploid versus diploid yeast. Nature 455(7217):1251–1254
78. De Pourcq K, De Schutter K, Callewaert N (2010) Engineering of glycosylation in yeast and other fungi: current state and perspectives. Appl Microbiol Biotechnol 87(5):1617–1631
79. Demirbas A (2009) Biofuels securing the planet's future energy needs. Energy Convers Manag 50(9):2239–2249
80. Demirbas A (2009) Political, economic and environmental impacts of biofuels: a review. Appl Energy 86:S108–S117
81. Diamandis EP (2000) Sequencing with microarray technology–a powerful new tool for molecular diagnostics. Clin Chem 46(10):1523–1525
82. Doudna JA, Cech TR (2002) The chemical repertoire of natural ribozymes. Nature 418 (6894):222–228

83. Drinnenberg IA, Weinberg DE, Xie KT et al (2009) RNAi in budding yeast. Science 326 (5952):544–550
84. Dueber JE, Mirsky EA, Lim WA (2007) Engineering synthetic signaling proteins with ultrasensitive input/output control. Nat Biotechnol 25(6):660–662
85. Dueber JE, Wu GC, Malmirchegini GR et al (2009) Synthetic protein scaffolds provide modular control over metabolic flux. Nat Biotechnol 27(8):753–759
86. Dueber JE, Yeh BJ, Bhattacharyya RP et al (2004) Rewiring cell signaling: the logic and plasticity of eukaryotic protein circuitry. Curr Opin Struct Biol 14(6):690–699
87. Edwards JR, Ruparel H, Ju J (2005) Mass-spectrometry DNA sequencing. Mutat Res/ Fundam Mol Mech Mutagen 573(1–2):3–12
88. Elbeik T, Markowitz N, Nassos P et al (2004) Simultaneous runs of the Bayer VERSANT HIV-1 version 3.0 and HCV bDNA version 3.0 quantitative assays on the system 340 platform provide reliable quantitation and improved work flow. J Clin Microbiol 42 (7):3120–3127
89. Elbeik T, Surtihadi J, Destree M et al (2004) Multicenter evaluation of the performance characteristics of the bayer VERSANT HCV RNA 3.0 assay (bDNA). J Clin Microbiol 42 (2):563–569
90. Elena SF, Lenski RE (2003) Evolution experiments with microorganisms: the dynamics and genetic bases of adaptation. Nat Rev Genet 4(6):457–469
91. Ellis DI, Goodacre R (2006) Metabolic fingerprinting in disease diagnosis: biomedical applications of infrared and Raman spectroscopy. Analyst 131(8):875–885
92. Ellis T, Wang X, Collins JJ (2009) Diversity-based, model-guided construction of synthetic gene networks with predicted functions. Nat Biotechnol 27(5):465–471
93. Ennis BM, Marshall CT, Maddox IS et al (1986) Continuous product recovery by in-situ gas stripping/condensation during solvent production from whey permeate using Clostridium acetobutylicum. Biotechnol Lett 8(10):725–730
94. Enserink M (2005) Infectious diseases. Source of new hope against malaria is in short supply. Science 307(5706):33
95. Escobar JC, Lora ES, Venturini OJ et al (2009) Biofuels: environment, technology and food security. Renew Sustain Energy Rev 13(1275–1287)
96. Esener AA, Veerman T, Roels JA et al (1982) Modeling of bacterial growth; formulation and evaluation of a structured model. Biotechnol Bioeng 24(8):1749–1764
97. Eyidogan M, Ozsezen AN, Canakci M et al (2010) Impact of alcohol-gasoline fuel blends on the performance and combustion characteristics of an SI engine. Fuel 89(10):2713–2720
98. Ezeji T, Milne C, Price ND et al (2010) Achievements and perspectives to overcome the poor solvent resistance in acetone and butanol-producing microorganisms. Appl Microbiol Biotechnol 85(6):1697–1712
99. Ezeji TC, Qureshi N, Blaschek HP (2003) Production of acetone, butanol and ethanol by Clostridium beijerinckii BA101 and in situ recovery by gas stripping. World J Microbiol Biotechnol 19(6):595–603
100. Ezeji TC, Qureshi N, Blaschek HP (2004) Acetone butanol ethanol (ABE) production from concentrated substrate: reduction in substrate inhibition by fed-batch technique and product inhibition by gas stripping. Appl Microbiol Biotechnol 63(6):653–658
101. Ezeji TC, Qureshi N, Blaschek HP (2005) Continuous butanol fermentation and feed starch retrogradation: butanol fermentation sustainability using Clostridium beijerinckii BA101. J Biotechnol 115(2):179–187
102. Ezeji T, Qureshi N, Blaschek HP (2007) Butanol production from agricultural residues: impact of degradation products on clostridium beijerinckii growth and butanol fermentation. Biotechnol Bioeng 97(6):1460–1469
103. Ezeji T, Qureshi N, Blaschek HP (2007) Production of acetone-butanol-ethanol (ABE) in a continuous flow bioreactor using degermed corn and Clostridium beijerinckii. Process Biochem 42(1):34–39
104. Fabrizio P, Abelson J (1990) Two domains of yeast U6 small nuclear RNA required for both steps of nuclear precursor messenger RNA splicing. Science 250(4979):404–409

105. Feng X, Page L, Rubens J et al (2010) Bridging the gap between fluxomics and industrial biotechnology. J Biomed Biotechnol 2010:460717
106. Feng W, Raghuraman MK, Brewer BJ (2007) Mapping yeast origins of replication via single-stranded DNA detection. Methods 41(2):151–157
107. Ferea TL, Bowman BJ (1996) The vacuolar ATPase of Neurospora crassa is indispensable: inactivation of the vma-1 gene by repeat-induced point mutation. Genetics 143(1):147–154
108. Ficarro SB, McCleland ML, Stukenberg PT et al (2002) Phosphoproteome analysis by mass spectrometry and its application to Saccharomyces cerevisiae. Nat Biotechnol 20(3):301–305
109. Finos L, Farcomeni A (2011) k-FWER control without p -value adjustment, with application to detection of genetic determinants of multiple sclerosis in Italian twins. Biometrics 67 (1):174–181
110. Fischer CR, Tseng HC, Tai M et al (2010) Assessment of heterologous butyrate and butanol pathway activity by measurement of intracellular pathway intermediates in recombinant Escherichia coli. Appl Microbiol Biotechnol 88(1):265–275
111. Gao H, Zhuo Y, Ashforth E et al (2010) Engineering of a genome-reduced host: practical application of synthetic biology in the overproduction of desired secondary metabolites. Protein Cell 1(7):621–626
112. Gaynor JW, Campbell BJ, Cosstick R (2010) RNA interference: a chemist's perspective. Chem Soc Rev 39(11):4169–4184
113. Georgiou G, Valax P (1996) Expression of correctly folded proteins in Escherichia coli. Curr Opin Biotechnol 7(2):190–197
114. Gianchandani EP, Chavali AK, Papin JA (2010) The application of flux balance analysis in systems biology. Wiley Interdiscip Rev Syst Biol Med 2(3):372–382
115. Gibson DG, Benders GA, Andrews-Pfannkoch C et al (2008) Complete chemical synthesis, assembly, and cloning of a Mycoplasma genitalium genome. Science 319(5867):1215–1220
116. Gibson DG, Glass JI, Lartigue C et al (2010) Creation of a bacterial cell controlled by a chemically synthesized genome. Science 329(5987):52–56
117. Giga-Hama Y, Tohda H, Takegawa K et al (2007) Schizosaccharomyces pombe minimum genome factory. Biotechnol Appl Biochem 46(Pt 3):147–155
118. Gilmore JM, Washburn MP (2010) Advances in shotgun proteomics and the analysis of membrane proteomes. J Proteomics 73(11):2078–2091
119. Glass JI, Assad-Garcia N, Alperovich N et al (2006) Essential genes of a minimal bacterium. Proc Natl Acad Sci U S A 103(2):425–430
120. Glover III WR (2007) Systems and methods of analyzing nucleic acid polymers and related components. USA Patent
121. Goffeau A, Barrell BG, Bussey H et al (1996) Life with 6000 genes. Science 274(5287):546, 563–567
122. Gorg A, Weiss W, Dunn MJ (2004) Current two-dimensional electrophoresis technology for proteomics. Proteomics 4(12):3665–3685
123. Gowda M, Li H, Alessi J et al (2006) Robust analysis of 5′-transcript ends (5′-RATE): a novel technique for transcriptome analysis and genome annotation. Nucleic Acids Res 34(19):e126
124. Graslund S, Sagemark J, Berglund H et al (2008) The use of systematic N- and C-terminal deletions to promote production and structural studies of recombinant proteins. Protein Expr Purif 58(2):210–221
125. Grate D, Wilson C (2001) Inducible regulation of the S. cerevisiae cell cycle mediated by an RNA aptamer-ligand complex. Bioorg Med Chem 9(10):2565–2570
126. Grossmann J, Roschitzki B, Panse C et al (2010) Implementation and evaluation of relative and absolute quantification in shotgun proteomics with label-free methods. J Proteomics 73 (9):1740–1746
127. Guadalupe Medina V, Almering MJ, van Maris AJ et al (2010) Elimination of glycerol production in anaerobic cultures of a Saccharomyces cerevisiae strain engineered to use acetic acid as an electron acceptor. Appl Environ Microbiol 76(1):190–195
128. Gustafsson C, Govindarajan S, Minshull J (2004) Codon bias and heterologous protein expression. Trends Biotechnol 22(7):346–353

129. Gygi SP, Rist B, Gerber SA et al (1999) Quantitative analysis of complex protein mixtures using isotope-coded affinity tags. Nat Biotechnol 17(10):994–999
130. Hahn-Hagerdal B, Karhumaa K, Fonseca C et al (2007) Towards industrial pentose-fermenting yeast strains. Appl Microbiol Biotechnol 74(5):937–953
131. Hanlon SE, Rizzo JM, Tatomer DC et al (2011) The Stress Response Factors Yap6, Cin5, Phd1, and Skn7 Direct Targeting of the Conserved Co-Repressor Tup1-Ssn6 in *S. cerevisiae*. PLoS One 6(4):e19060
132. Harris LM, Desai RP, Welker NE et al (2000) Characterization of recombinant strains of the Clostridium acetobutylicum butyrate kinase inactivation mutant: need for new phenomenological models for solventogenesis and butanol inhibition? Biotechnol Bioeng 67(1):1–11
133. Hartner FS, Ruth C, Langenegger D et al (2008) Promoter library designed for fine-tuned gene expression in Pichia pastoris. Nucleic Acids Res 36(12):e76
134. Harvey I, Garneau P, Pelletier J (2002) Inhibition of translation by RNA-small molecule interactions. RNA 8(4):452–463
135. Haseloff J, Ajioka J (2009) Synthetic biology: history, challenges and prospects. J R Soc Interface 6(Suppl 4):S389–S391
136. Hasunuma T, Sanda T, Yamada R et al (2011) Metabolic pathway engineering based on metabolomics confers acetic and formic acid tolerance to a recombinant xylose-fermenting strain of Saccharomyces cerevisiae. Microb Cell Fact 10(1):2
137. Haubold B, Wiehe T (2004) Comparative genomics: methods and applications. Naturwissenschaften 91(9):405–421
138. Herbert D, Phipps PJ, Strange RE (1971) Chemical analysis of microbial cells. In: Norris JR, Ribbons DW (eds) Methods in microbiology, vol 5, Part 2. Academic Press, London, pp 209–344
139. Herrgard MJ, Fong SS, Palsson BO (2006) Identification of genome-scale metabolic network models using experimentally measured flux profiles. PLoS Comput Biol 2(7):e72
140. Hershey JWB, Mathews M, Sonenberg N et al (1996) Translational control, vol 30, Cold spring harbor monograph series. Cold Spring Harbor Laboratory Press, Plainview
141. Hicks WM, Yamaguchi M, Haber JE (2011) Real-time analysis of double-strand DNA break repair by homologous recombination. Proc Natl Acad Sci U S A 108(8):3108–3115
142. Hoefnagel MH, Starrenburg MJ, Martens DE et al (2002) Metabolic engineering of lactic acid bacteria, the combined approach: kinetic modelling, metabolic control and experimental analysis. Microbiology 148(Pt 4):1003–1013
143. Højer-Pedersen J, Smedsgaard J, Nielsen J (2008) The yeast metabolome addressed by electrospray ionization mass spectrometry: initiation of a mass spectral library and its applications for metabolic footprinting by direct infusion mass spectrometry. Metabolomics 4(4):393–405
144. Hong WK, Kim CH, Heo SY et al (2010) Enhanced production of ethanol from glycerol by engineered Hansenula polymorpha expressing pyruvate decarboxylase and aldehyde dehydrogenase genes from Zymomonas mobilis. Biotechnol Lett 32(8):1077–1082
145. Hong WK, Kim CH, Heo SY et al (2011) 1,3-Propandiol production by engineered Hansenula polymorpha expressing dha genes from Klebsiella pneumoniae. Bioprocess Biosyst Eng 34(2):231–236
146. Hong K-K, Vongsangnak W, Vemuri GN et al (2011) Unravelling evolutionary strategies of yeast for improving galactose utilization through integrated systems level analysis. Proc Natl Acad Sci U S A 108(29):12179–12184
147. Hu Q, Noll RJ, Li H et al (2005) The Orbitrap: a new mass spectrometer. J Mass Spectrom 40 (4):430–443
148. Hu H, Qian J, Chu J et al (2009) DNA shuffling of methionine adenosyltransferase gene leads to improved S-adenosyl-L-methionine production in Pichia pastoris. J Biotechnol 141 (3–4):97–103
149. Huang B, Guo J, Yi B et al (2008) Heterologous production of secondary metabolites as pharmaceuticals in Saccharomyces cerevisiae. Biotechnol Lett 30(7):1121–1137

150. Huang WC, Ramey DE, Yang ST (2004) Continuous production of butanol by Clostridium acetobutylicum immobilized in a fibrous bed bioreactor. Appl Biochem Biotechnol 113–116:887–898
151. Hughes TR, Mao M, Jones AR et al (2001) Expression profiling using microarrays fabricated by an ink-jet oligonucleotide synthesizer. Nat Biotechnol 19(4):342–347
152. Insenser MR, Hernáez ML, Nombela C et al (2010) Gel and gel-free proteomics to identify Saccharomyces cerevisiae cell surface proteins. J Proteomics 73(6):1183–1195
153. Inui M, Suda M, Kimura S et al (2008) Expression of Clostridium acetobutylicum butanol synthetic genes in Escherichia coli. Appl Microbiol Biotechnol 77(6):1305–1316
154. Isaacs FJ, Collins JJ (2005) Plug-and-play with RNA. Nat Biotechnol 23(3):306–307
155. Isaacs FJ, Dwyer DJ, Collins JJ (2006) RNA synthetic biology. Nat Biotechnol 24(5):545–554
156. Isaacs FJ, Dwyer DJ, Ding C et al (2004) Engineered riboregulators enable post-transcriptional control of gene expression. Nat Biotechnol 22(7):841–847
157. Ishihama Y, Oda Y, Tabata T et al (2005) Exponentially modified protein abundance index (emPAI) for estimation of absolute protein amount in proteomics by the number of sequenced peptides per protein. Mol Cell Proteomics 4(9):1265–1272
158. Jansen RP (2001) mRNA localization: message on the move. Nat Rev Mol Cell Biol 2 (4):247–256
159. Jarboe LR, Zhang X, Wang X et al (2010) Metabolic engineering for production of biorenewable fuels and chemicals: contributions of synthetic biology. J Biomed Biotechnol 2010:761042
160. Jeffries TW (2006) Engineering yeasts for xylose metabolism. Curr Opin Biotechnol 17 (3):320–326
161. Jensen PR, Hammer K (1998) The sequence of spacers between the consensus sequences modulates the strength of prokaryotic promoters. Appl Environ Microbiol 64(1):82–87
162. Jeppsson M, Johansson B, Jensen PR et al (2003) The level of glucose-6-phosphate dehydrogenase activity strongly influences xylose fermentation and inhibitor sensitivity in recombinant Saccharomyces cerevisiae strains. Yeast 20(15):1263–1272
163. Ji Y, Zhang B, Van SF et al (2001) Identification of critical staphylococcal genes using conditional phenotypes generated by antisense RNA. Science 293(5538):2266–2269
164. Jiang Y, Xu C, Dong F et al (2009) Disruption of the acetoacetate decarboxylase gene in solvent-producing Clostridium acetobutylicum increases the butanol ratio. Metab Eng 11 (4–5):284–291
165. Jöbses IML, Egberts GTC, van Baalen A et al (1985) Mathematical modelling of growth and substrate conversion of Zymomonas mobilis at 30 and 35 °C. Biotechnol Bioeng 27 (7):984–995
166. Juneau K, Palm C, Miranda M et al (2007) High-density yeast-tiling array reveals previously undiscovered introns and extensive regulation of meiotic splicing. Proc Natl Acad Sci U S A 104(5):1522–1527
167. Kasianowicz JJ, Brandin E, Branton D et al (1996) Characterization of individual polynucleotide molecules using a membrane channel. Proc Natl Acad Sci U S A 93(24):13770–13773
168. Keasling JD (2008) Synthetic biology for synthetic chemistry. ACS Chem Biol 3(1):64–76
169. Keasling JD (2010) Manufacturing molecules through metabolic engineering. Science 330 (6009):1355–1358
170. Khongto B, Laoteng K, Tongta A (2010) Fermentation process development of recombinant Hansenula polymorpha for gamma-linolenic acid production. J Microbiol Biotechnol 20 (11):1555–1562
171. Kim DS, Gusti V, Pillai SG et al (2005) An artificial riboswitch for controlling pre-mRNA splicing. RNA 11(11):1667–1677
172. Kim I-K, Roldão A, Siewers V et al (2011) A systems-level approach for metabolic engineering of yeast cell factories. FEMS Yeast Res 12(2):228–248
173. Kim SH, Shin DH, Liu J et al (2005) Structural genomics of minimal organisms and protein fold space. J Struct Funct Genomics 6(2–3):63–70

174. Kind S, Jeong WK, Schroder H et al (2010) Identification and elimination of the competing N-acetyldiaminopentane pathway for improved production of diaminopentane by Corynebacterium glutamicum. Appl Environ Microbiol 76(15):5175–5180
175. Kirpekar F, Nordhoff E, Larsen LK et al (1998) DNA sequence analysis by MALDI mass spectrometry. Nucleic Acids Res 26(11):2554–2559
176. Kitney RI (2007) Synthetic biology – engineering biologically-based devices and systems. In: IFMBE proceedings. Springer, Berlin/Heidelberg, pp 1138–1139
177. Klimacek M, Krahulec S, Sauer U et al (2010) Limitations in xylose-fermenting Saccharomyces cerevisiae, made evident through comprehensive metabolite profiling and thermodynamic analysis. Appl Environ Microbiol 76(22):7566–7574
178. Knoshaug EP, Zhang M (2009) Butanol tolerance in a selection of microorganisms. Appl Biochem Biotechnol 153(1–3):13–20
179. Kobayashi K, Ehrlich SD, Albertini A et al (2003) Essential Bacillus subtilis genes. Proc Natl Acad Sci U S A 100(8):4678–4683
180. Koonin E (2000) How many genes can make a cell: the minimal-gene-Set concept. Annu Rev Genomics Hum Genet 1:18
181. Koonin EV (2003) Comparative genomics, minimal gene-sets and the last universal common ancestor. Nat Rev Microbiol 1(2):127–136
182. Kotter P, Weigand JE, Meyer B et al (2009) A fast and efficient translational control system for conditional expression of yeast genes. Nucleic Acids Res 37(18):e120
183. Krivoruchko A, Siewers V, Nielsen J (2011) Opportunities for yeast metabolic engineering: lessons from synthetic biology. Biotechnol J 6(3):262–276
184. Kümmel A, Ewald JC, Fendt S-M et al (2010) Differential glucose repression in common yeast strains in response to HXK2 deletion. FEMS Yeast Res 10(3):322–332
185. Kuyper M, Toirkens MJ, Diderich JA et al (2005) Evolutionary engineering of mixed-sugar utilization by a xylose-fermenting Saccharomyces cerevisiae strain. FEMS Yeast Res 5 (10):925–934
186. Kvitek DJ, Sherlock G (2011) Reciprocal sign epistasis between frequently experimentally evolved adaptive mutations causes a rugged fitness landscape. PLoS Genet 7(4)
187. Kwok R (2010) Five hard truths for synthetic biology. Nature 463(7279):288–290
188. Larbi NB, Jefferies C (2009) 2D-DIGE: comparative proteomics of cellular signalling pathways. Methods Mol Biol 517:105–132
189. Lee SM, Cho MO, Park CH et al (2008) Continuous butanol production using suspended and immobilized Clostridium beijerinckii NCIMB 8052 with supplementary butyrate. Energy Fuel 22(5):3459–3464
190. Lee S, Kim P-J, Jeong H (2011) Global organization of protein complexome in the yeast Saccharomyces cerevisiae. BMC Syst Biol 5(1):126
191. Lee SJ, Lee DY, Kim TY et al (2005) Metabolic engineering of Escherichia coli for enhanced production of succinic acid, based on genome comparison and in silico gene knockout simulation. Appl Environ Microbiol 71(12):7880–7887
192. Lee DK, Park JW, Kim YJ et al (2003) Toward a functional annotation of the human genome using artificial transcription factors. Genome Res 13(12):2708–2716
193. Lee TI, Rinaldi NJ, Robert F et al (2002) Transcriptional regulatory networks in Saccharomyces cerevisiae. Science 298(5594):799–804
194. Lefrancois P, Euskirchen G, Auerbach R et al (2009) Efficient yeast ChIP-Seq using multiplex short-read DNA sequencing. BMC Genomics 10(1):37
195. Leon C, Garcia-Canas V, Gonzalez R et al (2011) Fast and sensitive detection of genetically modified yeasts in wine. J Chromatogr A 1218(42):7550–7556
196. Liew ST, Arbakariya A, Rosfarizan M et al (2006) Production of solvent (acetone-butanol-ethanol) in continuous fermentation by clostridium saccharobutylicum DSM 13864 using gelatinised sago starch as a carbon source. Malays J Microbiol 2(2):42–50
197. Liu L, Agren R, Bordel S et al (2010) Use of genome-scale metabolic models for understanding microbial physiology. FEBS Lett 584(12):2556–2564

198. Liu S, Bischoff KM, Qureshi N et al (2010) Functional expression of the thiolase gene thl from Clostridium beijerinckii P260 in Lactococcus lactis and Lactobacillus buchneri. N Biotechnol 27(4):283–288
199. Liu E, Hu Y (2010) Construction of a xylose-fermenting Saccharomyces cerevisiae strain by combined approaches of genetic engineering, chemical mutagenesis and evolutionary adaptation. Biochem Eng J 48(2):204–210
200. Liu H, Liu K, Yan M et al (2011) GTME for improved adaptation of Saccharomyces cerevisiae to corn cob acid hydrolysate. Appl Biochem Biotechnol 164(7):1150–1159
201. Liu HL, Wang WC (2003) Protein engineering to improve the thermostability of glucoamylase from Aspergillus awamori based on molecular dynamics simulations. Protein Eng 16(1):19–25
202. Liu H, Yan M, Lai C et al (2010) gTME for improved xylose fermentation of Saccharomyces cerevisiae. Appl Biochem Biotechnol 160(2):574–582
203. Lombardi A, Bursomanno S, Lopardo T et al (2010) Pichia pastoris as a host for secretion of toxic saporin chimeras. FASEB J 24(1):253–265
204. Lorenz A, Fuchs J, Trelles-Sticken E et al (2002) Spatial organisation and behaviour of the parental chromosome sets in the nuclei of Saccharomyces cerevisiae x S. paradoxus hybrids. J Cell Sci 115(Pt 19):3829–3835
205. Lu C, Jeffries T (2007) Shuffling of promoters for multiple genes to optimize xylose fermentation in an engineered Saccharomyces cerevisiae strain. Appl Environ Microbiol 73 (19):6072–6077
206. Luetz S, Giver L, Lalonde J (2008) Engineered enzymes for chemical production. Biotechnol Bioeng 101(4):647–653
207. Lynch SA, Desai SK, Sajja HK et al (2007) A high-throughput screen for synthetic riboswitches reveals mechanistic insights into their function. Chem Biol 14(2):173–184
208. Lynch M, Sung W, Morris K et al (2008) A genome-wide view of the spectrum of spontaneous mutations in yeast. Proc Natl Acad Sci U S A 105(27):9272–9277
209. Ma M, Liu L (2010) Quantitative transcription dynamic analysis reveals candidate genes and key regulators for ethanol tolerance in Saccharomyces cerevisiae. BMC Microbiol 10(1):169
210. Madsen KM, Udatha GD, Semba S et al (2011) Linking genotype and phenotype of Saccharomyces cerevisiae strains reveals metabolic engineering targets and leads to triterpene hyper-producers. PLoS One 6(3):e14763
211. Mahadevan R, Edwards JS, Doyle FJ 3rd (2002) Dynamic flux balance analysis of diauxic growth in Escherichia coli. Biophys J 83(3):1331–1340
212. Maier T, Guell M, Serrano L (2009) Correlation of mRNA and protein in complex biological samples. FEBS Lett 583(24):3966–3973
213. Malmstrom J, Beck M, Schmidt A et al (2009) Proteome-wide cellular protein concentrations of the human pathogen Leptospira interrogans. Nature 460(7256):762–765
214. Marchisio MA, Rudolf F (2011) Synthetic biosensing systems. Int J Biochem Cell Biol 43 (3):310–319
215. Martins FS, Elian SDA, Vieira AT et al (2011) Oral treatment with Saccharomyces cerevisiae strain UFMG 905 modulates immune responses and interferes with signal pathways involved in the activation of inflammation in a murine model of typhoid fever. Int J Med Microbiol 301 (4):359–364
216. Matsuda F, Furusawa C, Kondo T et al (2011) Engineering strategy of yeast metabolism for higher alcohol production. Microb Cell Fact 10:70
217. Matsuda T, Kigawa T, Koshiba S et al (2006) Cell-free synthesis of zinc-binding proteins. J Struct Funct Genomics 7(2):93–100
218. Mellado MC, Peixoto C, Cruz PE et al (2008) Purification of recombinant rotavirus VP7 glycoprotein for the study of in vitro rotavirus-like particles assembly. J Chromatogr B Analyt Technol Biomed Life Sci 874(1–2):89–94
219. Meng F, Cargile BJ, Patrie SM et al (2002) Processing complex mixtures of intact proteins for direct analysis by mass spectrometry. Anal Chem 74(13):2923–2929

220. Mermelstein LD, Papoutsakis ET, Petersen DJ et al (1993) Metabolic engineering of Clostridium acetobutylicum ATCC 824 for increased solvent production by enhancement of acetone formation enzyme activities using a synthetic acetone operon. Biotechnol Bioeng 42(9):1053–1060
221. Merritt J, DiTonno JR, Mitra RD et al (2003) Parallel competition analysis of Saccharomyces cerevisiae strains differing by a single base using polymerase colonies. Nucleic Acids Res 31 (15):e84
222. Miclet E, Stoven V, Michels PAM et al (2001) NMR spectroscopic analysis of the first two steps of the pentose-phosphate pathway elucidates the role of 6-phosphogluconolactonase. J Biol Chem 276(37):34840–34846
223. Mignone F, Gissi C, Liuni S et al (2002) Untranslated regions of mRNAs. Genome Biol 3(3): REVIEWS0004
224. Miller I, Crawford J, Gianazza E (2006) Protein stains for proteomic applications: which, when, why? Proteomics 6(20):5385–5408
225. Moazed D (2009) Small RNAs in transcriptional gene silencing and genome defence. Nature 457(7228):413–420
226. Morimoto T, Kadoya R, Endo K et al (2008) Enhanced recombinant protein productivity by genome reduction in Bacillus subtilis. DNA Res 15(2):73–81
227. Murakami K, Tao E, Ito Y et al (2007) Large scale deletions in the Saccharomyces cerevisiae genome create strains with altered regulation of carbon metabolism. Appl Microbiol Biotechnol 75(3):589–597
228. Murphy KF, Balazsi G, Collins JJ (2007) Combinatorial promoter design for engineering noisy gene expression. Proc Natl Acad Sci U S A 104(31):12726–12731
229. Mushegian AR, Koonin EV (1996) A minimal gene set for cellular life derived by comparison of complete bacterial genomes. Proc Natl Acad Sci U S A 93(19):10268–10273
230. Mutschlechner O, Swoboda H, Gapes JR (2000) Continuous two-stage ABE-fermentation using Clostridium beijerinckii NRRL B592 operating with a growth rate in the first stage vessel close to its maximal value. J Mol Microbiol Biotechnol 2(1):101–105
231. Na D, Kim TY, Lee SY (2010) Construction and optimization of synthetic pathways in metabolic engineering. Curr Opin Microbiol 13(3):363–370
232. Nagar B, Hantschel O, Young MA et al (2003) Structural basis for the autoinhibition of c-Abl tyrosine kinase. Cell 112(6):859–871
233. Nair RV, Green EM, Watson DE et al (1999) Regulation of the sol locus genes for butanol and acetone formation in Clostridium acetobutylicum ATCC 824 by a putative transcriptional repressor. J Bacteriol 181(1):319–330
234. Namjoshi AA, Hu WS, Ramkrishna D (2003) Unveiling steady-state multiplicity in hybridoma cultures: the cybernetic approach. Biotechnol Bioeng 81(1):80–91
235. Napoli F, Olivieri G, Russo ME et al (2010) Butanol production by Clostridium acetobutylicum in a continuous packed bed reactor. J Ind Microbiol Biotechnol 37(6):603–608
236. Ndukum J, Fonseca LL, Santos H et al (2011) Statistical inference methods for sparse biological time series data. BMC Syst Biol 5:57
237. Nevoigt E (2008) Progress in metabolic engineering of Saccharomyces cerevisiae. Microbiol Mol Biol Rev 72(3):379–412
238. Nevoigt E, Fischer C, Mucha O et al (2007) Engineering promoter regulation. Biotechnol Bioeng 96(3):550–558
239. Nevoigt E, Kohnke J, Fischer CR et al (2006) Engineering of promoter replacement cassettes for fine-tuning of gene expression in Saccharomyces cerevisiae. Appl Environ Microbiol 72 (8):5266–5273
240. Ng P, Tan JJ, Ooi HS et al (2006) Multiplex sequencing of paired-end ditags (MS-PET): a strategy for the ultra-high-throughput analysis of transcriptomes and genomes. Nucleic Acids Res 34(12):e84
241. Ng P, Wei CL, Sung WK et al (2005) Gene identification signature (GIS) analysis for transcriptome characterization and genome annotation. Nat Methods 2(2):105–111

242 Nielsen J, Jewett MC (2008) Impact of systems biology on metabolic engineering of saccharomyces cerevisiae. FEMS Yeast Res 8(1):122–131
243. Nielsen KL, Hogh AL, Emmersen J (2006) DeepSAGE–digital transcriptomics with high sensitivity, simple experimental protocol and multiplexing of samples. Nucleic Acids Res 34 (19):e133
244. Nielsen DR, Leonard E, Yoon SH et al (2009) Engineering alternative butanol production platforms in heterologous bacteria. Metab Eng 11(4–5):262–273
245. Nigam PS, Singh A (2011) Production of liquid biofuels from renewable resources. Prog Energ Combust Sci 37(1):52–68
246. Norbeck J, Påhlman A-K, Akhtar N et al (1996) Purification and characterization of two isoenzymes of DL-glycerol-3-phosphatase from saccharomyces cerevisiae. J Biol Chem 271 (23):13875–13881
247. Olivares-Hernandez R, Bordel S, Nielsen J (2011) Codon usage variability determines the correlation between proteome and transcriptome fold changes. BMC Syst Biol 5:33
248. Olivares-Hernández R, Usaite R, Nielsen J (2010) Integrative analysis using proteome and transcriptome data from yeast to unravel regulatory patterns at post-transcriptional level. Biotechnol Bioeng 107(5):865–875
249. Ong SE, Blagoev B, Kratchmarova I et al (2002) Stable isotope labeling by amino acids in cell culture, SILAC, as a simple and accurate approach to expression proteomics. Mol Cell Proteomics 1(5):376–386
250. Otero JM, Olsson L, Nielsen J (2007) Metabolic engineering of Saccharomyces cerevisiae microbial cell factories for succinic acid production. J Biotechnol 131(2, Supplement):S205
251. Otero JM, Vongsangnak W, Asadollahi MA et al (2010) Whole genome sequencing of Saccharomyces cerevisiae: from genotype to phenotype for improved metabolic engineering applications. BMC Genomics 11:723
252. Papworth M, Kolasinska P, Minczuk M (2006) Designer zinc-finger proteins and their applications. Gene 366(1):27–38
253. Parawira W, Tekere M (2011) Biotechnological strategies to overcome inhibitors in lignocellulose hydrolysates for ethanol production: review. Crit Rev Biotechnol 31(1):20–31
254. Pardo C, Hoose SA, Pondugula S et al (2009) DNA methyltransferase probing of chromatin structure within populations and on single molecules. Methods Mol Biol 523:41–65
255. Parekh SR, Parekh RS, Wayman M (1988) Ethanol and butanol production by fermentation of enzymatically saccharified SO2-prehydrolysed lignocellulosics. Enzyme Microb Technol 10(11):660–668
256. Park KS, Kim JS (2006) Engineering of GAL1 promoter-driven expression system with artificial transcription factors. Biochem Biophys Res Commun 351(2):412–417
257. Park JM, Kim TY, Lee SY (2009) Constraints-based genome-scale metabolic simulation for systems metabolic engineering. Biotechnol Adv 27(6):979–988
258. Park JH, Kim TY, Lee KH et al (2011) Fed-batch culture of Escherichia coli for L-valine production based on in silico flux response analysis. Biotechnol Bioeng 108(4):934–946
259. Park KS, Lee DK, Lee H et al (2003) Phenotypic alteration of eukaryotic cells using randomized libraries of artificial transcription factors. Nat Biotechnol 21(10):1208–1214
260. Park CH, Okos MR, Wankat PC (1989) Acetone-butanol-ethanol (ABE) fermentation in an immobilized cell trickle bed reactor. Biotechnol Bioeng 34(1):18–29
261. Park KS, Seol W, Yang HY et al (2005) Identification and use of zinc finger transcription factors that increase production of recombinant proteins in yeast and mammalian cells. Biotechnol Prog 21(3):664–670
262. Partow S, Siewers V, Bjorn S et al (2010) Characterization of different promoters for designing a new expression vector in Saccharomyces cerevisiae. Yeast 27(11):955–964
263. Pavlik P, Simon M, Schuster T et al (1993) The glycerol kinase (GUT1) gene of Saccharomyces cerevisiae: cloning and characterization. Curr Genet 24(1–2):21–25
264. Petersson A, Almeida JRM, Modig T et al (2006) A 5-hydroxymethyl furfural reducing enzyme encoded by the Saccharomyces cerevisiae ADH6 gene conveys HMF tolerance. Yeast 23(6):455–464

265. Pfleger BF, Pitera DJ, Smolke CD et al (2006) Combinatorial engineering of intergenic regions in operons tunes expression of multiple genes. Nat Biotechnol 24(8):1027–1032
266. Pharkya P, Burgard AP, Maranas CD (2004) OptStrain: a computational framework for redesign of microbial production systems. Genome Res 14(11):2367–2376
267. Pharkya P, Maranas CD (2006) An optimization framework for identifying reaction activation/inhibition or elimination candidates for overproduction in microbial systems. Metab Eng 8(1):1–13
268. Phugare S, Patil P, Govindwar S et al (2010) Exploitation of yeast biomass generated as a waste product of distillery industry for remediation of textile industry effluent. Int Biodeter Biodegr 64(8):716–726
269. Posfai G, Plunkett G 3rd, Feher T et al (2006) Emergent properties of reduced-genome Escherichia coli. Science 312(5776):1044–1046
270. Price JV, Cech TR (1985) Coupling of Tetrahymena ribosomal RNA splicing to beta-galactosidase expression in Escherichia coli. Science 228(4700):719–722
271. Pscheidt B, Glieder A (2008) Yeast cell factories for fine chemical and API production. Microb Cell Fact 7:25
272. Quek LE, Wittmann C, Nielsen LK et al (2009) OpenFLUX: efficient modelling software for 13 C-based metabolic flux analysis. Microb Cell Fact 8:25
273. Qureshi N, Blaschek HP (2001) Recent advances in ABE fermentation: hyper-butanol producing Clostridium beijerinckii BA101. J Ind Microbiol Biotechnol 27(5):287–291
274. Qureshi N, Ezeji TC, Ebener J et al (2008) Butanol production by Clostridium beijerinckii. Part I: use of acid and enzyme hydrolyzed corn fiber. Bioresour Technol 99(13):5915–5922
275. Qureshi N, Li X-L, Hughes S et al (2006) Butanol production from corn fiber xylan using clostridium acetobutylicum. Biotechnol Prog 22(3):673–680
276. Qureshi N, Maddox IS (1995) Continuous production of acetone-butanol-ethanol using immobilized cells of clostridium-acetobutylicum and integration with product removal by liquid-liquid-extraction. J Ferment Bioeng 80(2):185–189
277. Qureshi N, Saha BC, Cotta MA (2008) Butanol production from wheat straw by simultaneous saccharification and fermentation using Clostridium beijerinckii: part II–Fed-batch fermentation. Biomass Bioenergy 32(2):176–183
278. Qureshi N, Saha BC, Dien B et al (2010) Production of butanol (a biofuel) from agricultural residues: part I - use of barley straw hydrolysate. Biomass Bioenergy 34(4):559–565
279. Qureshi N, Saha BC, Hector RE et al (2010) Production of butanol (a biofuel) from agricultural residues: part II - use of corn stover and switchgrass hydrolysates. Biomass Bioenergy 34(4):566–571
280. Ramon A, Marin M (2011) Advances in the production of membrane proteins in Pichia pastoris. Biotechnol J 6(6):700–706
281. Raspor P, Zupan J, ČAdeŽ N (2007) Validation of yeast identification by in Silico Rflp. J Rapid Methods Autom Microbiol 15(3):267–281
282. Ratcliffe RG, Shachar-Hill Y (2005) Revealing metabolic phenotypes in plants: inputs from NMR analysis. Biol Rev Camb Philos Soc 80(1):27–43
283. Ratcliffe RG, Shachar-Hill Y (2006) Measuring multiple fluxes through plant metabolic networks. Plant J 45(4):490–511
284. Redding-Johanson AM, Batth TS, Chan R et al (2011) Targeted proteomics for metabolic pathway optimization: application to terpene production. Metab Eng 13(2):194–203
285. Ren B, Robert F, Wyrick J et al (2000) Genome-wide location and function of DNA binding proteins. Science 290(5500):2306–2309
286. Reschiglian P, Moon MH (2008) Flow field-flow fractionation: a pre-analytical method for proteomics. J Proteomics 71(3):265–276
287. Rinaudo K, Bleris L, Maddamsetti R et al (2007) A universal RNAi-based logic evaluator that operates in mammalian cells. Nat Biotechnol 25(7):795–801
288. Ro DK, Ouellet M, Paradise EM et al (2008) Induction of multiple pleiotropic drug resistance genes in yeast engineered to produce an increased level of anti-malarial drug precursor, artemisinic acid. BMC Biotechnol 8:83

289. Ro DK, Paradise EM, Ouellet M et al (2006) Production of the antimalarial drug precursor artemisinic acid in engineered yeast. Nature 440(7086):940–943
290. Roepstorff P (1997) Mass spectrometry in protein studies from genome to function. Curr Opin Biotechnol 8(1):6–13
291. Ronaghi M, Karamohamed S, Pettersson B et al (1996) Real-time DNA sequencing using detection of pyrophosphate release. Anal Biochem 242(1):84–89
292. Ross PL, Huang YN, Marchese JN et al (2004) Multiplexed protein quantitation in Saccharomyces cerevisiae using amine-reactive isobaric tagging reagents. Mol Cell Proteomics 3 (12):1154–1169
293. Roth RJ, Acton N (1989) A simple conversion of artemisinic acid into artemisinin. J Nat Prod 52(5):1183–1185
294. Rubin-Pitel SB, Zhao H (2006) Recent advances in biocatalysis by directed enzyme evolution. Comb Chem High Throughput Screen 9(4):247–257
295. Ruder WC, Lu T, Collins JJ (2011) Synthetic biology moving into the clinic. Science 333 (6047):1248–1252
296. Sachs AB, Sarnow P, Hentze MW (1997) Starting at the beginning, middle, and end: translation initiation in eukaryotes. Cell 89(6):831–838
297. Saito H, Inoue T (2009) Synthetic biology with RNA motifs. Int J Biochem Cell Biol 41(2):398–404
298. Salusjarvi L, Kankainen M, Soliymani R et al (2008) Regulation of xylose metabolism in recombinant Saccharomyces cerevisiae. Microb Cell Fact 7:18
299. Sampaio PN, Sousa L, Calado CRC et al (2011) Use of chemometrics in the selection of a Saccharomyces cerevisiae expression system for recombinant cyprosin B production. Biotechnol Lett 33(11):2111–2119
300. Sanda T, Hasunuma T, Matsuda F et al (2011) Repeated-batch fermentation of lignocellulosic hydrolysate to ethanol using a hybrid Saccharomyces cerevisiae strain metabolically engineered for tolerance to acetic and formic acids. Bioresour Technol 102(17):7917–7924
301. Sanger F, Nicklen S, Coulson AR (1977) DNA sequencing with chain-terminating inhibitors. Proc Natl Acad Sci U S A 74(12):5463–5467
302. Sansonetti S, Hobley TJ, Calabro V et al (2011) A biochemically structured model for ethanol fermentation by Kluyveromyces marxianus: A batch fermentation and kinetic study. Bioresour Technol 102(16):7513–7520
303. Satyanarayana T, Kunze G (2009) Yeast biotechnology: diversity and applications. Springer, Dordrecht
304. Schilling CH, Edwards JS, Letscher D et al (2000) Combining pathway analysis with flux balance analysis for the comprehensive study of metabolic systems. Biotechnol Bioeng 71 (4):286–306
305. Schmidt M (2009) Synthetic biology: the technoscience and its societal consequences. Springer, Dordrecht/New York
306. Schwartz DC, Cantor CR (1984) Separation of yeast chromosome-sized DNAs by pulsed field gradient gel electrophoresis. Cell 37(1):67–75
307. Seino A, Yanagida Y, Aizawa M et al (2005) Translational control by internal ribosome entry site in Saccharomyces cerevisiae. Biochim Biophys Acta 1681(2–3):166–174
308. Sera T (2009) Zinc-finger-based artificial transcription factors and their applications. Adv Drug Deliv Rev 61(7–8):513–526
309. Shen CR, Lan EI, Dekishima Y et al (2011) Driving forces enable high-titer anaerobic 1-butanol synthesis in Escherichia coli. Appl Environ Microbiol 77(9):2905–2915
310. Shiraki T, Kondo S, Katayama S et al (2003) Cap analysis gene expression for high-throughput analysis of transcriptional starting point and identification of promoter usage. Proc Natl Acad Sci U S A 100(26):15776–15781
311. Shlomi T, Berkman O, Ruppin E (2005) Regulatory on/off minimization of metabolic flux changes after genetic perturbations. Proc Natl Acad Sci U S A 102(21):7695–7700
312. Sillers R, Chow A, Tracy B et al (2008) Metabolic engineering of the non-sporulating, non-solventogenic Clostridium acetobutylicum strain M5 to produce butanol without acetone

demonstrate the robustness of the acid-formation pathways and the importance of the electron balance. Metab Eng 10(6):321–332
313. Skotheim RI, Thomassen GO, Eken M et al (2009) A universal assay for detection of oncogenic fusion transcripts by oligo microarray analysis. Mol Cancer 8:5
314. Smith LM, Sanders JZ, Kaiser RJ et al (1986) Fluorescence detection in automated DNA sequence analysis. Nature 321(6071):674–679
315. Smolke CD, Carrier TA, Keasling JD (2000) Coordinated, differential expression of two genes through directed mRNA cleavage and stabilization by secondary structures. Appl Environ Microbiol 66(12):5399–5405
316. Smolke CD, Keasling JD (2002) Effect of gene location, mRNA secondary structures, and RNase sites on expression of two genes in an engineered operon. Biotechnol Bioeng 80 (7):762–776
317. Steen EJ, Chan R, Prasad N et al (2008) Metabolic engineering of Saccharomyces cerevisiae for the production of n-butanol. Microb Cell Fact 7:36
318. Sterner R, Merkl R, Raushel FM (2008) Computational design of enzymes. Chem Biol 15(5):421–423
319. Storey JD, Tibshirani R (2003) Statistical methods for identifying differentially expressed genes in DNA microarrays. Methods Mol Biol 224:149–157
320. Subramanian A, Tamayo P, Mootha VK et al (2005) Gene set enrichment analysis: a knowledge-based approach for interpreting genome-wide expression profiles. Proc Natl Acad Sci U S A 102(43):15545–15550
321. Suen WC, Zhang N, Xiao L et al (2004) Improved activity and thermostability of Candida antarctica lipase B by DNA family shuffling. Protein Eng Des Sel 17(2):133–140
322. Suess B, Hanson S, Berens C et al (2003) Conditional gene expression by controlling translation with tetracycline-binding aptamers. Nucleic Acids Res 31(7):1853–1858
323. Suzuki N, Nonaka H, Tsuge Y et al (2005) Multiple large segment deletion method for Corynebacterium glutamicum. Appl Microbiol Biotechnol 69(2):151–161
324. Syed QUA, Nadeem M, Nelofer R (2008) Enhanced butanol production by mutant strains of clostridium acetobutylicum in molasses medium. Turk J Biochem 33(1):25–30
325. SynBERC. http://64.13.253.69/parts
326. Tang WL, Zhao H (2009) Industrial biotechnology: tools and applications. Biotechnol J 4 (12):1725–1739
327. Tashiro Y, Shinto H, Hayashi M et al (2007) Novel high-efficient butanol production from butyrate by non-growing Clostridium saccharoperbutylacetonicum N1-4 (ATCC 13564) with methyl viologen. J Biosci Bioeng 104(3):238–240
328. Tashiro Y, Takeda K, Kobayashi G et al (2004) High butanol production by Clostridium saccharoperbutylacetonicum N1-4 in fed-batch culture with pH-Stat continuous butyric acid and glucose feeding method. J Biosci Bioeng 98(4):263–268
329. Teixeira AP, Portugal CA, Carinhas N et al (2009) In situ 2D fluorometry and chemometric monitoring of mammalian cell cultures. Biotechnol Bioeng 102(4):1098–1106
330. Teixeira MC, Raposo LR, Mira NP et al (2009) Genome-wide identification of Saccharomyces cerevisiae genes required for maximal tolerance to ethanol. Appl Environ Microbiol 75 (18):5761–5772
331. Ten years of synergy (2010) Nature 463(7279):269–270
332. Thompson SR, Gulyas KD, Sarnow P (2001) Internal initiation in Saccharomyces cerevisiae mediated by an initiator tRNA/eIF2-independent internal ribosome entry site element. Proc Natl Acad Sci U S A 98(23):12972–12977
333. Tobe VO, Taylor SL, Nickerson DA (1996) Single-well genotyping of diallelic sequence variations by a two-color ELISA-based oligonucleotide ligation assay. Nucleic Acids Res 24(19):3728–3732
334. Tochigi Y, Sato N, Sahara T et al (2010) Sensitive and convenient yeast reporter assay for high-throughput analysis by using a secretory luciferase from Cypridina noctiluca. Anal Chem 82(13):5768–5776

335. Tomas CA, Welker NE, Papoutsakis ET (2003) Overexpression of groESL in Clostridium acetobutylicum results in increased solvent production and tolerance, prolonged metabolism, and changes in the cell's transcriptional program. Appl Environ Microbiol 69(8):4951–4965
336. Topp S, Reynoso CM, Seeliger JC et al (2010) Synthetic riboswitches that induce gene expression in diverse bacterial species. Appl Environ Microbiol 76(23):7881–7884
337. Toussaint M, Wellinger RJ, Conconi A (2010) Differential participation of homologous recombination and nucleotide excision repair in yeast survival to ultraviolet light radiation. Mutat Res/Genet Toxicol Environ Mutagen 698(1–2):52–59
338. Ubiyvovk VM, Ananin VM, Malyshev AY et al (2011) Optimization of glutathione production in batch and fed-batch cultures by the wild-type and recombinant strains of the methylotrophic yeast Hansenula polymorpha DL-1. BMC Biotechnol 11:8
339. Usaite R, Jewett MC, Oliveira AP et al (2009) Reconstruction of the yeast Snf1 kinase regulatory network reveals its role as a global energy regulator. Mol Syst Biol 5:319
340. van den Hombergh JP, van de Vondervoort PJ, Fraissinet-Tachet L et al (1997) Aspergillus as a host for heterologous protein production: the problem of proteases. Trends Biotechnol 15(7):256–263
341. van der Velden AW, Thomas AA (1999) The role of the 5′ untranslated region of an mRNA in translation regulation during development. Int J Biochem Cell Biol 31(1):87–106
342. Van Der Westhuizen A, Jones DT, Woods DR (1982) Autolytic Activity and Butanol Tolerance of Clostridium acetobutylicum. Appl Environ Microbiol 44(6):1277–1281
343. van Dijk EL, Chen CL, d/'Aubenton-Carafa Y et al (2011) XUTs are a class of Xrn1-sensitive antisense regulatory non-coding RNA in yeast. Nature 475(7354):114–117
344. Velculescu VE, Zhang L, Zhou W et al (1997) Characterization of the yeast transcriptome. Cell 88(2):243–251
345. Vieira HL, Estevao C, Roldao A et al (2005) Triple layered rotavirus VLP production: kinetics of vector replication, mRNA stability and recombinant protein production. J Biotechnol 120(1):72–82
346. Villas-Boas SG, Moxley JF, Akesson M et al (2005) High-throughput metabolic state analysis: the missing link in integrated functional genomics of yeasts. Biochem J 388(Pt 2):669–677
347. Wang Y, Chu J, Zhuang Y et al (2009) Industrial bioprocess control and optimization in the context of systems biotechnology. Biotechnol Adv 27(6):989–995
348. Wang Z, Gerstein M, Snyder M (2009) RNA-Seq: a revolutionary tool for transcriptomics. Nat Rev Genet 10(1):57–63
349. Watanabe S, Abu Saleh A, Pack SP et al (2007) Ethanol production from xylose by recombinant Saccharomyces cerevisiae expressing protein-engineered NADH-preferring xylose reductase from Pichia stipitis. Microbiology 153(Pt 9):3044–3054
350. Watanabe S, Kodaki T, Makino K (2004) Various mutations by using yeast gene for protein engineering. Nucleic Acids Symp Ser 48:197–198
351. Watanabe S, Saleh AA, Pack SP et al (2007) Ethanol production from xylose by recombinant Saccharomyces cerevisiae expressing protein engineered NADP + -dependent xylitol dehydrogenase. J Biotechnol 130(3):316–319
352. Weiss B, Davidkova G, Zhou LW (1999) Antisense RNA gene therapy for studying and modulating biological processes. Cell Mol Life Sci 55(3):334–358
353. Werstuck G, Green MR (1998) Controlling gene expression in living cells through small molecule-RNA interactions. Science 282(5387):296–298
354. Westers H, Dorenbos R, van Dijl JM et al (2003) Genome engineering reveals large dispensable regions in Bacillus subtilis. Mol Biol Evol 20(12):2076–2090
355. Winkler WC, Breaker RR (2005) Regulation of bacterial gene expression by riboswitches. Annu Rev Microbiol 59:487–517
356. Winkler WC, Cohen-Chalamish S, Breaker RR (2002) An mRNA structure that controls gene expression by binding FMN. Proc Natl Acad Sci U S A 99(25):15908–15913
357. Winkler WC, Nahvi A, Roth A et al (2004) Control of gene expression by a natural metabolite-responsive ribozyme. Nature 428(6980):281–286
358. Winston F (2001) EMS and UV Mutagenesis in Yeast. Curr Protoc Mol Biol. Wiley

359. Winzeler EA, Shoemaker DD, Astromoff A et al (1999) Functional characterization of the S. cerevisiae genome by gene deletion and parallel analysis. Science 285(5429):901–906
360. Wisselink HW, Toirkens MJ, Wu Q et al (2009) Novel evolutionary engineering approach for accelerated utilization of glucose, xylose, and arabinose mixtures by engineered Saccharomyces cerevisiae strains. Appl Environ Microbiol 75(4):907–914
361. Wolf JJ, Dowell RD, Mahony S et al (2010) Feed-forward regulation of a cell fate determinant by an RNA-binding protein generates asymmetry in yeast. Genetics 185(2):513–522
362. Wright J, Bellissimi E, de Hulster E et al (2011) Batch and continuous culture-based selection strategies for acetic acid tolerance in xylose-fermenting Saccharomyces cerevisiae. FEMS Yeast Res 11:299–306
363. Wu L, Mashego MR, van Dam JC et al (2005) Quantitative analysis of the microbial metabolome by isotope dilution mass spectrometry using uniformly 13 C-labeled cell extracts as internal standards. Anal Biochem 336(2):164–171
364. Wu D, Wang Q, Assary RS et al (2011) A computational approach to design and evaluate enzymatic reaction pathways: application to 1-butanol production from pyruvate. J Chem Inf Model 51(7):1634–1647
365. Xia M, Farkas T, Jiang X (2007) Norovirus capsid protein expressed in yeast forms virus-like particles and stimulates systemic and mucosal immunity in mice following an oral administration of raw yeast extracts. J Med Virol 79(1):74–83
366. Xia X, Holcik M (2009) Strong eukaryotic IRESs have weak secondary structure. PLoS One 4(1):1–3
367. Yazawa H, Kamisaka Y, Kimura K et al (2011) Efficient accumulation of oleic acid in Saccharomyces cerevisiae caused by expression of rat elongase 2 gene (rELO2) and its contribution to tolerance to alcohols. Appl Microbiol Biotechnol 91(6):1593–1600
368. Yen L, Svendsen J, Lee JS et al (2004) Exogenous control of mammalian gene expression through modulation of RNA self-cleavage. Nature 431(7007):471–476.
369. Young E, Lee SM, Alper H (2010) Optimizing pentose utilization in yeast: the need for novel tools and approaches. Biotechnol Biofuels 3:24
370. Young JD, Walther JL, Antoniewicz MR et al (2008) An elementary metabolite unit (EMU) based method of isotopically nonstationary flux analysis. Biotechnol Bioeng 99(3):686–699
371. Yu KO, Jung J, Kim SW et al (2011) Synthesis of FAEEs from glycerol in engineered Saccharomyces cerevisiae using endogenously produced ethanol by heterologous expression of an unspecific bacterial acyltransferase. Biotechnol Bioeng 109(1):110–115
372. Zamboni N, Sauer U (2009) Novel biological insights through metabolomics and 13 C-flux analysis. Curr Opin Microbiol 12(5):553–558
373. Zarif BR, Azin M, Amirmozafari N (2011) Increasing the bioethanol yield in the presence of furfural via mutation of a native strain of Saccharomyces cerevisiae. Afr J Microbiol Res 5 (6):651–656
374. Zhang L, Chang S, Wang J (2010) How to make a minimal genome for synthetic minimal cell. Protein Cell 1(5):8
375. Zhang L, Chang S, Wang J (2011) Synthetic biology: from the first synthetic cell to see its current situation and future development. Chinese Sci Bull 56(3):9
376. Zhang J, Lau MW, Ferre-D'Amare AR (2010) Ribozymes and riboswitches: modulation of RNA function by small molecules. Biochemistry 49(43):9123–9131
377. Zhang Y, Thiele I, Weekes D et al (2009) Three-dimensional structural view of the central metabolic network of Thermotoga maritima. Science 325(5947):1544–1549
378. Zhang J, Vaga S, Chumnanpuen P et al (2011) Mapping the interaction of Snf1 with TORC1 in Saccharomyces cerevisiae. Mol Syst Biol 7:545
379. Zhang J, Vemuri G, Nielsen J (2010) Systems biology of energy homeostasis in yeast. Curr Opin Microbiol 13(3):382–388
380. Zhang Y, Yu Z, Xin L et al (2010) Expression of the hemagglutinin and neuramidinase gene of influenza A virus H1N1 in Pichia methanolica. Sheng Wu Gong Cheng Xue Bao 26(8):1068–1073
381. Zhao L, Wang J, Zhou J et al (2011) Modification of carbon flux in Sacchromyces cerevisiae to improve L-lactic acid production. Wei Sheng Wu Xue Bao 51(1):50–58

Chapter 10
Design of Superior Cell Factories for a Sustainable Biorefinery By Synthetic Bioengineering

Tomohisa Hasunuma, Fumio Matsuda, and Akihiko Kondo

Contents

Abstract To build an energy and material secure future, we must pioneer the next generation of renewable fuels and chemicals using environmentally benign production processes. Since biomass represents an abundant carbon-neutral

T. Hasunuma
Organization of Advanced Science and Technology, Kobe University, 1-1 Rokkodai, Nada-ku, Kobe 657-8501, Japan

Precursory Research for Embryonic Science and Technology (PRESTO), Japans Science and Technology Agency (JST), 3-5 Sanbancho, Chiyoda-ku, Tokyo 113-0027, Japan

F. Matsuda
Organization of Advanced Science and Technology, Kobe University, 1-1 Rokkodai, Nada-ku, Kobe 657-8501, Japan

A. Kondo (✉)
Department of Chemical Science and Engineering, Graduate School of Engineering, Kobe University, 1-1 Rokkoda, Nada-ku, Kobe 657-8501, Japan
e-mail: akondo@kobe-u.ac.jp

C. Wittmann and S.Y. Lee (eds.), *Systems Metabolic Engineering*,
DOI 10.1007/978-94-007-4534-6_10, © Springer Science+Business Media Dordrecht 2012

renewable resource for the production of biofuels, numerous environmental and social benefits could result from the replacement of petroleum-based transport fuels with bioethanol converted from biomass. One of the key technologies for the development of biorefineries is cell surface engineering, which is a powerful tool for engineering and functionalizing many organisms. Using the technology, various kinds of functional proteins, such as enzymes, can be expressed on the cell surface without loss of cell activity. The display of amylolytic and cellulolytic enzymes on the surface of *Saccharomyces cerevisiae* has accomplished direct ethanol production from starchy and cellulosic biomass. Moreover, the display of hemicellulase on the surface of *S. cerevisiae* that has a xylose-assimilating metabolic pathway has enabled production of ethanol from hemicellulosic materials. Furthermore, reutilization of the cell-surface engineered yeast has an advantage in the reduction of enzyme cost, which enables reuse of enzymes on the cell surface by collecting the cells. Thus, cell surface engineering is a promising technology for the development of a consolidated bioprocess by integrating enzyme production, saccharification and fermentation. Regardless of the biomass hydrolysis, metabolic engineering of microorganisms is emphasized for the efficient production of ethanol from biomass. Specifically, lignocellulosic hydrolysates contain high concentrations of inhibitors that negatively affect metabolism and ethanol yields. To circumvent these difficulties, robust *S. cerevisiae* strains that efficiently ferment mixtures of hexose and pentose sugars in the presence of various chemical contexts for industrial ethanol production should be constructed through metabolic engineering approaches. A combination of a cell-surface displayed enzyme system and an intracellular metabolic engineering system is a very effective approach for developing cells with improved fermentation ability for industrial applications. The technology (synthetic bioengineering) will open up new applications of cell factories to industrially important processes.

Keywords Cell factory • Sustainability • Biorefinery • Synthetic bioengineering • Bioethanol • Biofuels • Surface engineering • Metabolic engineering • Metabolomics • Tolerance • Renewable resources • *Saccharomyces cerevisiae* • Lignocellulosic biomass • Saccharification • Sugars • Industrial feedstocks • Consolidated bioprocessing (CBP) • Whole-cell biocatalyst • Glucoamylase • Trichoderma • Yeast • Pretreatment • Microbial inhibitors

10.1 Introduction

Bioenergy represents the utilization of biomass as starting material for the production of sustainable fuels and chemicals [1]. Environmental concerns and the depletion of oil reserves have resulted in governmental actions and incentives to establish greater energy independence by promoting research on environmentally benign and sustainable biofuels, such as bioethanol and biodiesel. Ethanol is the most widely used biofuel, either as a fuel or as a gasoline blend. World ethanol production has reached approximately 51,000 million liters, with the United States and Brazil being the foremost producers [2]. On average, 73% of the ethanol produced worldwide corresponds to fuel ethanol, 17% to beverage ethanol and 10% to industrial ethanol [3].

Fuel ethanol can be obtained from sucrose directly or from starchy and lignocellulosic biomass via several steps. Although the process complexity depends on the type of feedstock, the spectrum of designed and implemented technologies ranges from simple conversion of sugars by fermentation to multistage conversion of lignocellulosic biomass to ethanol [3]. The main feedstock for bioethanol production is sugarcane in the form of either cane juice or molasses (by-products of sugar mills) and starchy biomass. Approximately 79% of the ethanol in Brazil is produced from sugarcane juice and the remainder is from cane molasses [4]. The most widely used microorganism for ethanol fermentation is *Saccharomyces cerevisiae* due to its ability to hydrolyze cane sucrose into fermentable sugars. Furthermore, *S. cerevisiae* has many advantages as an ethanol producer, such as fast sugar consumption, high ethanol yield from glucose, and high resistance to ethanol. Among bacteria, *Zymomonas mobilis* is credited as a high ethanol yielding strain, producing up to 97% of the theoretical yield. However, its range of fermentable substrates is too narrow (glucose, fructose and sucrose) [5] and bacteria also produce undesirable compounds, such as sorbitol, which reduce the efficiency of the conversion of sucrose to ethanol [6].

Starchy and lignocellulosic materials are gaining importance as potential substrates for bioethanol production. However, because the yeast *S. cerevisiae* cannot directly utilize starch and lignocelluloses as carbon sources, several strategies have been developed for the expression of biomass degrading enzymes [7–11] using recombinant *S. cerevisiae* strains. Simultaneous saccharification and fermentation with engineered microbes is the key factor for achieving process integration to produce bioethanol cost effectively.

Recently, the development of cell surface engineering of *S. cerevisiae* has opened a new approach to the issue [12–20]. This chapter attempts to present a comprehensive overview of the developing technologies for biorefineries through yeast cell surface engineering and metabolic pathway engineering for the production of bioethanol from various biomass resources.

10.2 Cell Surface Engineering for Bioethanol Production

10.2.1 Yeast Cell-Surface Engineering System

In a cell-surface engineering system established using the yeast *S. cerevisiae*, so-called arming yeasts are constructed that are armed with biocatalysts in the forms of enzyme, functional proteins, antibodies, and combinatorial protein libraries [17]. Among many advantages of the system, in which proteins are genetically displayed on the cell surface, are easy reproduction of the displayed biocatalysts and easy separation of product from catalyst.

The cell surface is a functional interface between the inside and outside of the cell. Some surface proteins extend across the plasma membrane, and others are bound by noncovalent interactions to the cell-surface components. Cells have systems for anchoring surface-specific proteins and for confining surface proteins to particular domains on the cell surface. In biotechnology, the cell surface can be exploited by making use of known mechanisms for the transport of proteins to the cell surface. In particular, *S. cerevisiae* is useful as a host for genetic engineering, since it allows the folding and glycosylation of expressed heterologous eukaryotic proteins and can be genetically manipulated. Moreover, the yeast can be cultivated to a high density in an inexpensive medium, so that the display of enzymes on yeast cell surface has several applications in bioconversion processes.

Many glucoamylase-extractable proteins on the yeast cell surface, for example, agglutinin (Agα1 and Aga1), flocculin Flo1, Sed1, Cwp1, Cwp2, Tip1, and Tir1/Srp1 have glycosylphosphatidylinositol (GPI) anchors that play important roles in the expression of cell-surface proteins [21, 22]. GPI anchored proteins contain hydrophobic peptides at their C-termini. After the completion of protein synthesis, the precursor protein remains anchored in the endoplasmic reticulum (ER) membrane by the hydrophobic C-terminal sequence, with the rest of the protein in the ER lumen. Within less than a minute, the hydrophobic C-terminal sequence is cleaved at this site and concomitantly replaced with a GPI anchor, presumably by the action of a transamidase [17].

Among the GPI anchor proteins, α-agglutinin and flocculin anchors are suitable for the expression of hydrolytic enzymes. The molecular information on α-agglutinin is utilized to target the heterologous proteins of biotechnological importance to the outermost glycoprotein layer of the cell wall. In the α-agglutinin system, the C-terminal half of the α-agglutinin containing the GPI anchor attachment signal has been used to anchor heterologous proteins on the yeast cell surface [23]. In the flocculin system, two types of cell-surface display methods have been developed. In one system, C-terminal region of Flo1p, which contains a GPI-attachment signal, is used, while the second system attempts to utilize the ability of the flocculation functional domain of Flo1p to create a novel surface display apparatus [17].

10.2.2 Consolidated Bioprocessing Using Yeast Cell-Surface Engineering

The conventional method of bioethanol production from starchy or lignocellulosic biomass is a complex multistep process. Since wild-type *S. cerevisiae* strains have no ability to degrade polysaccharide biomass, the bioconversion of these raw materials to ethanol consists of three stages: pretreatment, hydrolysis by an enzyme complex, and fermentation [24]. Hydrolysis and fermentation can be carried out concurrently in a

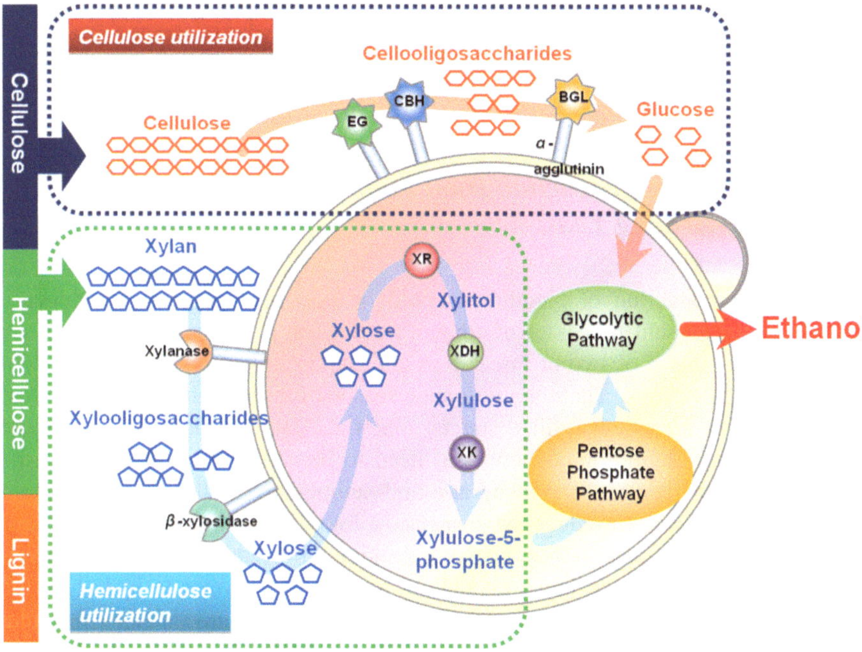

Fig. 10.1 Direct ethanol production from lignocellulosic materials using an arming *S. cerevisiae* strain displaying cellulolytic and hemicellulolytic enzymes (endoglucanase (*EG*), cellobiohydrolase (*CBH*), β-glucosidase (*BGL*), xylanase and β-xylosidase) on the cell surface and expressing xylose-assimilating enzymes (xylose reductase (*XR*), xylitol dehydrogenase (*XDH*) and xylulokinase (*XK*)) in the cell

process known as simultaneous saccharification and fermentation (SSF) [25]. SSF reduces equipment costs by performing the hydrolysis and fermentation in a single reactor and eliminates the need for expensive materials capable of withstanding strong acids or other chemicals. Additionally, SSF increases the hydrolysis rate by reducing product inhibition, as the sugars are rapidly consumed by yeast during SSF. A major challenge in improving the SSF process is matching the temperature conditions required for optimum performance of the enzyme and the fermenting microorganism, because the optimum temperature for cellulase enzymes is higher than can be tolerated by most yeast used for industrial ethanol production [26, 27].

Recently, a new method known as consolidated bioprocessing (CBP), which combines enzyme production, cellulose saccharification, and fermentation into a single process, has been proposed that would greatly enhance the cost-effectiveness of bioethanol production [24, 25, 28]. The recent advancement of yeast cell-surface engineering provided the tools for the display of amylolytic, cellulolytic and xylanolytic enzymes, which allow the utilization of a yeast whole-cell biocatalyst for direct ethanol production from starchy and lignocellulosic materials (Fig. 10.1). Moreover, combining enzyme display technology and metabolic pathway integration for simultaneous assimilation of pentose and hexose could provide a simple fermentation strategy without any additional enzymes [29].

10.2.3 Ethanol Production from Starchy Biomass by Arming Yeast

Large-scale production of bioethanol could be possible from starchy biomass. However, the present processes of ethanol production from starchy materials via fermentation consist of two or three steps and require improvements for efficient production at low cost. There are two main reasons for the high cost. One is that, since *S. cerevisiae* cannot utilize starchy materials, large amounts of amylolytic enzymes (glucoamylase and α-amylase) need to be added. The other is that the starchy materials need to be cooked at a high temperature (140–180°C) to obtain a high ethanol yield. To reduce the energy cost for cooking, Matsumoto et al. [30] reported non-cooking and low-temperature cooking fermentation systems. However, it is still necessary to add large amounts of amylolytic enzymes to hydrolyze the starch to glucose. The following section describes the bioconversion of different starch-containing materials by cell-surface engineered yeast strains.

Surface expression of amylolytic enzymes was initiated by the pioneering work of Murai et al. [18]. They reported a strategy for development of recombinant *S. cerevisiae* strains displaying amylolytic enzymes on the cell surface. *Rhizopus oryzae* glucoamylase, an exo-type amylolytic enzyme cleaving α-1,4-linked and α-1,6-linked glucose from starch, was displayed as a fusion with α-agglutinin using the secretion signal peptide of *R. oryzae* glucoamylase on the cell surface of *S. cerevisiae* strain MT8-1. The anchoring on the cell wall was verified by immunofluorescence labeling with anti-glucoamylase IgG [18, 19].

Kondo et al. [13] used a flocculating yeast for the surface expression of glucoamylase because such a strain has an advantage in the recovery of cells with displayed enzymes by flocculation. The display of glucoamylase did not negatively affect the flocculation ability of the yeast during growth and ethanol fermentation phases. Moreover, the activity of the glucoamylase displayed on the surface of flocculent yeast was similar to that displayed on non-flocculent yeast. The recombinant glucoamylase-displaying flocculent yeast strain maintained a high ethanol production rate (approximately 0.6–0.7 g L^{-1} h^{-1}) during repeated utilization of surface engineered yeast cells for fermentation over 300 h. The display of glucoamylase has another advantage in reducing the glucose concentration in the fermentation medium, which minimizes the risk of contamination.

On the other hand, the display of only glucoamylase led to the accumulation of insoluble starch during fed-batch fermentation because of the lack of a liquefying α-amylase. In order to overcome this problem, Shigechi et al. [31] developed two recombinant yeast strains: one co-displaying glucoamylase and α-amylase on the cell surface, and the other displaying glucoamylase and secreting α-amylase into the culture medium. In fed-batch fermentations of soluble potato starch as the carbon source using these recombinant yeasts co-expressing sequential amylolytic enzymes, ethanol production reached 60 g L^{-1} after 100 h under anaerobic conditions. Moreover, ethanol production from low-temperature cooked starch has several advantages over the conventional high-temperature cooking (140–180°C) process [30], because the high-temperature cooking requires high energy and the addition of large amounts of amylolytic enzymes. Shigechi et al. [32] directly produced ethanol from corn starch cooked at 80°C in a single step using a recombinant yeast strain co-displaying both glucoamylase from *R. oryzae* and α-amylase from *Bacillus stearothermophilus*. α-Amylase hydrolyzing α-1,4-linkages of starch in a random fashion played an important role in the co-operative and sequential decomposition of the starch, which led to the efficient production of ethanol. Moreover, the maximum ethanol titer, ethanol-production rate and substrate consumption rate in the fermentation from low-temperature cooked starch were almost the same as those from high-temperature cooked starch. Using the co-displaying strain, the yield of ethanol produced was 0.50 g per gram of carbohydrate consumed, which corresponds to 97.2% of the theoretical yield.

The isolation of amylase from the lactic acid bacterium *Streptococcus bovis* [33] opened new horizons for the hydrolysis of raw starch. Shigechi et al. [34] developed a novel non-cooking fermentation system for direct ethanol production with a yeast strain that co-displayed *R. oryzae* glucoamylase and *S. bovis* α-amylase using α-agglutinin and Flo1p as anchor proteins. They reported that α-amylase activity depends on the anchor protein, whereas glucoamylase activity is anchor independent. The use of the Flo1 anchor led to 40 times higher α-amylase activity than the α-agglutinin anchor. Since several α-amylases have a starch binding domain at their C-terminal region [35], the co-display of glucoamylase and α-amylase using α-agglutinin and Flo1p as anchors would be effective for ethanol production from raw corn starch [36–38]. The yield in terms of grams of ethanol produced per gram of sugar consumed was 0.44 g g^{-1}, which corresponds to 86.5% of the theoretical yield [34].

10.2.4 Ethanol Production from Lignocellulosic Biomass By Arming Yeast

Recently, lignocellulosic biomass such as agricultural waste, forestry residues, waste paper, and industrial wastes has been recognized as an inexpensive and abundant source of sugar for the production of sustainable transportation fuel. Lignocellulose is composed of cellulose (insoluble fibers of β-1,4-glucan), hemicellulose (non-cellulosic polysaccharides, including β-1,4-xylan, xyloglucan, glucoxylan), and lignin (a complex of phenolic compounds) [39, 40].

The filamentous fungus *Trichoderma reesei* is a well-known cellulolytic organism that secretes a large number of cellulolytic enzymes, including endoglucanase (EG), cellobiohydrolase (CBH), and β-glucosidase (BGL) [41, 42]. EGs act randomly against the amorphous region of the cellulose chain to produce reducing and non-reducing ends for CBHs that produce cellobiose from reducing or non-reducing ends of crystalline cellulose. Cellulose chains are thus efficiently degraded to soluble cellobiose and cellooligosaccharides by the endo-exo synergism of EG and CBH [43–45]. In the last step of enzymatic cellulose degradation, cellooligosaccharides are hydrolyzed to glucose by β-glucosidase. Exo-exo synergism between two cellobiohydrolases has also been reported [46]. The expression of these cellulolytic enzymes in bacteria [47, 48] and yeast [49–51] has been expected to be a way of reducing the cost of cellulase production and other pretreatments in ethanol production from lignocellulosic materials [52–54]. Recently, some researchers have developed ethanologenic bacteria [47, 48, 55] and yeast [49] for the production of cellulosic ethanol. For instance, the recombinant *Klebsiella oxytoca* SZ21 was able to produce ethanol directly from amorphous cellulose, although the yield was low [48]. Apart from *Trichoderma*, the anaerobic bacteria *Clostridium thermocellum* and *Clostridium cellulovorans* produce a cellulosome complex consisting of cellulase and hemicellulase organized on the cell surface [56, 57]. Besides these species, the mesophilic cellulolytic *Clostridium cellulolyticum* has been widely exploited for its cellulosome [58, 59].

For cellulose conversion to ethanol, enzymatic conversion of the substrate to fermentable sugars is followed by fermentation by a microbe, such as yeast. Recently, CBP has been proposed to greatly enhance the cost-effectiveness of bioethanol production (Fig. 10.2) [24, 25]. An ideal microorganism for CBP would be capable of simultaneous cellulose saccharification and ethanol production from sugars. Yeast, particularly *S. cerevisiae*, is the first choice for industrial ethanol production, because it has many advantages as an ethanol producer, such as fast sugar consumption, high ethanol yield form glucose, and higher resistance to ethanol and other compounds present in lignocellulosic hydrolysates than bacteria [60, 61]. Thus, heterologous production of cellulases has been pursued with this yeast [62, 63].

Fujita, et al. [64] attempted to convert a cellulosic β-glucan from barley into ethanol by constructing cellulose-degrading yeast cells that co-display *Aspergillus aculeatus* β-glucosidase 1 (BGL1) and *T. reesei* endoglucanase II (EGII). β-glucan is a linear, soluble polysaccharide composed of an average of 1,200 glucose residues joined by approximately 70% β-1,4-glycosidic linkages and approximately 30% β-1,3-glycosidic linkages. The recombinant yeast showed growth in synthetic medium containing β-glucan as a sole carbon source and was able to directly ferment 45 g L^{-1} β-glucan to produce 16.5 g L^{-1} ethanol within 50 h. The yield in terms of grams of ethanol produced per gram of carbohydrate utilized was 0.48 g g^{-1}, which corresponds to 93.3% of the theoretical yield. Furthermore, Fujita et al. [65] reported simultaneous saccharification and fermentation of a phosphoric acid-swollen cellulose (PASC) to ethanol with the use of a recombinant yeast strain co-displaying three cellulolytic enzymes, *T. reesei* EGII and cellobiohydrolase II (CBHII), and *A. aculeatus* BGL1. The yeast strain successfully

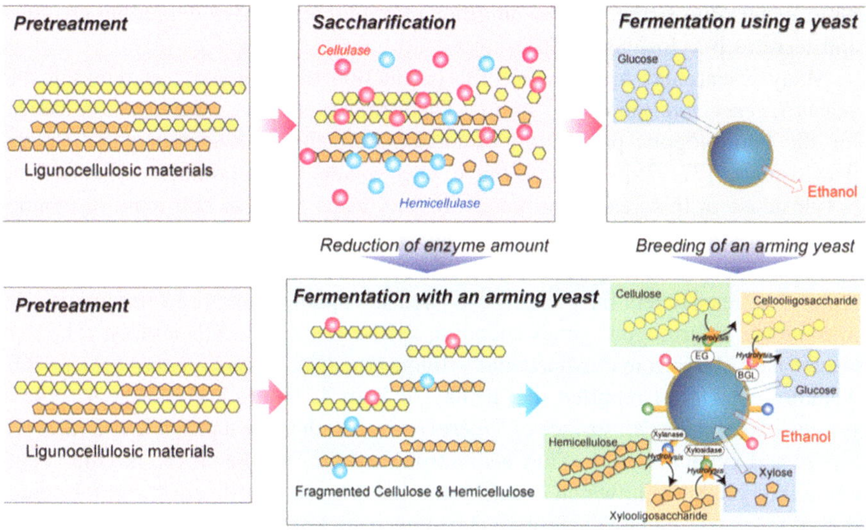

Fig. 10.2 Process improvement using a cell surface engineered *S. cerevisiae* strain

produced ethanol from PASC as a sole carbon source. The yield of ethanol was 0.45 g g^{-1}, which corresponds to 88.5% of the theoretical yield.

Recently, cell surface engineering was applied to a thermotolerant strain of the yeast *Kluyveromyces marxianus* strain for the display of cellulolytic enzymes on the cell surface [27]. One of the major drawbacks in CBP is the optimum temperature for saccharification and fermentation. The saccharification with cellulolytic enzymes is best done around 50°C, while most fermenting microbes have an optimum temperature for ethanol fermentation between 28°C and 37°C. Because the pretreatment of lignocellulose is at high temperature, the saccharification and fermentation temperature should match the previous process of CBP. In a recent report [27], the strain grew well at a temperature as high as 48°C, at which it was able to produce ethanol from the cellulosic material β-glucan with a yield of 0.47 g ethanol per gram of consumed carbohydrate. This study supports the development of CBP for bioethanol production.

β-1,4-Xylan, the most abundant component of hemicellulose, is a complex polysaccharide consisting of a backbone of β-1,4-linked xylopyranoside partially substituted with acetyl, glucuronosyl, and arabinosyl side chains. For the efficient utilization of biomass, the conversion of xylan as well as cellulose is required. Xylan is hydrolyzed to xylooligosaccharides by endo-β-xylanase, after which β-xylosidase hydrolyzes xylooligosaccharides to xylose [66, 67]. Many bacterial and fungal species are able to utilize xylan as a carbon source [68]. *Trichoderma* spp. and *Aspergillus* spp. secrete large amounts of xylan-degrading enzymes [69, 70]. Specifically, *T. reesei* secretes the two major inducible endo-β-xylanases xylanase I (XYNI) and xylanase II (XYNII). XYNII represents more than 50% of the total xylanolytic activity of *T. reesei* cultivated on xylan [71]. The industrial koji mold *Aspergillus oryzae* is also an efficient producer of xylan degrading enzymes and secretes β-xylosidase [72].

Many researchers have studied the production of xylanolytic enzymes in the yeasts *S. cerevisiae* and *Pichia stipitis*. The former has been used as a host organism for the heterologous production of bacterial or fungal xylanases [73–76] and β-xylosidase [77, 78]; attempts have been made to co-produce xylanase and β-xylosidase in this yeast as a means of converting xylan to cell mass or ethanol through simultaneous saccharification and fermentation [79]. However, wild-type *S. cerevisiae* is unable to utilize xylose for growth or fermentation. Accordingly, a number of researchers have engineered yeasts capable of xylose fermentation by introducing heterologous genes encoding xylose reductase (XR) and xylitol dehydrogenase (XDH) from *P. stipitis* and xylulokinase (XK) from *S. cerevisiae* [80–83]. Katahira et al. [84] reported the display of both *T. reesei* XYNII and *A. oryzae* β-xylosidase on the cell surface of *S. cerevisiae*, which made it possible to hydrolyze birchwood xylan to xylose. They also introduced XR, XDH and XK into the xylan-degrading yeast strain, which enabled simultaneous saccharification and fermentation of xylan to ethanol. Xylanolytic enzymes were also expressed in *P. stipitis* with the ability to ferment xylose [85–87]. The direct production of ethanol from xylan was performed using a recombinant *P. stipitis* strain secreting *Cryptococcus albidus* xylanase, although ethanol productivity and yield were not reported [87].

Lignocellulosic material is often decomposed by acid treatment to release not only glucose, but also various monosaccharides, such as xylose, mannose, galactose, and arabinose, and oligosaccharides. Therefore, microorganisms that are able to ferment these sugars are required for the industrial production of ethanol. Although the fermentation of xylose and cellobiose is simultaneously performed by the thermotolerant methylotrophic yeast *Hansenula polymorpha* [88], the ethanol yield and productivity are insufficient. Katahira et al. [89] constructed a recombinant *Saccharomyces* strain with xylose-fermenting ability by introducing genes of *P. stipitis* for XR and XDH and *S. cerevisiae* XK for intracellular expression, as well as cellooligosaccharide-degrading ability by introducing a fusion gene of *A. aculeatus* BGL1 and the 3′ half of α-agglutinin for cell-surface display. Using the recombinant strain, ethanol was produced from a lignocellulosic hydrolysate obtained by concentrated sulfuric acid pretreatment of wood chips.

10.3 Metabolic Pathway Engineering for the Improvement of Ethanol Production from Lignocellulosic Materials

10.3.1 Inhibitor Tolerance

To exploit lignocellulosic materials for fuel ethanol production, the improvement of not only fermentation capacity, but also tolerance to compounds present in the hydrolysates is required. Specifically, lignocellulosic hydrolysates contain a variety of toxic compounds that negatively affect microbial growth, metabolism and ethanol yield, in contrast to sugarcane- or starch-derived feedstocks. Harsh conditions used in the pretreatment of the raw lignocellulosic material release inhibitors including weak organic acids, furan derivatives, and phenolics [90, 91]. In addition, acetic acid released during solubilization and hydrolysis of hemicellulase [92] is usually found at a high concentration in the hydrolysate. Although levels of acetate depend on the type of biomass and the pretreatment method, concentrations typically range from 1 to 10 g L^{-1} in the hydrolysate [90]. Formic acid is typically present at lower concentrations than acetic acid, but is more toxic to *S. cerevisiae* than acetic acid [93, 94]. Other toxic weak acids, for which the concentrations in the hydrolysate are rarely reported, are present at even lower concentrations than formic acid. Other microbial inhibitors [e.g., furfural, hydroxymethylfurfural (HMF), vanillin and syringaldehyde] are also produced by decomposition of sugars and lignin during the biomass pretreatment. Therefore, detoxification procedures for lignocellulosic hydrolysates are under development [95]; however, practical large-scale detoxification is technically complex and adds cost to the fermentation process.

To improve ethanol production in the presence of fermentation inhibitors, a metabolic engineering approach identifying enzymes responsible for the inhibitor tolerance of *S. cerevisiae* strains was launched [60, 91]. Overexpression of homologous or heterologous genes encoding the enzymes has detoxified specific inhibitors such as furan derivatives and phenolics [96–98]. However, metabolic engineering strategies that have addressed yeast tolerance to weak acids in lignocellulosic hydrolysates are rare.

Although the mechanism of inhibition by weak acids is not easily elucidated, the inhibitory effect of weak acids has been ascribed to uncoupling and intracellular anion accumulation [90, 92]. The undissociated form of weak acids can diffuse from the fermentation medium across the plasma membrane into the cell and dissociate there due to the higher intracellular pH, thus decreasing the cytosolic pH. In addition, intracellular accumulation of the anionic species may contribute to weak acid toxicity [99, 100]. If the anionic form of acetic acid is captured in the cells, undissociated acid in the medium will diffuse into the cell until equilibrium is reached. Thus, enhanced basic research is expected to identify further target genes for improved acid tolerance.

10.3.2 *Improving Inhibitor Tolerance of* S. cerevisiae *Through Metabolic Pathway Engineering*

The metabolic profiling technique [101] can be a powerful tool to gain insight into functional biology. Comprehensive analysis of a wide range of metabolites from cellular extracts by high-sensitivity mass spectrometry makes it well suited for the identification of metabolic compounds that are important for specific biological questions [102–104]. Accordingly, metabolomics analysis can be used for designing novel strategies to develop more efficient cell capabilities through genetic engineering.

Recently, the effect of acetic acid on xylose fermentation was determined through a metabolomic approach by both capillary electrophoresis-mass spectrometry and gas chromatography-mass spectrometry, which revealed that the accumulation of metabolites involved in the non-oxidative pentose phosphate pathway (PPP) [e.g., sedoheptulose 7-phosphate, ribulose 5-phosphate, ribose 5-phosphate and erythrose 4-phosphate] increased with an increase in acetic acid concentration in a recombinant xylose-fermenting *S. cerevisiae* strain [105]. Transaldolase (TAL) and transketolase (TKL) have been considered rate-limiting enzymes in the non-oxidative PPP, which is supported by several genome-scale and enzymatic analyses of mutant strains with improved xylose metabolism, in which *TAL1*, *TKL1*, or both genes have been found to be upregulated [106–110]. Consequently, *TAL1* and *TKL1* from *S. cerevisiae* were overexpressed in the strain, which successfully conferred increased ethanol productivity in the presence of acetic and formic acids in the xylose-fermenting yeast.

The biochemical and molecular basis of yeast stress tolerance has been investigated by a global gene expression analysis using a DNA microarray [111]. Evolutionarily engineered yeast that tolerates furfural and 5-HMF showed enhanced expression of *ZWF1*, *GND1*, *GND2*, and *TDH1*, genes involved in NAD(P)H regeneration steps in PPP and glycolysis pathway [112]. Furthermore, screening of a single gene deletion mutant library revealed that PPP genes *ZWF1*, *GND1*, *RPE1*, and *TKL1* were required for furfural tolerance in *S. cerevisiae* [97]. These reports support PPP being intimately associated with yeast stress tolerance to fermentation inhibitors.

The identification of an effective target for engineering requires an understanding of innate mechanisms for stress tolerance. For this purpose, microarray analysis has been applied to survey the transcriptional response of *S. cerevisiae* to stress, a strategy that has uncovered many unpredictable stress responses [113, 114]. For example, an analysis of *S. cerevisiae* under ethanol stress conditions revealed novel stress responsive genes such as tryptophan biosynthesis-related genes as candidates for tolerance engineering [115, 116]. The microarray analysis of xylose-fermenting *S. cerevisiae* under different formic acid concentrations revealed that the expression of 7 genes (*FDH1, FDH2, ALD5, HXK2, VCX1, GPD2,* and *HXT4*) was upregulated in response to increasing formic acid dosage [117]. The list of upregulated genes suggested a response of *S. cerevisiae* to formic acid, and the major defense mechanism is likely to be upregulation of formate dehydrogenase (FDH) activity. In order to consume NADH produced by the formic acid breakdown reaction, transcription of two NAD-dependent reductases, alcohol dehydrogenase (*ALD5*), and glycerol 3-phosphate dehydrogenase (*GPD2*), were induced in an attempt to maintain the redox balance. Genes in the initial steps for sugar assimilation, encoding a hexose transporter (*HXT4*), and hexokinase (*HXK2*), were also upregulated to produce more ATP. The quantification of formic acid in the medium indicated that the innate activity of FDH was too weak to detoxify formic acid.

To reinforce the capability for formic acid breakdown, the *FDH1* gene was also overexpressed in the xylose-fermenting recombinant yeast strain. This modification allowed the yeast to rapidly decompose excess formic acid. The fermentation profile indicated that the production of xylitol and glycerol, major by-products in xylose fermentation, was not affected by the upregulation of FDH activity.

The inhibition of yeast fermentation by weak acids is clearly a significant hurdle for the development of ethanol production from lignocellulose. However, there are only a few reported studies on strategies to overcome this inhibition [60]. During the past few years, global gene expression analysis and proteomic analysis have been performed under conditions mimicking the authentic ethanol fermentation process, presenting dynamic views of gene expression and protein biosynthesis in the yeast stress response [111]. As a part of systems biology, metabolomics and transcriptomics discussed here, and fluxomics represent the distal readout of cellular status, in contrast to genomics and proteomics, and closely reflect the activities of the cell at a functional level. An important challenge for metabolic engineering is identification of gene targets that have material importance for improvement of cellular function. Metabolomics and transcriptomics would facilitate metabolic engineering to optimize microorganisms for industrial ethanol production from lignocellulosic feedstocks through white biotechnology.

10.4 Conclusion and Future Outlook

By displaying novel enzymes on the cell surface, efficient whole cell biocatalysts are constructed. The displayed enzymes are regarded as a kind of self-immobilized enzyme on the cell surface. As shown in the case of ethanol production, a combination of a cell-surface displayed enzyme system and an intracellular metabolic system is a very effective approach to develop cells with novel metabolic ability for industrial applications. Since cell surface engineering is now available in various microbial cells [118–121], this approach will be applicable to various bioconversion processes. By combining the modification of the cell surface and engineering of the intracellular metabolic pathway, various types of whole cell biocatalysts could be prepared.

Recent advances in omics techniques, including genomics, transcriptomics, and metabolomics, have made it possible to generate information required for further engineering [97, 105, 122, 123]. As mentioned in this chapter, metabolomics and transcriptomics are powerful tools to gain insight into the effects of environmental perturbation on microbial metabolism and then to design rational strategies to confer tolerance to the stress through metabolic engineering. Comparative genome analysis using re-sequenced genome information could identify causal genes and polymorphisms responsible for stress tolerance or specific metabolic functions in non-model microbes. Transcriptome and metabolome analysis becomes a useful diagnostic tool to infer bottlenecks or problems in the cell process that should be engineered to improve microbe performance. By employing omics tools as well as advanced cell engineering technologies, a super-yeast with extreme performance for cost-effective bioethanol production may be designed and constructed.

References

1. Bungay HR (1981) Energy, The Biomass Options. Wiley, New York
2. Renewable Fuels Association. Industry statistics. 2007. http://www.eth-anolrfa.org/industry/statistics/(September 2007)
3. Sánchez OJ, Cardona CA (2008) Trends in biotechnological production of fuel ethanol from different feedstocks. Bioresour Technol 99:5270–5295
4. Wilkie AC, Riedesel KJ, Owend JM (2000) Stilage characterization and anaerobic treatment of ethanol stillage from conventional and cellulosic feed stocks. Biomass Bioenerg 19:63–102
5. Classen PAM, van Lier JB, Lopez Contreras AM, van Neil EWJ, Sijtsma L, Stams AJM, de Vries SS, Weusthuis RA (1999) Utilization of biomass for the supply of energy carriers. Appl Microbiol Biotechnol 52:741–755
6. Lee WG, Huang C (2000) Modelling of ethanol fermentation using *Zymomonas mobilis* ATCC 10988 grown on the media containing glucose and fructose. Biochem Eng J 4:217–227
7. Ashikari T, Kunisaki N, Matsumoto T, Amachi T, Yoshizumi H (1989) Direct fermentation of raw corn starch to ethanol by yeast transformants containing a modified *Rhizopus* glucoamylase gene. Appl Microbiol Biotechnol 32:129–133
8. Inlow D, McRae J, Ben-Bassat A (1988) Fermentation of corn starch to ethanol with genetically engineered yeast. Biotechnol Bioeng 32:227–234
9. Birol G, Onsan ZI, Kirdar B, Oliver SG (1998) Ethanol production and fermentation characteristics of recombinant *Saccharomyces cerevisiae* strains grown on starch. Enzyme Microb Technol 22:672–677
10. de Moreas L, Astolfi-Filho S, Oliver SG (1995) Development of yeast strains for the efficient utilization of starch: evaluation of constructs that express alpha amylase and glucoamylase separately or as bifunctional fusion proteins. Appl Microbiol Biotechnol 43:1067–1076
11. Eksteen JM, van Rensburg P, Cordero Otero RR, Pretorius IS (2003) Starch fermentation by recombinant *Saccharomyces cerevisiae* strains expressing the α-amylase and glucoamylase genes from *Lipomyces kononenkoae* and *Saccharomyces fibuligera*. Biotechnol Bioeng 84:639–646
12. Boder ET, Wittrup KD (1997) Yeast surface display for screening combinatorial polypeptide libraries. Nat Biotechnol 15:553–557
13. Kondo A, Shigechi H, Abe M, Uyama K, Matsumoto T, Takahashi S, Ueda M, Tanaka A, Kishimoto M, Fukuda H (2002) High-level ethanol production from starch by a flocculent *Saccharomyces cerevisiae* strain displaying cell-surface glucoamylase. Appl Microbiol Biotechnol 58:291–296
14. Murai T, Ueda M, Kawaguchi T, Arai M, Tanaka A (1998) Assimilation of cellooligosaccharides by a cell surface-engineered yeast expressing β-glucosidase and carboxymethylcellulase from *Aspergillus aculeatus*. Appl Environ Microbiol 64:4857–4861
15. Nakamura Y, Shibasaki S, Ueda M, Tanaka A, Fukuda H, Kondo A (2001) Development of novel whole-cell immunoadsorbents by yeast surface display of the IgG-binding domain. Appl Microbiol Biotechnol 57:500–505
16. Schreuder MP, Mooren ATA, Toschka HY, Verrips CT, Klis FM (1996) Immobilizing proteins on the surface of yeast cells. Trends Biotechnol 14:115–120
17. Kondo A, Ueda M (2004) Yeast cell-surface display-applications of molecular display. Appl Microbiol Biotechnol 64:28–40

18. Murai T, Ueda M, Yamamura Y, Atomi H, Shibasaki Y, Kamasawa N, Osumi M, Amachi T, Tanaka A (1997) Construction of a starch utilizing yeast by cell surface engineering. Appl Environ Microbiol 63:1362–1366
19. Ueda M, Murai T, Shibasaki Y, Kamasawa N, Osumi M, Tanaka A (1998) Molecular breeding of polysaccharide-utilizing yeast cells by surface engineering. Ann N Y Acad Sci 13(864):528–537
20. Ueda M, Tanaka A (2000) Genetic immobilization of proteins on the yeast cell surface. Biotechnol Adv 18:121–140
21. Roy A, Lu CF, Marykwas DL, Lipke PN, Kurjan J (1991) The AGA1 product is involved in cell surface attachment of the *Saccharomyces cerevisiae* cell adhesion glycoprotein a-agglutinin. Mol Cell Biol 11:4196–4206
22. Watari J, Takata Y, Ogawa M, Sahara H, Koshino M, Onnela ML, Airaksinen U, Jaatinen R, Penttila M, Keranen S (1994) Molecular cloning and analysis of the yeast flocculation gene FLO1. Yeast 10:211–225
23. Cappellaro C, Hauser K, Mrŝa V, Watzele M, Watzele G, Gruber C, Tanner W (1991) *Saccharomyces cerevisiae* a- and α-agglutinin: characterization of their molecular interaction. EMBO J 10:4081–4088
24. Lynd LR, van Zyl WH, McBride JE, Laser M (2005) Consolidated bioprocessing of cellulosic biomass: an update. Curr Opin Biotechnol 16:577–583
25. Xu Q, Singh A, Himmel ME (2009) Perspectives and new directions for the production of bioethanol using consolidated bioprocessing of lignocellulose. Curr Opin Biotechnol 20:364–371
26. Kiran Sree N, Sridhar M, Suresh K, Banat IM, Venkateswar Rao L (2000) Isolation of thermotolerant, osmotolerant, flocculating *Saccharomyces cerevisiae* for ethanol production. Bioresour Technol 72:43–46
27. Yanase S, Hasunuma T, Yamada R, Tanaka T, Ogino C, Fukuda H, Kondo A (2010) Direct ethanol production from cellulosic materials at high temperature using the thermotolerant yeast *Kluyveromyces marxianus* displaying cellulolytic enazymes. Appl Microbiol Biotechnol 88:381–388
28. Cardona CA, Sánchez OJ (2007) Fuel ethanol production: process design trends and integration opportunities. Bioresour Technol 98:2415–2457
29. Sakamoto T, Hasunuma T, Hori Y, Yamada R, Kondo A (2011) Direct ethanol production from hemicellulosic materials of rice straw by use of an engineered yeast strain codisplaying three types of hemicellulolytic enzymes on the surface of xylose-utilizing *Saccharomyces cerevisiae* cells. J Biotechnol. doi:10.1016/j.biotec.2011.06.025
30. Matsumoto N, Fukunishi O, Miyanaga M, Kakihara K, Nakajima E, Yoshizumi H (1982) Industrialization of a non-cooking system for alcoholic fermentation from grains. Agric Boil Chem 46:1549–1558
31. Shigechi H, Uyama K, Fujita T, Matsumoto T, Ueda M, Tanaka A, Fukuda H, Kondo A (2002) Efficient ethanol production from starch through development of novel flocculent yeast strains displaying glucoamylase and co-displaying or secreting α-amylase. J Mol Cat B 17:179–187
32. Shigechi H, Fujita Y, Koh J, Ueda M, Fukuda H, Kondo A (2000) Energy saving direct ethanol production from low temperature cooked corn starch using a cell surface engineered yeast strain co-displaying glucoamylase and α-amylase. Biochem Eng J 350:477–484
33. Satoh E, Niimura Y, Uchimura T, Kozaki M, Komagata K (1993) Molecular cloning and expression of two α-amylase genes from *Streptococcus bovis* 148 in *Escherichia coli*. Appl Environ Microbiol 59:3669–3673
34. Shigechi H, Koh J, Fujita Y, Matsumoto T, Bito Y, Ueda M, Satoh E, Fukuda H, Kondo A (2004) Direct ethanol production from raw corn starch via fermentation by use of novel surface engineered yeast strain co-displaying glucoamylase and α-amylase. Appl Environ Microbiol 70:5037–5040

35. Lo HF, Lin LL, Chiang WY, Chie MC, Hsu WH, Chang CT (2002) Deletion analysis of the C-terminal region of the α-amylase of *Bacillus* sp. strain TS-23. Arch Microbiol 178:115–123
36. Yamada R, Tanaka T, Ogino C, Fukuda H, Kondo A (2010) Novel strategy for yeast construction using δ-integration and cell fusion to efficiently produce ethanol from raw starch. Appl Microbiol Biotechnol 85:1491–1498
37. Yamakawa S, Yamada R, Tanaka T, Ogino C, Kondo A (2010) Repeated batch fermentation from raw starch using a maltose transporter and amylase expressing diploid yeast strain. Appl Microbiol Biotechnol 87:109–115
38. Yamada R, Bito Y, Adachi T, Tanaka T, Ogino C, Fukuda H, Kondo A (2009) Efficient production of ethanol from raw starch by a mated diploid *Saccharomyces cerevisiae* with integrated α-amylase and glucoamylase genes. Enzyme Microb Technol 44:344–349
39. Aristidou A, Penttilä M (2000) Metabolic engineering applications to renewable resource utilization. Curr Opin Biotechnol 11:187–198
40. Thomson JA (1993) Molecular biology of xylan degradation. FEMS Microbiol Rev 10:65–82
41. Dashtban M, Schraft H, Qin H (2009) Fungal bioconversion of lignocellulosic residues; opportunities & perspectives. Int J Biol Sci 5:578–595
42. Percival Zhang Y-H, Himmel ME, Mielenz JR (2006) Outlook for cellulase improvement: screening and selection strategies. Biotechnol Adv 24:452–481
43. Medve J, Karlsson J, Lee D, Tjerneld F (1998) Hydrolysis of microcrystalline cellulose by cellobiohydrolase I and endoglucanase II from *Trichoderma reesei*: adsorption, sugar production pattern, and synergism of the enzymes. Biotechnol Bioeng 59:621–634
44. Teeri TT (1997) Crystalline cellulose degradation: new insight into the function of cellobiohydrolases. Trends Biotechnol 15:160–167
45. Woodward J (1991) Synergism in cellulose systems. Bioresour Technol 36:67–75
46. Medve J, Ståhlberg J, Tjerneld F (1994) Adsorption and synergism of cellobiohydrolase I and II of *Trichoderma reesei* during hydrolysis of microcrystalline cellulose. Biotechnol Bioeng 44:1064–1073
47. Wood BE, Ingram LO (1992) Ethanol production from cellobiose, amorphous cellulose, and crystalline cellulose by recombinant *Klebsiella oxytoca* containing chromosomally integrated *Zymomonas mobilis* genes for ethanol production and plasmids expressing thermostable cellulase genes from *Clostridium thermocellum*. Appl Environ Microbiol 58:2103–2110
48. Zhou S, Ingram LO (2001) Simultaneous saccharification and fermentation of amorphous cellulose to ethanol by recombinant *Klebsiella oxytoca* SZ21 without supplemental cellulose. Biotechnol Lett 23:1455–1462
49. Cho KM, Yoo YJ (1999) Novel SSF process for ethanol production from microcrystalline cellulose using the δ-integrated recombinant yeast, *L2612GC*. J Microbiol Biotechnol 9:340–345
50. van Rensburg P, van Zyl WH, Pretorius IS (1996) Co-expression of a *Phanerochaete chrysosporium* cellobiohydrolase gene and a *Butyrivibrio fibrisolvens* endo-β-1,4-glucanase gene in *Saccharomyces cerevisiae*. Curr Genet 30:246–250
51. van Rensburg P, van Zyl WH, Pretorius IS (1998) Engineering yeast for efficient cellulose degradation. Yeast 14:67–76
52. Lynd LR, Wyman CE, Gerngross TU (1999) Biocommodity engineering. Biotechnol Prog 15:777–793
53. Sun Y, Cheng J (2002) Hydrolysis of lignocellulosic materials for ethanol production: a review. Bioresour Technol 83:1–11
54. Zaldivar J, Nielsen J, Olsson L (2001) Fuel ethanol production from lignocellulose: a challenge for metabolic engineering and process integration. Appl Microbiol Biotechnol 56:17–34
55. Guedon E, Desvaux M, Petitdemange H (2002) Improvement of cellulolytic properties of *Clostridium cellulolyticum* by metabolic engineering. Appl Environ Microbiol 68:53–58
56. Doi RH, Tamaru Y (2001) The *Clostridium cellulovorans* cellulosome: an enzyme complex with plant cell wall degrading activity. Chem Rec 1:24–32

57. Shoham Y, Lamed R, Bayer EA (1999) The cellulosome concept as an efficient microbial strategy for the degradation of insoluble polysaccharides. Trends Microbiol 7:275–281
58. Desvaux M (2005) The cellulosome of *Clostridium cellulolyticum*. Enzyme Microb Technol 37:373–385
59. Desvaux M (2005) *Clostridium cellulolyticum*: model organism of mesophilic cellulolytic clostridia. FEMS Microbiol Rev 29:741–764
60. Nevoigt E (2008) Progress in metabolic engineering of *Saccharomyces cerevisiae*. Microbiol Mol Biol Rev 72:379–412
61. Olsson L, Hahn-Hägerdal B (1993) Fermentative performance of bacteria and yeasts in lignocellulose hydrolysates. Process Biochem 28:249–257
62. Den Haan R, Rose SH, Lynd LR, van Zyl WH (2007) Hydrolysis and fermentation of amorphous cellulose by recombinant *Saccharomyces cerevisiae*. Metab Eng 9:87–94
63. Lynd LR, Weimer PJ, van Zyl WH, Pretorius IS (2002) Microbial cellulose utilization: fundamentals and biotechnology. Microbiol Mol Biol Rev 66:506–577
64. Fujita Y, Takahashi S, Ueda M, Tanaka A, Okada H, Morikawa Y, Kawaguchi T, Arai M, Fukuda H, Kondo A (2002) Direct and efficient production of ethanol from cellulosic material with a yeast strain displaying cellulolytic enzymes. Appl Environ Microbiol 68:5136–5141
65. Fujita Y, Ito J, Ueda M, Fukuda H, Kondo A (2004) Synergistic saccharification, and direct fermentation to ethanol of amorphous cellulose by use of engineered yeast strain codisplaying three types of cellulolytic enzyme. Appl Environ Microbiol 70:1207–1212
66. Biely P (1985) Microbial xylanolytic systems. Trends Biotechnol 3:286–290
67. Kulkarni N, Shendye A, Rao M (1999) Molecular and biotechnological aspects of xylanases. FEMS Microbiol Rev 23:411–456
68. Jeffries TW (1983) Utilization of xylose by bacteria, yeasts, and fungi. Adv Biochem Eng 27:1–32
69. Herrmann MC, Vrsanska M, Jurickova M, Hirsch J, Biely P, Kubicek CP (1997) The β-D-xylosidase of *Trichoderma reesei* is a multifunctional β-D-xylan xylohydrolase. Biochem J 321:375–381
70. De Vries RP, Kester HCM, Poulsen CH, Benen JAE, Visser J (2000) Synergy between enzymes from *Aspergillus* involved in the degradation of plant cell wall polysaccharides. Carbohydr Res 327:401–410
71. Törrönen A, Harkki A, Rouvinen J (1994) Three-dimensional structure of endo-1,4-β-xylanase II from *Trichoderma reesei*: two conformational states in the active site. EMBO J 13:2493–2501
72. Kitamoto N, Yoshino S, Ohmiya K, Tsukagoshi N (1999) Sequence analysis, overexpression, and antisense inhibition of a β-xylosidase gene, *xylA*, from *Aspergillus oryzae* KBN616. Appl Environ Microbiol 65:20–24
73. La Grange DC, Pretorius IS, van Zyl WH (1996) Expression of a *Trichoderma reesei* β-xylanase gene (*XYN2*) in *Saccharomyces cerevisiae*. Appl Environ Microbiol 62:1036–1044
74. Li XL, Lijungdahl LG (1996) Expression of *Aureobasidium pullulans xynA* in, and secretion of the xylanase from, *Saccharomyces cerevisiae*. Appl Environ Microbiol 62:209–213
75. Moreau A, Durand S, Morosoli R (1992) Secretion of a *Cryptococcus albidus* xylanase in *Saccharomyces cerevisiae*. Gene 116:109–113
76. Pérez-Gonzalez JA, De Graff LH, Visser J, Ramón D (1996) Molecular cloning and expression in *Saccharomyces cerevisiae* of two *Aspergillus nidulans* xylanase genes. Appl Environ Microbiol 62:2179–2182
77. La Grange DC, Pretorius IS, van Zyl WH (1997) Cloning of the *Bacillus pumilus* β-xylosidase gene (*xynB*) and its expression in *Saccharomyces cerevisiae*. Appl Microbiol Biotechnol 47:262–266
78. Margolles-Clark E, Tenkanen M, Nakari-Setälä T, Penttilä M (1996) Cloning of genes encoding α-L-arabinofuranosidase and β-xylosidase from *Trichoderma reesei* by expression in *Saccharomyces cerevisiae*. Appl Environ Microbiol 62:3840–3846

79. La Grange DC, Pretorius IS, Claeyssens M, van Zyl WH (2001) Degradation of xylan to D-xylose by recombinant *Saccharomyces cerevisiae* coexpressing the *Aspergillus niger* β-xylosidase (*xlnD*) and the *Trichoderma reesei* xylanase II (*xyn2*) genes. Appl Environ Microbiol 67:5512–5519
80. Eliasson A, Christensson C, Wahlbom CF, Hahn-Hägerdal B (2000) Anaerobic xylose fermentation by recombinant *Saccharomyces cerevisiae* carrying *XYL1*, *XYL2*, and *XKS1* in mineral medium chemostat cultures. Appl Environ Microbiol 66:3381–3386
81. Ho NWY, Chen Z, Brainard AP (1998) Genetically engineered *Saccharomyces* yeast capable of effective cofermentation of glucose and xylose. Appl Environ Microbiol 64:1852–1859
82. Jeppsson M, Johansson B, Hahn-Hägerdal B, Gorwa-Grauslund MF (2002) Reduced oxidative pentose phosphate pathway flux in recombinant xylose-utilizing *Saccharomyces cerevisiae* strains improves the ethanol yield from xylose. Appl Environ Microbiol 68:1604–1609
83. Johansson B, Christensson C, Hobley T, Hahn-Hägerdal B (2001) Xylulokinase overexpression in two strains of *Saccharomyces cerevisiae* also expressing xylose reductase and xylitol dehydrogenase and its effect on fermentation of xylose and lignocellulosic hydrolysate. Appl Environ Microbiol 67:4249–4255
84. Katahira S, Fujita Y, Mizukie A, Fukuda H, Kondo A (2004) Construction of a xylan fermenting yeast strain through codisplay of xylanolytic enzymes on the surface of xylose utilizing *Saccharomyces cerevisiae* cells. Appl Environ Microbiol 70:5407–5414
85. Den Haan R, van Zyl WH (2001) Differential expression of the *Trichoderma reesei* β-xylanase II (*xyn2*) gene in the xylose-fermenting yeast *Pichia stipitis*. Appl Microbiol Biotechnol 57:521–527
86. Den Haan R, van Zyl WH (2003) Enhanced xylan degradation and utilisation by *Pichia stipitis* overproducing fungal xylanolytic enzymes. Enzyme Microb Technol 33:620–628
87. Morosoli R, Zalce E, Durand S (1993) Secretion of a *Cryptococcus albidus* xylanase in *Pichia stipitis* resulting in a xylan fermenting transformant. Curr Genet 24:94–99
88. Ryabova OB, Chmil OM, Sibirny AA (2003) Xylose and cellobiose fermentation to ethanol by the thermotolerant methylotrophic yeast *Hansenula polymorpha*. FEMS Yeast Res 4:157–164
89. Katahira S, Mizuike A, Fukuda H, Kondo A (2006) Ethanol from lignocellulosic hydrolysate by a recombinant xylose- and cellooligosachharide-assimilating yeast strain. Appl Mircobiol Biotechnol 72:1136–1143
90. Almeida JRM, Modig T, Petersson A, Hahn-Hägerdal B, Lidén G, Gorwa-Grauslund MF (2007) Increased tolerance and conversion of inhibitors in lignocellulosic hydrolysates by *Saccharomyces cerevisiae*. J Chem Technol Biotechnol 82:340–349
91. van Maris AJA, Abbott DA, Bellissimi E, van den Brink J, Kuyper M, Luttik MAH, Wisselink HW, Scheffers WA, van Dijken JP, Pronk JT (2006) Alcoholic fermentation of carbon sources in biomass hydrolysates by *Saccharomyces cerevisiae*: current status. Antonie Van Leeuwenhoek 90:391–418
92. Palmqvist E, Hahn-Hägerdal B (2000) Fermentation of lignocellulosic hydrolysates. II: inhibitors and mechamisms of inhibition. Bioresour Technol 74:25–33
93. García-Aparicio MP, Ballesteros I, Gonzárez A, Oliva JM, Ballesteros M, Negro MJ (2006) Effect of inhibitors released during steam-explosion pretreatment of barley straw on enzymatic hydrolysis. Appl Biochem Biotechnol 129:278–288
94. Martin C, Alriksson B, Sjöde A, Nilvebrant N-O, Jönsson LJ (2007) Dilute sulfuric acid pretreatment of agricultural and agro-industrial residues for ethanol production. Appl Biochem Biotechnol 137:339–352
95. Klinke HB, Thomsen AB, Ahring AB (2004) Inhibition of ethanol-producing yeast and bacteria by degradation products produced during pre-treatment of biomass. Appl Microbiol Biotechnol 66:10–26
96. Almeida JRM, Bertilsson M, Hahn-Hägerdal B, Lidén G, Gorwa-Grauslund M-F (2009) Carbon fluxes of xylose-consuming *Saccharomyces cerevisiae* strains are affected differently by NADH and NADPH usage in HMF reduction. Appl Microbiol Biotechnol 84:751–761

97. Gorsich SW, Dien BS, Nichols NN, Slininger PJ, Liu ZL, Skory CD (2006) Tolerance to furfural-induced stress is associated with pentose phosphate pathway genes *ZWF1*, *GND1*, *RPE1*, and *TKL1* in *Saccharomyces cerevisiae*. Appl Microbiol Biotechnol 71:339–349
98. Larrson S, Cassland P, Jönsson LJ (2001) Development of a *Saccharomyces cerevisiae* strain with enhanced resistance to phenolic fermentation inhibitors in lignocellulose hydrolysates by heterologous expression of laccase. Appl Environ Microbiol 67:1163–1170
99. Helle SS, Murray A, Lam J, Cameron DR, Duff SJB (2004) Xylose fermentation by genetically modified *Saccharomyces cerevisiae* 259ST in spent sulfite lilquor. Bioresour Technol 92:163–171
100. Russel JB (1992) Another explanation for the toxicity of fermentation acids at low pH: anion accumulation versus uncoupling. J Appl Bacteriol 73:363–370
101. Oldiges M, Lütz S, Pflug S, Schroer K, Stein N, Wiendahl C (2007) Metabolomics: current state and evolving methodologies and tools. Appl Microbiol Biotechnol 76:495–511
102. Baran R, Reindl W, Northern TR (2009) Mass spectrometry based metabolomics and enzymatic assays for functional genomics. Curr Opin Microbiol 12:547–552
103. Garcia DE, Baidoo EE, Benke PI, Pingitore F, Tang YJ, Villa S, Keasling JD (2008) Separation and mass spectrometry in microbial metabolomics. Curr Opin Microbiol 11:233–239
104. Yoshida S, Imoto J, Minato T, Oouchi R, Sugihara M, Imai T, Ishiguro T, Mizutani S, Tomita M, Soga T, Yoshimoto H (2008) Development of bottom-fermenting *Saccharomyces* strains that produce high SO_2 levels, using integrated metabolome and transcriptome analysis. Appl Environ Microbiol 74:2787–2796
105. Hasunuma T, Sanda T, Yamada R, Yoshimura K, Ishii J, Kondo A (2011) Metabolic pathway engineering based on metabolomics confers acetic and formic acid tolerance to a recombinant xylose-fermenting strain of *Saccharomyces cerevisiae*. Microb Cell Fact 10:2
106. Sonderegger M, Jeppsson M, Hahn-Hägerdal B, Sauer U (2004) Molecular basis for anaerobic growth of *Saccharomyces cerevisiae* on xylose, investigated by global gene expression and metabolic flux analysis. Appl Environ Microbiol 70:2307–2317
107. Jin Y-S, Alper H, Yang Y-T, Stephanopoulos G (2005) Improvement of xylose uptake and ethanol production in recombinant *Saccharomyces cerevisiae* through an inverse metabolic engineering. Appl Environ Microbiol 71:8249–8256
108. Pitkänen J-P, Rintala E, Aristidou A, Ruohonen L, Penttilä M (2005) Xylose chemostat isolates of Saccharomyces cerevisiae show altered metabolite and enzyme levels compared with xylose, glucose, and ethanol metabolism of the original strain. Appl Microbiol Biotechnol 67:827–837
109. Toivari MH, Salusjärvi L, Ruohonen L, Penttilä M (2004) Endogenous xylose pathway in *Saccharomyces cerevisiae*. Appl Environ Microbiol 70:3681–3686
110. Wahlbom CF, Cordero Otero RR, van Zyl WH, Hahn-Hägerdal B, Jönsson LJ (2003) Molecular analysis of a *Saccharomyces cerevisiae* mutant with improved ability to utilize xylose shows enhanced expression of proteins involved in transport, initial xylose metabolism and the pentose phosphate pathway. Appl Environ Microbiol 69:740–746
111. Zhao XQ, Bai FW (2009) Mechanism of yeast stress tolerance and its manipulation for efficient fuel ethanol production. J Biotechnol 144:23–30
112. Liu ZL, Ma M, Song M (2009) Evolutionarily engineered ethanologenic yeast detoxifies lignocellulosic biomass conversion inhibitors by reprogrammed pathways. Mol Genet Genomics 282:233–244
113. Li BZ, Yuan YJ (2010) Transcriptome shifts in response to furfural and acetic acid in *Saccharomyces cerevisiae*. Appl Microbiol Biotechnol 86:1915–1924
114. Schüller C, Mamnun YM, Mollapour M, Krapf G, Schuster M, Bauer BE, Piper PW, Kuchler K (2004) Global phenotypic analysis and transcriptional profiling defines the weak acid stress response regulon in *Saccharomyces cerevisiae*. Mol Biol Cell 15:706–720
115. Hirasawa T, Yoshikawa K, Nakakura Y, Nagahisa K, Furusawa C, Katakura Y, Shimizu H, Shioya S (2007) Identification of target genes conferring ethanol stress tolerance to *Saccharomyces cerevisiae* based on DNA microarray data analysis. J Biotechnol 131:34–44

116. Yoshikawa K, Tanaka T, Furusawa C, Nagahisa K, Hirasawa T, Shimizu H (2009) Comprehensive phenotypic analysis for identification of genes affecting growth under ethanol stress in *Saccharomyces cerevisiae*. FEMS Yeast Res 9:32–44
117. Hasunuma T, Sung K-M, Sanda T, Yoshimura K, Matsuda F, Kondo A (2011) Efficient fermentation of xylose to ethanol at high formic acid concentrations by metabolically engineered *Saccharomyces cerevisiae*. Appl Microbiol Biotechnol 90:997–1004
118. Okano K, Zhang Q, Kimura S, Narita J, Tanaka T, Fukuda H, Kondo A (2008) System using tandem repeats of the cA peptidoglycan-binding domain from *Lactococcus lactis* for display of both N- and C-terminal fusions on cell surfaces of lactic acid bacteria. Appl Environ Microbiol 74:1117–1123
119. Narita J, Okano K, Tateno T, Tanino T, Sewaki T, Sung MH, Fukuda H, Kondo A (2006) Display of active enzymes on the cell surface of *Escherichia coli* using PgsA anchor protein and their application to bioconversion. Appl Microbiol Biotechnol 70:564–572
120. Tateno T, Hatada K, Tanaka T, Fukuda H, Kondo A (2009) Development of novel cell surface display in *Corynebacterium glutamicum* using porin. Appl Microbiol Biotechnol 84:733–739
121. Tabuchi S, Ito J, Adachi T, Ishida H, Hata Y, Okazaki F, Tanaka T, Ogino C, Kondo A (2010) Display of both N- and C-terminal target fusion proteins on the *Aspergillus oryzae* cell suface using a chitin-binding module. Appl Microbiol Biotechnol 87:1783–1789
122. Heer D, Heine D, Sauer U (2009) Resistance of *Saccharomyces cerevisiae* to high concentrations of furfural is based on NADPH-dependent reduction by at least two oxidoreductase. Appl Environ Microbiol 75:7631–7638
123. Mira NP, Palma M, Guerreiro JF, Sá-Correia I (2010) Genome-wide identification of *Saccharomyces cerevisiae* genes required for tolerance to acetic acid. Microb Cell Fact 9:79

Chapter 11
Systems-Level Analysis of Cancer Metabolism

Paulo A. Gameiro, Christian M. Metallo, and Gregory Stephanopoulos

Contents

Abstract The complexity of cancer requires systems-level approaches to examine uncontrolled proliferation, with many analytical tools now providing massive information on distinct cellular processes. In contrast to the genetic anchors founded in cancer biology that underpin tumor suppressors and oncogenes as

P.A. Gameiro
Department of Chemical Engineering, Massachusetts Institute of Technology, Cambridge, MA 02139, USA

Department of Life Sciences, University of Coimbra, Coimbra, Portugal

Department of Medicine, Massachusetts General Hospital, Boston, MA 02114, USA

C.M. Metallo • G. Stephanopoulos (✉)
Department of Chemical Engineering, Massachusetts Institute of Technology, Cambridge, MA 02139, USA
e-mail: gregstep@mit.edu

C. Wittmann and S.Y. Lee (eds.), *Systems Metabolic Engineering*,
DOI 10.1007/978-94-007-4534-6_11,

units of malignant function, we now see a shift of attention towards metabolism. This trend calls for the increased use of stable isotopic tracers to dissect effects in metabolic fluxes that arise from gene deregulation. When combined with analytical techniques such as mass spectrometry or nuclear magnetic resonance (NMR) and computational tools to interpret such datasets, isotopic tracers can allow for the determination of various metabolic events involved in tumorigenesis at a fine resolution. As such, the interplay between fluxes and signaling warrants a thorough investigation that will lead to targeted therapies rooted on metabolic targets. This chapter describes stable isotopic methods to determine fluxes and identify switches, illustrating how metabolic activity can be quantitatively interpreted to address fundamental questions in cancer.

Keywords Systems-level analysis • Cancer metabolism • Cell proliferation • Tumor repressor • Oncogenes • Malignent function • Isotopic tracers • Nuclear magnetic resonance (NMR) • Metabolic flux analysis • Anabolic switch • Metabolic targets • Lipid synthesis • Nucleic acid synthesis • Protein synthesis • Cell culture • Tumorgenesis • Metabolic markers • Metabolomics • Mass spectrometry • Isotopic enrichment • Experimental design • Turnover rates • Metabolic steady-state • Pulse-chase strategy • Isotopomer distribution • Elemental tracers • Non-targeted tracer fate detection (NTFD) • Warburg effect • Metabolic switch • Splice isoform • Pyruvate kinase • Glutamine • Metabolic reprogramming • Catabolic arrest

Abbreviations

2HG	R(2)-2-hydroxyglutarate
AA	Aminoacid
ACL	ATP citrate lyase
AMP	Adenosine 5′-monophosphate
AMPK	AMP-activated protein kinase
ATP	Adenosine 5′-triphosphate
dN	Deoxyribonucleoside
DNNS	*De novo* nucleoside synthesis
dR	deoxyribose (dR)
EI	Electron impact
EMU	Elementary metabolic unit
ETC	Electron transport chain
FH	Fumarate hydratase
FSR	Fractional synthesis rate
GAP	Glyceraldehyde 3-phosphate
GC	Gas Chromatography
HK-II	Hexokinase II
IDH	Isocitrate dehydrogenase
ISA	Isopotomer spectral analysis

LC	Liquid Chromatography
LDH-A	Lactate dehydrogenase A
MFA	Metabolic flux analysis
MID	Mass isotopomer distribution
m/z	Mass-to-charge ratio
NEAA	Non-essential amino acid
NMFA	Nonstationary metabolic flux analysis
NTFDA	Non-targeted tracer fate detection
OXPHOS	Oxidative phosphorylation
PC	Pyruvate carboxylase
PDH	Pyruvate dehydrogenase
PEP	Phosphoenolpyruvate
PGAM1	Phosphoglycerate mutase
PK	Pyruvate kinase
PPP	Pentose phosphate pathway
TBDMS	*Tert*-butyldimethylsilyl
TCA	Tricarboxylic acid
TGF-β	Transforming growth factor β
TKT	Transketolase
TMS	Trimethylsilyl
VDAC	Voltage-dependent anionic channel

11.1 Introduction

The metabolism of cells reflects their function. Different cell types display distinct metabolic features which depend on their differentiation program and specialization. Cancer cells show an altered metabolic phenotype on their own, which is presumably optimum for uncontrolled proliferation. Cellular metabolism regained attention in the last decade with the development of tools that enable system-wide inquiries of protein function via analysis of metabolites [52]. Defined as the comprehensive analysis of all measurable metabolites under a given set of conditions, metabolomics is employed to monitor metabolic phenotypes, which entails the determination of metabolite concentrations to ultimately uncover metabolic markers of disease [36]. Metabolomics is commonly referred as metabonomics when it investigates perturbations of the metabolome in response to genetic modification or pathophysiological stimuli (diet and toxins); a term that was coined by Jeremy Nicholson [86]. This tool has been extensively applied to analyze metabolic phenotypes of normal and transformed cells, tissues and biofluids [5, 28, 110] and represents a minimally invasive means for early diagnosis and prognosis in cancer [42].

A major shortcoming of metabolomics is that changes in static concentrations may arise from perturbations in various nodes or pathways within a reaction network. Without further knowledge on the activity of enzymatic reactions, the causal pathway to a specific metabolic outcome remains elusive. For example, an increase in lactate production can arise from glycolysis, glutamine oxidation or pentose phosphate pathway (PPP). Since metabolite pool sizes do not directly capture the dynamics of metabolic systems, pathway tracing is not possible. As stated by Alan Saghatelian, "When you find a new metabolite, you can't just knock it out like a gene or a protein; you have to figure out the pathways that make it and break it down, and what it binds to" [4]. In this context, stable isotopes represent the most practical means for tracing metabolic pathways, wherein isotopically-enriched molecules (substrates) are used to follow their intracellular fate via metabolism. Provided that knowledge of atom transitions is available, stable isotopes inform on the contribution of various pathways to metabolite enrichment. In turn, such information can be used to ascertain or estimate metabolic fluxes, which provide the ultimate representation of pathway activity. Thus, examining the isotopic fate of substrates in the metabolic network provides a quantitative measure of what metabolic pathways are activated by a specific growth signal, mutation or drug [64]. In essence, stable isotopes in conjunction with metabolic flux analysis (MFA) are the reagents to investigate metabolism in any biological system [104].

The motivation for determining a flux stems from its proximity to the phenotype. Should mutations in proto-oncogenes promote proliferation, they ultimately affect the reactions that produce macromolecules and energy. Thus, the flux, or the rate of substrate consumption per cell per unit time, captures multiple levels of organization, including changes in metabolite concentrations, covalent modifications, protein interactions and gene expression; hence it represents a powerful metric to provide insights into the functioning of cellular processes [48, 88]. Enzymatic reactions are linked by the same cofactor molecules and control mechanisms e.g. phosphorylation, acetylation [62], with multiple pathways producing and degrading common metabolites. As a result, it is impossible to elucidate the role of biochemical pathways through the kinetic characterization of isolated reactions *in vitro*. In this context, fluxes provide the "top-down" dimension that integrates the expression of enzymes and metabolic control into a unit of cellular activity. Therefore, MFA and pathway tracing are powerful tools in elucidating the metabolic effects driven by the multitude of oncogenic signals that promote tumorigenesis. This warrants the use of stable isotopes to decipher cancer metabolism, as it can potentially define therapies that are orthogonal to the strategies currently used in the clinic [9].

11.2 Determination of Isotopic Enrichment

In classical biochemistry, radioisotopes marked the era of pathway identification by teaching us the individual steps comprising metabolic networks. Pioneering work with radioisotopes in conjugation with chemical analysis elucidated the topology of the metabolic networks and led to the discovery of the pathways we so well know today: for instance, Glycolysis and Krebs cycle [63]. As pointed out by Koshland, we moved from an era of pathway regulation, which shed light into the allosterism and enzyme kinetics, to the era of pathway quantification [62]. This is largely due to the implementation of stable isotopic tracers, as these reagents allowed metabolic engineers to introduce metabolic fluxes as readouts of biological function [48]. Furthermore, while avoiding the safety issues associated with radioisotopes, stable isotopes provide information on the abundance of labeled as well as unlabeled molecules. (For a review on stable isotopes and radioisotopes, see [7, 58]).

Compounds enriched with stable isotopes can be detected by nuclear magnetic resonance (NMR) spectroscopy and mass spectrometry (MS). The former detects the magnetic spins of atomic nuclei. Since the magnetic states of an individual isotope atom are influenced by the magnetic properties of bonded atoms, isotopes afford molecules a NMR signature that discriminates isotopic content as well as position of enrichment. For a molecule with n carbon atoms, there are 2^n isotopomers [107] and the relative abundance of each isotopomer is informative of tracer distribution. The great advantage of NMR is its non-invasive nature which is suited for applications in humans [43, 53]. Since ^{12}C is not active in NMR, enrichment with ^{13}C is necessary to detect carbons in metabolites with enough sensitivity. In the context of isotopic studies, it is worth carrying this rationale further; pathway tracing requires asymmetry in labeling, so one would not benefit if nature had selected ^{13}C as the most abundant carbon isotope. If it were the case, one would use ^{12}C tracers to introduce isotopic asymmetry in the naturally labeled ^{13}C metabolites. This highlights that the availability of stable isotopes endows scientists the power to study metabolite transformations without the need of an external marker, so we should use them. Although NMR-based methods are limited to metabolites occurring in the mM range of cellular concentrations, its chemical specificity enables detailed isotopomer analysis of abundant metabolites such as glutamate [72] and other amino and organic acids [137]. For example, (i) glycolytic/pentose-phosphate cycling, and (ii) pyruvate dehydrogrenase (PDH)/pyruvate carboxylase (PC) fluxes can be accurately resolved from the labeling patterns of lactate and glutamate using ^{13}C glucose tracers, respectively [16, 67, 81].

In most implementations, MS–based methods are carried out by coupling a separation technique to a mass analyzer through different ionization interfaces. Typically, separation is achieved by gas chromatography (GC) or liquid chromatography (LC), in which metabolites are retained for different periods of time resulting in a temporal resolution that is expressed as a retention time or retention index. GC-MS is a widely implemented technique for the analysis of volatile and thermally stable metabolites [44]. In brief, metabolites in a complex mixture are separated in a column coupled to a MS detection system that generates ions through electron impact (EI) or chemical ionization. EI is almost exclusively used as it is highly reproducible and suitable for library-based identification, which can be exploited in non-targeted isotopic studies [51]. A fragmentation pattern specific to the metabolite structure is produced, and a mass spectrum is recorded where fragments are sorted based on their mass-to-charge ratio (m/z). Several reviews on MS principles, methodologies and applications in metabolism are available [6, 31, 44]. GC-MS offers high chromatographic resolution and sensitivity for amino and organic acids, which comprise a significant part of the molecules in central carbon metabolism. A disadvantage of GC-MS is its restriction to metabolites that are either volatile or rendered volatile by derivatization. Silylation with *tert*-butyldimethylsilylation (TBDMS) is an established method to reduce the polarity of amino and organic acids through an anhydrous reaction that forms ester derivatives [2, 92]. A disadvantage of TBDMS is that it significantly increases the molecular mass of its derivatives which compromises the detection of metabolites containing several derivatizable groups e.g. sugar phosphate metabolites [44]. For these metabolites, trimethylsilyl (TMS) is an alternative method. However, TMS is more sensitive to hydrolysis and the quantification of specific phosphorylated metabolites may be limited [92]. LC-MS and LC-MS/MS have become more popular recently for detecting isotopic enrichment in a broader range of metabolites [124]. Overall, LC-MS/MS complements GC-MS methods in the ability to determine isotopic enrichment in phosphorylated metabolites more effectively [6].

Mass spectrometry detects molecules that differ in mass, as determined by mass isotopomers. Because quadrupole mass spectrometers do not resolve exact masses, measured mass isotopomers may comprise molecules that differ in their isotopic composition. For example, ^{15}N and ^{13}C-labeled glutamine are different isotopologues, but they are detected by MS as the same mass isotopomer [50, 107]. In practical terms, an output in the form of mass isotopomer distribution (MID) distinguishes populations of molecules that have a different number of neutrons, wherein each mass isotopomer comprise isotopologues sharing a nominal mass. Although no positional enrichment is directly measured by MS, it is often possible to retrieve such information via analysis of the molecular fragmentation. For a detailed analysis of amino acid fragmentation patterns arising in GC-MS, please see Antoniewicz et al. [2]. Mass isotopomers are generally represented by the mass *m* plus the number of *non-abundant isotopes*. The fraction mass + 0 (or M + 0) denotes the mass of molecules comprised by atoms of the most common isotope present naturally. The increase in mass (M + 1, M + 2, M + 3, etc) refers to molecules enriched with stable isotopes from the tracer or from the natural abundance of the element, e.g. 1.1% for ^{13}C. Since the MID pattern of a metabolite readily indicates the abundance of each mass isotopomer relative to the total, it reflects the activity of the reactions metabolizing the tracer.

A complication associated with MID patterns is the natural abundance of ^{13}C, ^{2}H, ^{17}O, ^{18}O, ^{15}N and ^{29}Si atoms, among others. In isotopic studies using TBDMS, the natural abundance of atoms in the derivatization group contributes to the isotopic labeling in the analyte, so measured MIDs need to be corrected for the labeled atoms that are not derived from the tracer [35]. For example, glutamine labeled with ^{13}C in the five carbon atoms (or [U-$^{13}C_5$] glutamine, in which "U" denotes uniformly) is a widely used tracer for MFA in mammalian cells. When A549 cells (a human lung carcinoma line) are labeled with 98% [U-$^{13}C_5$] glutamine, prior to correction, MID depicts a M + 5 glutamine fractional abundance of approximately 60% which is a consequence of abundant M + 6 and M + 7 fractions in the population of molecules. By applying a correction method adapted from Fernandez' et al. [35], a MID is determined which closely resembles that expected of the original tracer, with M + 5

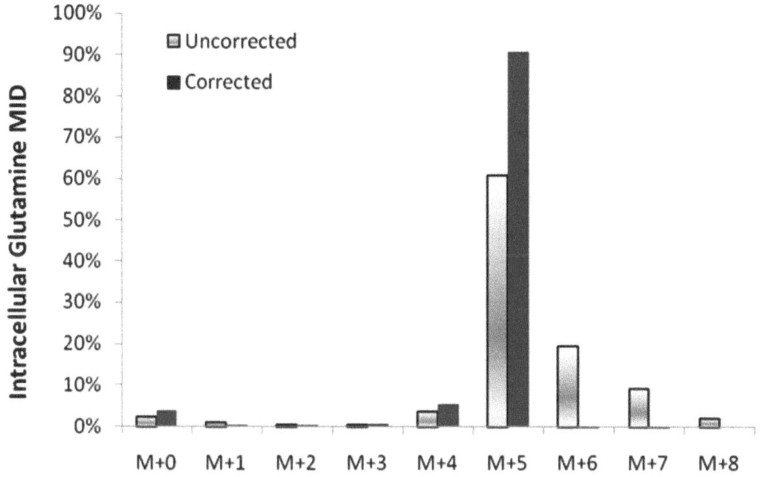

Fig. 11.1 Mass isotopomer distribution (*MID*) of intracellular glutamine in a lung carcinoma line (A549) treated with [U-$^{13}C_5$] (99%) glutamine. The glutamine MID was obtained by GC-MS analysis of a cell extract. The figure displays the TBDMS-derivatized glutamine fragment 431 m/z. The correction algorithm accounts for the natural abundance of C, H, O, N, P, Si, S and F atom. Because glutamine is mostly a substrate in mammalian cells, it appears almost as labeled as the tracer

comprising approximately 90% of the intracellular glutamine (Fig. 11.1). Also note that the M + 4 peak arising from incomplete atomic enrichment in the tracer and the M + 0 abundance from protein turnover are readily detected.

11.3 Metabolic Flux Analysis

MFA effectively employs MID data to resolve fluxes in a metabolic network through the use of stable isotopes. Well known by metabolic engineers to determine flux changes in response to modulated enzyme expression, MFA enables the evaluation of network rigidity to ultimately enhance the productivity of a desired metabolite [104, 112]. Throughout the 1990s, MFA was employed to determine metabolic fluxes in batch cultures of microorganisms and hybridomas [73, 112, 114, 146]. Typically, a labeled substrate (e.g. ^{13}C-enriched glucose) of known composition is applied to a biological system and the labeling in metabolite pools is measured. Meanwhile, simulated measurements are generated with some initial flux "guess" values and a stoichiometric matrix outlined by the reaction network and atom transitions. Fluxes are ultimately estimated by iteratively adjusting a flux vector by minimizing the difference (or residuals) between simulated and measured MIDs. As isotopic measurements usually outnumber the free fluxes or degrees of freedom, the system is overdetermined [131]. In conjunction with independently determined substrate uptake and product secretion rates, fluxes in mol per cell per time unit can be generated using the MFA approach [111]. The rationale, methodology and troubleshooting of MFA have been extensively reviewed in the literature [37, 89, 100, 119, 131, 144]. Several frameworks for MFA applications were developed to simulate the isotopic labeling of metabolites within the model network, with the atom and isotopomer matrix mapping (AMM and IMM) representing breakthroughs in the field [107, 147]. More recently, the elementary metabolic unit (EMU) decomposition method was introduced, dramatically decreasing flux estimation times by reducing the set of isotopomers/cumomers (2^n) considered in such calculations [3]. In addition, the determination of the sensitivities of the simulated measurements (i.e. MIDs or isotopic distribution) with respect to fluxes generates flux confidence intervals, which are of paramount importance to interpret MFA results [1]. While widely applied in microorganisms for biotechnological purposes, MFA has been increasingly applied to cultured human cells [38, 78, 82, 91, 98, 137]. As such, MFA is emerging as an important systems-level tool for identifying metabolic targets in cancer [9].

MFA is based on the assumption that measured metabolites are at isotopic and metabolic steady-state; that is, labeling patterns and metabolite concentrations do not vary over time. Therefore, the measured isotopic distribution is strictly a function of tracer(s) composition, network topology, pathway activity and the presence on endogenous sources. The metabolic steady-state condition ensures that changes of metabolite concentrations do not lead to unsynchronized tracer incorporation which would invalidate comparisons of isotopic distributions. In contrast to microorganisms, the complex medium used to culture mammalian cells contains multiple nutrients, e.g. non-essential aminoacids (NEAA) that contribute with unlabeled carbons. As NEAA can potentially be consumed and produced, changes in medium composition during cultivation may affect the efflux/influx balance, which shifts metabolism and introduces non-stable isotopic behavior [29]. The concentration of main carbon sources, i.e. glucose and glutamine, and accumulation of toxic products are also important variables to consider. As reported by Deshpande et al. in batch cultures of CHO cells, optimized versions of culture media may be required to guarantee metabolic and isotopic steady-state for valid MFA [29]. Surprisingly, no characterization has been reported for adherent cultured cells scrutinizing the effects of media composition, serum types and other variables on the isotopic steady-state condition.

An important variant of isotopic studies is the non-stationary MFA (NMFA) that uses transient isotopic data and intracellular pool sizes to estimate fluxes [132]. Despite the requirement of efficient rapid sampling [90], NMFA is ideal for determining fluxes in biological systems where the steady-state assumption is vulnerable, such as slow-growing or quiescent mammalian cells [71]. When the kinetics of fatty acids, DNA and protein metabolism in mammalian cells may require time-prohibitive experiments, NMFA represents a powerful method to determine biosynthetic fluxes. Young et al. employed NMFA to determine the flux distribution in brown adipocytes [142]. Maier et al. reported the effect of statins in the cholesterol synthesis pathway in primary hepatocytes [70]. As illustrated in this chapter, other methods have also been developed to quantitatively trace slow-turnover pathways.

Furthermore, the capability of NMFA to quantify metabolite levels makes it suited for metabolic control analysis applications in determining the control exerted by specific reactions on overall pathway activity, i.e. the flux control coefficients, and the local effect of changes of a drug or metabolite in the rates of specific reactions, i.e. elasticity coefficients [34, 56]. In the study of Maier et al. [70], the authors used the information of NMFA to derive flux control coefficients for the target enzyme 3-hydroxy-3-methyl-glutaryl-coenzyme A reductase. Altogether, these studies underpin the value of stationary and dynamic MFA to identify target fluxes of specific drugs and other effectors as well as the significance of changing these fluxes – via flux control coefficients [15].

11.4 The Impact of Tracers

The experimental design is critical to maximize the flux information in isotopic studies. In particular, the choice of tracer dictates the measured MID of a metabolite, thereby strongly influencing the quality of the data. Because a wide collection of stable isotopes is now available, the labeling strategy is a challenging component in MFA applications. The knowledge of atom transitions [103] suffices to select the best tracer in some cases. For example, [1,2 $^{13}C_2$] glucose is *a priori* optimal to probe flux split ratios into the PPP versus anaerobic glycolysis; the oxidative decarboxylation of 6-phosphogluconate into D-ribulose 5-phosphate ultimately produces singly ^{13}C-labeled molecules of glyceraldehyde 3-phosphate (GAP) through the non-oxidative branch of PPP whereas glycolytic flux forms M + 2 GAP, so the relative activity of these two pathways can be ascertained by the isotopic distribution of the downstream metabolites pyruvate and lactate [81, 122]. Likewise, mixtures of [U-$^{13}C_6$] glucose and unlabeled glucose can effectively resolve anaplerotic fluxes and TCA cycle in microbial systems that typically rely on a single carbon source [37]. Conversely, in mammalian cells, both glucose and glutamine heavily feed carbon into the TCA cycle and compartmentalization exists for many metabolites, which confound the *a priori* tracer selection process. Therefore, the labeling strategy in complex systems requires a quantitative evaluation of which tracers are more informative of specific subnetworks. If we want meaningful flux estimates, we must quantitatively address the question: what tracer should we use?

The effectiveness of a set of 18 ^{13}C glucose and glutamine tracers was recently tested in a carcinoma cell line to accurately estimate fluxes in central carbon metabolism [80]. The method was based on the uncertainty of a specific tracer to estimate fluxes of interest; tracers generating narrower confidence intervals associated with each individual flux were considered more useful. Non-linear 95% confidence intervals were generated using the platform *Metran*, in which the residual disagreement between observed and simulated MIDs is assessed as a

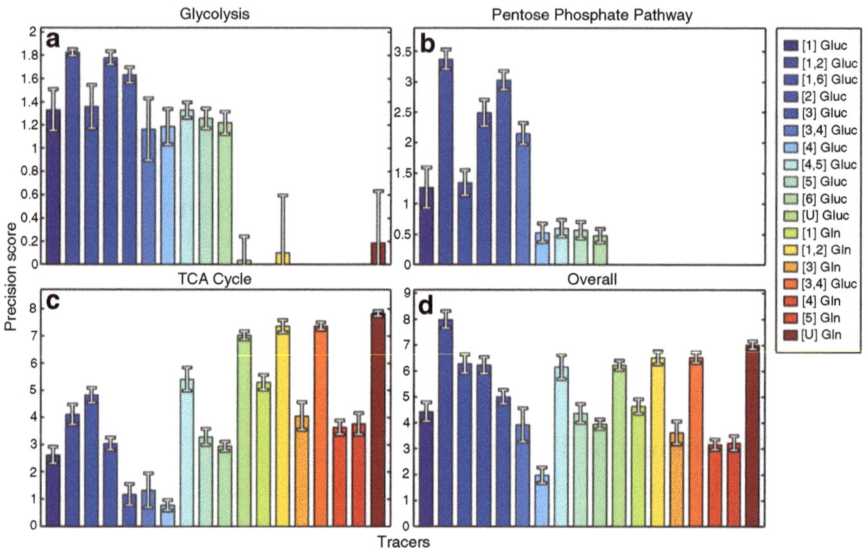

Fig. 11.2 Metallo et al. [80] Optimal ^{13}C tracers for the analysis of pathways in central carbon metabolism. [1] Gluc depicts ^{13}C-labeled glucose in the 1 carbon; [U]Gln denotes ^{13}C - "uniformly" labeled glutamine in all carbons. The precision score demonstrates that (**a**) Glycolysis and (**b**) pentose pathways are accurately determined using [1,2] Gluc. (**c**) Several glutamine tracers and [U]Gluc scored the highest for **c**. TCA cycle, and (**d**) the overall best tracers were [1,2] Gluc and [U]Gln due to the precision in resolving fluxes of glycolysis/pentose phosphate pathway and TCA cycle, respectively

function of the estimated flux values; large variations of an estimated flux *i* that result in small variations in the residuals indicate that this flux *i* is poorly defined [1]. While optimal tracers were previously evaluated for other biological systems, Metallo et al. demonstrated the utility of specific ^{13}C tracers to conduct MFA in a two-substrate cancer network. Figure 11.2 depicts the optimal ^{13}C tracers for subnetworks of interest in central carbon metabolism, including glycolysis, PPP and TCA cycle. It should be noted that surprising results may emerge from a thorough selection of the tracer. Although *a priori* choice may well indicate [U-$^{13}C_6$] glucose as appropriate to probe intermediary metabolism, multiple labeled glutamine ^{13}C tracers generated equally good or better estimates for anaplerotic and TCA cycle fluxes. Because glucose and glutamine fuel the TCA cycle with two and four carbons, respectively (neglecting PC activity), glutamine tracers are optimal because they allow ^{13}C to be more heavily propagated throughout the pathway via TCA cycling and scrambling at the succinate/fumarate node, whereas doubly labeled metabolites from glucose are more easily diluted. Additionally, [1,2-$^{13}C_2$] glucose was the overall best tracer for the full network, which is due to its high precision to determine glycolysis and PPP fluxes. As specific cells certainly make use of fluxes to different extents, systematic analyses of other cancer cell lines and optimal ^{2}H and ^{15}N tracers represent high priority studies.

11.5 Synthesis of Macromolecules

Macromolecules are duplicated during cell cycle; hence biosynthetic fluxes must accompany proliferation. One of the kinetic features of proteins, nucleic acids and lipids is that turnover rates are typically slow and pool sizes are large. As a result, the enrichment of a protein with atoms from the tracer is significantly diluted by abundant unlabeled proteins already present in the biological system and endogenous pools of amino acid (AA) precursors. As described herein, this paradigm led to the development of mathematical methods to distinguish between pre-existing and newly synthesized macromolecules.

The tracer incorporation method is traditionally used to determine the fractional synthesis rate (FSR) of a product. In the simplest case, the turnover rate of a single-pool product is represented by a single constant K (in units of time $^{-1}$) that equals the FSR at metabolic steady-state [94]. FSR represents the fraction of product pool that turns over in a given period of time [134]. Upon a constant infusion, the appearance of the tracer in the product will increase monoexponentially as a function of time (E_t) until the enrichment in the product eventually approaches the enrichment of the precursor pool, E_P. FSR can be derived from the enrichment of the product (at a time t), according to

$$E^t = E^P(1 - e^{-Kt}) \tag{11.1}$$

$$FSR = \frac{E^{t2} - E^{t1}}{E^P \bullet (t^2 - t^1)} \tag{11.2}$$

E_P denotes the precursor enrichment and E_t represents the enrichment of the product-bound precursor at a time *t*, i.e. tracer/tracee ratio (Fig. 11.3) [94]. FSR can be calculated by dividing the initial rate of change of product enrichment by the average precursor enrichment during tracer incorporation (11.2). Therefore, one needs to know E_P to accurately determine FSR. For a review on the tracer incorporation method, see Patterson [94], Wolfe and Chinkles [134]. The isotopic content of the true precursor represents a key challenge in biosynthetic labeling experiments, and various strategies have been employed to account for this parameter. In fast-growing cells (e.g. microorganisms), the labeled substrate can be administered until the fractional enrichment of the precursor pool is set to 1.0, i.e. fully labeled, and the subsequent administration of unlabeled substrate allows the measurement of the turnover via chase of the degrading pre-labeled product, which is referred as "pulse-chase" strategy [8]. In mammalian cells or intact animals, the presence of multiple carbon sources makes it impracticable to attain fully labeled precursors in many applications. When the labeled fraction of the precursor pool is not complete, the isotopic content of the precursor pool can be measured; however, there is disagreement as to whether the precursor pool can be *truly* represented by isolating the biosynthetic precursor from cells or a tissue sample. For example, the enrichment of aminoacyl-transfer RNA should provide the best readout of the true

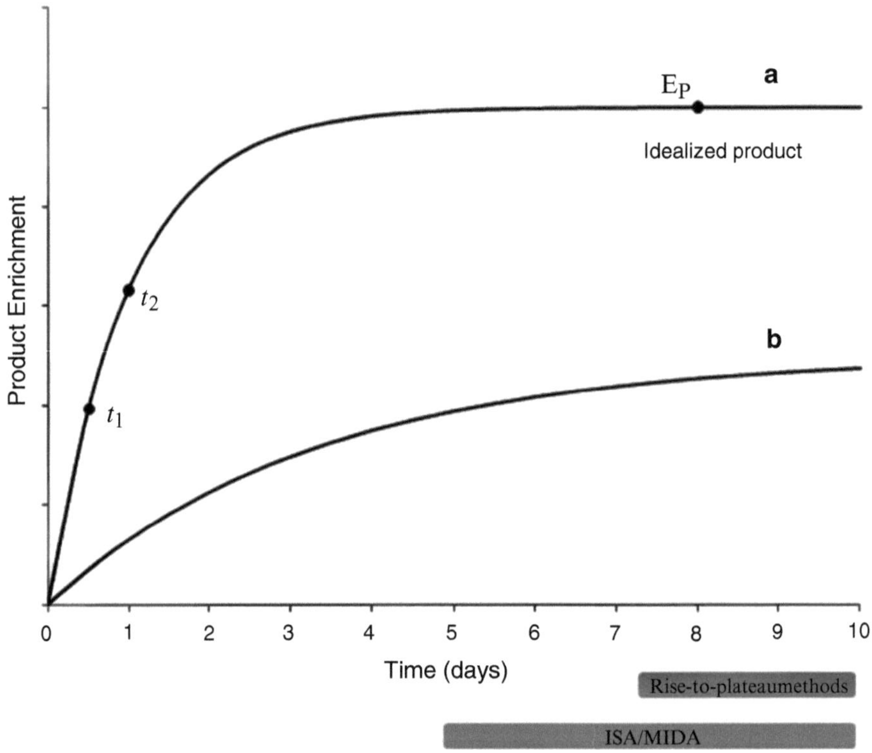

Fig. 11.3 The fractional enrichment of a product from labeled precursors is shown. The parameters to determine the fractional synthesis rate (FSR) are displayed. Following a constant tracer administration, the enrichment of a product (e.g. a protein) increases until the plateau phase, in which it equates the isotopic enrichment of the amino acid (AA) precursor (E_p). FSR can be determined by measuring the initial rate of protein enrichment and the steady-state enrichment of the AA precursor. (**a**) In an idealized protein of fast turnover ($K = 1$ day^{-1}), the product enrichment rapidly approaches E_P. (**b**) In a slow-turnover protein (e.g. collagen), "rise-to-plateau" approaches require prolonged tracer infusion to attain a steady isotopic enrichment in the product. The isotopomer spectral analysis/mass isotopomer distribution analysis (ISA/MIDA) method allows the FSR to be determined without reaching the plateau phase. Endogenous production of unlabelled precursors might be significant in diluting tracer incorporation, as emphasized in (**b**)

protein precursor [128], but the fast turnover might impose experimental difficulties to measure this precursor. Since only the subunits incorporated in the products at the site of product formation are the *true* precursors, the biosynthetic subunits are ultimately inaccessible to measurement [47, 49]. Alternatively, a labeled precursor can be continuously administered until the fractional enrichment in the product reaches an isotopic plateau that must equal the enrichment of the *true* precursor. This strategy avoids the need to measure the enrichment of the precursor, so that FSR can be determined by measuring the initial rate of increase and isotopic plateau enrichment of the product [74]. However, the "rise-to-plateau" approach might be cost and time-prohibitive when measuring slow-turnover proteins with labeled essential AAs. To circumvent this, ^2H_2O is a unique tracer because water equilibrates quickly across compartments, thereby allowing the FSR of macromolecules to be determined from the plateau isotopic enrichment of ^2H_2O in an accessible compartment [12]. However, the number of $^2H_2$ incorporated in biosynthesis is variable and needs to be determined.

As a consequence of the inability to accurately determine the precursor enrichment, state-of-the-art methods were developed to derive the isotopic content of both the precursor and the product through the analysis of the mass isotopomers of the product only [49, 59]. From the mathematical standpoint, the isotopic content of the precursor is not easily recovered from the product mass isotopomers because the sampled product is a mixture of pre-existing (naturally labeled) and newly synthesized molecules (tracer-enriched and naturally labeled); the system is characterized by two fluxes, i.e. contribution of tracer molecules to the precursor pool and FSR of the product during the labeling period, but it only contains one measurable variable, the mass isotopomers of the product [49]. Kelleher and Hellerstein independently developed a precursor-product mathematical relationship, the isotopomer spectral analysis (ISA) and mass isotopomer distribution analysis (MIDA) approaches, that predicts the appearance of each mass isotopomer of the product as a function of the two unknown fluxes, establishing a probability model that allows FSR and E_P to be determined without following the product enrichment until the plateau (Fig. 11.3).

11.5.1 Lipid Synthesis

The principle of ISA (Kelleher's terminology) showed proof of concept when applied to the *de novo* synthesis of fatty acids using ^{13}C acetate and ^2H_2O [30, 60]. There are two dimensionless parameters that are determined by the ISA approach: D is the fractional contribution of ^{13}C-enriched atoms of the tracer versus endogenous pathways (1-D) as sources of the precursor (e.g. lipogenic acetyl CoA); g(t) indicates the fraction of newly synthesized product (e.g. palmitate) in the sample determined by GC-MS. The model is reliant on the assumption that the system is at metabolic steady-state prior to tracer addition and that the mixture of precursors rapidly achieve metabolic and isotopic steady-state during the incubation with the tracer. The sampled product is not required to attain steady-state, and a function g(t) defines the fraction of sampled product that is newly synthesized during tracer incubation, as exemplified in Fig. 11.4 for lipogenesis. In essence, the ISA model estimates the appearance of each mass isotopomer (M + 0, M + 1, M + 2, etc) based on different D and g(t) values. The calculated isotopomer distribution is compared with the actual isotopic distribution of sampled palmitate (MID measured via GC-MS) and a fitting procedure outputs a best estimate for D and g(t) [59].

The ISA approach has great value in mammalian cells, in that it can determine the contribution of different carbon sources (e.g. glucose, glutamine, acetate) for

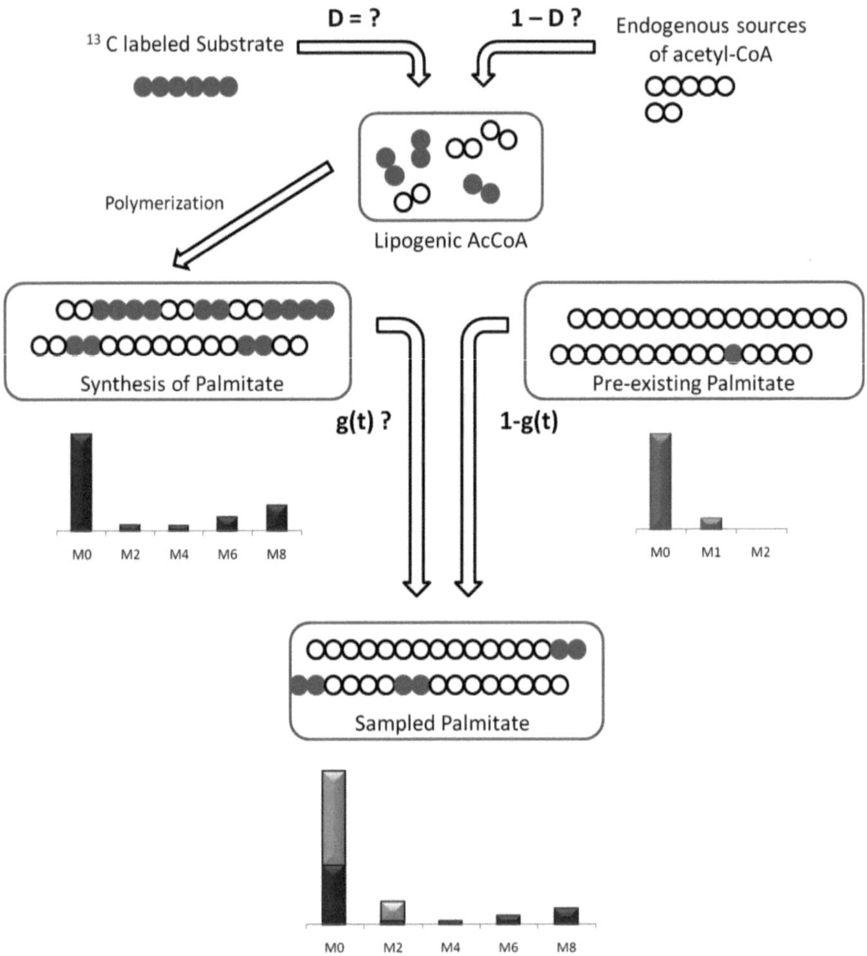

Fig. 11.4 The isotopomer spectral analysis (ISA) method. A [U-$_6{}^{13}$] Glucose tracer and naturally labeled endogenous sources (e.g. acetate, glutamine) feed carbon into the pool of lipogenic acetyl-CoA (*AcCoA*). The D parameter indicates the unknown fractional flux of [U-$_6{}^{13}$] Glucose to lipogenic AcCoA. During the incubation with the tracer, palmitate is synthesized from 8 molecules of AcCoA at the *true* site of precursor synthesis. Newly synthesized palmitate mixes with an existing pool (prior to tracer addition) and the fraction of *de novo* lipid synthesis in the sampled mixture after a period *t* is g(t). Each bar graph represents the mass isotopomer distribution (MID) of the product. Pre-existing palmitate has significant M + 1 which arises from naturally labeled elements. The parameters D and g(t) are estimated by minimizing the difference between predicted and measured MID in the sampled palmitate

the synthesis of fatty acids [140]. When employed with specific tracers, it allows the quantification of particular fluxes. Yoo et al. [141] employed ISA using [5-$^{13}C_1$] glutamine to quantify the contribution of glutamine to lipogenesis through the reductive carboxylation reaction of isocitrate dehydrogenase (IDH) enzymes in a brown adipocyte cell line. The authors showed that 90% of the glutamine-to-lipid flux was accounted by the reverse reaction of IDH rather than the common glutaminolysis pathway, underlining the value of the ISA in determining the biosynthetic activity of specific reactions. Overall, ISA provides a framework to measure biosynthetic rates of macromolecules despite of sub-cellular compartmentalization, insofar the biochemistry follows the polymerization principle with monomeric subunits (the precursors) condensing into a polymer (the product).

11.5.2 Nucleic Acid and Protein Synthesis

DNA replication represents a biochemical key step of cellular division [69], and is governed by the activity of two interacting pathways: (i) the *de novo* nucleoside synthesis (DNNS), and (ii) the deoxyribonucleoside (dN) salvage pathways. Traditional techniques to determine nucleic acid synthesis in cell culture are based upon the incorporation of [3H] thymidine (3HdT) and 5-bromodeoxyuridine (BrdU) into the DNA through the dN salvage pathway. In the seminal studies of Hellerstein and colleagues, deuterated water (^2H_2O) and [6,6-$^2H_2$] glucose were used to determine the rate of DNA synthesis via GS-MS analysis of the enrichment in the deoxyribose (dR) moiety of DNA [69, 84]. Whereas the contribution of the dN salvage pathway may be variable, the incorporation of ^2H into dR is a flux signature of DNA synthesis through the DNNS pathway (Fig. 11.5). There are seven entry points of ^2H from ^2H_2O in the dR carbon skeleton that are mediated by various pathways, including PPP, gluconeogenesis/glycolysis and ribonucleotide reductase [84]. Cells can be incubated with 5% ^2H_2O in the culturing medium, and the ^2H enrichment in derivatized purine dRs is measured via GC-MS after isolation and hydrolysis of DNA [13]. In contrast to other tracers, ^2H_2O is not subjected to compartmentalization which makes it highly quantitative for determining biosynthetic fluxes through the rise-to-plateau approach. Furthermore, such characteristics allow ^2H_2O to be effectively used *in vivo* for determination of cell proliferation [13]. Overall, the incorporation of ^2H into dRs represents an isotopic tracer-based assay to determine rates of cellular proliferation (Fig. 11.5) [48].

The DNNS pathway involves the production of the deoxyribose and purine (or pyrimidine) moieties, with the latter being synthesized via *de novo* and salvage pathways as well. Therefore, newly synthesized nucleic acids can also be traced through the enrichment of purines/pyrimidines. In particular, purines are more actively synthesized than pyrimidines during S-phase [69], making them a suitable analyte to measure nucleic acid synthesis. As the incorporation of ^2H from ^2H_2O into the base rings can also occur through the base salvage pathway, the determination of newly synthesized purines requires the use of other tracers. In this context, the purine ring contains nitrogen atoms derived from glutamine, glycine and aspartate, whereas glycine and the "1 carbon pool" provide the carbon skeleton (Fig. 11.5). As such, [$^{13}C_2$] glycine is an optimal tracer to determine the rate of nucleic acid synthesis through the incorporation of ^{13}C in the purine ring [96].

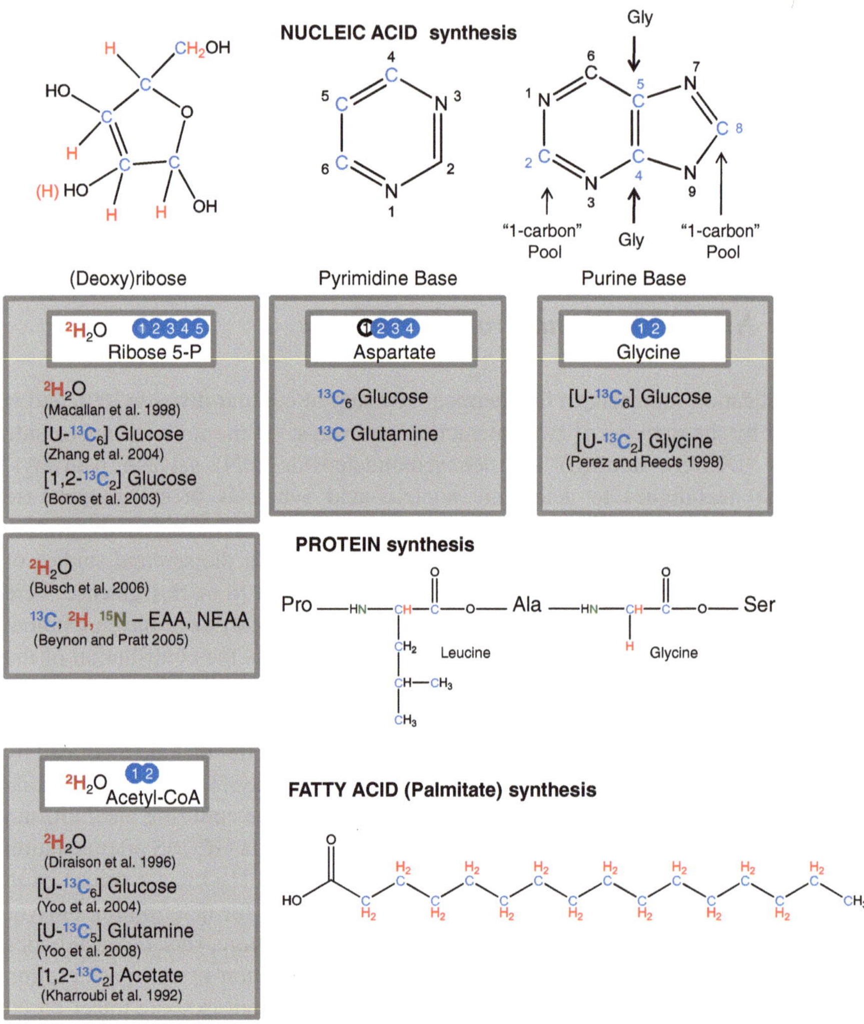

Fig. 11.5 Tracers used in the literature to trace biosynthetic pathways. Colors denote the potential sites of isotopic incorporation. White boxes display the biosynthetic precursor that is enriched by atoms from the tracer. The use of ^{13}C-glucose and glutamine tracers to probe the *de novo* synthesis of pyrimidines represent a hypothetical example [103]

Various elemental tracers (^{2}H, ^{13}C or ^{15}N) can be advantageous to probe the synthesis of nucleic acids and the choice will again affect the information ascertained on particular biosynthetic fluxes. During the DNNS pathway, the deoxyribose precursor 5-phosphoribosyl pyrophosphate combines with either free purines/pyrimidines (from the base salvage pathway) or base precursors (from the *de novo* base synthesis pathway). Whilst the *de novo* base synthesis pathway can be measured using ^{13}C or ^{15}N-glycine, ^{2}H and ^{13}C-labeled glucose inform on the total dR synthesis which is active whenever cells replicate DNA (Fig. 11.5) [145]. Additionally, ^{15}N or ^{13}C-labeled AAs may be more appropriate than ^{2}H-tracers to dually trace different biosynthetic pathways, i.e. nucleic acid and protein synthesis *in vivo* [96, 145]. In sum, ^{2}H is viewed more as an universal tracer to determine *in vivo* and *in vitro* turnover rates of macromolecules [48], whereas ^{13}C and ^{15}N-labeled substrates allow one to trace specific fluxes involving carbon and nitrogen transformations, respectively [33]. As a hypothetical example, the aspartate carbon skeleton is used for pyrimidine synthesis, and can be formed from glutamine (via oxidation to oxaloacetate followed by transamination) as well as from glucose-derived carbons (via PC and transamination). Hence, ^{13}C glutamine and glucose tracers in conjugation with the ISA method may resolve the relative contribution of these reactions for the production of pyrimidine bases. Remarkably, owing to the variety of tracers available, it emerges that many biosynthetic fluxes can now be determined, provided that analytical tools are available to translate isotopic data into meaningful flux information. Figure 11.5 depicts stable isotopes that have been employed to quantify the biosynthetic incorporation of tracer in macromolecules.

Retrospectively, MFA/ISA methods monitor genetic and proteomic information at the metabolic level, e.g. glucose oxidation or fatty acid synthesis, a perspective that is well embodied in the metabolic engineering field. It is now time to use these tools to elucidate the significance of gene deregulation and mutagenesis in cancer.

11.6 Non-targeted Pathway Tracing

In the theme of cancer biology, elucidating the metabolic changes that are brought about during tumorigenesis may require metabolic pathways to be examined at the global level. When the metabolic determinants of aberrant growth may possibly operate in distant nodes of metabolism and because the cellular metabolic network is immense, non-targeted pathway tracing allows one to discover target reactions and biomarkers that may not be captured by evaluating the labeling in a restricted set of analytes. Hiller et al. employed stable isotopes in combination with GC-MS to detect all observable metabolites that incorporate label from a specific isotopic tracer [51]. In essence, the non-targeted tracer fate detection (NTFD) method of Hiller's requires the GC-MS analysis of two independent metabolite extracts: one originated from cells treated with a mixture of labeled and unlabeled tracer and the other obtained from the treatment of cells with unlabeled tracer. Mass spectra from the mixture of labeled/unlabeled metabolites are then paired with the mass spectra of the unlabeled metabolites. A prominent step of NTFD is the subsequent calculation of the difference spectrum that produces a positive intensity for labeled fragments whereas unlabeled fragments are eliminated through subtraction. As a result, integration of the difference spectrum yields a peak for each labeled fragment that reflects isotopic enrichment and is used to calculate the correspondent MID.

Unlike targeted approaches that require the knowledge of the metabolite identity to compute MIDs, the mass spectra from unlabeled extracts in NTFD represent the internal standard for all the metabolites that are candidates for labeling detection. NTFD is an advance in the methodology of pathway tracing, in that it can map out fluxes interconnected in the catabolism of drugs or substrates that appear distant in the metabolic map. A prime example of an application of such tools is the phenomenon of mutations in metabolic enzymes observed in glioblastoma and leukemia that are associated with the production of the glutamine-derived "oncometabolite" R(2)-2-hydroxyglutarate (2HG) [138], as described herein. In this context, NTFD can simultaneously identify such "oncometabolites" while quantifying relative fluxes throughout metabolism which may be perturbed as a result.

11.7 Cell Culture as a Tool

In view of the fact that the majority of genetic biomarkers do not reach the clinical level [99], system-level metabolic analysis should be employed so as to complement previous efforts, e.g. discovering metabolic targets that may be used to combat drug resistance more effectively. On one hand, NMR and MS-based methods provide sensitive detection of metabolites levels and isotopic enrichment in cells, tissues and biofluids, with the latter representing an informative analyte of minimal invasiveness to humans [53]. On the other, the emerging notion in cancer research is that the intricate relationship between metabolic activity and signaling pathways is crucial to discover mechanisms of malignancy [14]. Since biofluids analysis is restricted to secreted metabolites, the determination of specific fluxes in central carbon metabolism may be limited. In this context, *in vitro* cell culture offers high level of control over experimental variables, such as combinatorial use of tracers, nutrient availability, oxygen supply, growth signals, gene expression and stress. Investigating fluxes that link oncogenic events to aberrant growth represents a method to identify metabolic targets, and the use of stable isotopes in monocultured cancer cells is therefore powerful in elucidating cellular metabolism because fluxes can be evaluated under a wide variety of conditions. Indeed, responses in cell culture cannot be directly extrapolated to *in vivo* conditions and fluxes are more realistically assessed in a tumor xenograft model in a host organism. However, our understanding of the interplay between fluxes and signaling is only in its infancy [66], and isotopic studies with cancer cell lines can pave the way to elucidate the fundamental biochemistry of proliferation. Table 11.1 depicts the standard conditions, reagents and steps to conduct isotopic experiments in cultured cells using GC-MS as the analytical platform.

Table 11.1 Experimental requirements for GC-MS – based isotopic studies in monocultured cells

	Labeled Medium	Specific Components	Labeling Time	Culture scale
Cell Culture	Ex. 1 [^{13}C]Glc (25 mM) + Gln (4 mM)	10% dialyzed Fetal Bovine Serum (NEAA – free)	Stationary: 6–24 h for intermediates Non-stationary: 3-7 days for macromolecules	6 well culture plate
	Reagents	*Outcome*		
A.1 Metabolite extraction	1V[b] Methanol : 1V[b] H2O	Quenching		
	2V[b] Chloroform	Deproteination and extraction[a]		
A.2 Metabolite Derivatization	TBDMS for polar fraction	Silylation of amino and organic and acids		
	Methylation Reagent	Fatty acid methyl esters		
A.3 GC-MS	**A.2**	MID		
B. Spent medium	Glc, Gln and Lactate analyzer	Extracellular Fluxes		
Metabolic Flux Analysis	**A** and **B**	Fluxes (mol. cell.$^{-1}$ time^{-1})		

Adapted from methods in Metallo et al. [80] and Yoo et al. [140]

[a]Separation of metabolites occurs. Upper phase (methanol: H_2O) contains metabolites for polar analysis; bottom phase (chloroform) contains metabolites for non-polar analysis (e.g. fatty acids, cholesterols). Derivatization is normally preceded by evaporation

[b]Volumes added

See abbreviations list for other acronyms

11.8 Introduction to Cancer Metabolism

During proliferation, a cell undergoes fundamental changes in metabolism prior to division. One of the metabolic effects associated with cancer is a robust increase in glucose uptake and a preferential use of glycolysis over oxidative phosphorylation (OXPHOS) even under the presence of adequate oxygen – *aerobic glycolysis* [126]. This is termed the *Warburg effect* and the repertoire of findings shows that avid glycolysis is a metabolic shift that confers many proliferation advantages to cancer cells [26, 32, 40, 61, 85, 93, 97, 117].

The increase in aerobic glycolysis is linked to the increased expression of hexokinase II (HK-II) and consequent binding to the mitochondria through the voltage-dependent anionic channel (VDAC) [95]. Pedersen and coworkers demonstrated that the transformation of normal hepatocytes to hepatoma cells is associated with an interaction between HK-II and VDAC that reduces the sensitivity of HK-II to glucose-6-phosphate inhibition and prevents apoptosis [41, 75]. The *Warburg effect* and the mechanistic connection to a mitochondrial-bound isoform of HK represents the basis of the positron-emission tomography (PET) imaging technique that is currently used to detect tumors [20].

Aerobic glycolysis was the first evidence for a metabolic switch associated with cancer, though Warburg's work has been accompanied by a tortuous history due to his statement that cancer originates from defective OXPHOS followed by an increase in glycolysis to cope with ATP deficiency [125]. While several somatic mutations in mitochondrial DNA are reported for primary human tumors [17], mounting evidences suggest that many cancer cells have functional OXPHOS [39], exhibit active TCA cycle [78], and require electron transport chain (ETC) for Kras-mediated tumorigenicity [129]. Furthermore, ATP citrate lyase (ACL) is important for tumor growth, suggesting that cancer cells actively re-route pyruvate to the mitochondrion [46]. In this case, a "truncated" version of TCA cycle ensures efficient glucose-to-lipid flux, whereby citrate is transported to the cytosol and converted by ACL into oxaloacteate and acetyl-coA. Ultimately, cells must exhibit some functional mitochondrial metabolism, as pyrimidine synthesis is dependent on the activity of complex III [116] with TCA cycle providing the biosynthetic precursors. Remarkably, although studies have shed light into the overall role of mitochondria on bioenergetics of cancer cells [45, 135], no study has actually reported TCA cycle activities in functional versus OXPHOS-deficient cancer cells. Because cancer cells exhibit a great level of heterogeneity in their genomes which ultimately converge into a tumorigenic phenotype [113], elucidating the molecular basis of cancer requires a systematic investigation of fluxes, enzymes and oncogenes to identify critical reactions in tumorigenesis. Importantly, similar metabolic phenotypes are observed across cancer cell types, which suggest that common fluxes and pathways are activated to support proliferation in different cancer cells [26, 115]. As cancer research calls for novel therapeutic targets, stable isotopes can be applied to define additional metabolic signatures associated with oncogenes to decipher aberrant growth. This will be exemplified in the next section.

The Warburg effect has been reinterpreted by the recent discovery that aerobic glycolysis in cancer cells is determined by a switch in a splice isoform of the glycolytic enzyme pyruvate kinase (PK), which converts phosphoenolpyruvate (PEP) into pyruvate [18]. Whilst the M1 isoform of PK (PKM1) is expressed in most adult tissues, the M2 isoform (PKM2) is a splice variant that is expressed during embryonic development. Christofk et al. observed that PKM2 is predominantly expressed in cancer cells that exhibit aerobic glycolysis. Conversely, replacing the expression of PKM1 by PKM2 reversed the Warburg effect and diminished tumor xenograft growth, demonstrating the definite role of the PKM1-to-PKM2 switch in mediating aberrant growth. Cantley and coworkers have further demonstrated that the tetramer PKM2 is regulated by phosphorylation, upon which it is converted to the dimeric form commonly found in cancer cells [19]. Paradoxically, while the more active tetrameric form of PKM2 favors glycolysis and lactate formation, the less active dimeric form of PKM2 favors to the accumulation of upstream glycolytic intermediates [76]. Though the accumulation of triose-phosphates stimulates the synthesis of amino acids and lipids, this is incongruent with the increased lactate production that accompanies the expression of PKM2 in cancer cells. In this context, Luo et al. recently showed that PKM2 promotes the transactivation of the transcription factor hypoxia-inducible factor 1, which activates the expression of lactate dehydrogenase (LDH-A) thereby stimulating lactate production [68]. In addition, Vander Heiden et al. demonstrated that PEP, serving as a phosphate donor for the phosphorylation of phosphoglycerate mutase (PGAM1), can be converted to pyruvate in the absence of PK [121]. The expression of both PKM2 and PGAM1 therefore shift cancer cells toward anabolic metabolism by allowing high rates of glycolysis and lactate production decoupled from ATP formation (Fig. 11.6).

The metabolic reprogramming of other nutrients also accounts for the metabolic switch observed in cancer cells. Glutamine is highly abundant in the human body and a major *energy* source [83], being highly converted to lactate in some cancer cells [27, 85]. These findings stimulate the view that glucose and glutamine consumption is a hallmark of proliferating cells. In this aspect, while aerobic glycolysis ensures a fine tuning between anabolism and proliferation, less is known about the role of glutamine metabolism and coordination with aerobic glycolysis [21]. Stable isotopes and MFA can address some of these questions, as described herein.

11.9 Signaling & Metabolic Fluxes: Observability via Isotopic Tracers

11.9.1 The Switch to Anabolism

During tumorigenesis, cancer cells exhibit mutations in tumor suppressors to ensure a growth-permissive state that enforces the neoplastic process [123]. Among the

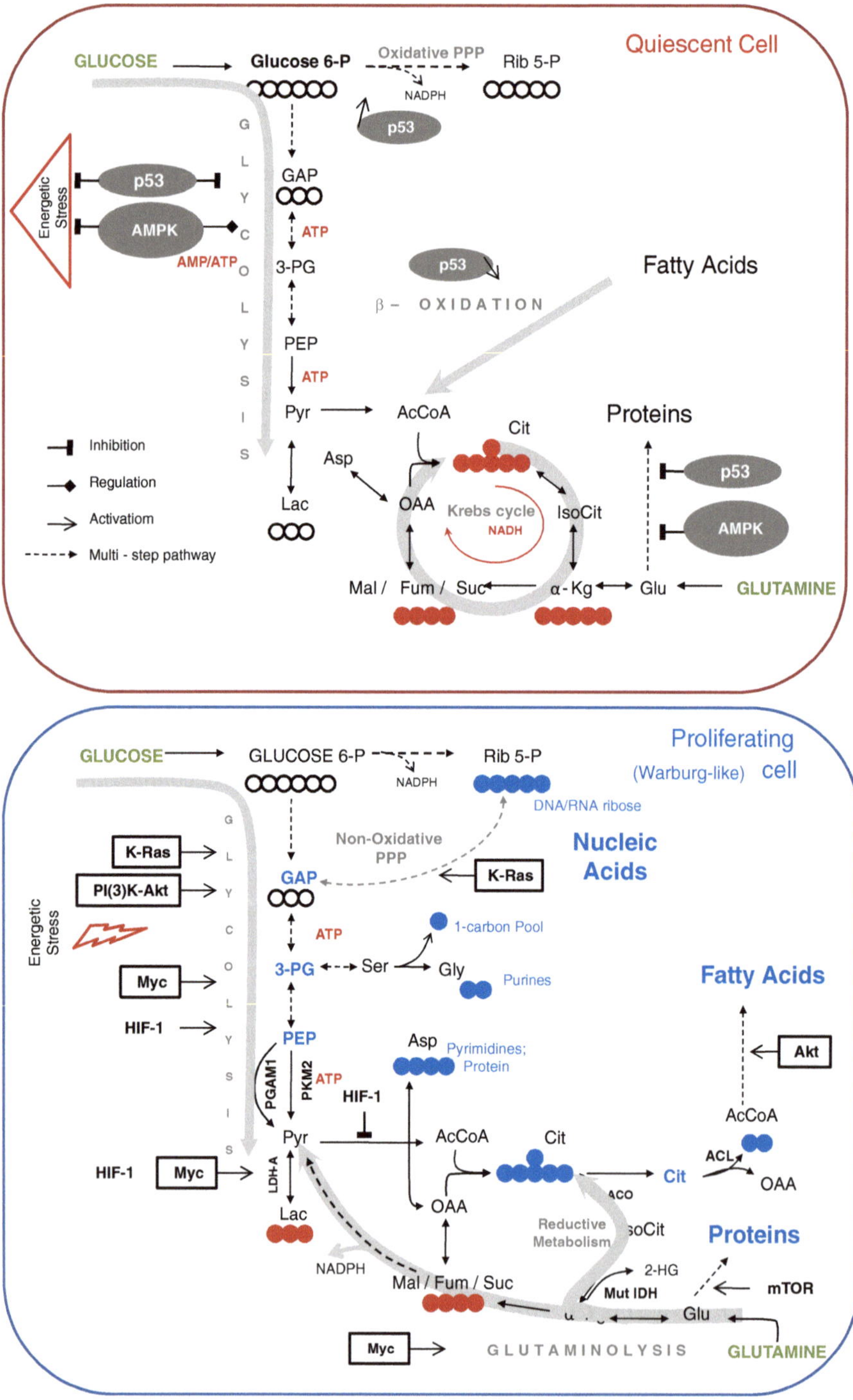

Fig. 11.6 Diagram depicting the metabolic pathways of the *catabolic arrest* and *anabolic switch* in a quiescent and a proliferating (cancer) cell, respectively. In a quiescent cell, pyruvate is actively diverted into the mitochondria, which in combination with β – oxidation, ensure active

most mutated tumor suppressor genes in cancer, p53 regulates many functions in the cell, including cell cycle and apoptosis. p53 also regulates the energetic status of cells through the stimulation of OXPHOS, β-oxidation, autophagy, and proteins that combat reactive oxygen species [54, 139]. In addition, the AMP-activated protein kinase (AMPK) is a central regulatory switch that activates glucose oxidation and decreases lipogenesis in response to unfavorable changes in AMP/ATP levels [108]. In essence, the AMPK-p53 pathway endow cells a metabolic adaptation to cope with energetic stresses through the regulation of ATP-producing pathways in central carbon metabolism. Therefore, while ensuring unchecked growth, loss of tumor suppressors reduces the metabolic plasticity to energetic insults or damage, which opens an avenue for therapeutic strategies focused on the energetic stress of cancer cells.

On the contrary, growth signals and mutations in proto-oncogenes enhance the metabolic potential of cancer cells in an autonomous fashion [109]. Blanch and colleagues have employed NMR-based MFA using ^{13}C glucose to show that estrogen-stimulated breast cancer cell proliferation increases the activity of PPP and glutamine uptake [38]. Boros et al. demonstrated using the [1,2-$^{13}C_2$] glucose tracer that the transformation of A549 lung carcinoma cells by transforming growth factor β (TGF-β) diverts glucose carbons to nucleic acid synthesis via the non-oxidative reactions of PPP [11]. The fact that the transformation of cancer cells involved the specific stimulation of non-oxidative PPP places the flux-controlling transketolase (TKT) enzyme as a potential target for cancers dependent of particular transforming factors and associated kinases [10]. Particularly, a mutated isoform of TKT is upregulated in colon and urothelial tumors [65], suggesting that anti-TKT drugs may be effective for cancer cells exhibiting TGF-β and kinase-stimulated non-oxidative PPP. Moreover, this underlines the biochemical significance of the PKM2 expression and *aerobic glycolysis* in promoting the anabolic switch through the accumulation of glycolytic intermediates. Likewise, the Ras-PI(3)K-Akt signaling strongly influences metabolism, and constitutive activation of this pathway increases glucose consumption through the activation of phosphofrutokinase-1 and HK-II [101]. Mutations in Ras and PI(3)K comprise one of the most common sets of mutations in human tumors, and ^{13}C glucose tracers can identify the metabolic effects of such oncogenic signals. As demonstrated with [1,2-$^{13}C_2$] glucose, codon-specific mutations of K-Ras differentially stimulate glycolytic

Fig. 11.6 (continued) TCA cycle, OXPHOS and efficient ATP production. P53 and AMPK regulate glycolysis at the phosphofrutokinase-1 level and through the maintenance of AMP:ATP ratio. In a cancer cell, PKM2 promotes the accumulation of glycolytic intermediates while stimulating lactate production. ACL mediates the "truncated" TCA cycle whereby citrate is mainly exported to the cytosol and converted to lipogenic AcCoA via ACL. The constitutive activation of the Ras-PI(3)K-Akt pathway increase glucose consumption and glycolysis; myc promotes glutaminolysis; IDH mediates the reductive metabolism of glutamine, and mutant IDH produces the "oncometabolite" 2-hydroxyglutarate. *Red* and *blue* markers denote carbons that are mostly directed to catabolic and anabolic pathways, respectively. *Blue* metabolites depict a hypothetical increase in pool sizes. The abbreviation for the enzymes is elucidated in the text

activity, oxidative PPP and pyruvate dehydrogenase (PDH) flux [122], remarkably suggesting that point mutations (in the same protein) associated with distinct aggressiveness also communicate differently with fluxes.

The activation of oncogenes affects the metabolism of other nutrients, with the up-regulation of glutamine catabolism being now recognized as an equally important metabolic switch of cancer cells [25]. Among its pleiotropic effects in cell cycle and metabolism, the transcription factor c-Myc (or Myc) stimulates glutamine uptake and glutaminolysis – the oxidation of glutamine to pyruvate via TCA cycle and mitochondrial $NADP^+$-malic enzyme [105] – through the repression of the micro RNAs miR-23a and miR-23b ([22] and references therein). Glutamine can potentially (i) be oxidized for energy, (ii) provide nitrogen atoms for nucleotide and hexosamine biosynthesis, (iii) fuel the TCA cycle with anabolic carbons, including aspartate and citrate, and (iv) be partially oxidized through glutaminolysis to produce lactate and NADPH [22]. Without a quantitative determination of the intracellular fate, one can neither ascertain the actual impact of Myc nor the pathways through which glutamine is metabolized in cancer cells. The metabolism of glutamine in cancer is reviewed in [25]. Meadows et al. [78] analyzed the isotopic enrichment of glutamate and glutamine from different combinations of ^{13}C glucose tracers to demonstrate that breast cancer cells exhibit higher irreversibility of glutamine-to-glutamate flux when compared to normal breast epithelial cells. Deberardinis et al. [27] showed the definite use of glutaminolysis in glioblastoma cells using [3-^{13}C] glutamine. Recently, making use of [U-$^{13}C_5$] glutamine as a tracer, Wise et al. reported that Myc is required to maintain the glutaminolytic phenotype and promote addiction to glutamine in transformed glial cells [133]. Together, these findings define glutaminolysis as a metabolic signature of glioblastoma cells that is driven by the oncogene Myc. Furthermore, Myc upregulates the transcription of ribonucleoproteins involved in the alternative splicing of PK, favoring the formation of PKM2 [24]. The complex regulation of aerobic glycolysis and glutaminolysis at different levels reflects how multiple oncogenes can converge for the activation of specific pathways that enforce anabolism (Fig. 11.6) [118]. The intricacy between signaling and metabolism thus warrants further isotopic studies to identify fluxes that are upregulated by specific oncogenes.

In retrospect, loss of tumor suppressors increases the sensitivity of cancer cells to energetic stresses, and current efforts in chemotherapy aim to target ATP-production pathways and DNA/RNA damage [102]. Conversely, the elucidation of the anabolic switch in cancer cells is providing a new mechanistic basis for the metabolic reprogramming that promotes aberrant growth [120]. As a result, the functional outcome of mutagenesis is now being elucidated at the level of metabolic fluxes (Fig. 11.6), a perspective that shall be encouraged to reach personalized therapy.

11.9.2 Tumorigenesis By Enzymes and Metabolites

Nutrient availability and growth signals must be coupled to permit division only when anabolic and energetic requirements are fulfilled [26]. Glutamine stimulates glucose uptake and cellular proliferation through the repression of the MondoA transcription factor [55], suggesting that carbon sources actively coordinate the uptake of one another through the regulation of genes. In addition, ACL activity was recently shown to be required for histone acetylation in a glucose-dependent manner [130], and intracellular levels of L-glutamine regulate mTOR activity through a membrane transport that involves influx of essential AAs [87]. In this context, how metabolic fluxes exert feedback control on signaling networks is becoming an active area of research [79]. While mutations in proto-oncogenes and tumor suppressors affect metabolic fluxes, nutrient availability can select for some of these mutations. As shown recently, glucose deprivation induces the development of K-Ras mutations [143]. This suggests that nutrients and intermediary metabolites regulate proteins and the genome, highlighting the notion that the interplay between mutagenesis and metabolism is a reciprocal interaction. Therefore, a turning point in cancer biology is that fluxes also drive tumorigenesis [77], so that nutrient uptake and signaling pathways are tuned for rapid proliferation.

The understanding of cancer metabolism has been reawakened by the recent discovery of mutations in enzymes. Mutations at a single arginine residue (R132) of the human cytosolic isocitrate dehydrogenase 1 (IDH1) are found in >70% of grade II-III gliomas and secondary glioblastomas, while mutations in the analogous amino acid (R172) of the isocitrate dehydrogenase 2 (IDH2) are reported in gliomas that lack mutations in IDH1 [136]. Mutually exclusive IDH1 and IDH2 mutations also occur in a significant subset of acute myeloid leukemias [127], representing a neomorphic event that promotes the $NADP^+$-dependent reduction of α-ketoglutarate to R(2)-2-hydroxyglutarate (2HG) [23, 127].

Intriguingly, IDH1 and IDH2 exist as homodimers but only heterozygous mutations are found in the tumors, which suggests that the wild-type allele of IDH1 and/or IDH2 may play a role in the proliferation of cancer cells (cancer research on IDH mutations is reviewed in [138]). Remarkably, the wild-type IDH2 contributes to the mitochondrial production of citrate from glutamine through the reductive carboxylation of α-ketoglutarate in the mitochondria [127]. Ward et al. employed [U-$^{13}C_5$] glutamine to detect TCA cycle activity and reverse IDH flux through the analysis of isotopic enrichment in the M + 4 and M + 5 citrate fractions, respectively. While oxidation of α-ketoglutarate via TCA cycle and subsequent condensation with unlabeled acetyl-CoA produces M + 4 citrate, the formation of M + 5 citrate from M + 5 α-ketoglutarate is indicative of reverse IDH reaction. The authors observed that siRNA-mediated knockdown of IDH2 reduced cell proliferation and decreased the labeling of M + 5 citrate while M + 4 citrate was unchanged, pointing to a specific role of IDH2 in mediating the glutamine-to-citrate flux through reductive carboxylation. It should be noted that glioblastoma cells undergo active glutaminolysis [27]. Since isotopic exchange via mitochondrial malic enzyme generates M + 3 pyruvate and malate, M + 5 citrate can also arise from the citrate synthase reaction using M + 2 acetyl-CoA (derived from M + 3 pyruvate) and M + 3 oxaloacetate molecules, which makes it an ambiguous readout of reductive carboxylation in cells that rely heavily on the glutaminolysis pathway. In addition, it is proposed that the direct interconversion of NADH and NADPH by the mitochondrial transhydrogenase favors the reductive carboxylation in the mitochondria [106], whereas IDH1-mediated production of NADPH is assumed to be required for biosynthesis [27]. To date, neither the potential role of cytosolic IDH1 to reductively produce citrate from α-ketoglutarate has been tested nor the hypothesized role of transhydrogenase experimentally validated. Furthermore, the activity of reductive carboxylation by IDH1 and IDH2 in mutant versus wild-type dimers is equally unknown and warrants isotopic studies.

Altogether, these findings illustrate how metabolic studies and stable isotopes enable the functional characterization of enzymes, i.e. while metabolomic approaches identified the "oncometabolite" 2HG as being produced by mutant IDH1/IDH2, Ward et al. used [U-$^{13}C_5$] glutamine to show that IDH2 contributes to the reductive conversion of α-ketoglutarate into citrate, which provides an explanation to the heterozygosity found in IDH2 mutations. Hence, pathway tracing using stable isotopes is powerful because it can elucidate cellular metabolism at a fine resolution. In combination with untargeted tools such as NTFD, it could identify other "oncometabolites" and pinpoint single reactions as potential targets in cancer [52].

11.10 Concluding Remark

The metabolic reprogramming of cancer cells is now a central paradigm for cancer biologists. In 1926, Warburg demonstrated that cancer cells exhibited aerobic glycolysis. Throughout the twentieth century, while biology was focused on growth factors and oncogenic signals, sight was lost on the significance of this metabolic transformation. The understanding of the cancer metabolism is now being rekindled by (1) the discovery of enzymatic mutations and isoform switches that connect metabolism to malignancy, (2) the mounting evidence that oncogenes and tumor suppressors intimately communicate with fluxes through signaling pathways and (3) system-level tools that can identify the outcome of gene deregulation at a fine level. In particular, isotopic studies should shed light into some of the feedback mechanisms that allow cancer cells to maximize fluxes and proliferate within disparate microenvironments. This represents a milestone in cancer biology that is fostering a new era of personalized therapies [57].

References

1. Antoniewicz MR, Kelleher JK, Stephanopoulos G (2006) Determination of confidence intervals of metabolic fluxes estimated from stable isotope measurements. Metab Eng 8 (4):324–337
2. Antoniewicz MR, Kelleher JK, Stephanopoulos G (2007) Accurate assessment of amino acid mass isotopomer distributions for metabolic flux analysis. Anal Chem 79(19):7554–7559

3. Antoniewicz MR, Kelleher JK, Stephanopoulos G (2007) Elementary metabolite units (EMU) – a novel framework for modeling isotopic distributions. Metab Eng 9(1):68–86
4. Baker M (2011) Metabolomics: from small molecules to big ideas. Nat Methods 8(2):117–121
5. Beckonert O, Keun HC, Ebbels TMD, Bundy J, Holmes E, Lindon JC, Nicholson JK (2007) Metabolic profiling, metabolomic and metabonomic procedures for NMR spectroscopy of urine, plasma, serum and tissue extracts. Nat Protoc 2(11):2692–2703
6. Bennett BD, Yuan J, Kimball EH, Rabinowitz JD (2008) Absolute quantitation of intracellular metabolite concentrations by an isotope ratio-based approach. Nat Protoc 3(8):1299–1311
7. Bequette BJ, Sunny NE, El-Kadi SW, Owens SL (2006) Application of stable isotopes and mass isotopomer distribution analysis to the study of intermediary metabolism of nutrients. J Anim Sci 84(13 suppl):50–59
8. Beynon RJ, Pratt JM (2005) Metabolic labeling of proteins for proteomics. Mol Cell Proteomics (MCP) 4(7):857–872
9. Boros LG (2005) Metabolic targeted therapy of cancer: current tracer technologies and future drug design strategies in the old metabolic network. Metabolomics 1(1):11–15
10. Boros LG, Brackett DJ, Harrigan GG (2003) Metabolic biomarker and kinase drug target discovery in cancer using stable isotope-based dynamic metabolic profiling (SIDMAP). Curr Cancer Drug Targets 3(6):445–453
11. Boros LG, Torday JS, Lim S, Bassilian S, Cascante M, Lee WN (2000) Transforming growth factor beta2 promotes glucose carbon incorporation into nucleic acid ribose through the nonoxidative pentose cycle in lung epithelial carcinoma cells. Cancer Res 60(5):1183–1185
12. Busch R, Kim Y-K, Neese RA, Schade-Serin V, Collins M, Awada M, Gardner JL, Beysen C, Marino ME, Misell LM, Hellerstein MK (2006) Measurement of protein turnover rates by heavy water labeling of nonessential amino acids. Biochimica et Biophysica Acta (BBA) – General Subjects 1760(5):730–744
13. Busch R, Neese RA, Awada M, Hayes GM, Hellerstein MK (2007) Measurement of cell proliferation by heavy water labeling. Nat Protoc 2(12):3045–3057
14. Cairns RA, Harris IS, Mak TW (2011) Regulation of cancer cell metabolism. Nat Rev Cancer 11:85–95. doi:10.1038/nrc2981
15. Cascante M, Boros LG, Comin-Anduix B, de Atauri P, Centelles JJ, Lee PWN (2002) Metabolic control analysis in drug discovery and disease. Nat Biotechnol 20(3):243–249
16. Chatham JC, Bouchard B, Des Rosiers C (2003) A comparison between NMR and GCMS 13 C-isotopomer analysis in cardiac metabolism. Mol Cell Biochem 249(1–2):105–112
17. Chatterjee A, Mambo E, Sidransky D (2006) Mitochondrial DNA mutations in human cancer. Oncogene 25(34):4663–4674
18. Christofk HR, Vander Heiden MG, Harris MH, Ramanathan A, Gerszten RE, Wei R, Fleming MD, Schreiber SL, Cantley LC (2008) The M2 splice isoform of pyruvate kinase is important for cancer metabolism and tumour growth. Nature 452(7184):230–233
19. Christofk HR, Vander Heiden MG, Wu N, Asara JM, Cantley LC (2008) Pyruvate kinase M2 is a phosphotyrosine-binding protein. Nature 452(7184):181–186
20. Czernin J, Phelps ME (2002) Positron emission tomography scanning: current and future applications. Annu Rev Med 53(1):89–112
21. Dang CV (2009) PKM2 Tyrosine phosphorylation and glutamine metabolism signal a different view of the Warburg effect. Sci Signal 2(97):pe75
22. Dang CV (2010) Glutaminolysis: supplying carbon or nitrogen, or both for cancer cells? Cell Cycle 9(19):3884–3886
23. Dang L, White DW, Gross S, Bennett BD, Bittinger MA, Driggers EM, Fantin VR, Jang HG, Jin S, Keenan MC, Marks KM, Prins RM, Ward PS, Yen KE, Liau LM, Rabinowitz JD, Cantley LC, Thompson CB, Vander Heiden MG, Su SM (2009) Cancer-associated IDH1 mutations produce 2-hydroxyglutarate. Nature 462(7274):739–744
24. David CJ, Chen M, Assanah M, Canoll P, Manley JL (2010) HnRNP proteins controlled by c-Myc deregulate pyruvate kinase mRNA splicing in cancer. Nature 463(7279):364–368

25. DeBerardinis RJ, Cheng T (2010) Q's next: the diverse functions of glutamine in metabolism, cell biology and cancer. Oncogene 29(3):313–324
26. DeBerardinis RJ, Lum JJ, Hatzivassiliou G, Thompson CB (2008) The biology of cancer: metabolic reprogramming fuels cell growth and proliferation. Cell Metab 7(1):11–20
27. DeBerardinis RJ, Mancuso A, Daikhin E, Nissim I, Yudkoff M, Wehrli S, Thompson CB (2007) Beyond aerobic glycolysis: transformed cells can engage in glutamine metabolism that exceeds the requirement for protein and nucleotide synthesis. Proc Natl Acad Sci U S A 104(49):19345–19350
28. Denkert C, Budczies J, Weichert W, Wohlgemuth G, Scholz M, Kind T, Niesporek S, Noske A, Buckendahl A, Dietel M, Fiehn O (2008) Metabolite profiling of human colon carcinoma – deregulation of TCA cycle and amino acid turnover. Mol Cancer 7:72
29. Deshpande R, Yang TH, Heinzle E (2009) Towards a metabolic and isotopic steady state in CHO batch cultures for reliable isotope-based metabolic profiling. Biotechnol J 4(2):247–263
30. Diraison F, Pachiaudi C, Beylot M (1996) In vivo measurement of plasma cholesterol and fatty acid synthesis with deuterated water: Determination of the average number of deuterium atoms incorporated. Metabolism 45(7):817–821
31. Dunn WB, Bailey NJC, Johnson HE (2005) Measuring the metabolome: current analytical technologies. Analyst 130(5):606–625
32. Elstrom RL, Bauer DE, Buzzai M, Karnauskas R, Harris MH, Plas DR, Zhuang H, Cinalli RM, Alavi A, Rudin CM, Thompson CB (2004) Akt stimulates aerobic glycolysis in cancer cells. Cancer Res 64(11):3892–3899. doi:10.1158/0008-5472.CAN-03-2904
33. Fan MZ, Chiba LI, Matzat PD, Yang X, Yin YL, Mine Y, Stein HH (2006) Measuring synthesis rates of nitrogen-containing polymers by using stable isotope tracers. J Anim Sci 84 (13 suppl):E79–E93
34. Fell DA (2004) Enzymes, metabolites and fluxes. J Exp Bot 56(410):267–272
35. Fernandez CA, Rosiers CD, Previs SF, David F, Brunengraber H (1996) Correction of 13 C mass isotopomer distributions for natural stable isotope abundance. J Mass Spectrom 31(3):255–262
36. Fiehn O, Kopka J, Dormann P, Altmann T, Trethewey RN, Willmitzer L (2000) Metabolite profiling for plant functional genomics. Nat Biotechnol 18(11):1157–1161
37. Fischer E, Zamboni N, Sauer U (2004) High-throughput metabolic flux analysis based on gas chromatography-mass spectrometry derived 13 C constraints. Anal Biochem 325(2):308–316
38. Forbes NS, Meadows AL, Clark DS, Blanch HW (2006) Estradiol stimulates the biosynthetic pathways of breast cancer cells: detection by metabolic flux analysis. Metab Eng 8(6):639–652
39. Funes JM, Quintero M, Henderson S, Martinez D, Qureshi U, Westwood C, Clements MO, Bourboulia D, Pedley RB, Moncada S, Boshoff C (2007) Transformation of human mesenchymal stem cells increases their dependency on oxidative phosphorylation for energy production. Proc Natl Acad Sci USA 104(15):6223–6228
40. Gatenby RA, Gillies RJ (2004) Why do cancers have high aerobic glycolysis? Nat Rev Cancer 4(11):891–899
41. Gogvadze V, Orrenius S, Zhivotovsky B (2008) Mitochondria in cancer cells: what is so special about them? Trends Cell Biol 18(4):165–173
42. Griffin JL, Shockcor JP (2004) Metabolic profiles of cancer cells. Nat Rev Cancer 4(7):551–561
43. Gruetter R, Adriany G, Choi I-Y, Henry P-G, Lei H, Öz G (2003) Localized in vivo13C NMR spectroscopy of the brain. NMR Biomed 16(6–7):313–338
44. Halket JM, Waterman D, Przyborowska AM, Patel RKP, Fraser PD, Bramley PM (2005) Chemical derivatization and mass spectral libraries in metabolic profiling by GC/MS and LC/MS/MS. J Exp Bot 56(410):219–243
45. Hao W, Chang C-PB, Tsao C-C, Xu J (2010) Oligomycin-induced bioenergetic adaptation in cancer cells with heterogeneous bioenergetic organization. J Biol Chem 285(17):12647–12654

46. Hatzivassiliou G, Zhao F, Bauer DE, Andreadis C, Shaw AN, Dhanak D, Hingorani SR, Tuveson DA, Thompson CB (2005) ATP citrate lyase inhibition can suppress tumor cell growth. Cancer Cell 8(4):311–321
47. Hellerstein M (2000) Methods for measuring polymerisation biosynthesis: three general solutions to the problem of the "true precursor". Diabetes Nutr Metab 13(1):46–60
48. Hellerstein MK (2004) New stable isotope-mass spectrometric techniques for measuring fluxes through intact metabolic pathways in mammalian systems: introduction of moving pictures into functional genomics and biochemical phenotyping. Metab Eng 6(1):85–100
49. Hellerstein MK, Neese RA (1992) Mass isotopomer distribution analysis: a technique for measuring biosynthesis and turnover of polymers. Am J Physiol 263(5 Pt 1):E988–E1001
50. Hellerstein MK, Neese RA (1999) Mass isotopomer distribution analysis at eight years: theoretical, analytic, and experimental considerations. Am J Physiol 276(6 Pt 1):E1146–E1170
51. Hiller K, Metallo CM, Kelleher JK, Stephanopoulos G (2010) Nontargeted elucidation of metabolic pathways using stable-isotope tracers and mass spectrometry. Anal Chem 82 (15):6621–6628
52. Hiller K, Metallo C, Stephanopoulos G (2011) Elucidation of cellular metabolism via metabolomics and stable-isotope assisted metabolomics. Curr Pharm Biotechnol 12(7):1075–1086
53. Jones JG, Garcia P, Barosa C, Delgado TC, Diogo L (2009) Hepatic anaplerotic outflow fluxes are redirected from gluconeogenesis to lactate synthesis in patients with Type 1a glycogen storage disease. Metab Eng 11(3):155–162
54. Jones RG, Thompson CB (2009) Tumor suppressors and cell metabolism: a recipe for cancer growth. Genes Dev 23(5):537–548
55. Kaadige MR, Looper RE, Kamalanaadhan S, Ayer DE (2009) Glutamine-dependent anapleurosis dictates glucose uptake and cell growth by regulating MondoA transcriptional activity. Proc Natl Acad Sci 106(35):14878–14883
56. Kacser H, Burns JA (1973) The control of flux. Symp Soc Exp Biol 27:65–104
57. Kaelin WG, Thompson CB (2010) Q&A: cancer: clues from cell metabolism. Nature 465 (7298):562–564
58. Kelleher JK (2001) Flux estimation using isotopic tracers: common ground for metabolic physiology and metabolic engineering. Metab Eng 3(2):100–110
59. Kelleher JK, Masterson TM (1992) Model equations for condensation biosynthesis using stable isotopes and radioisotopes. Am J Physiol 262(1 Pt 1):E118–E125
60. Kharroubi AT, Masterson TM, Aldaghlas TA, Kennedy KA, Kelleher JK (1992) Isotopomer spectral analysis of triglyceride fatty acid synthesis in 3 T3-L1 cells. Am J Physiol 263 (4 Pt 1):E667–E675
61. Kim J-w, Dang CV (2006) Cancer's molecular sweet tooth and the Warburg effect. Cancer Res 66(18):8927–8930
62. Koshland DE (1998) The era of pathway quantification. Science 280(5365):852–853
63. Kresge N, Simoni RD, Hill RL (2005) Otto Fritz Meyerhof and the elucidation of the glycolytic pathway. J Biol Chem 280(4):e3
64. Lane AN, Fan TWM, Higashi RM (2008) Stable isotope-assisted metabolomics in cancer research. IUBMB Life 60(2):124–129
65. Langbein S, Zerilli M, Az H, Staiger W, Rensch-Boschert K, Lukan N, Popa J, Ternullo MP, Steidler A, Weiss C, Grobholz R, Willeke F, Alken P, Stassi G, Schubert P, Coy JF (2006) Expression of transketolase TKTL1 predicts colon and urothelial cancer patient survival: Warburg effect reinterpreted. Br J Cancer 94(4):578–585
66. Levine AJ, Puzio-Kuter AM (2010) The control of the metabolic switch in cancers by oncogenes and tumor suppressor genes. Science 330(6009):1340–1344
67. Lu D, Mulder H, Zhao P, Burgess SC, Jensen MV, Kamzolova S, Newgard CB, Sherry AD (2002) 13 C NMR isotopomer analysis reveals a connection between pyruvate cycling and glucose-stimulated insulin secretion (GSIS). Proc Natl Acad Sci U S A 99(5):2708–2713
68. Luo W, Hu H, Chang R, Zhong J, Knabel M, O'Meally R, Cole RN, Pandey A, Semenza GL (2011) Cell 145(5):732–744

69. Macallan DC, Fullerton CA, Neese RA, Haddock K, Park SS, Hellerstein MK (1998) Measurement of cell proliferation by labeling of DNA with stable isotope-labeled glucose: studies in vitro, in animals, and in humans. Proc Natl Acad Sci USA 95(2):708–713
70. Maier K, Hofmann U, Bauer A, Niebel A, Vacun G, Reuss M, Mauch K (2009) Quantification of statin effects on hepatic cholesterol synthesis by transient 13 C-flux analysis. Metab Eng 11(4–5):292–309
71. Maier K, Hofmann U, Reuss M, Mauch K (2008) Identification of metabolic fluxes in hepatic cells from transient 13 C-labeling experiments: part II. Flux estimation. Biotechnol Bioeng 100(2):355–370
72. Malloy CR, Sherry AD, Jeffrey FM (1988) Evaluation of carbon flux and substrate selection through alternate pathways involving the citric acid cycle of the heart by 13 C NMR spectroscopy. J Biol Chem 263(15):6964–6971
73. Mancuso A, Sharfstein ST, Tucker SN, Clark DS, Blanch HW (1994) Examination of primary metabolic pathways in a murine hybridoma with carbon-13 nuclear magnetic resonance spectroscopy. Biotechnol Bioeng 44(5):563–585
74. Martini WZ, Chinkes DL, Wolfe RR (2004) Quantification of DNA synthesis from different pathways in cultured human fibroblasts and myocytes. Metabolism 53(1):128–133
75. Mathupala SP, Ko YH, Pedersen PL (2006) Hexokinase II: cancer's double-edged sword acting as both facilitator and gatekeeper of malignancy when bound to mitochondria. Oncogene 25(34):4777–4786
76. Mazurek S, Boschek CB, Hugo F, Eigenbrodt E (2005) Pyruvate kinase type M2 and its role in tumor growth and spreading. Semin Cancer Biol 15(4):300–308. doi:10.1016/j.semcancer.2005.04.009
77. McKnight SL (2010) On getting there from here. Science (New York, NY) 330 (6009):1338–1339
78. Meadows AL, Kong B, Berdichevsky M, Roy S, Rosiva R, Blanch HW, Clark DS (2008) Metabolic and morphological differences between rapidly proliferating cancerous and normal breast epithelial cells. Biotechnol Prog 24(2):334–341
79. Metallo CM, Vander Heiden MG (2010) Metabolism strikes back: metabolic flux regulates cell signaling. Genes Dev 24(24):2717–2722
80. Metallo CM, Walther JL, Stephanopoulos G (2009) Evaluation of 13 C isotopic tracers for metabolic flux analysis in mammalian cells. J Biotechnol 144(3):167–174
81. Miccheli A, Tomassini A, Puccetti C, Valerio M, Peluso G, Tuccillo F, Calvani M, Manetti C, Conti F (2006) Metabolic profiling by 13 C-NMR spectroscopy: [1,2-13 C2]glucose reveals a heterogeneous metabolism in human leukemia T cells. Biochimie 88(5):437–448
82. Munger J, Bennett BD, Parikh A, Feng X-J, McArdle J, Rabitz HA, Shenk T, Rabinowitz JD (2008) Systems-level metabolic flux profiling identifies fatty acid synthesis as a target for antiviral therapy. Nat Biotechnol 26(10):1179–1186
83. Neermann J, Wagner R (1996) Comparative analysis of glucose and glutamine metabolism in transformed mammalian cell lines, insect and primary liver cells. J Cell Physiol 166 (1):152–169
84. Neese RA, Siler SQ, Cesar D, Antelo F, Lee D, Misell L, Patel K, Tehrani S, Shah P, Hellerstein MK (2001) Advances in the stable isotope-mass spectrometric measurement of DNA synthesis and cell proliferation. Anal Biochem 298(2):189–195
85. Newsholme EA, Crabtree B, Ardawi MS (1985) The role of high rates of glycolysis and glutamine utilization in rapidly dividing cells. Biosci Rep 5(5):393–400
86. Nicholson JK, Connelly J, Lindon JC, Holmes E (2002) Metabonomics: a platform for studying drug toxicity and gene function. Nat Rev Drug Discov 1(2):153–161
87. Nicklin P, Bergman P, Zhang B, Triantafellow E, Wang H, Nyfeler B, Yang H, Hild M, Kung C, Wilson C, Myer VE, MacKeigan JP, Porter JA, Wang YK, Cantley LC, Finan PM, Murphy LO (2009) Bidirectional transport of amino acids regulates mTOR and autophagy. Cell 136 (3):521–534
88. Nielsen J (2003) It is all about metabolic fluxes. J Bacteriol 185:7031–7035

89. Niklas J, Schneider K, Heinzle E (2010) Metabolic flux analysis in eukaryotes. Curr Opin Biotechnol 21(1):63–69
90. Noack S, Nöh K, Moch M, Oldiges M, Wiechert W (2011) Stationary versus non-stationary 13 C-MFA: a comparison using a consistent dataset. J Biotechnol 154(2–3):179–190
91. Noguchi Y, Young JD, Aleman JO, Hansen ME, Kelleher JK, Stephanopoulos G (2009) Effect of anaplerotic fluxes and amino acid availability on hepatic lipoapoptosis. J Biol Chem 284(48):33425–33436. doi:10.1074/jbc.M109.049478
92. Ohie T, Fu X-w, Iga M, Kimura M, Yamaguchi S (2000) Gas chromatography-mass spectrometry with tert.-butyldimethylsilyl derivatization: use of the simplified sample preparations and the automated data system to screen for organic acidemias. J Chromatogr B Biomed Sci Appl 746(1):63–73
93. Ortega ÁD, Sánchez-Aragó M, Giner-Sánchez D, Sánchez-Cenizo L, Willers I, Cuezva JM (2009) Glucose avidity of carcinomas. Cancer Lett 276(2):125–135
94. Patterson BW (1997) Use of stable isotopically labeled tracers for studies of metabolic kinetics: an overview. Metabolism 46(3):322–329
95. Pedersen P (2007) Warburg, me and hexokinase 2: multiple discoveries of key molecular events underlying one of cancers' most common phenotypes, the "Warburg effect", i.e., elevated glycolysis in the presence of oxygen. J Bioenerg Biomembr 39(3):211–222
96. Perez JF, Reeds PJ (1998) A new stable isotope method enables the simultaneous measurement of nucleic acid and protein synthesis in vivo in mice. J Nutr 128(9):1562–1569
97. Pfeiffer T, Schuster S, Bonhoeffer S (2001) Cooperation and competition in the evolution of ATP-producing pathways. Science 292(5516):504–507
98. Portais J-C, Schuster R, Merle M, Canioni P (1993) Metabolic flux determination in C6 glioma cells using carbon-13 distribution upon [1-13 C]glucose incubation. Eur J Biochem 217(1):457–468
99. Poste G (2011) Bring on the biomarkers. Nature 469(7329):156–157
100. Quek L-E, Wittmann C, Nielsen L, Kromer J (2009) OpenFLUX: efficient modelling software for 13 C-based metabolic flux analysis. Microb Cell Fact 8(1):25
101. Robey RB, Hay N (2009) Is Akt the "Warburg kinase"?–Akt-energy metabolism interactions and oncogenesis. Semin Cancer Biol 19(1):25–31
102. Rodríguez-Enríquez S, Marín-Hernández A, Gallardo-Pérez JC, Carreño-Fuentes L, Moreno-Sánchez R (2009) Targeting of cancer energy metabolism. Mol Nutr Food Res 53(1):29–48
103. Salway JG (2004) Metabolism at a glance, 3rd edn. Blackwell, Malden
104. Sauer U (2006) Metabolic networks in motion: 13 C-based flux analysis. Mol Syst Biol 2:62. doi:10.1038/msb4100109
105. Sauer LA, Dauchy RT, Nagel WO, Morris HP (1980) Mitochondrial malic enzymes. Mitochondrial NAD(P) + -dependent malic enzyme activity and malate-dependent pyruvate formation are progression-linked in Morris hepatomas. J Biol Chem 255(9):3844–3848
106. Sazanov LA, Jackson JB (1994) Proton-translocating transhydrogenase and NAD- and NADP-linked isocitrate dehydrogenases operate in a substrate cycle which contributes to fine regulation of the tricarboxylic acid cycle activity in mitochondria. FEBS Lett 344 (2–3):109–116
107. Schmidt K, Carlsen M, Nielsen J, Villadsen J (1997) Modeling isotopomer distributions in biochemical networks using isotopomer mapping matrices. Biotechnol Bioeng 55(6):831–840
108. Shackelford DB, Shaw RJ (2009) The LKB1-AMPK pathway: metabolism and growth control in tumour suppression. Nat Rev Cancer 9(8):563–575
109. Shaw RJ, Cantley LC (2006) Ras, PI(3)K and mTOR signalling controls tumour cell growth. Nature 441(7092):424–430
110. Sreekumar A, Poisson LM, Rajendiran TM, Khan AP, Cao Q, Yu J, Laxman B, Mehra R, Lonigro RJ, Li Y, Nyati MK, Ahsan A, Kalyana-Sundaram S, Han B, Cao X, Byun J, Omenn GS, Ghosh D, Pennathur S, Alexander DC, Berger A, Shuster JR, Wei JT, Varambally S, Beecher C, Chinnaiyan AM (2009) Metabolomic profiles delineate potential role for sarcosine in prostate cancer progression. Nature 457(7231):910–914

111. Stephanopoulos G, Aristidou AA, Nielsen J (1998) Metabolic engineering: principles and methodologies. Academic, San Diego
112. Stephanopoulos G, Vallino JJ (1991) Network rigidity and metabolic engineering in metabolite overproduction. Science 252(5013):1675–1681
113. Stratton MR, Campbell PJ, Futreal PA (2009) The cancer genome. Nature 458 (7239):719–724. doi:10.1038/nature07943
114. Szyperski T (1995) Biosynthetically directed fractional 13 C-labeling of proteinogenic amino acids. An efficient analytical tool to investigate intermediary metabolism. Eur J Biochem/ FEBS 232(2):433–448
115. Tennant DA, Duran RV, Gottlieb E (2010) Targeting metabolic transformation for cancer therapy. Nat Rev Cancer 10(4):267–277
116. Tolstonog GV, Deppert W (2010) Metabolic sensing by p53: keeping the balance between life and death. Proc Natl Acad Sci 107(30):13193–13194
117. Tong X, Zhao F, Mancuso A, Gruber JJ, Thompson CB (2009) The glucose-responsive transcription factor ChREBP contributes to glucose-dependent anabolic synthesis and cell proliferation. Proc Natl Acad Sci USA 106(51):21660–21665. doi:10.1073/pnas.0911316106
118. Tong X, Zhao F, Thompson CB (2009) The molecular determinants of de novo nucleotide biosynthesis in cancer cells. Curr Opin Genet Dev 19(1):32–37. doi:10.1016/j.gde.2009.01.002
119. Van Winden W, Verheijen P, Heijnen S (2001) Possible pitfalls of flux calculations based on 13 C-labeling. Metab Eng 3(2):151–162
120. Vander Heiden MG, Cantley LC, Thompson CB (2009) Understanding the Warburg effect: the metabolic requirements of cell proliferation. Science 324(5930):1029–1033
121. Vander Heiden MG, Locasale JW, Swanson KD, Sharfi H, Heffron GJ, Amador-Noguez D, Christofk HR, Wagner G, Rabinowitz JD, Asara JM, Cantley LC (2010) Evidence for an alternative glycolytic pathway in rapidly proliferating cells. Science 329(5998):1492–1499
122. Vizan P, Boros LG, Figueras A, Capella G, Mangues R, Bassilian S, Lim S, Lee W-NP, Cascante M (2005) K-ras codon-specific mutations produce distinctive metabolic phenotypes in human fibroblasts. Cancer Res 65(13):5512–5515
123. Vogelstein B, Kinzler KW (2004) Cancer genes and the pathways they control. Nat Med 10 (8):789–799
124. Wamelink MMC, Struys EA, Huck JHJ, Roos B, van der Knaap MS, Jakobs C, Verhoeven NM (2005) Quantification of sugar phosphate intermediates of the pentose phosphate pathway by LC-MS/MS: application to two new inherited defects of metabolism. J Chromatogr B 823(1):18–25
125. Warburg O (1956) On respiratory impairment in cancer cells. Science 124(3215):269–270
126. Warburg O (1956) On the origin of cancer cells. Science 123(3191):309–314
127. Ward PS, Patel J, Wise DR, Abdel-Wahab O, Bennett BD, Coller HA, Cross JR, Fantin VR, Hedvat CV, Perl AE, Rabinowitz JD, Carroll M, Su SM, Sharp KA, Levine RL, Thompson CB (2010) The common feature of leukemia-associated IDH1 and IDH2 mutations is a neomorphic enzyme activity converting [alpha]-ketoglutarate to 2-hydroxyglutarate. Cancer Cell 17(3):225–234
128. Watt PW, Lindsay Y, Scrimgeour CM, Chien PA, Gibson JN, Taylor DJ, Rennie MJ (1991) Isolation of aminoacyl-tRNA and its labeling with stable-isotope tracers: use in studies of human tissue protein synthesis. Proc Natl Acad Sci USA 88(13):5892–5896
129. Weinberg F, Hamanaka R, Wheaton WW, Weinberg S, Joseph J, Lopez M, Kalyanaraman B, Mutlu GM, Budinger GRS, Chandel NS (2010) Mitochondrial metabolism and ROS generation are essential for Kras-mediated tumorigenicity. Proc Natl Acad Sci USA 107 (19):8788–8793
130. Wellen KE, Hatzivassiliou G, Sachdeva UM, Bui TV, Cross JR, Thompson CB (2009) ATP-citrate lyase links cellular metabolism to histone acetylation. Science 324(5930):1076–1080
131. Wiechert W (2001) 13 C metabolic flux analysis. Metab Eng 3(3):195–206
132. Wiechert W, Nöh K (2005) From stationary to instationary metabolic flux analysis. Adv Biochem Eng Biotechnol 92:145–172

133. Wise DR, DeBerardinis RJ, Mancuso A, Sayed N, Zhang X-Y, Pfeiffer HK, Nissim I, Daikhin E, Yudkoff M, McMahon SB, Thompson CB (2008) Myc regulates a transcriptional program that stimulates mitochondrial glutaminolysis and leads to glutamine addiction. Proc Natl Acad Sci USA 105(48):18782–18787
134. Wolfe RR, Chinkles DL (2004) Measurement of the synthesis of specific products. In: Isotope tracers in metabolic research: principles and practice of kinetic analysis, 2nd edn. Wiley, Hoboken, pp 325–360
135. Wu M, Neilson A, Swift AL, Moran R, Tamagnine J, Parslow D, Armistead S, Lemire K, Orrell J, Teich J, Chomicz S, Ferrick DA (2007) Multiparameter metabolic analysis reveals a close link between attenuated mitochondrial bioenergetic function and enhanced glycolysis dependency in human tumor cells. Am J Physiol Cell Physiol 292(1):C125–C136
136. Yan H, Parsons DW, Jin G, McLendon R, Rasheed BA, Yuan W, Kos I, Batinic-Haberle I, Jones S, Riggins GJ, Friedman H, Friedman A, Reardon D, Herndon J, Kinzler KW, Velculescu VE, Vogelstein B, Bigner DD (2009) IDH1 And IDH2 mutations in gliomas. N Engl J Med 360(8):765–773
137. Yang TH, Frick O, Heinzle E (2008) Hybrid optimization for 13 C metabolic flux analysis using systems parametrized by compactification. BMC Syst Biol 2(1):29
138. Yen KE, Bittinger MA, Su SM, Fantin VR (2010) Cancer-associated IDH mutations: biomarker and therapeutic opportunities. Oncogene 29(49):6409–6417
139. Yeung SJ, Pan J, Lee MH (2008) Roles of p53, Myc and HIF-1 in regulating glycolysis — the seventh hallmark of cancer. Cell Mol Life Sci 65(24):3981–3999
140. Yoo H, Stephanopoulos G, Kelleher JK (2004) Quantifying carbon sources for de novo lipogenesis in wild-type and IRS-1 knockout brown adipocytes. J Lipid Res 45(7):1324–1332
141. Yoo H, Antoniewicz MR, Stephanopoulos G, Kelleher JK (2008) J Biol Chem 283 (30):20621–20627
142. Young JD, Walther JL, Antoniewicz MR, Yoo H, Stephanopoulos G (2008) An elementary metabolite unit (EMU) based method of isotopically nonstationary flux analysis. Biotechnol Bioeng 99(3):686–699. doi:10.1002/bit.21632
143. Yun J, Rago C, Cheong I, Pagliarini R, Angenendt P, Rajagopalan H, Schmidt K, Willson JKV, Markowitz S, Zhou S, Diaz LA, Velculescu VE, Lengauer C, Kinzler KW, Vogelstein B, Papadopoulos N (2009) Glucose deprivation contributes to the development of KRAS pathway mutations in tumor cells. Science 325(5947):1555–1559
144. Zamboni N, Fendt S-M, Ruhl M, Sauer U (2009) ^{13}C-based metabolic flux analysis. Nat Protoc 4(6):878–892
145. Zhang X-j, Chinkes DL, Wu Z, Martini WZ, Wolfe RR (2004) Fractional synthesis rates of DNA and protein in rabbit skin are not correlated. J Nutr 134(9):2401–2406
146. Zupke C, Sinskey AJ, Stephanopoulos G (1995) Intracellular flux analysis applied to the effect of dissolved oxygen on hybridomas. Appl Microbiol Biotechnol 44(1–2):27–36
147. Zupke C, Stephanopoulos G (1994) Modeling of isotope distributions and intracellular fluxes in metabolic networks using atom mapping matrixes. Biotechnol Prog 10(5):489–498

Index

DOI 10.1007/978-94-007-4534-6, © Springer Science+Business Media Dordrecht 2012

MIX
Papier aus verantwortungsvollen Quellen
Paper from responsible sources
FSC® C105338

If you have any concerns about our products,
you can contact us on
ProductSafety@springernature.com

In case Publisher is established outside the EU,
the EU authorized representative is:
Springer Nature Customer Service Center GmbH
Europaplatz 3, 69115 Heidelberg, Germany

Printed by Libri Plureos GmbH
in Hamburg, Germany